K. Taira, K. Kataoka, T. Niidome (Eds.)

Non-viral Gene Therapy

Gene Design and Delivery

K. Taira, K. Kataoka, T. Niidome (Eds.)

Non-viral Gene Therapy

Gene Design and Delivery

With 163 Figures, Including 25 in Color

Springer

Kazunari Taira, Ph.D
Department of Chemistry and Biotechnology, Graduate School of Engineering
The University of Tokyo
7-3-1 Hongo, Bunkyo-ku, Tokyo 113-8656, Japan
and
Gene Function Research Center
National Institute of Advanced Industrial Science and Technology (AIST)
Central 4, 1-1-1 Higashi, Tsukuba Science City 305-8562, Japan

Kazunori Kataoka, Ph.D
Department of Materials Science and Engineering, Graduate School of Engineering
The University of Tokyo
7-3-1 Hongo, Bunkyo-ku, Tokyo 113-8656, Japan

Takuro Niidome, Ph.D
Department of Applied Chemistry, Faculty of Engineering
Kyushu University
6-10-1 Hakozaki, Higashi-ku, Fukuoka 812-8581, Japan

Library of Congress Control Number: 2005925229

ISBN 4-431-25122-7 Springer-Verlag Tokyo Berlin Heidelberg New York

Springer is a part of Springer Science+Business Media
springeronline.com

Typesetting: SNP Best-set Typesetter Ltd., Hong Kong
Printing and binding: Shinano Inc., Japan

Printed on acid-free paper

Foreword

Several years ago, when the discovery of catalytic RNA was recognized in a public manner, many people asked if new fields of therapy would soon be available. Although some tentative positive answers were given, nobody would say with certainty that RNA of various kinds was a truly promising means of altering gene expression. In fact, over the past decade, both our knowledge of RNAs with different functions and the utility of RNA in the inhibition or enhancement of gene expression have occurred with great drama. We proceeded in terms of possible therapeutic tools from RNase P and group I introns through "hammerhead" RNA enzymes, antisense technology, and more recently, to RNAi and its derivatives. A useful practical method of RNA delivery in animals will complete the picture. The diversity of RNA and the varied role of it inside cells and in therapy should be a tremendous challenge for young molecular biologists. This volume will make their task easier.

Sidney Altman
Sterling Professor of Molecular, Cellular & Developmental Biology, Nobel Laureate
Department of Molecular, Cellular and Developmental Biology
Yale University

Delivery of nucleic acids to cells in an animal remains a challenging problem. It is the major obstacle to success of therapeutic approaches using genes and oligonucleotides, including siRNAs.

Solutions found so far by chemists are satisfactory only for transfection of cells in culture. In vivo delivery is generally poor, except where large amounts of complexes can physically reach a large mitotic cell surface. In this context, like a gleam of hope, a non-viral vector was recently shown to successfully deliver a therapeutic gene in patients (see http://www.gtmb.org/volume8/html/18_Ohana/18._Ohana_et_al,_181-192.html).

Altogether, intracellular barriers are responsible for a vanishingly small 0.001% yield of the transfection "reaction"; desperately seeking chemists to improve yields!

Moreover, even the perfect transfection reagent would be useless if it could not physically diffuse to the cells within the target organ—desperately needing physicists and pharmacists.

Gene delivery indeed is transdisciplinary by essence, and tremendous improvements can still come from anyone. I remain firmly confident that "naïve" scientists attracted by this exciting and now mature topic will help us metamorphose gene delivery into gene therapy. This is the goal of this book.

Jean-Paul Behr
Laboratoire de Chimie Génétique
associé au C.N.R.S.-Université Louis Pasteur de Strasbourg
Faculté de Pharmacie

Preface

The potential use of genes as therapeutic agents has attracted attention as a novel approach to the treatment of severe diseases. In the case of inherited disorders, the introduction of a normal copy of the affected gene can be effective, as in the well-known case of gene therapy for severe combined immune deficiency due to adenosine deaminase (ADA) deficiency, in which the normal gene for ADA is used to treat the affected patient. For the treatment of acquired disorders, such as cancer and infectious diseases, effective potential strategies involve not only the introduction of a therapeutic gene, such as the gene for a cytokine or an antigen, but also the silencing of the expression of an abnormal gene, whose expression is enhanced in the tissue of the diseased part.

Subsequent to the revolutionary discovery of ribozymes by Sidney Altman and Thomas Cech in the early 1980s, RNAs were recognized as a class of potentially therapeutic nucleic acids that might be able to regulate the expression of viral and endogenous genes in a clinical setting. In 2000, we realized the importance of stimulating interactions between scientists who are involved in research on potentially therapeutic genes and those who are involved in studies of drug- and gene-delivery systems. For this reason, we helped to establish the Japanese Society of Gene Design and Delivery (http://www.gene-delivery.org/) in 2001. That same year, the new and powerful phenomenon known as RNA interference (RNAi) was reported. This intracellular phenomenon has been recognized in cells of organisms that include and are higher on the evolutionary scale than *Schizosaccharomyces pombe*. The potential suppressive effects of RNAi might be expected to be significantly greater than those of ribozymes, bringing the dream of nucleic acid drugs closer, provided that suitable delivery systems can be developed.

Gene-delivery strategies fall into two categories: viral and non-viral. More than 70% of current gene therapies are performed using the viral method. However, this method is plagued by serious problems, such as toxicity, immunogenicity, and problems associated with large-scale high-quality production. Therefore, even though non-viral methods are also associated with problems, such as the limited efficiency of gene delivery, future practical applications of genes to medicine will require the development of safe non-viral gene-delivery systems.

In addition to antisense technology and the use of ribozymes and decoys, RNAi has become a major tool in efforts to control gene expression in the fields of molecular biology, biotechnology, and medicine. Non-viral gene-delivery systems will provide the key to the use of both synthetic small interfering RNA (siRNA) and long-lasting expression vector-based siRNA. However, efficient systems for the delivery of the active RNAs and/or the corresponding DNA templates are sorely needed.

We have entitled this book *Non-viral Gene Therapy: Gene Design and Delivery*. Our conviction that gene design and gene delivery must be considered together for future practical applications of gene therapy is incorporated in this title. In the first part, several non-viral gene-delivery systems are described, with an emphasis on chemical and physical issues. In the second part, current strategies of controlled gene delivery, including in vivo systems, and the status of clinical trials are introduced. Current protocols for gene delivery in vitro and in vivo are included as the third part, and full details are supplied to allow the reader to apply the various protocols. Finally, the fourth part includes details of the most recent developments in RNA technologies and delivery systems.

We hope that this book will be of interest to researchers who are studying gene design and delivery, and that it will stimulate their research and the research of the next generation of experts in this field.

We are deeply indebted to all the contributors to this book; they graciously produced their manuscripts at very short notice. The publication of this book would have been impossible without the considerable efforts and patience of these contributors, the members of the Japanese Society of Gene Design and Delivery, and the staff of Springer-Verlag, Tokyo.

February 2005

Kazunari Taira
Kazunori Kataoka
Takuro Niidome
The Japanese Society of Gene Design and Delivery

Kazunari Taira (Co-editor)
The University of Tokyo &
National Institute of Advanced
Industrial Science and
Technology (AIST)

Kazunori Kataoka (Co-editor)
The University of Tokyo

Leaf Huang (Author, pp. 3–10)
University of Pittsburgh

Ernst Wagner (Author, pp. 237–245)
Ludwig-Maximilians-Universität

John Rossi (Author, pp. 465–475)
Beckman Research Institute of
the City of Hope

Alexander V. Kabanov
University of Nebraska
Medical Center

Michinori Kohara
The Tokyo Metropolitan Institute of
Medical Science

Georg Sczakiel
Universität zu Lübeck

If gene therapy is to become a successful and widely applicable tool in a clinical setting, there is no question that such an outcome will depend on extensive interactions between researchers who are studying potentially therapeutic genes and those who are studying drug- and gene-delivery systems. We hope that this book will be of interest to researchers in both areas, and that it will stimulate their research and the research of the next generation of experts, who will, perhaps, witness the application of this decade's research to the next decade's patients.

Contents

2. Controlled Gene Delivery

3. Current Protocols of Gene Delivery

4. Design of Genes Based on Current RNA Technology

1. Molecules for Gene Therapy

Recent Progress in Non-viral Gene Delivery

CHRISTINE C. CONWELL and LEAF HUANG

1 Introduction

Gene therapy provides a unique approach to medicine as it can be adapted towards the treatment of both inherited and acquired diseases. Gene delivery relies upon the encapsulation of a gene of interest, which is then ideally delivered to target cells. After uptake up by endocytosis, the DNA must be released into the cell so that transcription and translation may occur to produce the protein of interest. To achieve successful gene delivery, significant barriers must be overcome at each step of this process in order to optimize gene activity while minimizing the potential for inhibitory inflammatory responses.

Particular interest has been paid in recent years to the development of efficient non-viral vectors. Viral vectors (i.e., retroviruses and adenoviruses) may provide superior gene delivery to target cells compared to their non-viral counterparts, but viral vectors also come with the significantly increased risk of triggering a specific immune response, which under extreme circumstances could result in death (Lehrman 1999; Marshall 1999). Non-viral vectors may trigger an inflammatory response but are not likely to elicit specific recognition, making these types of vectors less hazardous in terms of antigen-specific immune responses. Although non-viral vectors are more appealing in this respect, there are several other factors that must be considered in vector design, including specific cell targeting, optimized uptake, and efficient intracellular release of the vector, in addition to minimizing the immune response. The development of novel non-viral vectors intended to optimize one or more of these aspects of gene delivery will be discussed briefly in this chapter.

2 Delivery of Naked DNA by Physical Methods

Naked plasmid DNA provides a promising mechanism for gene delivery as it is less immunogenic than most non-viral vectors currently used. Complications arise, however, in that naked plasmid DNA has no target-specific recognition and is more susceptible to nuclease degradation than encapsulated DNA. These issues limit the

Center for Pharmacogenetics, School of Pharmacy, University of Pittsburgh, 3501 Terrace St., Pittsburgh, PA 15261, USA quantity of naked DNA that is able to reach the target cells, thereby limiting the efficiency of gene expression. Much effort has been made in recent years to overcome these limitations by developing new methods to effectively deliver naked DNA to target cells.

Delivery of naked DNA has been promoted through the application of electric or magnetic fields to the target areas. Electroporation involves injection of the naked DNA, followed by the application of an electric field over the targeted tissue. The electric field increases the permeability of the cell membranes and allows for increased uptake of the naked DNA into the cells. Compared to injection alone, electroporation has been found to increase gene expression by several orders of magnitude (Ndoye et al. 2004; Wells 2004). Efforts have been made specifically to reduce tissue damage associated with electroporation techniques. Modifications such as the application of low voltages, pulses of high voltage followed by low voltage, and the development of new electrodes have been shown to increase gene uptake and reduce damage compared to previous methods (Bureau et al. 2000, 2004; Liu and Huang 2002; Zhang et al. 2002). Electroporation is particularly useful for delivering DNA to superficial areas, but it requires invasive surgical procedures to deliver DNA to specific organs. Magnetic fields have also been employed as a mechanism to deliver DNA to targeted areas. Magnetofection involves attaching DNA onto a magnetic nanoparticle coated with a cationic polymer (i.e., PEI) (Huth et al. 2004; Scherer et al. 2002). In vitro studies indicated that the presence of a magnetic field increased the rate of transfection from hours to minutes compared to the controls (Scherer et al. 2002).

Laser beam gene transduction (LBST) aids in the delivery of naked DNA by following injection with several short pulses from a laser beam directed towards the targeted area. As described by Ziera et al., delivery by LBST resulted in gene expression that was comparable to that induced by electroporation (Zeira et al. 2003). The advantage of this method compared to electroporation is that histological analysis revealed minimal tissue damage as a result of LBST, whereas tissues treated by electroporation had substantial damage.

Clinical ultrasonic methods have been found to increase gene expression in naked DNA delivery. Cavitation resulting from the ultrasonic waves increases the permeability of the cells, allowing for more efficient gene delivery. High- and low-level ultrasonic waves have been investigated for increasing DNA delivery to a variety of tissues, such as muscle and heart as well as to tumors (Wells 2004). Further studies of gene delivery by sonoporation have found that the presence of microbubbles, which act as contrast agents by reflecting the ultrasonic waves, promote efficient gene expression (Unger et al. 2004). Advances have also been made in target-specific delivery by incorporation of specific ligands onto the surface of the microbubbles.

The use of high pressure to deliver naked DNA has led to several new and modified methods for non-viral gene delivery, including modification of particle bombardment techniques (i.e., gene gun), jet injection, mechanical massage and hydrodynamic injection. Recent efforts have yielded the production of a modified gene gun in which DNA coated on gold particles is introduced into the target tissue by a high-pressure helium blast. Gene expression was increased several orders of magnitude and tissue penetration more than doubled compared to delivery by the conventional gene gun (Dileo et al. 2003b). Jet injection, by contrast, involves the injection of low volumes of DNA solution at high pressures. Gene expression has been observed at depths of up to 10 mm in targeted tissues after delivery by a high-speed jet injector and transfection efficiencies were similar to those observed using particle bombardment techniques (Walther et al. 2001). DNA delivery by particle bombardment and jet injection are limited to a relatively small target area compared to many other gene delivery methods.

Hydrodynamic injection allows DNA delivery to larger target regions and is not limited to superficial tissue (Zhang et al. 2001, 2004; Maruyama et al. 2002). This method involves high-pressure injection of a large volume of solution containing the DNA of interest. Gene expression was found to increase, particularly in hepatocytes, as a result of defects in the cells resulting from the high-pressure injection (Zhang et al. 2004). A main drawback to this technique is that the large injection volume can lead to side effects, such as high blood pressure and low heart rate, and possibly death. Alternatively, mechanical massage of the liver (MML) has recently been found to successfully deliver naked DNA to hepatocytes in a manner that has the potential for less lethal side effects (Liu et al. 2004a).

3 Cationic Polymer-Based Non-viral Vectors

Encapsulation of naked DNA by cationic polymers to form polyplexes is one of the primary non-viral vector systems that has been examined for the optimization of gene delivery. Generally, the cationic polymer first functions as a condensing agent, to collapse DNA into compact bundles of a size suitable for delivery. After the complex arrives at the target cells, the polymer must promote the release of DNA from the endosomal compartments into the cell. To accomplish these tasks, the cationic polymer must form a strong association with the DNA but release it under the appropriate conditions as well as have the ability to recognize specific cells. Polymers frequently used for DNA encapsulation and delivery include polyethylenimine (PEI), poly-L-lysine, cationic dendrimers, and arginine-rich proteins or peptides (i.e., protamines, HIV-TAT) (Pannier and Shea 2004; Wagner 2004). In recent years, efforts have been directed towards optimizing the polymer to become more efficient at these functions, often by increasing the biocompatibility, primarily through modifications to existing polymers.

Polyethlyenimines (PEI) have evolved into a prominent cationic polymer for gene delivery. PEI condenses plasmid DNA into compact particles and protects them from nuclease degradation during gene delivery (Ferrari et al. 1999). Once encapsulated into the endosome, PEI has the ability to act as a proton sponge, which destabilizes the endosomal compartment and allows release of the DNA into the cytoplasm (Kircheis et al. 2001). The ability to facilitate DNA release into the cell is a major asset, as release into the cytoplasm is one of the main barriers encountered in gene delivery. Recently, PEI has been modified to contain various degrees of branching and chain lengths, as well as by methylation of the polymer to a charged quaternary ammonium derivative. The quaternary ammonium derivative was found to decrease the toxicity of the molecule by up to four-fold (Brownlie et al. 2004).

The formation of novel block copolymers has also led to the development of more efficient, less toxic delivery agents. Polyethylene glycol (PEG) has been linked to PEI and has been shown to decrease toxicity by nearly ten-fold as compared to PEI alone.

By linking the polymers to a more biocompatible molecule (e.g. PEG), the copolymers are less toxic but maintain other positive characteristics originally observed with the cationic polymer alone. PEG has also been linked to poly-L-lysine, polyamidoamine and poly-histidine, among others (Ahn et al. 2004; Miyata et al. 2004; Putnam et al. 2003; Rackstraw et al. 2002). Furthermore, some novel copolymers have also been developed. The copolymer polyhydroxyproline (PHLP) has been shown to maintain sustained gene activity but with reduced toxicity due to the biodegradable nature of the molecule (Li and Huang 2004). The development of copolymers allows for the formation of novel structures that maintain the characteristics of the individual components, which in return allows for the formation of optimized polymers. Advances in the development of novel copolymers have provided promising options for biocompatible and efficient non-viral gene delivery vectors.

Polyplexes may also be modified by the addition of lipids to form lipopolyplexes. One of the more studied systems incorporates lipid with protamine molecules to encapsulate DNA and form complexes of lipid-protamine-DNA (LPD). LPD vectors have been found to have increased gene activity compared to lipid-based vectors (Li and Huang 1997; Li et al. 1998). These vectors have recently been adapted to contain a target ligand, asialofetuin (AF), which aids in delivery to cells containing a receptor for this molecule, particularly heptatocytes. The incorporation of the target ligand increased gene expression by nearly ten-fold over the lipid-based vector in the absence of ligand and polymer (Arangoa et al. 2003). Additionally, recent studies have shown that modified LPD vectors may be useful in cancer treatment (Dileo et al. 2003a; Whitmore et al. 2001). Thus, it is evident that LPD vectors are multifaceted and that further modifications of these vectors may provide successful gene delivery systems for a wide variety of treatments.

4 Lipid-Based Non-viral Vectors

Liposomes and DNA combine to form complexes referred to as lipoplexes. Since the initial investigations of lipoplexes nearly 20 years ago by Felgner et al., various cationic lipids and combinations of cationic and neutral lipids have been thoroughly examined for their potential as non-viral vectors (Felgner et al. 1987). In recent years, optimization of lipoplexes has focused primarily on improving the targeting of these vectors to specific cells as well as increasing the release of the DNA into the cytoplasm. These modifications in lipoplex composition have significantly improved vector association with specific cells as well as DNA expression by working to overcome several significant barriers limiting successful gene delivery.

Lipoplexes have been designed to target cell types that contain specific receptors or have a higher propensity to recognize certain molecules than other cells. Ligands recognized by the target cells are incorporated into the liposomes. The ability to target cancer cells has been the main focus of numerous studies, as in many cases the cells are not necessarily localized. Both folate and transferrin receptors are known to be over expressed on cancer cells, so vectors have been designed to be recognized by these receptors. By incorporating the ligands, vectors were shown to have an increased affinity for tumor tissue in both instances (Dauty et al. 2002; Xu et al. 2002; Zuber et al. 2003). Further investigation of ligand incorporation may allow for increased specificity for gene delivery, thereby increasing the gene expression observed in the target cells.

Improved targeting of vectors to specific cells increases gene expression, but lipoplex modifications have also been aimed towards increasing DNA release inside the cell. This aim has proven multifaceted, in that by working to modify vectors to breakdown intracellularly, new vectors tend to be more biocompatible and do not elicit as severe an immune response from the recipient. Many liposomes are being engineered such that under a particular condition (e.g. low pH, reducing environments) the vector will become defective and allow DNA to leak out of the complex (Asokan and Cho 2002; Guo and Szoka 2003; Thomas and Tirrell 2000; Yatvin et al. 1980). A primary example is novel lipids designed to contain cysteine residues that can form stable disulfide linkages (i.e. ligand binding) under most conditions, but are sensitive to reducing conditions (Guo and Szoka 2001, 2003; Dauty et al. 2001; Kwok et al. 2003). These vectors may contain a ligand that assists in the delivery of the vector to specific cells then, once the vector is released into the cytoplasm, the environment will reduce the disulfide linkages and allow the DNA to escape. Similar systems relying on the addition of moieties that are sensitive to intracellular conditions, which trigger the release of DNA have also been developed for cationic polymer based vectors (Kwok et al. 2003; McKenzie et al. 2000; Erbacher et al. 1999).

5 Toxicity of Non-viral Vectors

Non-viral vectors, although less lethal than viral vectors, still may elicit a strong, non-specific immune response. Toxicity frequently results from characteristics of the encapsulating polymer or lipid such as the length, saturation, or branching of the polymer (Ahn et al. 2004; Kramer et al. 2004). Efforts to reduce the toxicity of non-viral vectors have largely resulted in efforts to make the vectors more biodegradable and biocompatible. Many of the aforementioned systems (i.e. triggered release with disulfides, PEG copolymers) incorporated more biologically active components into the system, thereby reducing the immune response induced by delivery of the vector. Further efforts to reduce toxicity have involved the incorporation of molecules known to suppress the production of the cytokine, NF-κB. These "safeplexes" were able to maintain low levels of tumor necrosis factor (TNF-α) as compared to lipoplex alone, while achieving comparable levels of gene expression (Liu et al. 2004b). Another method explored by Tan et al. found to significantly reduce toxicity requires the sequential injection of liposome, then DNA as opposed to delivery of the lipoplex. By injecting the liposome prior to the plasmid, cytokine levels (IL-12, TNF-α) were reduced by greater than 80% compared to lipoplex (Tan et al. 2001). Thus, significant advances have been made towards decreasing the toxicity of these non-viral vectors.

Acknowledgment. The original work from this lab was supported by NIH grants CA74918, AI48851, DK44935, DK68556, and AR45925.

References

Ahn C, Chae S, Bae Y, Kim S (2004) Synthesis of biodegradable multi-block copolymers of poly(L-lysine) and poly(ethylene glycol) as a nonviral gene carrier. Journal of Controlled Release 97:567–574

Arangoa M, Duzgunes N, Tros de Ilarduya C (2003) Increased receptor-mediated gene delivery to the liver by protamine-enhanced-asiolofetuin-lipoplexes. Gene Therapy 10:5–14
Asokan A, Cho M (2002) Exploitation of intracellular pH gradients in the cellular delivery of macromolecules. Journal of Pharmaceutical Science 91:903–913
Brownlie A, Uchegbu I, Schatzlein A (2004) PEI-based vesicle-polymer hybrid gene delivery sustem with improved biocompatibility. International Journal of Pharmaceutics 274:41–52
Bureau M, Gehl K, Deleuze V, Mir L, Scherman D (2000) Importance of association between permeablization and electrophoretic forces for intramuscular DNA electrotransfer. Biochimica et Biophysica Acta 1474:353–359
Bureau M, Naimi S, Torero Ibad R, Seguin J, Georger C, Arnould E, Maton L, Blanche F, Delaere P, Scherman D (2004) Intramuscular plasmid DNA electrotransfer: Biodistribution and degradation. Biochimica et Biophysica Acta 1676:138–148
Dauty E, Remy J, Blessing T, Behr J (2001) Dimerizable cationic detergents with a low cmc condense plasmid DNA into nanometric particles and transfect cells in culture. Journal of the American Chemical Society 123:9227–9234
Dauty E, Remy J, Zuber G, Behr J (2002) Intracellular delivery of nanometric DNA particles via the folate receptor. Bioconjugate Chemistry 13:831–839
Dileo J, Banerjee R, Whitmore M, Nayak JV, Falo LD, Huang L (2003) Lipid-protamine-DNA-mediated antigen delivery to antigen presenting cells results in ehnahnced antitumor immune responses. Molecular Therapy 7:640–648
Dileo J, Miller T, Chesnoy S, Huang L (2003) Gene transfer to subdermal tissues via a new gene gun design. Human Gene Therapy 14:79–87
Erbacher P, Remy J, Behr J (1999) Gene transfer with synthetic virus-like particles via the integrin-mediated endocytosis pathway. Gene Therapy 6:138–145
Felgner P, Gadek T, Holm M, Roman R, Chan H, Wenz M, Northrop J, Ringold G, Danielsen M (1987) Lipofection: A highly efficient, lipid-mediated DNA-transfection procedure. Proceedings of the National Academy of Science USA 84:7413–7417
Ferrari S, Pettenazzo A, Garbati N, Zacchello F, Behr J, Scarpa M (1999) Polyethylenimine shows properties of interest for cystic fibrosis gene therapy. Biochimica et Biophysica Acta 1447:219–225
Guo X, Szoka FJ (2001) Steric stabilization of fusogenic liposomes by a low-pH sensitive PEG-diortho ester-lipid conjugate. Bioconjugate Chemistry 12:291–300
Guo X, Szoka FJ (2003) Chemical approaches to triggerable lipid vesicles for drug and gene delivery. Accounts of Chemical Research 36:335–341
Huth S, Lausier J, Gersting S, Rudolph C, Welsch U, Rosenecker J (2004) Insights into the mechanisms of magnetofection using PEI-based magnetofectins for gene transfer. Journal of Gene Medicine 6:923–936
Kircheis R, Wightman L, Wagner E (2001) Design and gene delivery activity of modified polyethylenimines. Advanced Drug Delivery Reviews 53:341–358
Kramer M, Stumbe J, Grimm G, Kaufmann B, Kruger U, Weber M, Haag R (2004) Dendritic polyamines: Simple access to new materials with defined treelike structures for application in nonviral gene delivery. Chembiochem 5:1081–1087
Kwok K, Park Y, Yang Y, McKenzie D, Liu Y, Rice K (2003) In vivo gene transfer using sulfhydryl cross-linked PEG-peptide/glycopeptide DNA co-condensates. Journal of Pharmaceutical Science 92:1174–1185
Lehrman S (1999) Virus treatment questioned after gene therapy death. Nature (London) 401:517–518
Li S, Huang L (1997) In vivo gene transfer via intravenous administration of cationic lipid-protamine-DNA (LPD) complexes. Gene Therapy 4:891–900
Li S, Rizzo MA, Bhattacharya S, Huang L (1998) Characterization of catinoi lipid-protamine-DNA (LPD) complexes for intravenous gene delivery. Gene Therapy 5:930–937

Li Z, Huang L (2004) Sustained delivery and expression of plasmid DNA based on biodegradable polyester, poly(D,L-lactide-co-4-hydroxyl-L-proline). Journal of Controlled Release 98:437–446

Liu F, Huang L (2002) A syringe electrode device for simultaneous injection of DNA and electrotransfer. Molecular Therapy 5:323–328

Liu F, Lei J, Vollmer R, Huang L (2004a) Mechanism of liver gene transfer by mechanical massage. Molecular Therapy 9:452–457

Liu F, Shollenberger L, Huang L (2004b) Non-immunostimulatory nonviral vectors. FASEB 18:1779–81

Marshall E (1999) Gene therapy death prompts review of adenovirus vector. Science 286:2244–2245

Maruyama H, Higuchi N, Nishikawa Y, Hirahara H, Iino N, Kameda S, Kawachi H, Yaoita E, Gejyo F, Miyazaki J (2002) Kidney targeted naked DNA transfer by retrograde renal vein injection in rats. Human Gene Therapy 13:455–468

McKenzie D, Kwok K, Rice K (2000) A potent new class of reductively activated peptide gene delivery agents. Journal of Biological Chemistry 275:9970–9977

Miyata K, Kakizawa Y, Nishiyama M, Harada A, Yamasaki Y, Koyama H, Katoaka K (2004) Block catiomer polyplexes with regulated densities of charge and disulfide cross-linking directed to enhance gene expresion. Journal of the American Chemical Society 126:2355–2361

Ndoye A, Merlin J, Leroux A, Dolivet G, Erbacher P, Behr J, Berg K, Guillemin F (2004) Enhanced gene transfer and cell death following p53 gene transfer using photochemical internalization of glucosylated PEI-DNA complexes. Journal of Gene Medicine 6

Pannier A, Shea L (2004) Controlled release systems for DNA delivery. Molecular Therapy 10:19–26

Putnam D, Zelikin A, Izumrudov V, Langer R (2003) Polyhistidine-PEG:DNA nanocomposites for gene delivery. Biomaterials 24:4425–4433

Rackstraw B, Stolnik S, Davis S, Bignotti F, Garnett M (2002) Development of multicomponent DNA delivery systems based upon poly(amidoamine)-PEG co-polymers. Biochimica et Biophysica Acta 1576:269–286

Scherer F, Anton M, Schillinger U, Henke J, Bergemann C, Kruger A, Gansbacher B, Plank C (2002) Magnetofection: Enhancing and targeting gene delivery by magnetic force in vitro and in vivo. Gene Therapy 9:102–109

Tan Y, Liu F, Li Z, Li S, Huang L (2001) Sequential injection of cationic liposome and plasmid DNA effectively transfects the lung with minimal inflammatory toxicity. Molecular Therapy 3:673–682

Thomas J, Tirrell D (2000) Polymer-induced leakage of cations from dioleoyl phosphatidylcholine and phosphatidylglycerol liposomes. Journal of Controlled Release 76:203–209

Unger E, Porter T, Culp W, Labell R, Matsunga T, Zutshi R (2004) Therapeutic applications of lipid-coated microbubbles. Advanced Drug Delivery Reviews 56:1291–1314

Wagner E (2004) Strategies to improve DNA polyplexes for in vivo gene transfer: Will "artificial viruses" be the answer? Pharmaceutical Research 21:8–14

Walther W, Stein U, Fichtner I, Malcherek L, Lemm M, Schlag P (2001) Nonviral in vivo gene delivery into tumor using a novel low volume jet-injection technology. Gene Therapy 8:173–180

Wells D (2004) Gene therapy progress and prospects: electroporation and other physical methods. Gene Therapy 11:1363–1369

Whitmore MM, Li S, Falo LJ, Huang, L. (2001) Systemic administration of LPD prepared with CpG oligonucleotides inhibits the growth of established pulmonary metastases by stimulating innate and acquired antitumore immune response. Cancer Immunology and Immunotherapy 50:503–514

Xu L, Huang C, Huang W, Tang W, Rait A, Yin Y, Cruz I, Xiang L, Pirollo K, Chang E (2002) Systemic tumor-targeted gene delivery by anti-transferrin receptor scFv-immunoliposomes. Molecular Cancer Therapeutics 1:337–346

Yatvin M, Kreutz W, Horwitz B, Shinitzky M (1980) pH-sensitive liposomes: Possible clinical implications. Science 210:1253–1255

Zeira E, Manevitch A, Khatchatouriants A, Pappo O, Hyam E, Darash-Yahana M, Tavor E, Honigman A, Lewis A, Galun E (2003) Femto-second infrared laser-an efficient and safe in vivo gene delivery system for prolonged expression. Molecular Therapy 8:342–350

Zhang G, Budker V, Williams P, Subbotin V, Wolff J (2001) Efficient expression of naked DNA delivered intraarterialy to limb muscles of nonhuman primates. Human Gene Therapy 12:427–438

Zhang G, Gao X, Song Y, Vollmer R, Stolz D, Gasiorowski J, Dean D, Liu D (2004) Hydroporation as the mechanism of hydrodynamic delivery. Gene Therapy 11:675–682

Zhang L, Nolan E, Kreitschitz S, Rabussay D (2002) Enhanced delivery of naked DNA to the skin by non-invasive in vivo electroporation. Biochimica et Biophysica Acta 1572:1–9

Zuber G, Zammut-Italiano L, Dauty E, Behr J (2003) Targeted gene delivery to cancer cells: Directed assembly of nanometric DNA particles coated with folic acid. Angewandte Chemie 42:2666–2699

Physical Chemistry of DNA-Carrier Complexes

KENICHI YOSHIKAWA[1] and YUKO YOSHIKAWA[2]

1 Ordered and Disordered Structures in Compact DNA

The condensation of giant DNA molecules has been actively studied not only in order to develop non-viral gene therapies but also to understand self-regulation of genetic activity in living cells (Lerman 1971; Bloomfield 1996). DNA molecules in viral capsids, bacterial nucleoids, and nuclei of eukaryotes occupy a volume 10^{-4} – 10^{-6} times less than when free in aqueous solution (Livolant 1991; Reich et al. 1994; Bloomfield 1996). Extensive studies on this drastic change in the conformation of DNA, known as DNA condensation, have actively been carried out; for reviews on DNA condensation, the articles of Bloomfield (1991, 1996) are recommended. Unfortunately, most studies dealing with DNA condensation have not clearly distinguished between transitions occurring in individual DNA molecules and those involving aggregation/precipitation of numbers of DNA molecules. This has been due to limitations in available experimental approaches to analyzing DNA condensation, such as light scattering, sedimentation, viscometry, linear dichroism, circular dichroism, and UV spectrometry. Although these methodologies provide useful information on ensembles of DNA molecules, it has been very difficult to gain precise insight into the conformational change of an individual DNA molecule. In discussing this problem, Bloomfield (1996) stated, "Generally, the term 'condensation' is reserved for situations in which the aggregate is of finite size and orderly morphology." Actually, in past literatures, DNA condensation was frequently interpreted as being a cooperative phenomenon, in other words, the transition was regarded as a series of continuous steps. For example, Widom and Baldwin (1983) concluded that "the transition for monomolecular condensation is diffuse", after careful measurement of DNA condensation using light scattering.

Owing to a series of studies on the conformational transition of single DNA molecules, it has recently established that the transition of individual large DNA molecules is of the all-or-none type, or largely discrete (Minagawa et al. 1994; Mel'nikov et al. 1995; K. Yoshikawa et al. 1996).

[1] Department of Physics, Graduate School of Science, Kyoto University, Kyoto 606-8502, Japan
[2] Department of Food and Nutrition, Nagoya Bunri College, Nagoya 451-0077, Japan

The transition between elongated and compact states in a polymer chain is generally referred to as a coil-globule transition (de Gennes 1979; Grosberg and Kholhlov 1994). The term "globule" means of spherical shape, suggesting that the compact state is liquid-drop-like. Following intense discussions and consideration of the single-chain behavior, current standard textbooks in polymer physics concurred on the diffuse nature of the coil-globule transition for any polymer chain (de Gennes 1979; Grosberg and Kholhlov 1994). However, based on the results of theoretical and experimental studies, it has become clear that single polymer chains of a semi-flexible nature are able to undergo all-or-none transitions (Yoshikawa 2001; Yoshikawa and Yoshikawa 2002).

In the following, the intrinsic nature of the folding transition of semi-flexible polymers is discussed. Generally speaking, semi-flexible polymers are considered to be stiff molecular chains with the contour length, or full length, being much larger than the persistence length. The latter is on the order of 50 nm, corresponding to ca. 170 base pairs (bp). If DNA molecules above the size of several tens of bases are considered, then DNA can be regarded as a semi-flexible polymer. DNA molecules below the size of several hundreds of base pairs, by contrast, behave as a rod and do not undergo abrupt transitions. This model provides a rough estimation of the density change between the elongated coil and folded compact states. For example, for a DNA molecule with 30 kilo bp (kbp), the diameter of the double-stranded DNA is $d \approx 2\,nm$ and the Kuhn length, λ, is 100 nm, i.e., double the persistence length. The size of the elongated state is given as $R_c \sim \lambda N^{3/5}$ (de Gennes 1979; Grosberg and Kholhlov 1994), while the size of the compact state is given as $R_g \sim (\lambda d^2 N)^{1/3}$. The density difference between the compact and elongated states is therefore evaluated as $\rho_g/\rho_c = (R_c/R_g)^3 \sim 10^5$. From this rough estimation it becomes clear that large DNA molecules, above the size of several tens of base pairs, exhibit significant difference between the elongated coil and folded compact states.

Most important in relation to the manner of the transition are the morphological changes in the compact state. As illustrated in Fig. 1, when the transition is diffuse or continuous, the final compact state exhibits a spherical morphology with liquid-drop-like properties, whereas when the transition is discrete, an ordered packed state, generating toroid and rod morphologies, results (Noguchi and Yoshikawa 1998).

Figure 2 shows the morphological variations in the compact state generated from a single T4 DNA molecule. As typical shapes, toroid and rod structures are frequently generated with various condensing agents, such as polyamines, PEG, and multivalent metal cations (Bloomfield 1996; Noguchi et al. 1996). At high concentrations of spermidine (3+), a giant toroid is generated, indicating the presence of a loosely packed

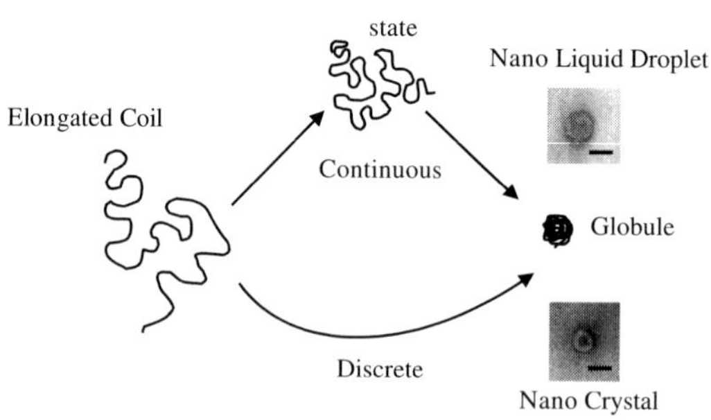

FIG. 1. The nature of the folding transition determines the final state of compaction. When the transition is continuous, the final state is a liquid-drop-like globule, whereas a nano-order crystalline state is obtained through a discrete transition. *Bar*, 0.1 μm

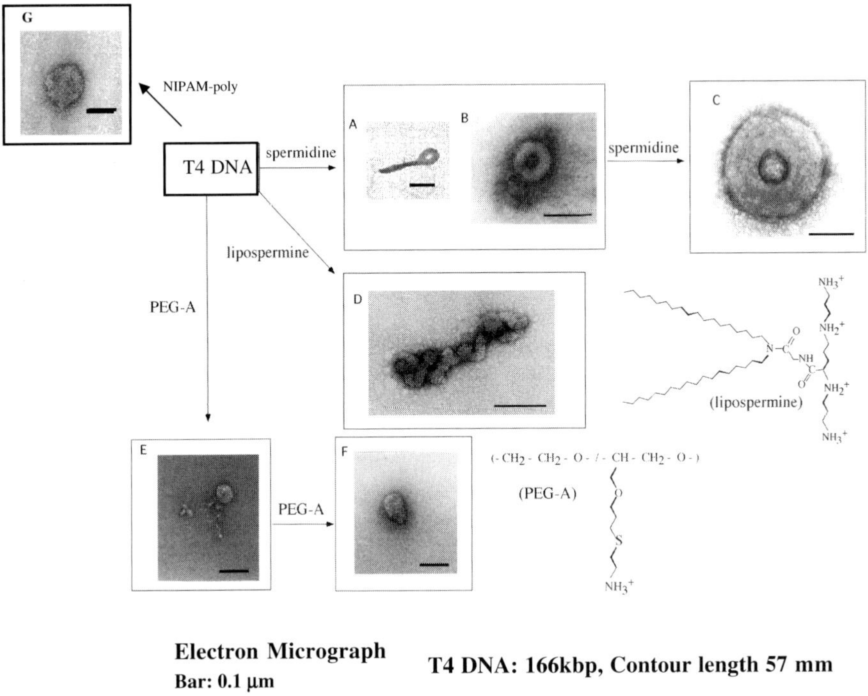

FIG. 2. Various kinds of morphologies generated from a single T4 DNA

state (Yoshikawa et al. 1999). At the top left of the figure, a complex of DNA with an intercalated polymer, exhibiting a spherical structure, is shown. It has been confirmed that, when the folding transition from the coil to the compact state is diffuse, the final compact structure is a liquid-drop-like globule (Iwataki 2000). It should be noted that the size of the spherical globule is significantly larger than that of the tightly packed state, such as yielding toroids and rods, as shown in Fig. 2. This implies that the spherical globule is in a swelled, loosely packed state. At the lower right of Fig. 2, a unique compact structure of the complex with lipospermine (Behr 1989), a potent artificial gene carrier, is shown. The DNA chain winds around a rod-like core and forms a spool-like structure (Y. Yoshikawa et al. 1996).

1.1 Effect of Charge on Compaction

DNA is a highly charged polymer with rather condensed arrays of highly acidic phosphate groups. Under physiological conditions, small counter-cations, such as sodium and potassium, are present at concentrations above mM. Thus, the degree of dissociation of the sodium or potassium salt of the phosphate moieties along DNA chain has to be considered.

Generally speaking, the degree of dissociation of a salt into free ions is described as the competition between the electrostatic energy and the free energy resulting from the translational entropy of ions. In aqueous solution, the electrostatic energy of 1 : 1 ions, i.e., monovalent cations and monovalent anions, is given as $U = -e^2/\varepsilon r$, where e

is the unit charge, ε is the dielectric constant and r is the distance between the charges. The free energy due to the freedom accompanied by the dissociation is around 1 k_BT per single species. Thus, a considerable amount of monovalent cations are condensed around DNA. According to the counter-ion condensation theory (Oosawa 1971; Manning 1978), about 70% of the negative charge of DNA is neutralized owing to condensation of the counter-cation. It is also believed that the degree of counter-ion condensation is insensitive to changes in the bulk concentrations of monovalent ions.

When a DNA molecule is tightly packed, almost all of the negative charge should be neutralized, accompanied by the enhancement of counter-ion condensation. More precisely, the volume part of the compact DNA is fully neutralized whereas negative charge remains on the surface (Yamasaki et al. 2001). It is obvious that an abrupt change in the degree of association of counter-ions is caused by the discrete transition between the elongated coil and folded compact states. In other words, the charge neutralization by cationic species has a significant contribution to the free energy change of the folding transition.

The folding transition of DNA induced by multivalent cations, such as spermidine (3+) and spermine (4+), is inhibited by monovalent ions. Consequently, at higher salt concentrations, larger amounts of multivalent cations are necessary to induce the compaction. However, in the folding transition induced in a crowding environment, the presence of salt is a necessary condition, i.e., salt is a promoting factor for compaction (Vasilevskaya et al. 1995).

1.2 The Difference Between DNA Compaction and Condensation

Up to quite recently, DNA condensation was interpreted as being a stepwise but continuous transition, a mixture of single-chain compaction and multiple-chain aggregation (Bloomfield 1996). Careful examination of the folding transition of individual DNA molecules using the experimental technique of single-chain observation has revealed that, for very large DNA molecules, the transition of individual chains is largely discrete (Vasilevskaya et al. 1995; K. Yoshikawa et al. 1996; Iwataki et al. 2004). The compact DNA thus generated loses almost all of the negative charge, except at the surface. Just after the transition, the compact particle thus behaves as a colloidal particle (Ichikawa et al. 2003). In the case of the folding transition induced by the addition of multivalent cations, the negative surface charge decreases with increasing multivalent ion concentration. As a result, large aggregates or precipitates are formed in the presence of a large excess of multivalent cations.

The negatively charged colloidal property of DNA occurring just after the transition implies that the individual compact state is stably solvable in aqueous solution. Indeed, it has been confirmed that, even after the forced contact of a pair of compact DNAs, these particles dissociate into individual DNA molecules as shown in Fig. 3 (Ichikawa et al. 2003). By contrast, when DNA is compacted with a cationic species containing a hydrophobic moiety, the compact DNA particles stick to each other (Ichikawa et al. 2003). Even in this case, however, the colloidal property is preserved, i.e., during Brownian motion individual DNA molecules tend to remain as individual particles.

Figure 2E shows the generation of an intra-chain segregated structure. The occurrence of such a segregated state is due to the electronic effect of DNA as a polyelectrolyte. When DNA is not tightly packed, the electronic charge remains in the volume part of the compact state. As the coulombic interaction is of long range, the swelled-compact state becomes destabilized. Thus, intra-chain segregation is induced when the compaction is not complete and there is remaining charge around the DNA chain (Yoshikawa et al. 2001; Takagi et al. 2001). Actually, it has been reported that the characteristic size of the segregation in a DNA chain can be controlled by changing the effective charge and the bulkiness of the condensation agents (Zinchenko 2003).

1.3 Higher-Order Structure of DNA and Genetic Activity

How does the higher-order structure affect the biological activity of genomic DNA? Figure 3 shows that compaction of very large, linear DNA completely inhibits the tran-

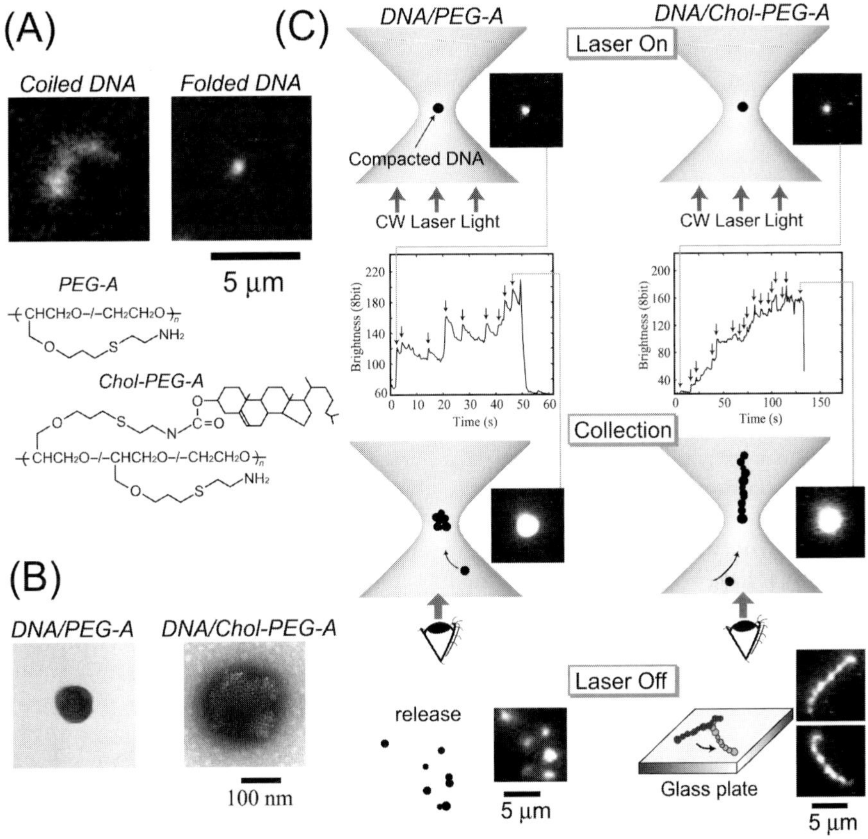

FIG. 3A–C. Colloidal property of compact DNA. As shown to the right of **C**, multiples of compact DNA are forced into contact with each other in the presence of laser tweezers; after turning off the laser, individual compact DNA molecules tend to dissolve each other. When a hydrophobic group is introduced to the compact DNA particle, the DNA molecules tend to stick to each other, as shown to the right of **C**. **A** Fluorescence microscopy images of elongated and coil DNA molecules. **B** Electron micrograph of the compact DNA molecules. Usual folded DNA (*left*), complex with a hydrophobic moiety (*right*) (Ichikawa et al. 2003)

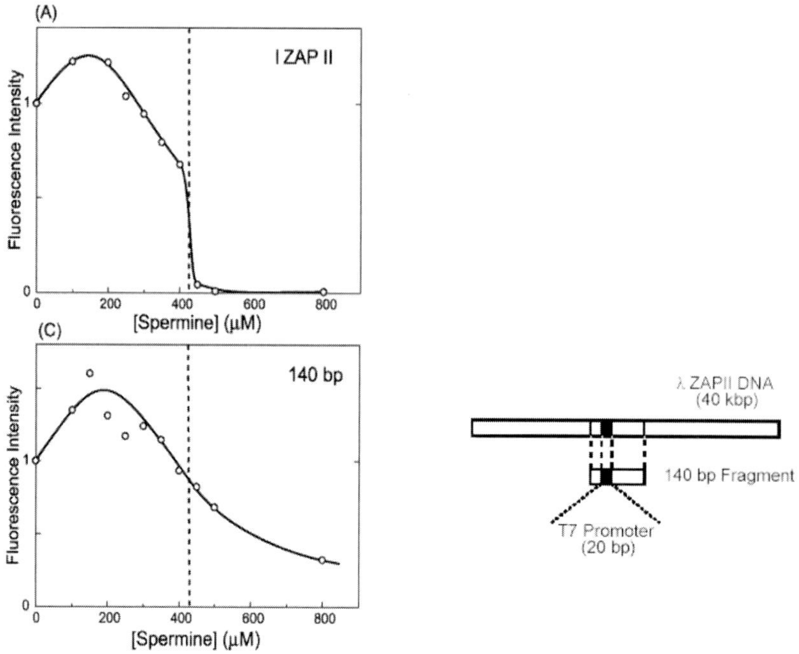

FIG. 4. The transcription of a very large DNA, λZAPII, is completely inhibited, whereas short-fragment DNA (140 bp) is inhibited only slightly (Tsumoto et al. 2003)

scriptional activity (Tsumoto et al. 2003), whereas short DNA molecules do not undergo significant changes on the transcription in response to a wide range of spermidine concentrations.

In the literature, some studies have reported promotion whereas others reported inhibition of enzymatic reactions accompanied by DNA condensation. Similarly, quite recently it has become clear that the folding transition of circular plasmid DNA with superhelicity is of a continuous nature. Even in the presence of condensing agents such as spermidine; this result is in contrast to the discrete nature of the transition of linear DNA.

Based on physicochemical insights into the folding transition of DNA, it has now become possible to propose a novel hypothesis regarding the relationship of the higher-order structure of DNA and its biological activity. When the compaction is complete and DNA molecule takes on ordered packing, there is complete inhibition of enzymatic action, such as transcription. However, when the folded structure is a swollen globule, enzymes can access DNA molecules. Indeed, folded DNA from circular plasmid exhibits high enzymatic sensitivity.

Based on studies of chromatin structure, it has been established that tightly packed and loosened parts coexist, except for during mitosis. As for the loosened region, both ends of the DNA chain are fixed to the scaffold of condensed chromatin. Thus, a loop structure is generally formed, suggesting a similar effect on circular DNA as described in the preceding section. This hypothesis can explain the on/off switching activation of a large number of genes during cell differentiation (Yoshikawa and Yoshikawa 2002).

It is highly expected that precise knowledge of DNA compaction will contribute to the development of non-viral vectors, and also to a full understanding of the mechanism of self-regulation of genetic activity in living cells (Yoshikawa 2002).

References

Behr JP, Demeneix B, Loeffler JP, Perez-Mutul J (1989) Efficient gene transfer into mammalian primary endocrine cells with lipopolyamine-coated DNA. Proc Natl Acad Sci USA 86:6982–6986

Bloomfield VA (1991) Condensation of DNA by multivalent cations: consideration on mechanism. Biopolymers 31:1471–1481

Bloomfield VA (1996) DNA condensation. Curr Opin Struct Biol 6:334–341

de Gennes PG (1979) Scaling Concepts in Polymer Physics, Cornell Univeristy Press, N. Y.

Grosberg A Yu, Kholhlov AR (1994) Statistical Physics of Macromolecules. AIP Press, N. Y.

Ichikawa M, Matsuzawa Y, Koyama Y, Yoshikawa K (2003) Langmuir 19:5444–5447

Iwataki T, Yoshikawa K, Kidoaki S, Umeno D, Kiji M, Maeda M (2000) Cooperativity vs. Phase Transition in a Giant Single DNA Molecule. J Am Chem Soc 122:9891–9896.

Iwataki T, Kidoaki S, Sakaue T, and Yoshikawa K (2004) Competition between compaction of single-chains and bundling of multiple. J Chem Phys 120:4004–4011

Lerman LS (1971) A transition to a compact form of DNA in polymer solutions. Proc Natl Acad Sci USA 68:1886–1890

Livolant F (1991) Ordered phases of DNA in vivo and in vitro. Physica A 176:117–137.

Manning GS (1978) The molecular theory of polyelectrolyte solutions with applications to the electrostatic properties of polynucleoides. Quarterly Rev Biophys 11:179–246

Mel'nikov SM, Serveyev VG, Yoshikawa K (1995) Discrete coil-globule transition of large DNA by cationic surfactant. J Am Chem Soc 117:2401–2408.

Minagawa K, Matsuzawa Y, Yoshikawa K, Khokhlov AR, Doi M (1994) Direct observation of the coil-globule transition in DNA molecules. Biopolymers 34:555–558

Noguchi H, Yoshikawa K (1998) Morphological Variation in a Collapsed Single Homopolymer Chain. J Chem Phys 109:5070–5077

Noguchi H, Saito S, Kidoaki S, Yoshikawa K (1996) Self Organized Nanostructure Constructed with a Single Polymer Chain. Chem Phys Lett 261:527–533

Oosawa F (1971) Polyelectrolytes. Marcel Dekker, New York

Reich Z, Ghirlando R, Minsky A (1991) Secondary conformational polymorphism of nucleic acids as a possible functional link between cellular parameters and DNA packaging processes. Biochemistry 30:7828–7836.

Takagi S, Tsumoto K, Yoshikawa K (2001) Intra-Molecular phase segregation in a single polyelectrolyte chain. J Chem Phys 114:6942–6949

Tsumoto K, Francois L, Yoshikawa K (2003) Biophys Chem 106:23–29

Vasilevskaya VV, Khokhlov A R, Matsuzawa Y, Yoshikawa K (1995) Collapse of single DNA in poly(ethylene glycol) solutions. J Chem Phys 102:6595–6602

Widom J, Baldwin RL (1983) Monomolecular condensation of DNA incuded by cobalt hexamine. Biopolymer 22:1595–1620

Yamasaki Y, Teramoto Y, Yoshikawa K (2001) Disappearance of the negative charge in giant DNA with a folding transition. Biophys J 80:2823–2832

Yoshikawa K (2001) Controlling the Higher-Order Structure of Giant DNA Molecules. Adv Dru Del Rev 52:235–244

Yoshikawa K (2002) Field Hypothesis on the Self-regulation of Gene Expression. J Biol Phys 28:701–712

Yoshikawa K, Takahashi M, Vasilevskaya VV, Khokhlov AR (1996) Large discrete transition in a single DNA molecule appears continuous in the ensemble. Phys Rev Lett 76: 3029–3031

Yoshikawa K, Yoshikawa Y (2002) Compaction and condensation of DNA, in *"Pharmaceutical Perspectives of Nucleic Acid-Based Therapeutics"*, eds., Mahato R I et. al., Taylor & Francis

Yoshikawa Y, Emi N, Kanbe T, Yoshikawa K, Saito H (1996) Folding and aggregation of DNA chains induced by complexation with lipospermie: formation of a nucleosome-like structure and network assembly. FEBS Lett 396:71–76

Yoshikawa Y, Velichko YS, Ichiba Y, Yoshikawa K (2001) Self-assembled pearling structure of long duplex DNA with histone H1. Eur J Biochem 268:2593–2599

Yoshikawa Y, Yoshikawa K, Kanbe T (1999) Formation of a giant toroid from long duplex DNA. Langmuir 15:4085–4088

Zinchenko AA, Sergeyev VG, Murata S, Yoshikawa K (2003) J Am Chem Soc 125:4414–4415

Liposomes

Kazuo Maruyama

1 Liposomal Drug Delivery Systems for Gene Delivery In Vivo

It was the German bacteriologist Paul Ehrlich who, in the late nineteenth century, coined the term "magic bullet," meaning a chemical that travels through the body and selectively kills diseased cells without harming neighboring healthy ones (Ehrlich 1956). Since then, many different approaches based on various physical and biochemical principles have been examined with the goal of developing systems with a therapeutically acceptable degree of target specificity (Poste and Kirsh 1983; Gupta 1990; Rowlinson-Busza and Epenetos 1992; Cummings and Smyth 1993).

Among the different approaches to drug delivery, immunoliposomes, using an antibody as a targeting ligand and a lipid vesicle, as carriers for both hydrophobic and hydrophilic drugs have attracted much attention. It has been demonstrated that the specific delivery of drugs to target cells is far more efficient with immunoliposomes than with liposomes lacking antibody (Wright and Huang 1989). The success of in vitro delivery to target cells using immunoliposomes (corresponding to type A in Fig. 1) has prompted similar experiments in vivo. However, targeting of immunoliposomes in vivo is far more complicated, as studies in vivo have revealed that bound antibodies lead to enhanced uptake of the immunoliposomes by the reticuloendothelial system (RES) (Aragnol and Leserman 1986; Derksen et al. 1988; Peeters et al. 1987), and that targeting efficiency depends on the antibody density on the immunoliposome surface (Maruyama et al. 1990). In addition, rapid uptake by the RES and endothelial barriers separating blood and tissues largely prevent immunoliposomes from reaching their target cells. Thus, highly efficient targeting and a relatively low level of RES uptake of the immunoliposomes are apparently mutually exclusive. As systemic administration is the most practical route for treatment, immunoliposomes overcoming these physiological barriers are highly desirable. The development of liposomes with RES-avoidance is a necessary first step in this direction.

School of Pharmaceutical Sciences, Teikyo University, Sagamiko, Tsukui-gun, Kanagawa 199-0195, Japan

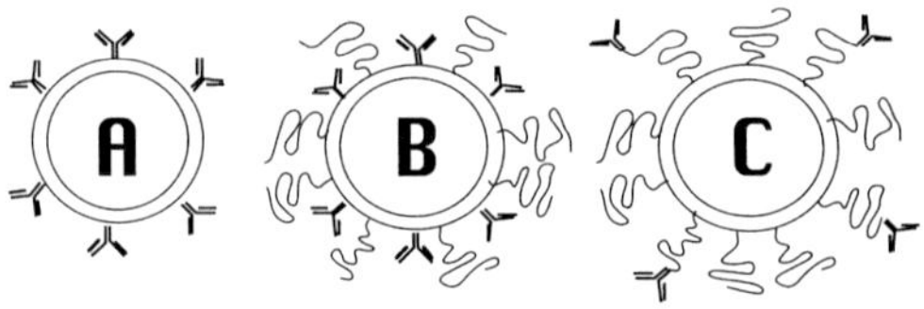

FIG. 1. Schematic illustration of the immobilization of antibody on liposomes. Type A: PEG-free immunoliposomes with antibody covalently linked to the short anchor *N*-glutaryl-phosphatidylethanolamine (NGPE); type B: PEG-immunoliposomes with antibody covalently linked to NGPE; type C: PEG-immunoliposomes with antibody attached to the distal terminal of distearoyl-*N*-(3-carboxypropionoyl poly(ethylene glycol) succinyl) phosphatidylethanolamine (DSPE-PEG-COOH), -maleimide or -hydorazide, so-called pendant-type PEG-immunoliposomes

A major development in the last few years has been the synthesis of liposomes with a prolonged circulation time in blood, commonly called long-circulating or sterically stabilized liposomes. Liposomes containing polyethylene glycol derivatives of phosphatidylethanolamine (PEG-lipid) (Klibanov et al. 1990; Blume and Cevc 1990; Allen et al. 1991; Maruyama et al. 1992) are not readily taken up by macrophages in the RES, and hence remain in the circulation for a relatively long period of time. Pharmacokinetic analysis and therapeutic studies in tumor-bearing mice, in which elevated liposome accumulation was shown, have demosntrated the potential of PEG-liposomes as drug carriers for use in cancer therapy (Lasic and Martin 1995; Gabizon and Papahadjopoulos 1988; Unezaki et al. 1995; Unezaki et al. 1994). Furthermore, results from preclinical studies with doxorubicin encapsulated into PEG-liposomes revealed an increased therapeutic efficacy compared to the free drug or drugs encapsulated in conventional liposomes. These new formulations of long-circulating liposomes should allow the development of immunoliposomes with both long survival times in the circulation and effective target recognition in vivo. PEG is very useful because of its ease of preparation, relatively low cost, controllable molecular weight and the ability to link it to lipids or protein (including antibody) by a variety of methods.

Given a suitable antibody with high specificity and affinity for the target antigen, the critical factor is the accessibility of target cells to the immunoliposomes. Efficient target binding of the injected immunoliposomes occurs only when the target cell is in the intravascular compartment or is accessible through leaky vascular structures. Thus, in terms of targeting drug delivery by immunoliposomes, two anatomical compartments can be considered. One is a readily accessible intravascular site, such as the vascular endothelial surface, leukemia cells, T cells, B cells or a thrombus. The other is a much less accessible extravascular site, such as a solid tumor, an infection site, or an inflammation site, where the vascular structure is leaky.

1.1 Targeting to a Readily Accessible Site

1.1.1 In Vivo Immunoliposome Targeting to the Lung Endothelial Surface

Immunoliposomes containing monoclonal antibodies specific for mouse pulmonary endothelium were used as a model system. The antibody 34A binds to a surface glycoprotein (gp112) expressed at high concentration in the mouse lung (Kennel et al. 1988). It has been demonstrated that mouse lung contains large amounts of gp112 (500 to 700 ng/mg protein), whereas other organs have very low (spleen, uterus) or undetectable (liver) levels. The anatomy and physiology of the lung favor its use as a model system for a number of reasons: all the blood circulates through the lung capillaries, the flow rate of blood in the lung is relatively slow for the purpose of oxygen and carbon dioxide exchange, and the largest capillary bed in the body is found in this organ. This convenient model has allowed us to investigate the effects on target binding of various physical parameters of immunoliposomes, such as the antibody-to-lipid ratio, liposome size, and injection dose.

Liposomes composed of egg phospahtidylcholine (PC), cholesterol (Chol) and *N*-glutaryl-phosphatidylethanolamine (NGPE) with a molar ratio of 5:5:1 were conjugated to the antibodies via NGPE (Maruyama et al. 1990). The resulting 34A-immunoliposomes (corresponding to type A in Fig. 1) were then injected into mice (Balb/c, 6–8 weeks old, male) via the tail vein.

Among the several parameters examined, the most important one seems to be the antibody-to-lipid ratio of the immunoliposomes. As shown in Table 1, immunoliposomes (Maruyama et al. 1990) with antibody 34A were prepared with various initial antibody-to-lipid ratios. The results clearly showed that there was a direct relationship between the antibody density of the immunoliposomes and the extent of their binding to the lung target. Approximately 60% of the injected dose was found in the lung within 15 min after injection when the immunoliposomes contained an average of 935 antibody molecules per liposome. Uptake by the liver and spleen followed an opposite trend from that of the lung, i.e. the accumulation of liposomes in the liver and spleen decreased as the antibody-to-lipid ratio increased, suggesting that uptake of immunoliposomes by the liver and spleen might limit the binding of immuno-

TABLE 1. Effect of antibody-to-lipid ratio of immunoliposomes on biodistribution. Immunoliposomes with different antibody-to-lipid ratio were prepared by employing different input ratios of antibody to lipids. The number of antibody molecules per liposome was calculated from the known conjugation efficiency for each preparation of immunoliposomes and the average diameter of the liposomes. The number of lipid molecules per liposome at a given diameter (~250 nm) was estimated according to the method of Enoch and Strittmatter (1979). Data represent the biodistribution of liposomes in the mice 15 min after liposome administration (i.v.). The percent of injected dose value for each ratio of antibody to lipid is the average ± S.D. of three mice. *BLP*, bare liposomes; *34A-LP*, 34A-immunoliposomes

Characteristic	BLP	34A-LP			
Initial antibody-to-lipid ratio (w/w)	—	1:50	1:10	1:5	1:1
Conjugation (%)	—	60	53	57	48
Size (nm)	250	224	236	234	247
No. antibody molecules/liposome	0	24	101	219	935
Organ	Percent injected dose				
Lung	0.4 ± 0.0	3.0 ± 0.3	20.0 ± 1.2	35.3 ± 2.4	59.8 ± 0.4
Blood	10.9 ± 0.7	7.9 ± 0.8	4.8 ± 0.2	2.9 ± 0.0	2.1 ± 0.1
Liver	49.7 ± 1.6	55.3 ± 1.7	41.7 ± 1.8	34.2 ± 2.5	12.1 ± 0.3
Kidney	1.2 ± 0.0	1.2 ± 0.1	1.0 ± 0.1	1.1 ± 0.0	1.3 ± 0.1
Spleen	4.3 ± 0.6	3.7 ± 1.0	3.3 ± 0.2	1.8 ± 0.3	0.6 ± 0.1

liposomes to the lung target. Thus, antibody density is an important factor for target binding in the blood.

1.1.2 The Influence of Prolonged Circulation Time of Liposomes on the Efficiency of Immunoliposome Targeting

The addition of PEG-lipid to liposomes can significantly prolong their circulation time in the blood when the conjugate is incorporated into the liposome membrane (Blume and Cevc 1990; Allen et al. 1991; Maruyama et al. 1992; Lasic and Martin 1995). The role of PEG in 34A-immunoliposome binding to the lung target was examined in the above animal model system. A series of PEG-immunoliposomes were designed, as shown in Fig. 1 (Maruyama et al. 1995). Type A consisted of PEG-free immunoliposomes with the antibody covalently linked to a short anchor. In type B PEG-immunoliposomes, the antibody was covalently linked to the short anchor such that antibody molecules and PEG molecules coexisted on the liposome surface. Type C consisted of a new type of PEG-immunoliposome, in which the antibody or antibody fragments were attached to the PEG terminus, forming so-called pendant-type PEG-immunoliposomes. The location of antibodies on the surface of the liposome can be clearly identified in the backscatter electron microscopy image (Takizawa et al. 1998). For the preparation of type C immunoliposomes, several functionalized PE-PEG derivatives, such as those bearing a carboxyl residue, maleimide residue, or hidorazide residue, at the end of the PEG molecule, were newly synthesized. Ideally, the coupling method should be both simple and rapid, producing a stable, non-toxic bond.

Plain liposomes for the preparation of type A or type C immunoliposomes were composed of egg PC and Chol (2:1, m/m) with 6 mol% NGPE or DSPE-PEG-COOH (distearoyl-*N*-(3-carboxypropionoyl poly(ethylene glycol) succinyl) phosphatidylethanolamine), each with an average molecular weight of 3,000. The plain liposomes for type B were composed of egg PC and Chol (2:1, m/m) with 6 mol% NGPE and 6 mol% DSPE-PEG with an average molecular weight of 3,000. Small unilamellar vesicles (SUVs) (90–130 nm in diameter) were prepared by the Reversed-phase evaporation vesicle (REV) method followed by extrusion through Nuclepore filters (0.1 μm). The carboxyl residues in the plain liposomes were activated and coupled to 34A antibodies.

In order to test the effects of antibody position and steric hindrance by the PEG chains on the behavior of immunoliposomes in vivo, three different types of 34A-immunoliposomes with similar numbers of antibody molecules per liposome were prepared and their targeting ability and biodistribution were evaluated in mice. As summarized in Table 2, type A, B and C 34A-immunoliposomes, containing approximately 35, 30 and 30 antibody molecules per liposome, respectively, were prepared and injected into mice via the tail vein (Fig. 2). Using the 34A-type A liposomes, with an average of 35 antibody molecules per liposome, 42.5% of the injected dose accumulated in the lung. The 34A-type B immunoliposomes showed a lower level of target binding and a significantly higher blood level than those of type A. In the case of 34A-type C, containing 30 antibody molecules per vesicle, the degree of target binding to the lung was 56.6% of the injected dose, 1.5-fold higher than that of type A. The target binding of 34A-type B is comparable with that of 34A-G_{M1}-immunoliposomes. Although long-circulating liposomes can also be obtained by incorporating PE-PEG

TABLE 2. Characteristics of 34A-immunoliposomes. Plain liposomes used for preparing type A or type C consisted of egg phospahtidylcholine:cholesterol (PC:Chol) (2:1, m/m) with 6 mol% *N*-glutaryl-phosphatidylethanolamine (NGPE) or distearoyl-*N*-(3-carboxypropionoyl poly(ethylene glycol) succinyl) phosphatidylethanolamine (DSPE-PEG-COOH) of average molecular weight 2000, respectively. The plain liposomes for type B were composed of egg PC:Chol (2:1, m/m) with 6 mol% NGPE and DSPE-PEG of average molecular weight of 2000. The average number of antibody molecules per liposome was estimated according to the method of Enoch and Strittmatter (1979)

	34A-type		
	A	B	C
PEG content (mol% of total lipid)	0	6	6
Mean diameter (nm)	121	111	122
Initial antibody:lipid ratio (w/w)	1.6	1.6	1.5
Conjugation efficiency (%)	35.6	31.8	24.8
Number antibody molecules/liposome	35	30	30

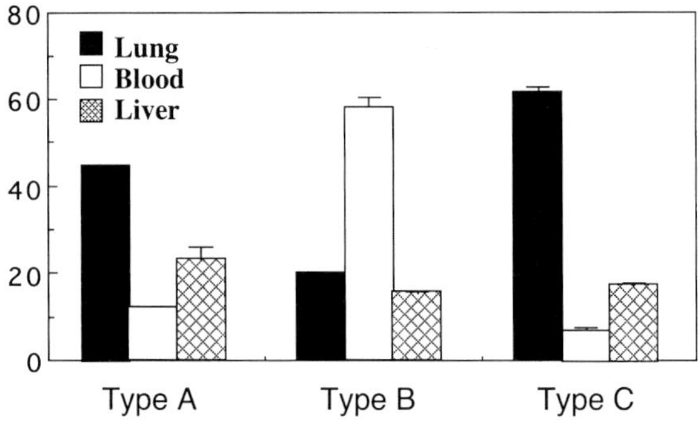

FIG. 2. Comparison of target binding to the lung and biodistribution (blood and liver) of the three different types of 34A-immunoliposomes. 34A-type A, B, and C conjugates contained approximately 35, 30, and 30 antibody molecules per liposome, respectively. Biodistribution (% of dose) was measured 30 min after intravenous injection. For details, see Table 2

derivatives in place of G_{M1}, inclusion of DSPE-PEG with an average molecular weight of 3000 reduced the target binding of 34A-type B. This effect depended on the chain length of PEG, suggesting that, although PEG prolongs the circulation time of immunoliposomes, it sterically hinders the binding of immunoliposomes to their target sites (Klibanov et al. 1991; Mori et al. 1991). Therefore, it was proposed that the use of longer-chain PEG with antibodies attached at the distal terminal of the PEG chain (type C) would afford immunoliposomes with both prolonged circulation time and good target binding (Klibanov et al. 1991; Klibanov and Huang 1992).

The efficiency of 34A-type C binding to the target as a function of the antibody content was evaluated A series of 34A-type C immunoliposomes was prepared with

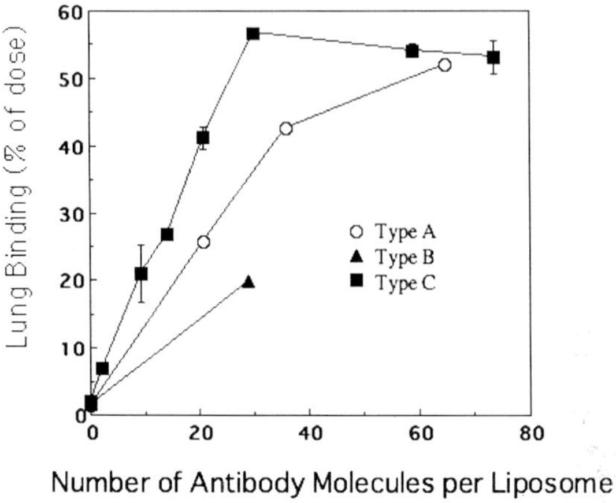

FIG. 3. Effect of antibody density on the lung targeting of 34A-type A, B, and C immunoliposomes. Lung binding (% of dose) was measured 30 min after injection

various initial antibody-to-lipid ratios. The final number of antibody molecules per liposome varied from 0 to 74, but the average sizes of the immunoliposomes were approximately the same, 90–130 nm in diameter. It is clear from the data in Fig. 3 that the efficiency of lung targeting was dependent on the antibody content of the immunoliposomes. Liposomes containing small numbers of antibody molecules per liposome accumulated in the lung at low levels and were retained in the blood at high levels. By contrast, linking an average of 74 antibody molecules per liposome on the PEG terminals resulted in the accumulation of 53% of the injected dose in the lung and only 7% in the blood. Lung binding reached a plateau at about 30 antibody molecules per liposome and a further increase in antibody content caused increased liver uptake. Thus, antibody density is an important factor for target binding even in the targeting of type C immunoliposomes. At low numbers of antibody molecules per liposome, such as 2 and 9 molecules, free PEG favors the evasion of RES uptake of the liposomes. This led to the high blood residence observed for 34A-type C immunoliposomes and a low efficiency of target binding due to the low antibody content. Type C liposomes showed higher immunotargetability than those of type A and type B at low antibody content (less than 30 antibody molecules per vesicle). Thus, type C is accumulated more effectively in the lung than the other immunoliposomes, in spite of the low antibody content.

1.1.3 Lymphoblastic Leukemia Cells

Patients with Philadelphia-chromosome-positive acute lymphoblastic leukemia (Ph^+ ALL) have a poor prognosis despite intensive therapeutic intervention. Recently, imatinib, a BCR-ABL tyrosine kinase inhibitor, has been proven to be an effective treatment for Ph^+ ALL, but nearly all patients rapidly acquire resistance. While high-dose imatinib administration might overcome this resistance, systemic toxicities would likely limit this approach. Therefore, a new delivery system allowing for the specific targeting of imatinib is needed. Since almost all Ph^+ ALL cells express CD19 on their surfaces, an immunoliposome carrying anti-CD19 antibody (CD19-type C) was developed (Harata et al. 2004). The internalization efficiency of CD19-type C approached

100% in all Ph+ ALL cells but was very low in CD19-negative cells. The cytocidal effect of imatinib-encapsulated CD19-type C on Ph+ ALL cell lines and primary leukemia cells from Ph+ ALL patients was much greater than that of imatinib with or without control liposomes. Importantly, imatinib- CD19-type C did not affect colony formation by CD34-positive hematopoietic cells, even at inhibitory concentrations of free imatinib. Taken together, these data clearly demonstrates that the imatinib- CD19-type C induces the specific and efficient death of Ph+ ALL cells. This new therapeutic approach may be an effective therapy for patients with Ph+ ALL, with fewer side effects than free imatinib.

2 Targeting to a Less Accessible Site (Solid Tumor)

2.1 The Influence of the Prolonged Circulation Time of Liposomes on Their Extravasation into Solid Tumor Tissue

Generally, the capillary permeability of the endothelial barrier in newly vascularized tumors is significantly greater than that of normal tissues outside the RES, which have continuous and nonfenestrated vascular endothelia, and the extravasation of macromolecules or liposomes is greatly limited (Jain and Gerlowski 1986). By contrast, the extravasation of circulating molecules from blood vessels to tumor tissue is a function of both local blood flow and microvascular permeability (Dvorak et al. 1988). In addition, since there is little or no lymphatic drainage in tumor tissues, accumulated macromolecules are retained in the tumor interstitium for a prolonged period of time. This phenomenon, termed the enhanced permeability and retention (EPR) effect, has been shown to occur universally in tumors (Maeda 1992; Maeda et al. 1992), and it is conceivable that long-circulating liposomes could take advantage of the EPR effect for efficient target binding in tumors.

The extravasation of liposomes of different sizes into solid tumors was examined in various tumor models in mice (Unezaki et al. 1995, 1996). As shown in Fig. 4, long-

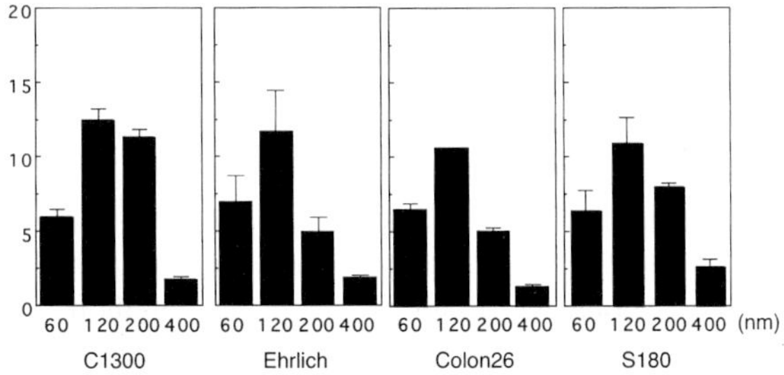

FIG. 4. Effect of liposomal diameter on the accumulation of PEG-liposomes in various implanted tumors in mice. PEG-liposomes composed of DSPC : Chol : DSPE-PEG (1 : 1 : 0.13, m/m) and containing ^{67}Ga-deferoxamine were intravenously injected into tumor-bearing mice. The mice were prepared by inoculating the following cells (1×10^7 cells) into the hind foot and used when the tumor had reached a diameter of 8 mm: mouse C-1300 neuroblastoma to A/J Sic mouse, mouse colon 26 adenocarcinoma to BALB/c mouse, mouse Ehrlich (1×10^7 cells) to ddY mouse or mouse sarcoma 180 to ddY mouse. Biodistribution was estimated at 6 h after injection

circulating liposomes composed of DSPC/Chol/DSPE-PEG (1:1:0.13, m/m), with an average diameter of 100–200 nm, accumulated efficiently in all tumor tissues examined. Clearly, liposome size is also an important factor for extravasation. Observations using fluorescence microscopy have shown that PEG-liposomes can indeed extravasate beyond the endothelial barrier, mainly through postcapillary venules (Unezaki et al. 1996; Huang et al. 1992; Ishida et al. 1999). Due to the increased circulation time of liposomes containing PE-PEG and the leaky structure of the microvasculature in solid tumor tissue, these liposomes accumulate preferentially in the tumor tissue. Thus, under physiological tumor conditions, only small liposomes ranging from 100–200 nm in diameter with a prolonged circulation half-life have a high probability of encountering the leaky vessels of the tumor tissue.

Recent studies have shown that PEG-liposomes encapsulating anthracyclines such as doxorubicin (DXR) exhibit improved therapeutic efficacy and reduced toxicity after i.v. injection into solid-tumor-bearing mice (Unezaki et al. 1995; Papahadjopoulos et al. 1991; Gabizon 1992).

2.2 Characterization of In Vivo Immunoliposome Targeting to Solid Tumors

In order to study whether immunoliposomes injected intravenously can extravasate into solid tumors and bind to tumor cells, we established a models system consisting of the monoclonal antibody 21B2, specific for human carcinoembryonic antigen (CEA), and mice bearing CEA-positive human gastric cancer strain MKN-45 (Uyama et al. 1994). Antibody 21B2 was isolated from BALB/c mice after immunization with human CEA antigen, purified from cells of a CEA-producing human gastric cancer line, MKN-45. Fab′ fragments of 21B2 were prepared by pepsin digestion of the antibody and 2-aminoethanethiol reduction of the F(ab′)2 fragments (Ishikawa et al. 1983). DPPE-PEG with a terminal maleimidyl group was synthesized for the preparation of Fab′-type C immunoliposomes (Maruyama et al. 1997a, 1997b). Two million MKN-45 cells were inoculated into the back of female BALB/c nu/nu mice. When the estimated tumor weight reached about 300 mg, the mice were used for experiments.

As shown in Fig. 5, PEG-Mal liposomes without the antibody showed prolonged residence in the circulation and low liver uptake, regardless of the presence of the terminal maleimidyl group. There were no marked differences in tissue distribution among liposomes containing DSPE-PEG, DSPE-PEG-COOH and DPPE-PEG-Mal, consistent with previous results (Rowlinson-Busza and Epenetos 1992; Aragnol and Leserman 1986; Derksen et al. 1988). The 21B2-type C conjugates, bearing approximately 51 whole antibody molecules per liposome, were rapidly cleared from the blood and were found entirely in the liver. In contrast, Fab′-type C conjugates, bearing approximately 517 Fab′ molecules per liposome, were retained longer in the circulation with a concomitant decrease in liver uptake compared with 21B2-type C. These results indicated that linkage of whole 21B2 antibodies to the PEG terminal enhanced RES uptake via a Fc-receptor-mediated mechanism (Aragnol and Leserman 1986; Derksen et al. 1988). This problem can be overcome by using the Fab′ fragment. In the case of Fab′-type C, the absence of the Fc portion and the presence of free PEG-Mal (not linked to the Fab′ fragment) may play a role in the prolonged circulation of the liposomes. Thus, the Fab′ fragment is much better than whole IgG in terms of

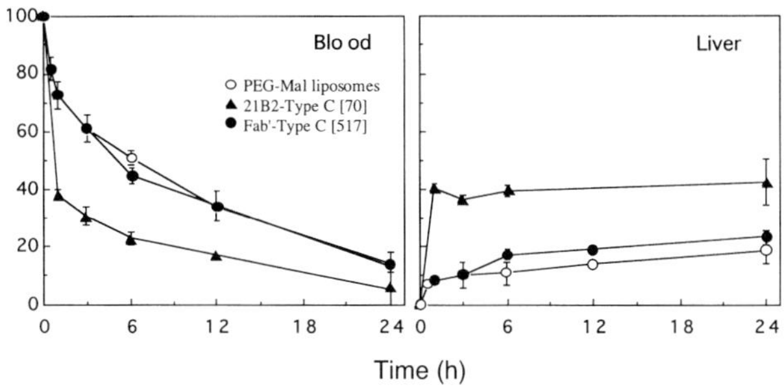

FIG. 5. Time course of blood residence and liver uptake of PEG-Mal liposomes, 21B2-type C [51] or Fab′-type C [517] in MKN-45-bearing BALB/c *nu/nu* mice. Two million MKN-45 cells were inoculated into the backs of female BALB/c *nu/nu* mice. The numbers in square brackets represent the average numbers of antibody or Fab′ molecules per liposome

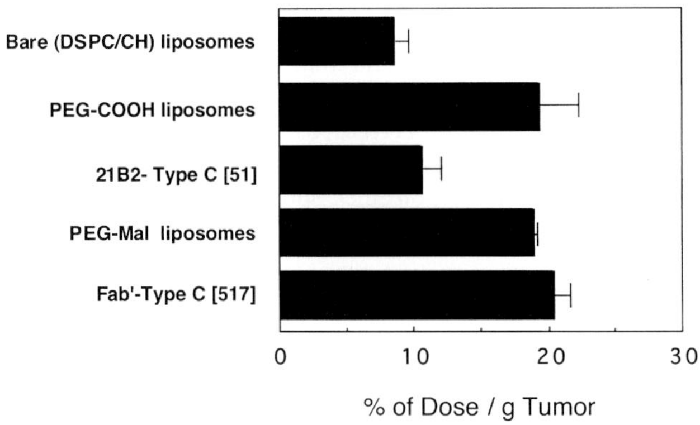

FIG. 6. Accumulation of immunoliposomes in solid tumors in MKN-45-bearing BALB/c *nu/nu* mice at 24 h after injection. Two million MKN-45 cells were inoculated into the backs of female BALB/c *nu/nu* mice. The numbers in square brackets represent the average numbers of antibody or Fab′ molecules per liposome

designing PEG-immunoliposomes with a prolonged circulation time. Furthermore, using the Fab′ fragment should greatly reduce the antigenicity.

The next question to arise was whether Fab′-type C could extravasate into solid tumor tissue and bind to tumor cells. Figure 6 shows the accumulation of Fab′-type C, 21B2-type C and comparable PEG-liposomes, with an average diameter of 100–130 nm, in MKN-45 solid tumor in mice at 24 h after injection. A relatively high accumulation was obtained with PEG-COOH liposomes, PEG-Mal liposomes and Fab′-type C immunoliposomes. These results clearly correlated with the prolonged circulation time. The accumulation rate of Fab′-type C was two-fold higher than that of 21B2-type C or bare liposomes, and equal to that of PEG-Mal and PEG-COOH liposomes.

2.3 Transferrin-Type C In Vivo Intracellular Targeting to Solid Tumors

Transferrin (TF) is a glycoprotein that transports ferric ion in the body. TF receptors are abundant in cancerous tissues and reflect the growth potential of the tumor. It is therefore reasonable to assume that TF receptors might be available as a target molecule for therapy (for review see Wagner et al. 1994). TF-receptor-mediated endocytosis is a normal physiological process by which TF delivers iron to cells (Huebers and Finch 1987; Aisen 1994). TF-type C conjugates, bearing approximately 25 TF molecules per liposome, readily bound to mouse Colon 26 cells in vitro, and were internalized by receptor-mediated endocytosis. Coupling of TF molecules did not cause enhanced RES uptake of liposomes, presumably because TF is an abundant serum glycoprotein. TF-PEG-liposomes had a prolonged residence time in the circulation and low RES uptake in Colon 26 tumor-bearing mice, resulting in enhanced extravasation of the liposomes into the solid tumor tissue. Electron microscopy studies in Colon 26 tumor-bearing mice revealed that the extravasated TF-PEG-liposomes were internalized into tumor cells by receptor-mediated endocytosis. TF-PEG-liposomes were taken up into endosomal-like intracellular vesicles, as visualized by transmission electron microscopy (data not shown) and maintained a high drug level in the tumor for over 72 h after injection. This high retention indicates cellular uptake of the extravasated TF-PEG-liposomes by TF-receptor-mediated endocytosis (Ishida et al. 2001; Maruyama et al. 2004). Thus, TF-PEG-liposomes are potential tools for in vivo cytosolic delivery of active macromolecules, such as genes or oligonucleotides.

3 Liposomal Gene Delivery Systems

The field of non-viral vector-mediated gene therapy, and particularly the use of cationic liposomes, has made great strides between the initial report by Felgner et al., in 1987, and their use in the world's first human gene-therapy clinical trial by Nabel et al. in 1992 (Nabel et al. 1993). Since then, various formulations of cationic liposomes have been developed for gene transfer and have been widely used for the in vitro transfection of eukaroytic cells (Li et al. 1999).

Cationic lipids are an interesting alternative to viral vector-mediated gene delivery in vitro and in vivo, and a large number of families of cationic vectors have been synthesized (Miller 1998). These vectors are easily prepared, and their production can be scaled-up. Most of them aim for in vitro gene delivery of plasmid DNA to give transient expression using cultured cells in culture Transfection kits for the in vitro delivery of genes or oligonucleotides into cells are important research reagents in many fields, including agriculture, medicine, and biotechnology, and are marketed worldwide.

Our goal is to deliver therapeutically active macromolecules as genes or oligonucleotides into the cytosol of target cells. The persisting key research issues is how to design the gene delivery systems to achieve significant cytosolic delivery of plasmic DNA. Many small drug molecules are readily taken up into cells by passive diffusion across cell membranes, but a growing number of pharmacologically active agents emerging from the field of biotechnology, such as proteins, peptides and nucleic acids

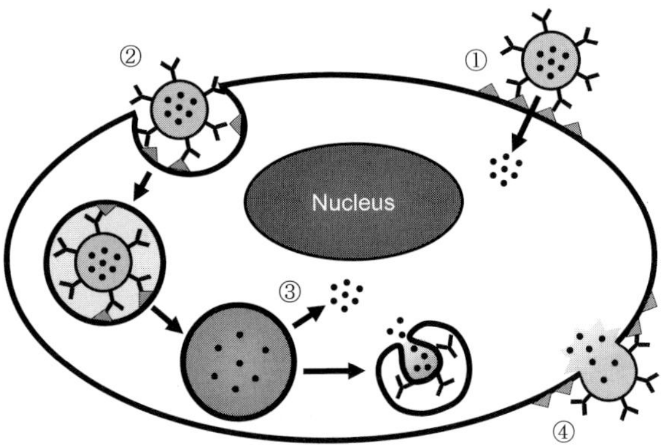

FIG. 7. Schematic representation of potential way by which targeted liposomes can achieve cytosolic drug delivery. *1* Transfer of lipophilic compounds from the liposomal bilayer to the plasma membrane. *2* Receptor-mediated endocytosis of immunoliposomes with subsequent cytosolic release of liberated drug molecules by passive diffusion out of the endosomes. *3* Cytosolic release of liberated drug molecules following low-pH-induced endosomal membrane destabilization or fusion. *4* Fusion of the liposomal bilayer with the plasma membrane

cannot easily pass biological membranes due to their unfavorable physicochemical characteristics. Therefore, cytosolic targeting of novel classes of drugs relies on delivery systems that allow translocation through cellular membranes.

The cytosolic delivery of liposome-entrapped drugs may be obtained in several different ways, as illustrated in Fig. 7. From these potential delivery routes, receptor-mediated endocytosis has been the focus of our research activities on the delivery of membrane-impermeable proteins and plasmid DNA (options 3 and 5, Fig. 7). This liposomal tactic involves: (1) specific targeting of liposomes to an internalizing receptor on the target cell population, leading to receptor-mediated endocytosis of ligand-directed liposomes, and (2) release of drugs from liposomes residing in endosomal compartments by membrane fusion/destabilization events triggered by the acidic environment within the endosomes.

3.1 Encapsulation of Plasmic DNA into Liposomes

When using liposomes as DNA carriers for in vivo gene therapy, an important research problem is how to encapsulate the DNA into liposome, since plasmid DNA is large. Improved formulations of cationic lipids have been used for the efficient delivery of DNA to cells in tissue culture (Felgner et al. 1987; Felgner et al. 1994). Much effort has also been devoted to the development of cationic lipids for the efficient delivery of nucleic acids in both animals and humans (Liu et al. 1997; Aksentijevich et al. 1996; Templeton et al. 1997).

Templeton et al. (1997) developed a unique cationic lipid structure, so-called BIV, using DOTAP, cholesterol and a novel formulation procedure. Nucleic acids are efficiently encapsulated between two bilamellar invaginated structures, BIVs. This procedure is different because it includes a brief, low-frequency sonication followed by manual extrusion through filters of decreasing pore size. The 0.1- and 0.2-μm filters consist of aluminum oxide membranes that contain a large number of pores per surface area, including evenly spaced and sized pores, and pores with straight channels. The bilamellar invaginated structures condense unusually large amounts of nucleic acids of any size. Furthermore, addition of DNA-condensing agents, including polymers, is not necessary. Encapsulation of nucleic acid by these bilamellar invaginated structures alone is spontaneous and immediate, and therefore cost-effective, requiring only one step of simple mixing. The extruded DOTAP:Chol-nucleic acid complexes are also large enough so that they are not cleared rapidly by Kupffer cells in the liver and yet extravasate across tight barriers and diffuse the target organ efficiently, which is critical for the gene to be expressed at a therapeutic level.

Gregoriadis and coworkers (Gregoriadis et al. 1998; Perrie and Gregoriadis 2000) have developed a dehydration-rehydration method to entrap plasmid DNA into liposomes. This is consistent with the notion that most of the DNA is incorporated within multilamellar vesicles and largely protected from nucleases by the bilayers. The potential of applying such liposome DNA carriers as oral delivery system for DNA vaccines was also assessed.

Wheeler et al. (1999) established "stabilized plasmid-lipid particles" (SPLP), produced by detergent dialysis employing a POPC/DODAC/PEG-CerC_{20} lipid mixture. The features of SPLP are the extended circulation lifetime of systemically administered gene therapy vectors and their accumulation at a distal tumor site. A detailed protocol for producing SPLP is provided in Fenske et al. (2002).

Liposome-protamine-DNA complexes (LPDs) for use as novel non-viral vectors, were developed by Huang's group on the basis of understanding the cellular and molecular barriers to cationic-lipid-based gene delivery systems (Fig. 8) (Sorgi et al. 1997; Li et al. 1998). Protamines are a group of small peptides (MW, 4000–4250) that play a unique role in condensing DNA to form a compact structure in the sperm and in delivering the sperm DNA into the nucleus of the egg after fertilization. They are highly positively charged because of their high arginine content. In addition, protamine sulfate is a USP and JP compound that is used clinically as an antidote to heparin-induced anti-coagulation. Thus, protamine sulfate may be a safe and efficient condensing agent in non-viral gene delivery systems. Since, however, it was found that positively charged liposomes are not suitable for active targeting to tumor cells, a novel formulation of anionic liposome-entrapped protamine-condensed DNA (LPD-II) was developed for gene targeting. The coupling of folate to LPD-II was optimized for systemic delivery of DNA to squamous cell carcinomas of the head and neck, and

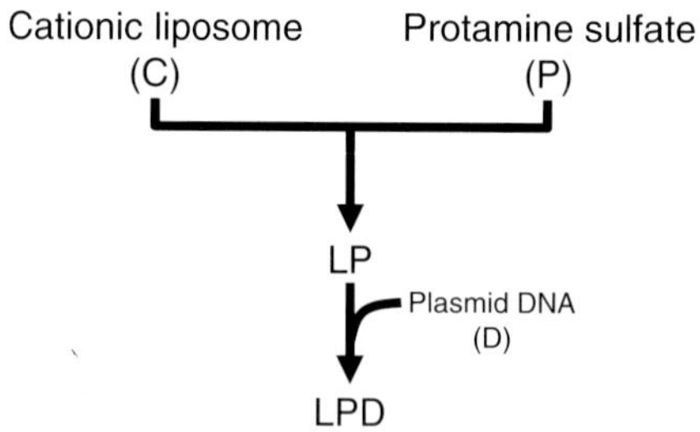

FIG. 8. Preparation of cationic liposome-protamine-DNA (LPD)

breast cancer xenografts (Xu et al. 1999). TF-coupled LPD-II liposomes were selectively targeted to myeroblast cells in vitro (Feero et al. 1997).

3.2 Liposomal Cytosolic Delivery

Receptor-mediated endocytosis would be expected to function as a pathway for the intracellular delivery of anticancer drugs in cancer chemotherapy as well as of nucleic acids in gene therapy (Harashima et al. 2001) (Fig. 7). However, once endocytosed, the control of intracellular trafficking is difficult, since it is under cellular regulation. Therefore, in order to optimize intracellular trafficking after receptor-mediated endocytosis to a target site, such as the cytosol or the nucleus, artificial sorting devices are required (Kamiya et al. 2001).

TF-receptor-mediated endocytosis as an entering pathway was chosen, these receptors are expressed in all nucleated cells in the body and in malignant cells, which have elevated receptor levels due to the high requirement of iron for growth. Once the TF-bearing liposomes are effectively internalized to the target cells, another sorting device is required to permit the liposomes, or their contents, to escape from endosomes to the cytosol. Based on results obtained with fusion peptides, as reported for viruses, GALA was examined, as the second sorting device. GALA is an artificially designed and pH-dependent fusogenic peptide composed of 30 amino acids (Subbarao et al. 1987; Nir and Nieva 2000). A highly efficient cytosolic release was achieved when the GALA peptide was present on the liposomal surfaces, which was attained by anchoring the GALA peptide to the liposomal membrane using a cholesterol derivative (Chol-GALA). Unexpectedly, there was also little escape from endosomes in the presence of GALA, which was introduced into the aqueous phase. Thus, intracellular trafficking can be controlled using TF-attached liposomes containing Chol-GALA via the TF-mediated endocytic pathway and can be applied for selective tumor targeting system (Kakudo et al. 2004).

References

Aisen P (1994) The transferrin receptor and the release of iron from transferrin. Adv Exp Med Biol 365:31–40

Aksentijevich L, Lunaridi-Iskandar Y, Gallo C, Gottesman M, Thrierry AR (1996) In vitro and liposome-mediated gene trasfer leads to human MDR1 expression in mouse bone marrow progenitor cell. Hum. Gene Ther 7:1111–1122

Allen TM, Hansen C, Martin F, Redemann C, Young AY (1991) Liposomes containing synthetic lipid derivatives of poly(ethylene glycol) show prolonged circulation half-lives *in vivo*. Biochim Biophys Acta 1066:29–36

Aragnol D, Leserman LD (1986) Immune clearance of liposomes inhibited by an anti-Fc receptor antibody *in vivo*. Proc. Natl Acad Set USA 83:2699–2703

Blume G, Cevc G (1990) Liposomes for sustained drug release *in vivo*. Biochim Biophys Acta 1029:91–97

Cummings J, Smyth JF (1993) Cytotoxic drug delivery. In: "Topics in phannacy, Vol. III." Florence AT Salole EG (Eds) Butterworth-Heinemann, Oxford, p 27–53

Derksen JTP, Morselt HWM, Scherphof GL (1988) Uptake and processing of immunoglobulin-coated liposomes by subpopulation of rat liver macrophages. Biochim Biophys Acta 971:127–136

Dvorak HF, Nagy JA, Dvorak JT (1988) Identification and characterization of the blood vessels of solid tumors that are leaky to circulating macromolecules. Am J Pathol 133:95–109

Ehrlich PA (1956) General review of the recent work in immunity. In "Immunology and cancer research." Himmelweit F (Ed.). Pergamon Press, London.

Enoch HG, Strittmatter P (1979) Formation and properties of 1000-Å-diameter, single-bilayer phospholipid vesicles. Proc Natl Acad Sci USA 76:145–149

Feero WG, Li S, Rosenblatt JD, Sirianni N, Morgan JE, Partridge TA, Huang L, Hoffman EP (1997) Selection and use of lighands for receptor-mediated gene delivery to myogenic cells. Gene Ther 4:664–674

Felgner JH, Kumar R, Sridhar CN, Wheeler CJ, Tsai YJ, Border R, Ramsey P, Martin M, Felgner PL (1994) Enhanced gene delivery and mechanism studies with a novel series of cationic lipid formulations. J Biol Chem 269:2550–2561

Felgner PL, Gadek TR, Holm M, Roman R, Chan HW, Wenz M, Northrop JP, Ringold GM, Danielsen M (1987) Lipofection: a highly efficient lipid-mediated DNA trancefection procedure. Proc. Natl Acid Sci USA 84:7413–7417

Fenske D, MacLachlan I, Cullis PR (2002) Stabilized plasmid-lipid particles: A systemic gene therapy vector method in engymology 346 ed by M I Phillips 36–71

Gabizon A (1992) Cancer Res 52:891–896

Gabizon A, Papahadjopoulos D (1988) Liposome formulations with prolonged circulation time in blood and enhanced uptake by tumors. Proc Natl Acad Sci USA 85:6949–6953

Gregoriadis G (1998) Genetic Vaccines: Strategies for Optimisation. Pharm Res 15:661–670

Gupta PK (1990) Drug targeting in cancer chemotherapy: a clinical perspective. J Pharm Sci 79:949–962

Harashima H, Shinohara Y, Kiwada H (2001) Intracellular control of gene trafficking using liposomes as drug carriers. Eur J Pharm Sci 13:85–89

Harata M, Soda Y, Tani K, Takizawa T, Maruyama K, Tojo A, Asano S (2004) CD19-targeting liposomes containing imatinib efficiently kill Philadelphia chromosome-positive acute Lymphoblastic leukemia cells. Blood 104:1442–1449

Huang SK, Lee KD, Hong K, Friend DS, Papahadjopoulos D (1992) Microscopic localization of sterically stabilized liposomes in colon carcinoma-bearing mice. Cancer Res 52:5135–5143

Huebers HA, Finch CA (1987) The physiology of transferrin and transferrin receptors. Physiol Rev 67:520–582

Ishida O, Maruyama K, Sasaki K, Iwatsuru M (1999) Size-dependent extravasation ant interstitial localization of polyethyleneglycol liposomes in solid tumor-bearing mice. Int J Pharm 190:49–56

Ishida O, Maruyama K, Tanahashi H, Iwatsuru M, Sasaki K, Eriguchi M, Yanagie H (2001) Liposomes bearing polyethyleneglycol-coupled transferrin with intracellular targeting property to the solid tumors *in vivo*. Pharm Res 18(7):1042–1048

Ishikawa E, Imagawa H, Hashida S, Yashitake S, Hamaguchi Y, Ueno T (1983) Enzyme labeling antibodies and their fragments for enzyme immunoassay and immunohistochemical staining. J Immunoassay 4:209–327

Jain RK, Gerlowski LE (1986) Extravascular transport in normal and tumor tissue. Crit Rev Oncol Hematol 5:115–170

Kakudo T, Chaki S, Futaki S, Nakase I, Akaji K, Kawakami T, Maruyama K, Kamiya H, Harashima H (2004) Transferrin-modified liposomes equipped with a pH-sensitive fusogenic peptide: An artificial viral-like delivery system. Biochemistry 43:5618–5628

Kamiya H, Tsuchiya H, Yamazaki J, Harashima H (2001) Intracellular trafficking and transgene expression of viral and non-viral gene vectors. Adv Drug Deliv Rev 52:153–164

Kennel SJ, Lankfold T, Hughes B, Hotchkiss JA (1988) Quantitation of a murine lung endothelial cell protein, P112, with a double monoclonal antibody assay. Laboratory Investigation 59:692–701

Klibanov A, Huang L (1992) Long-circulating liposomes: Development and perspectives. J Liposome Res 2:321–334
Klibanov A, Maruyama K, Beckerleg A, Torchilin VP, Huang L (1991) Activity of amphipathic poly(ethylene glycol) 5000 to prolong the circulation time of liposomes depends on the liposome size and is unfavorable for immunoliposome binding to target. Biochim Biophys Acta 1062:142–148
Klibanov A, Maruyama K, Torchilin VP, Huang L (1990) Amphipathic polyethyleneglycols effectively prolong the circulation time of liposomes. FEBS Lett 268:235–237
Lasic D, Martin F (1995) Stealth Liposomes. CRC Press, Inc Boca Raton
Li S, Rizzo MA, Bhattacharya S, Huang L (1998) Characterization of cationic lipid-protamine-DNA (LPD) complexes for intravenous gene delivery. Gene Therapy 5:930–937
Li S, Tseng WC, Stolz DB, Wu SP, Watkins SC, Huang L (1999) Dynamic changes in the characteristics of cationic lipidic vectors after exposure to mouse serum: implications for intravenous lipofection. Gene Therapy 6:858–594
Liu F, Qi H, Huang L, Liu D (1997) Factors controlling the efficiency of cationic lipid-mediated transfection in vivo via intravenous administration. Gene Ther 4:517–523
Maeda H (1992) The tumor blood vessel as an ideal target for macromolecular anticancer agents. J Control Release 19:315–324
Maeda H, Seymour LW, Miyamoto Y (1992) Conjugation of anticancer agents and polymers: advantages of macromolecular therapeutics *in vivo*. Bioconjug Chem 3:351–362
Maruyama K, Holmberg E, Kennel S, Klibanov A, Torchilin VP, Huang L (1990) Characterization of *in vivo* immunoliposomes targeting to pulmonary endothelium. J Pharm Sci 79:978–984
Maruyama K, Kennel SJ, Huang L (1990) Lipid composition is important for highly efficient target binding and retention of immunoliposomes. Proc Natl Acad Sci USA 87:5744–5748
Maruyama K, Takahashi N, Tagawa T, Nagaike K, Iwatsuru M (1997a) Immunoliposomes bearing polyethyleneglycol -coupled Fab′ fragment show prolonged circulation time and high extravasation into target solid tumors *in vivo*. 413:177–180
Maruyama K, Takizawa T, Takahashi N, Tagawa T, Nagaike K, Iwatsuru M (1997b) Advanced Drug Delivery Reviews 24:235–242
Maruyama K, Takizawa T, Yuda T, Kennel SJ, Huang L, Iwatsuru M (1995) Targetability of novel immunoliposomes modified with amphipathic polyethyleneglycols conjugated at their distal terminals to monoclonal antibodies. Biochim Biophys Acta 1234:74–80
Maruyama K, Yanagie H, Ishida O, Kasaoka S, Takizawa T, Utoguchi N, Shinohara A, Chiba M, Kobayashi H, Iwatsuru M, Eriguchi M (2004) Intracellular targeting of sodium mercaptoundecahydrododecaborate (BSH) to solid tumors by transferrin-PEG liposomes, for boron neutron-capture therapy (BNCT). J Controlled Release 98(2):195–207
Maruyama K, Yuda T, Okamoto S, Kojima S, Suginaka A, Iwatsuru M (1992) Prolonged circulation time *in vivo* of large unilamellar liposomes composed of distearoylphosphatidylcholine and cholesterol containing amphipathic poly(ethylene glycol). Biochim Biophys Acta 1128:44–49
Miller AD (1998) Cationic liposomes for gene therapy. Angew Chem Int Ed 37:1768–1785
Mori A, Klibanov A, Torchilin VP, Huang L (1991) Influence of the steric barrier activity of amphipathic poly(ethylene glycol) and ganglioside GM1 on the circulation time of liposomes and on the target binding of immunoliposomes *in vivo*. FEBS Let 284:263–266
Nabel GJ, Nabel EG Yang ZY, Fox BA, Plautz GE, Gao X, Huang L, Shu S, Gordon D (1993) Direct gene transfer with DNA-liposome complexes in melanoma: expression, biologic activity, and lack of toxicity. Proc. Natl Acad Sci USA 90:11307–11311
Nir S, Nieva JL (2000) Interactions of peptides with liposomes: pore formation and fusion. Prog Lipid Res 39:181–206

Papahadjopoulos D, Allen TM, Gabizon A, Mayhew E, Matthay K, Huang SK, Lee KD, Woodle MC, Lasic DD, Redemann C, Martin FJ (1991) Proc Natl Acad Sci USA 88: 11460–11464

Peeters PAM, Storm G, Crommelin DJA (1987) Liposomes *in vivo*: state of the art. Advanced Drug Delivery Review 1:249–266

Perrie Y, Gregoriadis G (2000) Liposome-Entrapped Plasmid DNA: Characterisation Studies. Biochim Biophys Acta 1475:125–132

Poste G, Kirsh R (1983) Site-specific (targeted) drug delivery in cancer therapy. Biotechnology 1:869–878

Rowlinson-Busza G, Epenetos AA (1992) Targeted delivery of biologic and other antineoplastic agents. Curr Opin Oncol 4:1142–1148

Sorgi FL, Bhattacharya S, Huang L (1997) Protamine sulfate enhances lipid mediated gene transfer. Gene Therapy 4:961–968

Subbarao NK, Parente RA, Szoka FC, Jr Nadasdi L, Pongracz K (1987) pH-dependent bilayer destabilization by an amphipathic peptide. Biochemistry 26:2964–2972

Takizawa T, Maruyama K, Iwatsuru M, Sasaki K (1998) Drug Delivery System 13:407–414

Templeton NS, Lasic DD, Frederik PM, Strey HH, Roberts DD, Pavlakis GN (1997) Improved DNA: liposome complexed systemic delivery and gene expressin. Nature Biotechnology 15:642–652

Unezaki S, Maruyama K, Hosoda J, Nagae I, Koyanagi Y, Nakata M, Ishida O, Iwatsuru M, Tsuchiya S (1996) Direct measurement of the extravasation of polyethyleneglycol-coated liposomes into solid tumor tissue by *in vivo* fluorescence microscopy. Int J Pharm 144:11–17

Unezaki S, Maruyama K, Ishida O, Suginaka A, Hosoda J, Iwatsuru M (1995) Enhanced tumor targeting and improved antitumor activity of doxorubicin by long-circulating liposomes containing amphipathic poly(ethylene glycol). Int J Pharm 126:41–48

Unezaki S, Maruyama K, Takahashi N, Koyama M, Yuda T, Suginaka A, Iwatsuru M (1994) Enhanced delivery and antitumor activity of doxorubicin using long-circulating thermosensitive liposomes containing amphipathic polyethyleneglycol in combination with local hyperthermia. Pharm Res 11:1180–1185

Uyama I, Kumai K, Yasuda T, Tagawa T, Ishibiki K, Kitajima M, Tadakuma T (1994) Improvement of therapeutic effect by using Fab′ fragment in the treatment of carcinoembryonic antigen-positive human solid tumors with adriamycin-entrapped immunoliposomes. Jpn J Cancer Res 85:434–440

Wagner E, Curiel D, Cotton M (1994) Delivery of drugs, proteins and genes into cells using transferrin as a ligand for receptor-mediated endocytosis. Advanced Drug Delivery Reviews 14:113–135

Wheeler JJ, Palmer L, Ossanlou M, MacLachlan I, Graham RW, Zhang YP, Hope MJ, Scherrer P, Cullis PR (1999) Gene Ther 6:271

Wright S, Huang L (1989) Antibody-directed liposomes as drug delivery vehicles. Advanced Drug Delivery Reviews 3:343–389

Xu L, Pirollo KF, Rait A, Murray A, Chang EH (1999) Systemic p53 Gene Therapy in Combination Rediation Result in Human Tumor Regression. Tumor-Targeting 4:92–114

Polymeric Gene Carriers

Tomoko Hashimoto and Tetsuji Yamaoka

1 Introduction

The concept of polymeric drugs for use as pharmaceuticals was first proposed around 1970 by Ringsdorf (1975). In this system, a water-soluble polymer is selected as a carrier material, to which organ-targeting moieties are bound along with the parent pharmaceuticals. The targeting moieties lead the drugs to the site of action by active targeting, hence the name, missile drugs. In this case, the main role of the carrier polymers is to prolong the half-life of the drug in the blood stream. The effects of molecular weight and electric charge of various water-soluble polymers, such as dextrans, pullulan, poly(ethylene glycol) (PEG), and poly(vinyl alcohol) (PVA), on the biological fate of the drugs have been reported (Yamaoka et al. 1995, 1994). Cationic macromolecules are known to affect on cellular interactions in vitro. Another role of polymeric carriers is to protect unstable drugs from various enzymes in the body. Protein and peptide drugs are susceptible to hydrolysis and would thus be digested by proteolytic enzymes in vivo.

Very recently, the genes coding for several peptide drugs have been attracting great attention with respect to curing genetic diseases, since protein drugs are very unstable in the body. In delivering genetic materials into cells or tissues, various water-soluble polymeric carriers have been selected as non-viral gene carriers because they are safer than viral vectors. However, the properties required for a gene carrier are quite different from those for conventional drug carries. Generally, polymeric gene carriers possess positively charged groups, which are known to condense DNA coils into 10^{-3}–10^{-4} of the original volume (De Smedt et al. 2000) by forming a polyion complex (PIC). PICs are formed via electrostatic interactions between the anionic phosphate groups of DNA and the cationic amino groups of polymeric gene carriers (Maes et al. 1967; Saymour 1999). These complexes of DNA and polycations are called "polyplexes". The electrostatic interaction of polyplexes is of increasing importance

Kyoto Institute of Technology, Matsugasaki, Sakyo-ku, Kyoto 606-8585, Japan not only in DNA condensing but also in maintaining the relaxed state of the polyplexes, which allows them to release DNA, as described later (Arigita et al. 1999; Kunath et al. 2003; Ruponen et al. 1999).

The other role of cationic polymers is to protect DNA from degradation by DNases existing in the body, which is one of the biggest objectives of the gene delivery system. Polyplexes of PLL, PEI, and polyamidoamine (PAMAM) dendrimer showed significant resistance against nuclease actions compared with free DNA. Protective interactive noncondensing (PINC) polymers, such as poly(*N*-vinyl pyrrolidone) (PVP) and PVA, form flexible polyplexes with DNA via hydrogen bonds (Mumper et al. 1998). Molecular modeling confirmed that PVP was located in the major groove of DNA. These PINC polymers protect DNA from nucleases and enhance intracellular uptake by interactions between the hydrophobic surface of polyplexes and the cell membrane. However, the strong protecting ability of the carriers does not necessarily lead to high levels of gene expression. Indeed, inverse effects between nuclease resistance and the rate of interexchange reactions of polyplexes (described in detail later) has been pointed out (Seymour 1997).

In the case of conventional polymeric drugs, the parent drug would simply be released from the carrier polymer by hydrolysis of the chemical bonds followed by diffusion into cells or to the site of action (Kopeek and Pohl 1988). In contrast, in the case of the gene delivery system, DNA molecules are diffuse with difficulty because of their huge molecular weights. Thus, carrier polymers continue to play an important role in the interaction with cell surfaces, in intracellular trafficking, and in transcription of transgene in the nucleus. In this chapter, the chemical structure, function, and mechanisms of various polymeric non-viral carriers for use in gene delivery systems in mammalian cells, in vitro and in vivo, are reviewed.

2 Various Polymeric Carriers

About 70% of the gene-therapeutic protocols available to date are based on viral vectors (http://www.wiley.co.uk/genetherapy/clinical/). Despite their high efficiency in vitro, clinical trials are often limited by several concerns, e.g. toxicity, immunogenicity, inflammatory properties, the limited size of the DNA, production and packing problems, and the high cost. In addition, an overwhelming immune reaction against adenovirus occurred in a patient at Pennsylvania University in 1999 (Marshall 2000, 1999) and a leukemia-like disease were reported in a French patient in 2002 (Marshall 2002; Kaiser 2003). As a result, non-viral vector-mediated systems have become of interest, because they are much safer, more cost-effective, and easier to manufacture than viral vector systems.

Some of the reasons for the few clinical trials using non-viral gene vectors are their toxicity (Choi et al. 1998), low transgene expression, and the tendency of the carrier/DNA complexes to aggregate in the blood (Dash et al. 1999). In addition, the detailed mechanisms of transgene expression following non-viral carrier-based gene transfer are not yet clear. The efficacy of transgene expression has been found to be clearly affected by the chemical structure of the cationic carriers, which indicates that the characteristics of the non-viral vectors would alter the efficacy of some process, such as internalization in cells and transgene expression.

2.1 Chemical Structure

The synthetic vectors described above are depicted in Fig. 1. All of them possess cationic charges and form PICs based on electrostatic interactions, but their transfection efficacies are quite different and depend on the chemical structure, charge density, and molecular weight of the polycation. For example, poly(L-arginine) and poly(L-lysine) are known to form complexes with quite different features (Liquier et al. 1975), and recently the former were found to lead to much higher levels of gene expression.

We have reported that polycations having abundant side-chain hydroxyl groups (or amide groups), such as poly(vinyl alcohol) dimethylaminoacetal (PVA3), are effective carriers with low cytotoxicity. Although the role of the hydroxyl or amide groups is still uncertain, they seem to strongly and effectively maintain the hydrophilic nature of the formed complexes thus preventing compaction of the complexes and ultimately allowing their dissociation. The methylene group length of the polyburene (PB) main chain (Fig. 1) also change the hydrophilic/hydrophobic balance (Aubin et al. 1994; Mita et al. 1977). The charge density of the polycations is the other important factors. In a comparison of DEAE-dextrans with different cationic group densities ranging from 20 to 55%, both transfection efficiency and cytotoxicity greatly decreased with decreased cationic group densities (Fig. 2). Other polycations, such as chitosan (Lee et al. 1998), polyethyleneimine (Godbey et al. 1999b; Pollard et al. 1998; Demeneix et al. 1998), and cationic polymethacrylate derivatives (van de Wetering et al. 1998;

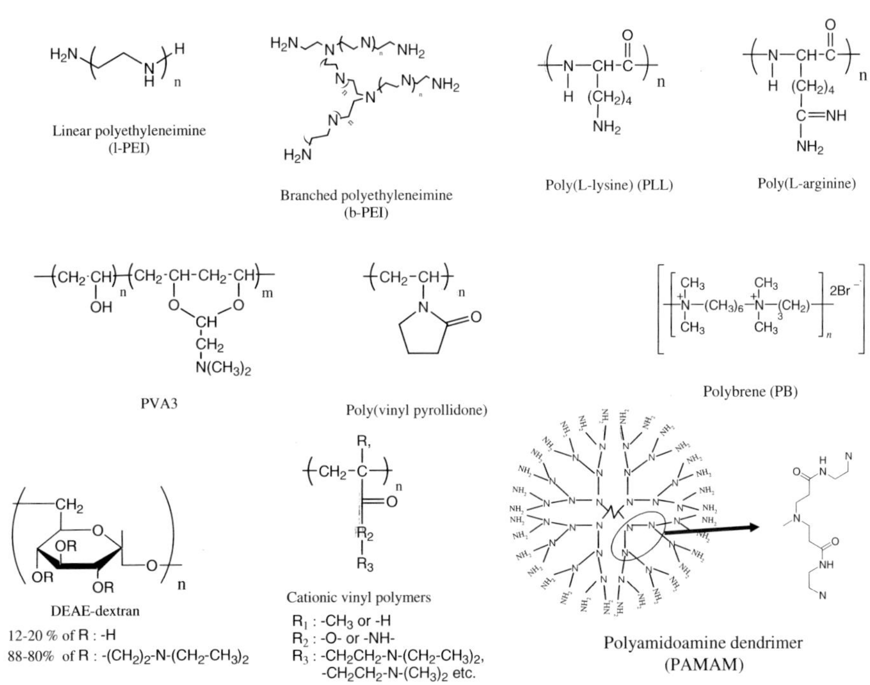

FIG. 1. Chemical structures of polycation-type non-viral gene carriers

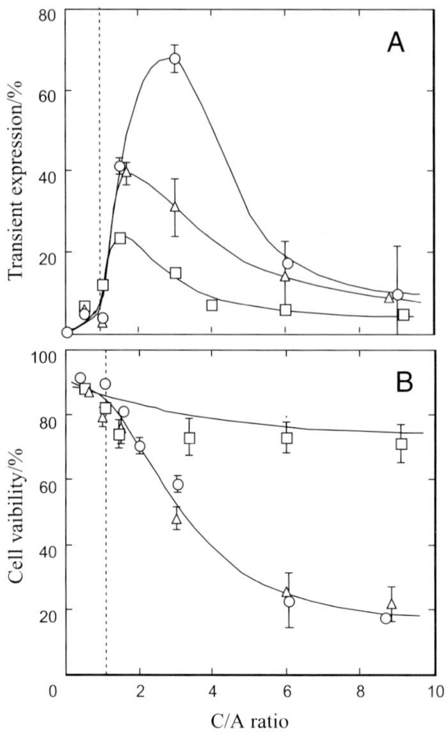

FIG. 2. **A** Transient expressio of *lacZ* introduced into COS-1 cells; **B** cytotoxicity after osmotic shock using DEAE-dextran with the substitution ratio for DEAE groups of: ○ 55, △ 31, and □ 22 %

Cherng et al. 1996) have also been proposed but there is little information on the correlation between their chemical structure and transfection efficacy. In 1995, Boussif et al. reported that polyethyleneimine (PEI) can be effectively used as a non-viral gene vector for the purpose of gene therapy (Boussif et al. 1995). Godbey et al. (1999a) reported the effect of molecular weight of PEI. Low-molecular-weight (1,800) PEI resulted in no gene expression but transfection efficacy increased with increasing molecular weight, with the most effective results obtained at 70,000.

2.2 Molecular Shape

When linear PEI and branched PEI were compared with respect to their transfection efficacy, the results were dependent on the adopted transfection procedure (Fig. 1) (Plank et al. 1999). They showed the efficiency of the branched structure, while Ohashi et al. reported more efficient gene transfer using linear PEI than branched PEI (Ohashi et al. 2001). Thus, further experiments are necessary to investigate the relationship between the molecular shape of polymers and gene expression.

In 1993, Haensler and Szoka reported on the effectiveness of hyper-branched polyamidoamine (PAMAM) dendrimers, a well-defined class of cascade polymers from methyl acrylate and ethylenediamine (Fig. 1) . They achieved excellent gene expression using heat-treated dendrimers (fractured dendrimers), which are a degraded form of the intact dendrimers, at the amide linkage (Tang et al. 1996) (see Chaps. 1–5).

Various amphiphilic block polymers, such as PEG ylated PLL, which self-organize into micelles in aqueous solution, have been used as carriers. DNA can form PICs with hydrophilic chains of block polymers. Micelles are formed from the core of these PICs surrounded by hydrophobic chains of block polymers (Erbacher et al. 1999).

3 Transfection Protocol Using Polyplexes

3.1 Various Inhibitors in In Vitro Gene Delivery

The roles of endosomes and lysosomes in gene transefer have been discussed in the literature (Wattiaux et al. 2000). Polyplexes are taken up via endocytosis and then travel through various intracellular pathway via endosomes and lysosomes. In order to avoid degradation of internalized polyplexes in these structures, the activity of the lysosomal enzymes must be suppressed by adding various endosomal or lysosomal inhibitors as shown in Table 1.

Weak bases, such as chloroquine and ammonium chloride, inhibit the acidication of endosomal or lysosomal environments as well as the degradation of polyplexes in endosomes and lysosomes (Maxfield 1982; Cotten et al. 1990; Choi et al. 1998). In a study comparing chloroquine and several other weak bases, it was found that only chloroquine enhances transgene expression, which may well be related to dissociation of the complexes (Erbacher et al. 1996). Bafilomycin A_1 and concanamycin A act as inhibitors of vacuolar ATPases known to block the endosomal proton pump. Photosensitizing compounds, such as AlPcS2a and TPPS2a, destabilize endosomes following photochemical reactions of these agents with visible light. Endosomal and lysosomal inhibitors thus improve the release of polyplexes from these organelles.

3.2 Mechanical Methods

Mechanical transfection, such as electroporation (Magin-Lachmann et al. 2004), microinjection (Zauner et al. 1999), and osmotic shock (Takai and Ohmori 1990; Okada and Rechsteiner 1982), are useful methods especially for studying transgene expression following endosomal escape. We use osmotic shock because, compared to microinjection, a larger number of cells can be easily treated and the polyplexes are directly delivered into the cytosolic compartment (Kimura et al. 2002). The method is applicable even for lymphoid cells. Briefly, cell suspensions are incubated with polyplexes for 30 min–1 h at 37°C, and a highly osmotic solution containing 1 M sucrose,

TABLE 1. Endosomal inhibitors

Inhibitors	Action	References
Chloroquine	Raises endosomal pH	(Erbacher et al. 1996; Cotten et al. 1990; Murphy et al. 1984)
Ammonium chloride Monensin, FCCP		(Erbacher et al. 1996; Fredericksen et al. 2002; Maxfield 1982)
Bafilomycin	ATPase inhibition	(Fredericksen et al. 2002; Drose and Altendorf 1997)
Fusogenic peptides	Membrane fusion	(Collins and Fabre 2004)

20% PEG 4000, 210 mM NaCl, and 70 mM Tris-HCl buffer (pH 7.3) is then added. After incubating the cells for a given period of time, they are rinsed twice with serum-free culture medium and exposed to hypotonic culture medium. After a 48-h incubation, transgene expression can be evaluated.

4 Biological Barriers

The mechanism of polyplex-mediated gene transfer is thought to follow the general endocytotic process. In order to lead to successfully high transgene expression, various biological barriers must be crossed: (1) interaction with the plasma membrane, (2) internalization, (3) escape from the endosome, (4) trafficking into the nucleus, and (5) dissociation of the complexes and/or DNA recognition by transcription factors (Fig. 3).

4.1 Internalization

Since positively charged complexes interact with the cell surface by an electrostatic interaction with anionic substances on the cell surface, such as sialic acid and proteoglycan, the zeta potential of polyplexes is important. Polyplexes composed of polycations with molecular weights of several thousands exhibit significantly higher zeta potentials (Ruponen et al. 1999; Jeong et al. 2001; Ahn et al. 2002; Wolfert et al. 1996; Toncheva et al. 1998; Cherng et al. 1996; Wolfert et al. 1999; Putnam et al. 2001; Howard et al. 2000), irrespective of the polycations used .

Internalization events also seem to be affected by polyplex size and zeta potential. The size of the polyplexes has been assessed by transmission electron microscopy

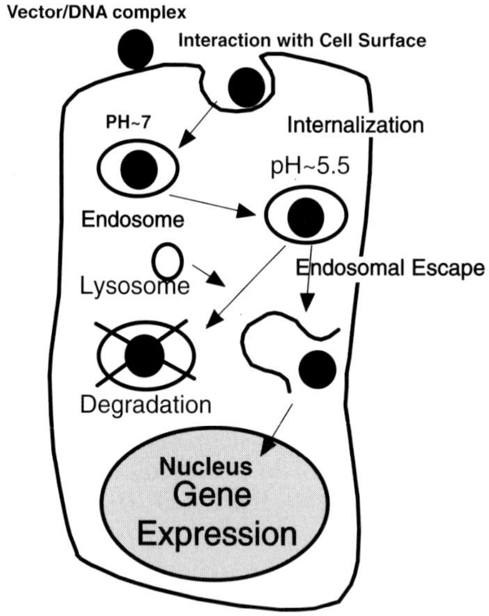

FIG. 3. Intracellular trafficking of polyplexes

(TEM) (Mannisto et al. 2002), atomic force microscopy (AFM) (Wolfert et al. 1996; Toncheva et al. 1998; Choi et al. 1999a), and dynamic light scattering (DLS) (Jeong et al. 2001; Cherng et al. 1996; Ogris et al. 1999). Generally, aggregation of the complexes makes their internalization difficult because of their large size. Tang and Szoka (1997) studied the aggregation properties of linear PLL, intact PAMAM dendrimers, fractured dendrimers, and branched PEI in forming complexes with DNA. These polycations formed similar complexes in terms of size and zeta potential but high-level gene expression was induced only by fractured dendrimers and branched PEI, due to the stability of the complexes and absence of aggregation (Tang and Szoka 1997). Despite these investigation, the correlation between transfection efficiency and size of the polyplexes remains unclear. Aggregation of the polyplexes in serum-containing medium or in blood is also a problem. In order to decrease both aggregation and the toxicity of polyplexes, conjugation with PEG has been proposed (Mannisto et al. 2002; Choi et al. 1999b). For example, the solubility of PEG ylated PEI and DNA polyplexes at higher therapeutic concentration was not only improved without aggregation but the in vivo toxicity was also reduced (Ogris et al. 1999; Kursa et al. 2003).

4.2 Receptor-Mediated Gene Delivery

In 1987, Wu et al. developed a system for targeting foreign genes to hepatocytes, which possess a unique receptor that binds and internalizes galactose-terminal asialoglycoproteins, through receptor-mediated endocytosis. It was shown that asialoorosomucoid-PLL carriers delivered pSV2-CAT plasmid DNA specifically to HepG2 hepatoma cells but not to other receptor (–) cell lines (Wu and Wu 1987, 1988a, b). Consequently, site-specific gene delivery has received much attention, especially in vivo direct gene transfer using various biologically active moieties, such as sugar (Midoux et al. 1993; Erbacher et al. 1996, 1997, 1995; Wu and Wu 1988a, b), transferrin (Wagner et al. 1991a,b), and LDL (Table 2).

Some types of cell, such as nonadherent primary hematopoietic cells, are well known to be difficult or almost impossible to transfect with foreign genes linked to conventional carriers because the endocytotic activity of these cells is quite low. Birnstiel and coworkers developed a system in which transferrin served as ligand and named the system "transferrinfection" (Wagner et al. 1991). The authors synthesized tranferrin-PLL conjugates using various molecular weights of PLL and different modification ratios of transferrin. A strong correlation was found between DNA condensation, evaluated using electron microscopy, and cellular DNA uptake.

Other candidates for receptor-mediated gene delivery are the receptors for integrin (Erbacher et al. 1999), insulin (Rosenkranz et al. 1992), and some growth factors (Fisher et al. 2000). Interestingly, polycations bound to VEGF (vascular endothelial growth factor) could not deliver DNA into nucleus but bFGF (basic fibroblast growth factor) could. The PEI derivatives conjugated to the integrin-binding peptide CYGGRGDTP via a disulfide bridge led to transgene expression in integrin-expressing epithelial cells (Hela) and fibroblasts (MRC5) at an expression level 10- to 100-fold higher than obtained with PEI. The advantage of receptor-mediated endocytosis is not just the cell-type specificity of the gene transfer but also the controlled intracellular trafficking of the complexes (Erbacher et al. 1999).

TABLE 2. Biological signal-mediated gene therapy

Barriers	Ligand	Base polymer	References
Plasma membrane	Galactose	PLL	(Wu and Wu 1987, 1988a, b; Nishikawa et al. 1998; Zanta et al. 1997)
		Vinyl polymer	(Lim et al. 2000)
	Lactose	PLL / PLL-PEG	(Erbacher et al. 1997; Midoux et al. 1993; Klink et al. 2001) / (Choi et al. 1999b)
	Folate	PLL / PLL-PEG	(Mislick et al. 1995; Leamon et al. 1999)
		PEI	(Guo and Lee 1999)
	Transfferin	PLL / PLL-PEG / PEI-PEG	(Wagner et al. 1991a,b 1990; Cotten et al. 1990;) / (Ogris et al. 1999; Vinogradov et al. 1999; Kursa et al. 2003)
	RGD	PLL	(Harbottle et al. 1998)
		PEI	(Erbacher et al. 1999)
	LDL	PLL	(Kim et al. 1998)
		PEI	(Furgeson et al. 2003)
Nuclear membrane	NLS	PLL	(Chan and Jans 1999; Chan et al. 1999)
		None	(Fritz et al. 1996; Balicki et al. 2002)

RGD; Arg-Gly-Asp tripeptide
LDL; low density lipoprotein
NLS; nuclear localization signal
PLL; poly(L-lysine)
PEG; poly(ethylene glycol)
PEI; polyethyleneimine

4.3 Endosomal Escape

The internal pH of endosomes containing polyplexes gradually decreases to about 5.5. Then, the endosomes fuse with lysosomes, resulting in the formation of secondary lysosomes, in which the incorporated DNA is normally hydrolyzed by lysosomal enzymes. DNA digestion at this step is one of the biggest barriers to effective gene transfer. When a foreign gene is transferred by microinjection or osmotic shock (Takai and Ohmori 1990), by which the transgene is compulsorily delivered into the cytosol, gene expression is generally much higher than obtained using the coculture method because there is no lysosomal digestion. If a non-viral vector has the ability to disrupt or fuse with the endosomal membranes, transferred foreign gene can escape from the endosome into the cytosol, and effective transgene expression should occur.

One promising strategy to release internalized complexes from the endosome is osmotic endosomal disruption. In 1995, Boussif et al. pointed out a novel mechanism, the "proton sponge hypothesis", which resulted in high-level expression of a transgene introduced into the cell using PEI as vector (Boussif et al. 1995). At pH 5.5.–7, PEI has a greater buffering capacity than PLL and other polycations. When PEI is used as carrier and is internalized into the endosome with DNA, a larger amount of H^+ should influx into the endosome, thus reducing the pH and raising the internal osmotic pressure, resulting in osmotic rupture of the endosome. In fact, fluorescence imaging showed that there was no overlap of PEI and labeled lysosomes while PLL was found together with lysosome when labeled PEI/DNA or PLL/DNA polyplexes were transfected into cells (Godbey et al. 2000; Remy-Kristensen et al. 2001). Recently, it was reported that DNA with PEI (linear or branched) showed rapid emdosomal escape (Itaka 2003). Hennink studied the effect of pKa of the cationic groups on their transfection efficacy using various cationic vinyl polymers (Zuidam et al. 2000; van de Wetering et al. 1999). The pKa ranges from 7.5 (for poly(2-(dimethylamino)ethyl methacrylate)) to 8.8 [for poly(3-(dimethylamino)propyl methacrylamide)], and the lower the pKa the higher the amount of gene expression. The pKa of the cationic groups is also influenced by their arrangement based on the polymer effect of the adjacent charged groups (van de Wetering et al. 1999).

4.4 Nuclear Transport

Transgenes must be transported to nucleus by some means. When lipoplexes are injected into the nucleus, gene expression is strongly suppressed by the cationic lipids, while polyplexes lead to strong gene expression after nuclear microinjection (Pollard et al. 1998). These results indicate that the intracellular trafficking and gene expression mechanisms for polyplexes and lipoplexes differ from each other.

One possible mechanism is the transportation through the nuclear membrane pore but this seems unlikely because the sizes of the complexes are too large, being usually around 100 nm. Another possibility is the accumulation of the complexes during the mitotic event accompanying nuclear membrane disappearance. Zauner et al. compared the role of mitosis in the transfection of confluent, contact-inhibited primary human cells using polyplexes and lipoplexes. It was shown that lipoplexes cannot lead to high-level gene expression at the confluent stage but that polyplexes can (Zauner et al. 1999). Godbey et al. (1999b) reported another mechanism of nuclear transport for PEI/DNA complexes. They suggested a mechanism in which the polyplexes come into contact with phospholipids of the endosome; the membrane then becomes permeabilized and bursts due to osmotic swelling, resulting in the polyplexes becoming coated with the phospholipids. The coated complexes could then enter the nucleus via fusion with the nuclear envelope.

In order to enhance trafficking through the nuclear pore, several nuclear localization signals (NLS) were utilized (Garcia-Bustos et al. 1991; Yoneda 1997). NLSs are oligopeptides mainly composed of cationic residues; they are 5–20 amino acids long and different sequences have been found in many species. NLS bound to PLL has been evaluated by many researchers as a gene carrier and was shown to be effective (Table 2) (Chan and Jans 1999).

4.5 Transcription of the Transgene

The delivered polyplexes may require disassembly in order to be transcribed but electrostatic polyplexes dissociate with difficulty under normal physiological conditions. In contrast, it is also possible that the DNA in the polyplexes is recognized without prior disassembly. In any case, the polyplex should possess adequate characteristics allowing these events. The tendency towards polyplex dissociation can be estimated by adding another polyanion, such as heparin sulfate, polyvinyl sulfate, and heparin, into the polyplex suspension (Erbacher et al. 1999; Wolfert et al. 1996; Ruponen et al. 1999). Kabanov et al. reported that under these conditions a polyion interexchange reaction occurs, resulting in free DNA release when an adequate amount of polyanion is added to polyplex suspensions (Vinogradov et al. 1998). The interexchange reaction of the complexes depends on the kind of polyanions added, such as poly(vinyl sulfonate) (Katayose and Kataoka 1998), poly(aspartic acid), or glycosaminoglycans (GAGs) (Mannisto et al. 2002). DNA was more easily released from pDMAEMA than from poly[(trimethylamino)ethyl methacrylate chloride)] (pTMAEMA) and the former showed high transfection efficiency than the latter (Arigita et al. 1999).

We reported that only cationic polymers containing nonionic hydrophilic groups lead to higher amounts of gene expression (Yamaoka et al. 1994). These groups seem to impart a hydrophilic nature to the complexes since they are not involved in complex formation. Gene expression in these complexes was higher even in an in vitro transcription/translation system using rabbit reticulocyte lysate. Based on this observation, we have attempted to improve the carrier ability of PLL (Yamaoka et al. 2000; Kimura et al. 2002). As shown in Fig. 4, PtmLS, which is a PLL derivative containing 25 mol % serine residues and quaternary ammonium groups, greatly enhanced transgene expression. PtmLS complexes were easily disassembled and were also transcribed in an in vitro translation system. Recently, low-molecular-weight PEI (LMW-PEI) was reported to be a promising carrier for gene transfer under in vitro and in vivo conditions compared with high-molecular-weight PEI (HMW-PEI). Poly-

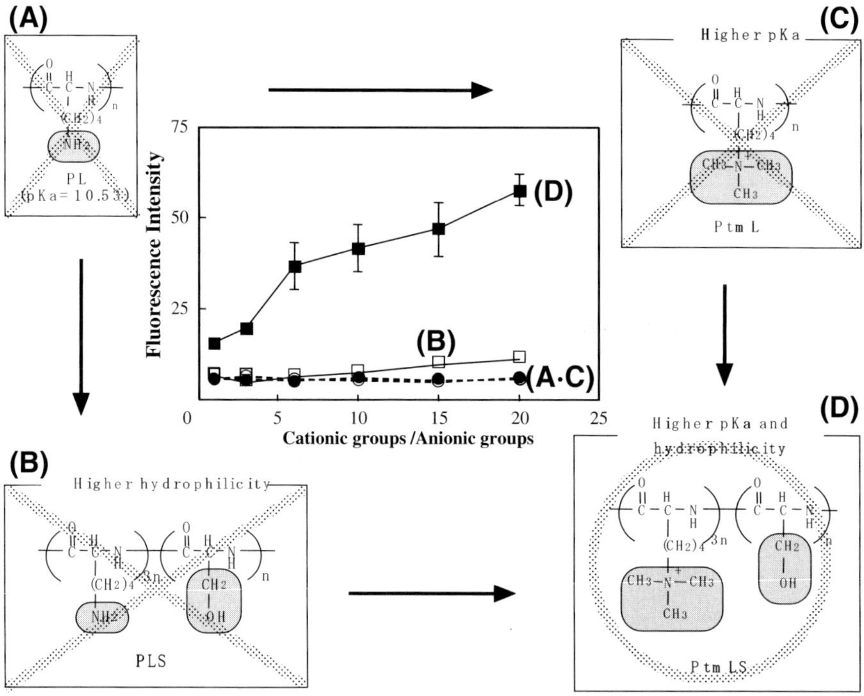

FIG. 4. Fluorescence intensity of EGFP expressed by COS-1 cells 40 h after osmotic shock of pEGFP using ○ PL, • PLS, □ PtmL, and ■ PtmLS

plexes formed from LMW-PEI also showed significantly reduced condensation and were reported to induce an higher transfection efficiency (Kunath et al. 2003).

5 Conclusions

The physicochemical features of the polyplexes are extremely important for designing effective non-viral carriers. Recently, several new systems have been described but are not well tested with respect to intracellular trafficking and transcription event. For example, non-condensing polyplexes are advantageous for transcription but, at the same time, are disadvantageous for nuclease digestion of DNA. In order to design and develop, new effective carriers, each step of intracellular trafficking route should be analyzed quantitatively.

References

Ahn CH, Chae SY, Bae YH, Kim SW (2002) Biodegradable poly(ethylenimine) for plasmid DNA delivery. J Control Release 80:273–282

Arigita C, Zuidam NJ, Crommelin DJ, Hennink WE (1999) Association and dissociation characteristics of polymer/DNA complexes used for gene delivery. Pharm Res 16: 1534–1541

Aubin RA, Weinfeld M, Mirzayans R, Paterson MC (1994) Polybrene/DMSO-assisted gene transfer. Generating stable transfectants with nanogram amounts of DNA. Mol Biotechnol 1:29–48

Balicki D, Putnam CD, Scaria PV, Beutler E (2002) Structure and function correlation in histone H2A peptide-mediated gene transfer. Proc Natl Acad Sci USA 99:7467–7471

Boussif O, Lezoualc'h F, Zanta MA, Mergny MD, Scherman D, Demeneix B, Behr JP (1995) A versatile vector for gene and oligonucleotide transfer into cells in culture and *in vivo*: polyethylenimine. Proc Natl Acad Sci USA 92:7297–7301

Chan CK, Jans DA (1999) Enhancement of polylysine-mediated transferrinfection by nuclear localization sequences: polylysine does not function as a nuclear localization sequence. Hum Gene Ther 10:1695–702

Cherng JY, van de Wetering P, Talsma H, Crommelin DJ, Hennink WE (1996) Effect of size and serum proteins on transfection efficiency of poly ((2-dimethylamino)ethyl methacrylate)-plasmid nanoparticles. Pharm Res 13:1038–1042

Choi JS, Lee EJ, Choi YH, Jeong YJ, Park JS (1999a) Poly(ethylene glycol)-block-poly(L-lysine) dendrimer: novel linear polymer/dendrimer block copolymer forming a spherical water-soluble polyionic complex with DNA. Bioconjug Chem 10:62–65

Choi YH, Liu F, Choi JS, Kim SW, Park JS (1999b) Characterization of a targeted gene carrier, lactose-polyethylene glycol-grafted poly-L-lysine and its complex with plasmid DNA. Hum Gene Ther 10:2657–2665

Choi YH, Liu F, Kim JS, Choi YK, Park JS, Kim SW (1998) Polyethylene glycol-grafted poly-L-lysine as polymeric gene carrier. J Control Release 54:39–48

Collins L, Fabre JW (2004) A synthetic peptide vector system for optimal gene delivery to corneal endothelium. J Gene Med 6:185–194

Cotten M, Langle-Rouault F, Kirlappos H, Wagner E, Mechtler K, Zenke M, Beug H, Birnstiel ML (1990) Transferrin-polycation-mediated introduction of DNA into human leukemic cells: stimulation by agents that affect the survival of transfected DNA or modulate transferrin receptor levels. Proc Natl Acad Sci USA 87:4033–4037

Dash PR, Read ML, Barrett LB, Wolfert MA, Seymour LW (1999) Factors affecting blood clearance and *in vivo* distribution of polyelectrolyte complexes for gene delivery. Gene Ther 6:643–650

De Smedt SC, Demeester J, Hennink WE (2000) Cationic polymer based gene delivery systems. Pharm Res 17:113–126

Demeneix B, Behr J, Boussif O, Zanta MA, Abdallah B, Remy J (1998) Gene transfer with lipospermines and polyethylenimines. Adv Drug Deliv Rev 30:85–95

Drose S, Altendorf K (1997) Bafilomycins and concanamycins as inhibitors of V-ATPases and P-ATPases. J Exp Biol 200 (Pt 1):1–8

Erbacher P, Roche AC, Monsigny M, Midoux P (1995) Glycosylated polylysine/DNA complexes: gene transfer efficiency in relation with the size and the sugar substitution level of glycosylated polylysines and with the plasmid size. Bioconjug Chem 6:401–410

Erbacher P, Roche AC, Monsigny M, Midoux P (1996) Putative role of chloroquine in gene transfer into a human hepatoma cell line by DNA/lactosylated polylysine complexes. Exp Cell Res 225:186–194

Erbacher P, Roche AC, Monsigny M, Midoux P (1997) The reduction of the positive charges of polylysine by partial gluconoylation increases the transfection efficiency of polylysine/DNA complexes. Biochim Biophys Acta 1324:27–36

Erbacher P, Remy JS, Behr JP (1999) Gene transfer with synthetic virus-like particles via the integrin-mediated endocytosis pathway. Gene Ther 6:138–145

Fisher KD, Ulbrich K, Subr V, Ward CM, Mautner V, Blakey D, Seymour LW (2000) A versatile system for receptor-mediated gene delivery permits increased entry of DNA into target cells, enhanced delivery to the nucleus and elevated rates of transgene expression. Gene Ther 7:1337–1343

Fredericksen BL, Wei BL, Yao J, Luo T, Garcia JV (2002) Inhibition of endosomal/lysosomal degradation increases the infectivity of human immunodeficiency virus. J Virol 76: 11440–11446

Fritz JD, Herweijer H, Zhang G, Wolff JA (1996) Gene transfer into mammalian cells using histone-condensed plasmid DNA. Hum Gene Ther 7:1395–1404

Furgeson DY CW, Yockman JW, Kim SW (2003) Modified Linear Polyethylenimine-Cholesterol Comjugates for DNA Complexation. Bioconjug Chem 14:840–847

Garcia-Bustos J, Heitman J, Hall MN (1991) Nuclear protein localization. Biochim Biophys Acta 1071:83–101

Godbey WT, Wu KK, Mikos AG (1999a) Size matters: molecular weight affects the efficiency of poly(ethylenimine) as a gene delivery vehicle. J Biomed Mater Res 45:268–275

Godbey WT, Wu KK, Mikos AG (1999b) Tracking the intracellular path of poly(ethylenimine)/DNA complexes for gene delivery. Proc Natl Acad Sci USA 96:5177–5181

Godbey WT, Barry MA, Saggau P, Wu KK, Mikos AG (2000) Poly(ethylenimine)-mediated transfection: a new paradigm for gene delivery. J Biomed Mater Res 51:321–328

Guo W, Lee RL (1999) Receptor-targeted gene delivery via folate-conjugated polyethylenimine. AAPS PharmSci 1:E19

Haensler J, Szoka Jr FC (1993) Polyamidoamine cascade polymers mediate efficient transfection of cells in culture. Bioconjug Chem 4:372–379

Harbottle RP, Cooper RG, Hart SL, Ladhoff A, McKay T, Knight AM, Wagner E, Miller AD, Coutelle C (1998) An RGD-oligolysine peptide: a prototype construct for integrin-mediated gene delivery. Hum Gene Ther 9:1037–1047

Howard KA, Dash PR, Read ML, Ward K, Tomkins LM, Nazarova O, Ulbrich K, Seymour LW (2000) Influence of hydrophilicity of cationic polymers on the biophysical properties of polyelectrolyte complexes formed by self-assembly with DNA. Biochim Biophys Acta 1475:245–255

Itaka (2003) In situ single cell observation by fluorescence resonance energy transfer reveals fast intra-cytoplasmic delivery and easy release of plasmid DNA complexed with linear polyethylenimine. J gene medicine 6:76–84

Jeong JH, Song SH, Lim DW, Lee H, Park TG (2001) DNA transfection using linear poly (ethylenimine) prepared by controlled acid hydrolysis of poly(2-ethyl-2-oxazoline). J Control Release 73:391–399

Kaiser J (2003) Gene therapy. Seeking the cause of induced leukemias in X-SCID trial. Science 299:495

Katayose S, Kataoka K (1998) Remarkable increase in nuclease resistance of plasmid DNA through supramolecular assembly with poly(ethylene glycol)-poly(L-lysine) block copolymer. J Pharm Sci 87:160–163

Kim JS, Kim BI, Maruyama A, Akaike T, Kim SW (1998) A new non-viral DNA delivery vector: the terplex system. J Control Release 53:175–182

Kimura T, Yamaoka Y, Iwase R, Murakami A (2002) Effect of physicochemical properties of polyplexes composed od chemically modified PL derivatives on transfection efficiency *in vitro*. Macromolecular Bioscience 2:437–446

Klink DT, Chao S, Glick MC, Scanlin TF (2001) Nuclear translocation of lactosylated poly-L-lysine/cDNA complex in cystic fibrosis airway epithelial cells. Mol Ther 3:831–841

Kopeek VuaJ, Pohl MBaVK J (1988) Cleavage of oligopeptide side-chains in N-(2-hydroxypropyl)meth-acrylamide copolymers by mixtures of lysosomal enzymes. J Control Release 8:133–140

Kunath K vHA, Fischer D, Petersen H, Bickel U, Voigt K, Kissel T (2003) Low-molecular-weight polyethylenimine as a non-viral vector for DNA delivery: comparison of physicochemical properties, transfection efficiency and *in vivo* distribution with high-molecular-weight polyethylenimine. J Control Release 89:113–125

Kursa M, Walker GF, Roessler V, Ogris M, Roedl W, Kircheis R, Wagner E (2003) Novel shielded transferrin-polyethylene glycol-polyethylenimine/DNA complexes for systemic tumor-targeted gene transfer. Bioconjug Chem 14:222–231

Leamon CP, Weigl D, Hendren RW (1999) Folate copolymer-mediated transfection of cultured cells. Bioconjug Chem 10:947–957

Lee KY, Kwon IC, Kim YH, Jo WH, Jeong SY (1998) Preparation of chitosan self-aggregates as a gene delivery system. J Control Release 51:213–220

Lim DW, Yeom YI, Park TG (2000) Poly(DMAEMA-NVP)-β-PEG-galactose as gene delivery vector for hepatocytes. Bioconjug Chem 11:688–695

Liquier J, Pinot-Lafaix M, Taillandier E, Brahms J (1975) Infrared linear dichroism investigations of deoxyribonucleic acid complexes with poly(L-arginine) and poly(L-lysine). Biochemistry 14:4191–4197

Maes R, Sedwick W, Vaheri A (1967) Interaction between DEAE-dextran and nucleic acids. Biochim Biophys Acta 134:269–276

Magin-Lachmann C, Kotzamanis G, D'Aiuto L, Cooke H, Huxley C, Wagner E (2004) In vitro and in vivo delivery of intact BAC DNA-comparison of different methods. J Gene Med 6:195–209

Mannisto M, Vanderkerken S, Toncheva V, Elomaa M, Ruponen M, Schacht E, Urtti A (2002) Structure-activity relationships of poly(L-lysines): effects of pegylation and molecular shape on physicochemical and biological properties in gene delivery. J Control Release 83:169

Marshall E (1999) Gene therapy death prompts review of adenovirus vector. Science 286: 2244–2245

Marshall E (2000) FDA halts all gene therapy trials at Penn. Science 287:565, 567

Marshall E (2002) Clinical research. Gene therapy a suspect in leukemia-like disease. Science 298:34–35

Maxfield FR (1982) Weak bases and ionophores rapidly and reversibly raise the pH of endocytic vesicles in cultured mouse fibroblasts. J Cell Biol 95:676–681

Midoux P, Mendes C, Legrand A, Raimond J, Mayer R, Monsigny M, Roche AC (1993) Specific gene transfer mediated by lactosylated poly-L-lysine into hepatoma cells. Nucleic Acids Res 21:871–878

Mislick KA, Baldeschwieler JD, Kayyem JF, Meade TJ (1995) Transfection of folate-polylysine DNA complexes: evidence for lysosomal delivery. Bioconjug Chem 6:512–515
Mita K, Zama M, Ichimura S (1977) Effect of charge density of cationic polyelectrolytes on complex formation with DNA. Biopolymers 16:1993–2004
Mumper RJ, Wang J, Klakamp SL, Nitta H, Anwer K, Tagliaferri F, Rolland AP (1998) Protective interactive noncondensing (PINC) polymers for enhanced plasmid distribution and expression in rat skeletal muscle. J Control Release 52:191–203
Murphy RF, Powers S, Cantor CR (1984) Endosome pH measured in single cells by dual fluorescence flow cytometry: rapid acidification of insulin to pH 6. J Cell Biol 98: 1757–1762
Nishikawa M, Takemura S, Takakura Y, Hashida M (1998) Targeted delivery of plasmid DNA to hepatocytes *in vivo*: optimization of the pharmacokinetics of plasmid DNA/galactosylated poly(L-lysine) complexes by controlling their physicochemical properties. J Pharmacol Exp Ther 287:408–415
Ogris M, Brunner S, Schuller S, Kircheis R, Wagner E (1999) PEGylated DNA/transferrin-PEI complexes: reduced interaction with blood components, extended circulation in blood and potential for systemic gene delivery. Gene Ther 6:595–605
Ohashi S, Kubo T, Ikeda T, Arai Y, Takahashi K, Hirasawa Y, Takigawa M, Satoh E, Imanishi J, Mazda O (2001) Cationic polymer-mediated genetic transduction into cultured human chondrosarcoma-derived HCS-2/8 cells. J Orthop Sci 6:75–81
Okada CY, Rechsteiner M (1982) Introduction of macromolecules into cultured mammalian cells by osmotic lysis of pinocytic vesicles. Cell 29:33–41
Plank C, Tang MX, Wolfe AR, Szoka Jr FC (1999) Branched cationic peptides for gene delivery: role of type and number of cationic residues in formation and *in vitro* activity of DNA polyplexes. Hum Gene Ther 10:319–332
Pollard H, Remy JS, Loussouarn G, Demolombe S, Behr JP, Escande D (1998) Polyethylenimine but not cationic lipids promotes transgene delivery to the nucleus in mammalian cells. J Biol Chem 273:7507–7511
Prasmickaite L, Hogset A, Berg K (2002) The role of the cell cycle on the efficiency of photochemical gene transfection. Biochim Biophys Acta 1570:210–218
Putnam D, Gentry CA, Pack DW, Langer R (2001) Polymer-based gene delivery with low cytotoxicity by a unique balance of side-chain termini. Proc Natl Acad Sci USA 98: 1200–1205
Remy-Kristensen A, Clamme JP, Vuilleumier C, Kuhry JG, Mely Y (2001) Role of endocytosis in the transfection of L929 fibroblasts by polyethylenimine/DNA complexes. Biochim Biophys Acta 1514:21–32
Ringsdorf H (1975) Structure and properties of pharmacologically active polymers. J Poly Sci Symp 51:135–153
Rosenkranz AA, Yachmenev SV, Jans DA, Serebryakova NV, Murav'ev VI, Peters R, Sobolev AS (1992) Receptor-mediated endocytosis and nuclear transport of a transfecting DNA construct. Exp Cell Res 199:323–329
Ruponen M, Yla-Herttuala S, Urtti A (1999) Interactions of polymeric and liposomal gene delivery systems with extracellular glycosaminoglycans: physicochemical and transfection studies. Biochim Biophys Acta 1415:331–341
Seymour LW (1997) Synthetic polymers for vectorial delivery of DNA: characterisation of polymer-DNA complexes by photon correlation spectroscopy and stability to nuclease degradation and disruption by polyanions *in vitro*. Journal of Controlled Release 48: 269–276
Seymour LW (1999) Turbidometric analysis of polyelectrolyte complexes formed between poly(L-lysine) and DNA. Colloids and Surfaces B 16:253–260
Takai T, Ohmori H (1990) DNA transfection of mouse lymphoid cells by the combination of DEAE-dextran-mediated DNA uptake and osmotic shock procedure. Biochim Biophys Acta 1048:105–109

Tang MX, Rademann CT, Szoka FC (1996) In vitro gene delivery by degraded polyamidoamine dendrimers. Bioconj Chem 6:703–714

Tang MX, Szoka FC (1997) The influence of polymer structure on the interactions of cationic polymers with DNA and morphology of the resulting complexes. Gene Ther 4:823–832

Toncheva V, Wolfert MA, Dash PR, Oupicky D, Ulbrich K, Seymour LW, Schacht EH (1998) Novel vectors for gene delivery formed by self-assembly of DNA with poly(L-lysine) grafted with hydrophilic polymers. Biochim Biophys Acta 1380:354–368

van de Wetering P, Cherng JY, Talsma H, Crommelin DJ, Hennink WE (1998) 2-(Dimethylamino)ethyl methacrylate based (co)polymers as gene transfer agents. J Control Release 53:145–153

van de Wetering P, Moret EE, Schuurmans-Nieuwenbroek NM, van Steenbergen MJ, Hennink WE (1999) Structure-activity relationships of water-soluble cationic methacrylate/methacrylamide polymers for non-viral gene delivery. Bioconjug Chem 10:589–597

Vinogradov S, Batrakova E, Li S, Kabanov A (1999) Polyion complex micelles with protein-modified corona for receptor-mediated delivery of oligonucleotides into cells. Bioconjug Chem 10:851–860

Vinogradov SV, Bronich TK, Kabanov AV (1998) Self-assembly of polyamine-poly(ethylene glycol) copolymers with phosphorothioate oligonucleotides. Bioconjug Chem 9: 805–812

Wagner E, Cotton M, Foisner R, Birnstiel ML (1991) Transferrin-polycation-DNA complexes: the effect of polycations on the structure of the complex and DNA delivery to cells. Proc Natl Acad Sci USA 88:4255–4259

Wagner E, Cotton M, Mechtler K, Kirlappos H, Birnstiel ML (1991) DNA-binding transferrin conjugates as functional gene-delivery agents: synthesis by linkage of polylysine or ethidium homodimer to the transferrin carbohydrate moiety. Bioconjug Chem 2:226–231

Wagner E, Zenke M, Cotton M, Beug H, Birnstiel ML (1990) Transferrin-polycation conjugates as carriers for DNA uptake into cells. Proc Natl Acad Sci USA 87:3410–3414

Wall DA, Wilson G, Hubbard AL (1980) The galactose-specific recognition system of mammalian liver: the route of ligand internalization in rat hepatocytes. Cell 21:79–93

Wattiaux R, Laurent N, Wattiaux-De Coninck S, Jadot M (2000) Endosomes, lysosomes: their implication in gene transfer. Adv Drug Deliv Rev 41:201–208

Wink T, Beer J, Hennink WE, Bult A, van Bennekom WP(1999) Interaction between plasmid DNA and cationic polymers studied by surface plasmon resonance spectrometry. Anal Chem 71:801–805

Wolfert MA, Dash PR, Nazarova O, Oupicky D, Seymour LW, Smart S, Strohalm J, Ulbrich K (1999) Polyelectrolyte vectors for gene delivery: influence of cationic polymer on biophysical properties of complexes formed with DNA. Bioconjug Chem 10:993–1004

Wolfert MA, Schacht EH, Toncheva V, Ulbrich K, Nazarova O, Seymour LW (1996) Characterization of vectors for gene therapy formed by self-assembly of DNA with synthetic block co-polymers. Hum Gene Ther 7:2123–2133

Wu GY, Wu CH (1987) Receptor-mediated *in vitro* gene transformation by a soluble DNA carrier system. J Biol Chem 262:4429–4432

Wu GY, Wu CH (1988a) Evidence for targeted gene delivery to Hep G2 hepatoma cells *in vitro*. Biochemistry 27:887–892

Wu GY, Wu CH (1988b) Receptor-mediated gene delivery and expression *in vivo*. J Biol Chem 263:14621–14624

Yamaoka T, Iwata H, Hamada N, Ide H, Kimura Y (1994) Transfection of foreign gene to mammalian cells using low-toxic cationic polymers. Nucleic Acids Symposium Series 31:229–230

Yamaoka T, Kimura T, Iwase R, Murakami A (2000) Enhanced expression of foreign gene transferred to mammalian cells *in vitro* using chemically modified poly(L-lysine)s as gene carriers. Chemistry Letters:118–119

Yamaoka T, Tabata Y, Ikada Y (1994) Distribution and tissue uptake of poly(ethylene glycol) with different molecular weights after intravenous administration to mice. J Pharm Sci 83:601–606

Yamaoka T, Tabata Y, Ikada Y (1995) Comparison of body distribution of poly(vinyl alcohol) with other water-soluble polymers after intravenous administration. J Pharm Pharmacol 47:479–486

Yoneda Y (1997) How proteins are transported from cytoplasm to the nucleus. J Biochem (Tokyo) 121:811–817

Zanta MA, Boussif O, Adib A, Behr JP (1997) *In vitro* gene delivery to hepatocytes with galactosylated polyethylenimine. Bioconjug Chem 8:839–844

Zauner W, Brunner S, Buschle M, Ogris M, Wagner E (1999) Differential behaviour of lipid based and polycation based gene transfer systems in transfecting primary human fibroblasts: a potential role of polylysine in nuclear transport. Biochim Biophys Acta 1428:57–67

Zenke M, Steinlein P, Wagner E, Cotton M, Beug H, Birnstiel ML (1990) Receptor-mediated endocytosis of transferrin-polycation conjugates: an efficient way to introduce DNA into hematopoietic cells. Proc Natl Acad Sci USA 87:3655–3659

Zuidam NJ, Posthuma G, de Vries ET, Crommelin DJ, Hennink WE, Storm G (2000) Effects of physicochemical characteristics of poly(2-(dimethylamino)ethyl methacrylate)-based polyplexes on cellular association and internalization. J Drug Target 8:51–66

Development of a Supramolecular Nanocarrier for Gene Delivery Based on Cationic Block Copolymers

Yuichi Yamasaki[1], Woo-Dong Jang[1], Nobuhiro Nishiyama[2], Naoki Kanayama[3], and Kazunori Kataoka[1,2]

1 Introduction

The potential to treat various life-threatening diseases by gene therapy has been experimentally validated by recent progress in molecular biology, biotechnology, and functional analyses of results from numerous preclinical studies. In a broad sense, gene therapy is defined as one that uses nucleic-acid-based drugs, including plasmid DNA (pDNA), antisense olidonucleotides, ribozymes, siRNA, and peptide nucleic acids.. However, some nucleic-acid-based drugs are unstable and cannot be delivered into target tissues without the help of a delivery systems. During the last 15 years, clinical trials of human gene therapy have been carried out in over 4,000 patients, but serious side effects have been reported in several patients due to the inherent toxicity of viral gene vectors (Marshall 1999, 2002). Although the high therapeutic efficiency of viral gene vector has been proven, the lack of safety remains a critical problem. Therefore, the establishment of a safe and effective gene therapy based on non-viral gene carrier systems has been strongly advocated.

In order to realize safe and effective non-viral gene therapy, the construction of a reliable gene carrier with minimal toxicity is necessary. Particularly for systemic administration, therapeutic genes need to be encapsulated inside of such carriers, which must also be small enough to circulate in the blood compartment. Although direct injection of naked pDNA can achieve a certain therapeutic efficacy for muscle and heart tissues, it results in the rapid clearance, within several minutes, of pDNA from the blood circulation (Harada-Shiba et al. 2002). Thus, the prevention of pDNA excretion and degradation in the blood compartment as well as the resistance of pDNA to both nuclease attack and reticular endothelial system recognition are also mandatory. The first step in this process is the compartmentalization of therapeutic pDNA into a nano-container that is suitable for blood circulation. This process of DNA

[1] Department of Materials Science and Engineering, Graduate School of Engineering, The University of Tokyo, Tokyo, Japan
[2] Center for Disease Biology and Integrative Medicine, Graduate School of Medicine, The University of Tokyo, Tokyo, Japan
[3] Department of Materials Science and Engineering, School of Engineering, Nagoya University, Aichi, Japan

compaction is closely related to the encapsulation of viral DNA into the capsid head. In addition to components of the viral system, several other proteins are known to induce DNA compaction. For example, a chromatin structure resulting from the electrostatic attractions between the lysine residues around histone octamer and DNA is essential for mammalian DNA packaging into the cell nucleus. These viral and nuclear compaction processes are referred to as DNA condensation and involve a shift in the DNA from an elongated state to a compact globular one. Using cationic reagents, DNA can be condensed in vitro into a form that is suitable for DNA delivery by non-viral gene carriers.

Among the different kinds of condensing agents, poly(L-lysine) (PLL) is one of the best known, since it can be treated as a model compound of basic proteins such as histones. In addition, PLL-induced DNA condensation is the most widely used model for studying chromosome structure. The morphology and size of PLL-induced DNA condensates have been investigated using transmission electron microscopy (TEM) and light scattering (LS). In TEM images, PLL-induced DNA condensates have often been observed as highly ordered toroidal or rod-like structures with a typical size of around 100 nm. LS studies showed that monomolecular DNA condensates were generated when an extra-dilute DNA solution was employed. However, under charge neutralization conditions, multi-molecular DNA condensates were often generated due to the formation of secondary aggregates, which caused the formation of large precipitates. By mixing an excess amount of PLL with DNA to form positively charged DNA condensates, the formation of aggregates was prevented. However, while the colloidal stability of the condensates was improved by this procedure, the cytotoxicity was enhanced by the presence of excess positive charges. Thus, PLL-induced DNA condensates were assessed only in preliminary studies using cultured cells as their systemic administration is not feasible.

Similar results are also expected for almost all of the cationic homopolymer systems. While much effort has been devoted to designing different types of cationic homopolymers, a construct that can be used for in vivo therapy remains elusive. It has therefore been concluded that the homopolymer system is appropriate for examining the fundamental properties of complexation and condensation, but is not suitable for actual clinical use. For this reason, we developed a cationic block copolymer that consists of hydrophilic poly(ethylene glycol) (PEG) and PLL chains. This cationic block copolymer, PEG-PLL, effectively produces a pDNA condensate through the formation of polyion complex (PIC) micelles that are surrounded by a hydrophilic layer. This arrangement leads to increased colloidal stability in a protein-containing medium and is thus expected to allow prolon ged circulation in the blood. In this chapter, the development of a block-copolymer-based pDNA delivery system is discussed in detail. The topics include: syntheses of PEG-PLL and other types of cationic block copolymers, the difference between PLL and PEG-PLL, and functionalization, as based on molecular design and ligand attachment.

2 Preparation of Block Copolymers as Gene Carriers

For the preparation of PIC micelles as an effective non-viral gene carrier, the molecular design of the block copolymer is very important. In this section, the characteristics and synthesis of block copolymers are briefly described. These carriers contain a polycationic segment that acts as a counterpart to anionic DNA in the formation of PICs, while a non-ionic segment prevents aggregation of the polycation-DNA complex.

2.1 Nonionic Segment

The requirements for the non-ionic segment are: high solubility in aqueous medium, flexibility, low toxicity, and low immunogenicity. In numerous non-ionic polymers, only limited examples successfully satisfied these criteria. PEG is one of the most widely used polymers in the synthesis of biomaterials due to its strong hydration ability and conformational flexibility, in addition to its remarkable biocompatibility (Harris and Zalipsky, 1997). Poly(*N*-(2-hydroxypropyl)methacrylamide) (PHPMA) (Oupicky et al. 1999; Vinogradov et al. 1999; Bohdanecky and Kopecek 1974) and dextran (Toncheva et al. 1998) also can be used as steric stabilizers of the DNA/polycation complex. The advantage of PHPMA is that its side chain can be modified in order to design functionalized DNA/polycation complexes (Dash et al. 2000). Dextran is a naturally occurring polysaccharide and has high biocompatibility. Comparative studies of dextran and PEG showed that the latter is more effective in preventing protein adsorption due to its large molecular exclusion volume and strong hydration ability (Osterberg et al. 1995).

2.2 Cationic Segment

The cationic segment of the block copolymer works as a counterpart to anionic DNA in the formation of PIC. To achieve maximum transfection efficiency, the relationship between the cationic segment structure and the desired function should be clarified. The stability and the dissociation behavior of DNA/polycation complexes is mainly controlled by the charge density and flexibility of the cationic portion. Additionally, protonation of the polycationic segment should facilitate the escape of the complex from the lysosome to the cytoplasm, i.e., the proton sponge effect (Boussif et al. 1995). Among the various cationic polymers, PLL has been the most widely used as the polycationic segment of block copolymers (Katayose and Kataoka, 1997; Harada and Kataoka, 1995). PLL has primary amino groups at the side-chains lysine residues, which can be utilized to introduce various functional groups, such as ligand moieties (Zauner et al. 1998), hydrophobic groups (Kim et al. 1998), thiol groups (Kakizawa et al. 1999), and histidyl moieties (Roufai and Midoux 2001). Recently, polyethyleneimine (PEI)-introduced block copolymers have been developed (Akiyama et al. 2000a). PEI has attracted increasing attention because of the relatively low pK_a values of its amino groups, which provide a high buffering capacity (Boussif et al. 1995). Similarly, poly(dimethylaminoethyl methacrylate) (PAMA) (Kataoka et al. 1999), poly(2-(trimethylammonio)ethyl methacrylate) (PTMAEM), and poly[*N*′-(2-amino)ethyl-*N*-methacryloylglycinamide] (poly[Ma-Gly-NH-$(CH_2)_2$-NH_2]) segments (Konak et al. 1999) have been conjugated to non-ionic polymers as DNA carriers and the transfection behaviors of the complexes have been evaluated. Polycationic dendrimers (Choi et al. 1999, 2000) have also been explored as gene-carrying cationic segments; when introduced to the PEG terminal and PIC micelles were successfully formed.

2.3 Synthesis of Block Copolymers

Two different methods have been used for the synthesis of block copolymers: (1) polymerization of the cationic segments from a non-ionic macro-initiator, and (2) coupling of the two different polymers. Recent progress in the polymerization technique has enabled the introduction of various functional groups at the chain ends of PEG (hetero-bifunctional PEG). Heterobifunctional PEG was used as a precursor of block copolymers and was prepared by ring-opening polymerization of ethylene oxide using an initiator bearing protected functional groups, such as an acetal (Kataoka et al. 1999; Nagasaki and Kataoka, 1998). The (-terminal group of the PEG was easily modified to amino, hydroxyl, sulfonyl, and disulfide groups (Akiyama et al. 2000b). Amino-terminated PEG can be used as a macro-initiator of N-carboxylic anhydride derivatives of amino acids (Katayose and Kataoka 1997; Harada and Kataoka 1995). For example, N-carboxylic anhydride of ε-(benzyloxycarbonyl)-L-lysine [lysine(Z)] was polymerized using amino-terminated PEG as a macro-initiator followed by deprotection of the side-chain amino groups of the lysine residues, yielding PEG-PLL block copolymers. PEG-PLL thus obtained has a very narrow molecular weight distribution, and the degree of polymerization is easily controlled by the initial ratio of monomer to initiator.

Similarly, PEG-PEI block copolymers were synthesized by the oxazoline polymerization initiated from the methanesulfonyl end of PEG, followed by the alkaline deprotection of amide moieties (Akiyama et al. 2000b). PEG-PAMA was synthesized by the sequential addition polymerization of dimethylaminoethyl methacrylate (AMA) from the active anionic terminal of PEG (Kataoka et al. 1999). Diblock or barbell-like triblock copolymers of PEG and PLL dendrimer were generated by stepwise liquid-phase peptide synthesis from the terminal amino group of PEG (Choi et al. 1999, 2000).

In addition, ligand moieties, such as sugar and peptides, can be introduced at the initiation terminal (α-terminal) of the hetero-bifunctional PEG segment to facilitate receptor-mediated endocytosis of polyplex micelles. For example, when potassium 3,3-diethoxypropanolate is used as an initiator of ethylene oxide, PEG segments will have an acetal group at their α-terminal that can easily be converted into an aldehyde in order to introduce ligand moieties (Kataoka et al. 1999).

The coupling reaction of cationic polymers and hydrophilic polymers is also applicable to the synthesis of block copolymers for use as DNA carriers. For example, PTMAEM was conjugated with PHPMA to yield PHPMA-PTMAEM (Konak et al. 1999). In this case, PTMAEM, which has quaternary ammonium groups on its side chains, has a terminal functionality of primary amino group that can be conjugated with the carboxylic acid terminal of PHPMA. PHPMA-Poly[Ma-Gly-NH-(CH2)2-NH2] was also synthesized by a similar approach. The terminal carboxylic-acid functionality of hydrophilic polymers is usually activated to react with the amino groups of cationic polymers by using *N*-hydroxy succinimide (Oupicky et al. 1999), carbodiimide (Toncheva et al. 1998), or thionyl chloride (Choi et al. 1998). Thus, block copolymers with primary amino groups on their side chains can be produced by protecting the amino groups before the conjugation reaction. Instead of the amide bond, urethane (Konak et al. 1999), thioether and disulfide bonds (Vinogradov et al. 1999) are also applicable to the synthesis of block copolymers, depending on the terminal functionalities.

It is also feasible to couple hydrophilic polymers onto the polyplexes of DNA and cationic polymers (Choi et al. 1998; Oupicky et al. 2000; Blessing et al. 2001). The conjugation reaction takes place at the cationic polymers located near the surface of the polyplex. If these have reactive amino groups on their side chains, such as is the case for PLL and PEI, the polymers obtained should have the grafted architecture.

3 Polylysine-Based Non-viral Gene Delivery Systems

PLL is the most extensively studied cationic polymer for constructing non-viral gene carriers. It has a comparatively high pK_a (~10.0) and is positively charged at physiological pH, providing the ability to mask the negative charge of pDNA, which it packages into a small particle (<200 nm) that can be effectively taken up by cells. Also, PLL/pDNA polyplexes protect pDNA from hydrolytic and enzymatic degradation, and positively-charged polyplexes prepared at non-stoichiometric mixing ratios exhibited facilitated cellular uptake via adsorptive endocytosis. The physicochemical and biological properties of PLL/pDNA polyplexes depend on the molecular weight of PLL, and the polyplexes from PLL with a relatively low molecular weight (~5,000) show inefficient gene transfer.

The drawback of PLL/pDNA polyplexes in general is their comparatively low transfection efficiency. While PLL/pDNA polyplexes are internalized into intracellular vesicles (e.g., endosomes), their ability to escape from the vesicles to the cytoplasm is substantially low, accounting for the insufficient gene transfer. The addition of endosomolytic agents, such as chloroquine, considerably increased the transfection efficiency of PLL/pDNA polyplexes (Erbacher et al. 1996). Also, the gene transfer efficiency can be improved by modifying PLL with inactivated virus particles, membrane-disruptive polymers (e.g., oligohistidine), and peptides (e.g., HA-2) (see below).

Low cytotoxicity is a prerequisite for in vivo gene delivery. It is well-known that the cationic nature of the polyplexes causes significant cyototoxicity. The use of PLL derivatives with ester and disulfide linkages in the backbone, which are cleavable in the intracellular compartment, significantly decreased the cytotoxicity of the polyplexes, since they were gradually degraded to low-cytotoxicity constituent oligomers or monomers (Lim et al. 2000; Oupicky et al. 2001). Moreover, these biodegradable PLL derivatives had a higher transfection efficiency than PLL, owing to the improved intracellular release of pDNA.

Receptor-mediated gene delivery based on PLL has been extensively studied, since the ε-amino groups in PLL are available for the attachment of various pilot molecules. Indeed, PLL was modified with a variety of agents, such as antibodies, peptides, sugars, folate, and transferrin, for cell-type specific transfection (Zauner et al. 1998). In many cases, however, cell type-specific transfection was obtained only with the aid of endosomolytic agents.

A large number of *in* vivo applications have been reported for PLL/pDNA polyplexes. The potential use of the poplyplexes for in vivo gene delivery was originally demonstrated for asialoorosomucoid (AsOR)-conjugated PLL for liver targeting (Wu et al. 1988). The PLL/pDNA polyplexes exhibited a relatively short residence time in the bloodstream, although the circulating properties vary according to the molecular weights of PLL and the mixing ratios of PLL to DNA (Nishikawa et al. 1998; Ward et al. 2001). A large portion of PLL/pDNA polyplexes accumulates in the liver following their intravenous injection. PLL/pDNA polyplexes (especially those consisting of low-molecular-weight PLL) activate the complement system, resulting in their accumulation in Kupffer cells in the liver. Meanwhile, gene expression in the liver was observed using hepatocye-targeted PLL/pDNA polyplexes (Zauner et al. 1998).

Surface modification of PLL/pDNA polyplexes with hydrophilic polymers, such as PEG, is a promising way to improve the colloidal stability and biocompatibility of the polyplexes. A typical core-shell type polyplex results from PEG-PLL block copolymers (Itaka et al. 2003). PEG-PLL and pDNA formed polyplex micelles of approximately 90 nm. The PIC core consisted of PLL and the single pDNA molecule was covered with dense PEG palisades. PEG-PLL/pDNA polyplex micelles showed high serum tolerability due to decreased non-specific interactions with biological components (i.e., the condensed state of pDNA in the PIC core as well as the cellular uptake and transfection efficiency of the polyplex micelles were not impaired by incubation with serum.), whereas conventional non-viral vectors, such as PLL/pDNA and lipoplexes, attenuate their efficacy in serum (Itaka et al. 2002, 2003). When PEG-PLL/pDNA polyplex micelles were intravenously injected, pDNA was observed in the blood circulation even after 3 h (Harada-Shiba et al. 2002). Thus, PEG-PLL/pDNA polyplex micelles show excellent blood-circulating properties. In order to achieve cell-type-specific in vivo gene delivery, the PLL/pDNA polyplexes with PEG palisades were modified with pilot molecules (Nah et al. 2002). This approach will be pursued with the goal of developing non-viral gene vectors.

Despite their surface modification with hydrophilic polymers, the polyplexes might be destabilized under harsh in vivo conditions, especially by polyelectrolyte exchange reactions with negatively charged biomolecules. Therefore, disulfide cross-linkings were introduced into PLL and resulted in further stabilization of the polyplexes. The disulfide bonds are relatively stable in the extracellular milieu, whereas they are cleavable in intracellular reductive environments. Hence, the cross-linked polyplex micelles from thiolated PEG-PLL efficiently released pDNA in response to reductive conditions mimicking those of the intracellular environment, but showed remarkable stability against counter-polyanion exchange under extracellular non-reductive conditions. Consequently, cross-linked polyplex micelles with optimized cationic charge and disulfide cross-linking densities showed more efficient gene transfer than non-cross-linked polyplex micelles (Miyata et al. 2004). It was also reported that introduction of disulfide cross-linkings into PLL/pDNA coated with PEG increased the amount of pDNA that circulated in the blood compartment from 6% to 40% of the initial doses at 30 min post-i.v. injection (Oupicky et al. 2001). Thus, polyplexes based on PLL are equipped with various functions that offer great potential for in vivo gene therapy.

4 Environmentally Responsive Polycation-Based Non-viral Gene Carriers

Most DNA/polycation complexes (polyplexes) adsorb to the cell membrane and are subsequently taken up by the cell via endocytosis. The endosomes fuse with lysosomes, where the pH of the endosomes/lysosomes is about 5. Nucleases in the lysosomes are activated by this acidic pH and eventually attack the polyplexes. Thus, it is widely recognized that the release of polyplexes from the endosomes is a critical issue in enhancing the efficiency of gene transfer (Wiethoff and Middaugh 2003).

In order to accelerate the endosomal escape of polyplexes, various strategies have been developed. For example, the effects of endosomotropic agents, such as chloroquine, were studied using various types of polyplexes in a variety of cell lines (Wattiaux et al. 2000). Although the in vitro transfection activity of polyplexes was significantly improved by chloroquine, the use of endosomotropic agents is not suitable for clinical application because of their strong toxicity.

This problem has been resolved by the development of environmentally responsive carriers, especially those that are pH-responsive, to enhance the endosomal escape efficiency of polyplexes. In 1995, Boussif et al. reported that PEI disrupts the endosomal membrane through the *proton sponge effect* (Boussif et al. 1995). Under physiological conditions, the amino groups of PEI are only partially protonated ($pK_{a,app}$ = 5~6) (Suh et al. 1994). Consequently, after endocytosis of PEI/DNA polyplexes, PEI buffers the endosomal acidification, which is accompanied by an accumulation of protons in endosomes coupled to a simultaneous influx of chloride anions. It is hypothesized that the combination of the swelling of the PEI conformation and the osmotic swelling of the endosome leads to destabilization of the endosome and, consequently, the release of its contents into the cytoplasm. Because of this membrane disruption activity, PEI-based polyplexes generally do not require the addition of endosomotropic agents for transfection. Furthermore, it is assumed that the proton sponge effect is not only valid for PEI but is also more generally applicable to polymers containing amino groups with a pK_a comparable to or below physiological pH, for example, PAMA (van de Wetering et al. 1999). Incorporation of histidine residues into polymer side chains also enhances transfection activity, because these residues are substantially protonated at endosomal pH and disrupt the endosomal membrane by the proton sponge effect (Midoux and Monsigny 1999). Thus, polycations with a low pK_a are promising agents to achieve effective gene transfer.

However, polyplexes based on PEI and PAMA exhibit positively charged characteristic (the ζ potential is in the order of several tens of mV), and these induce certain problems particularly in vivo, for example, non-specific interaction with non-target tissues and blood components, such as serum proteins. In order to overcome such problems, the conjugation of polycations with non-ionic and hydrophilic polymers, such as PEG (Katayose and Kataoka 1997), dextran (Maruyama et al. 1997) and PHPMA (Toncheva et al. 1998), to form block or grafted copolymers has been examined. By conjugating polycations with non-ionic and hydrophilic polymers, the surfaces of the polyplexes are eventually covered with a non-ionic and hydrophilic shell, so that they are electrically neutral and their colloidal stability is substantially increases compared to the nonconjugated polyplexes. In the case of PEI, the graft length and amount of PEG side chain strongly affect the shape of the polyplex and its surface charge. The stability and cytotoxicity of polyplexes improved with increasing graft length and amount of PEG side chain. However, the longer PEG side chains also interrupted gene transfer to cells, due to the greater steric hindrance and the reduced interaction with the negatively charged cell membrane (Petersen et al. 2002). A similar tendency was reported in the case of PAMA (Dubruel and Schacht 2000). Nonetheless,

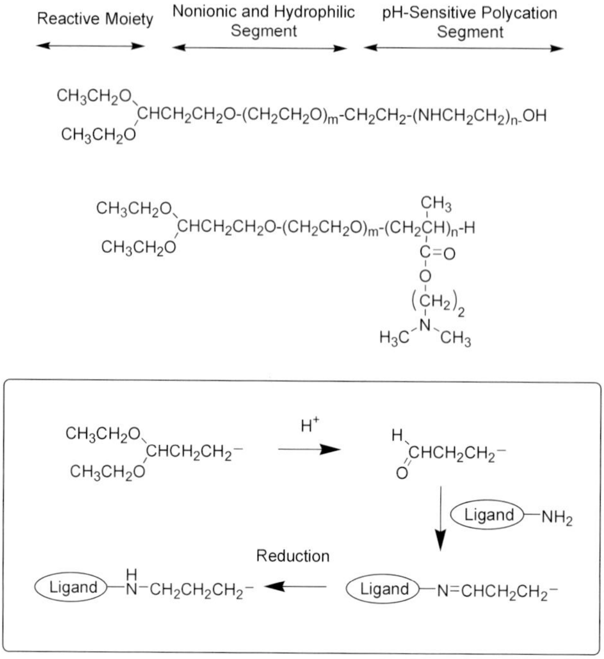

FIG. 1. Chemical structure of end-functionalized PEG-polycation block copolymers

the attachment of cell-specific ligands to the surfaces of polyplexes is a promising method to achieve effective gene transfer. For example, end-functionalized AB-type block copolymers of PEG and liner-PEI or PAMA have been successfully synthesized, (Fig. 1) (Akiyama et al. 2000b; Kataoka et al. 1999), and PEG-PAMA block-copolymer-based polyplex micelle to which lactose had been attached showed substantially higher transfection efficiency of HepG2 cells than polyplex micelles without lactose moieties (Wakebayashi et al. 2004). Murthy et al. (2003) reported a new strategy for the design and synthesis of pH-responsive polyplexes. A relatively hydrophobic, membrane-disruptive cationic polymer backbone was conjugated with PEG as the side chain via acid-labile linkages carrying the targeting unit to promote cell-specific uptake. By lowering the pH, PEG side chains were eliminated from the polymer backbone, which then disrupted the endosomal membrane to facilitate translocation of the polyplex from the endosomal compartment to the cytoplasm.

The incorporation of the pH-responsive ("fusogenic") peptides derived from viruses and artificial amphiphilic peptide to polyplexes has been described (Plank et al. 1994). These peptides change conformation upon lowering of the pH, become more hydrophobic; this, in turn, increases their affinity for phospholipid membranes and ultimately may solubilize (disrupt) the endosomal membrane.

5 Conclusions

In this chapter, recent trends in the development of supramolecular nanocarriers of genes and related compounds, based on cationic copolymers, was highlighted. PEG is widely used as the non-ionic and hydrophilic parts of these non-viral gene carriers, as it has been approved by the FDA for use in pharmaceutical formulations. The PEGylation of proteins and peptide drugs has led to an improvement of their pharmacokinetics, proving the validity of using PEG to improve drug efficacy. Similarly, it is expected that PEGylation of gene carriers offers great promise for their optimization. Recent progress in the development of novel synthetic backbones for various kinds of PEG conjugates as well as preparation procedures for sophisticated polyplexes further extend the applicability of PEGylated gene carriers. PEG conjugates possessing moieties that are responsive to changes in their chemical and biological environment act as environmentally responsive polyplexes that can release therapeutic genes in response to external stimuli. Surface modification of polyplexes using ligands attached to the distal end of the PEG segment significantly enhances the internalization efficiency of polyplexes into the cytoplasm. In addition, various functional counterparts to the PEG segment are also available due to the recent development of regulated polymerization techniques. Currently, numerous studies are focusing on polyplex-based PEG-polycation copolymers with respect to their safety and gene transfection efficiency. Investigations of block- and graft-copolymer-based gene delivery systems have progressed rapidly, and is expected to lead to progress in engineered syntheses of PEG-based block and graft copolymers.

Acknowledgments. This work was supported by the Core Research Program for Evolutional Science and Technology (CREST) from Japan Science and Technology Corporation (JST) and by Special Coordination Funds for Promoting Science and Technology from the Ministry of Education, Culture, Sports, Science and Technology of Japan.

References

Akiyama Y, Harada A, Nagasaki Y, Kataoka K (2000a) Synthesis of poly(ethylene glycol)-block-poly(ethylenimine) possessing an acetal group at the PEG end. Macromolecules 33:5841–5845

Akiyama Y, Otsuka H, Nagasaki Y, Kato M, Kataoka K (2000b) Selective synthesis of heterobifunctional poly(ethylene glycol) derivatives containing both mercapto and acetal terminals. Bioconjug Chem 11:947–50

Blessing T, Kursa M, Holzhauzer R, Kircheis R, Wagner E (2001) Different strategies for formation of PEGylated EGF-conjugated PEI/DNA complexes for targeted gene delivery. Bioconjug Chem 12:529–537

Boussif O, Lezoualc'h F, Zanta MA, Mergny MD, Scherman D, Demeneix B, Behr JP (1995) A versatile vector for gene and oligonucleotide transfer into cells in culture and *in vivo*: polyethylenimine. Proc Natl Acad Sci USA 92:7297–301

Choi JS, Joo DK, Kim CH, Kim K, Park JS (2000) Synthesis of a barbell-like triblock copolymer, poly(L-lysine) dendrimer-block-poly(ethylene glycol)-block-poly(L-lysine) dendrimer, and its self-assembly with plasmid DNA. J Am Chem Soc 122:474–480

Choi JS, Lee EJ, Choi YH, Jeong YJ, Park JS (1999) Poly(ethylene glycol)-*block*-poly(L-lysine) dendrimer: novel linear polymer/dendrimer block copolymer forming a spherical water-soluble polyionic complex with DNA. Bioconjug Chem 10:62–65

Choi YH, Liu F, Kim J-S, Choi YK, Park JS, Kim SW (1998) Polyethylene glycol-grafted poly-L-lysine as polymeric gene carrier. J Control Release 54:39–48

Dash PR, Read NL, Fisher KD, Howard KA, Wolfert M, Pupicky D, Subr V, Strohalm J, Ulbrich K, Seymour LW (2000) Decreased binding to proteins and cells of polymeric gene delivery vectors surface modified with a multivalent hydrophilic polymer and retargeting through attachment of transferring. J Biol Chem 275:3793–3802

Dubruel P, Schacht EH (2000) Effect of polyethylene oxide blocks or grafts on the physicochemical properties of poly(2-*N*-(dimethylaminoethyl) methacrylate) DNA complexes. J Bioact Compat Polym 15:279–296

Erbacher P, Roche AC, Monsigny M, Midoux P (1996) Putative role of chloroquine in gene transfer into a human hepatoma cell line by DNA/lactosylated polylysine complexes. Exp Cell Res 225:186–194

Harada A, Kataoka K (1995) Formation of Polyion complex in an aqueous milieu from a pair of oppositely charged block copolymers with poly(ethylene glycol) segments. Macromolecules 28:5294–5299

Harada-Shiba M, Yamauchi K, Harada A, Takamisawa I, Shimokado K, Kataoka K (2002) Polyion complex micelles as vectors in gene therapy—pharmacokinetics and *in vivo* gene transfer. Gene Ther 9:407–414

Harris JM, Zalipsky S (Eds.) (1997) Poly(ethylene glycol): Chemistry and Biological Applications, American Chemical Society, Washington, DC

Itaka K, Harada A, Nakamura K, Kawaguchi H, Kataoka K (2002) Evaluation by fluorescence resonance energy transfer of the stability of nonviral gene delivery vectors under physiological conditions. Biomacromolecules 3:841–845

Itaka K, Yamauchi K, Harada A, Nakamura K, Kawaguchi H, Kataoka K (2003) Polyion complex micelles from plasmid DNA and poly(ethylene glycol)-poly(L-lysine) block copolymer as serum-tolerable polyplex system: physiological properties of micelles relevant to gene transfection efficiency. Biomaterials 24:4495–4506

Kakizawa Y, Harada A, Kataoka K (1999) Environment-sensitive stabilization of core-shell structured polyion complex micelle by reversible cross-linking of the core through disulfide bond. J Am Chem Soc 121:11247–11248

Kataoka K, Harada A, Wakebayashi D, Nagasaki Y (1999) Polyion complex micelles with reactive aldehyde groups on their surface from plasmid DNA and end-functionalized charged block copolymers. Macromolecules 32:6892–6894

Katayose S, Kataoka K (1997) Water-soluble polyion complex associates of DNA and poly(ethylene glycol)-poly(L-lysine) block copolymer, Bioconjug. Chem 8:702–707

Kim JS, Kim BI, Maruyama A, Akaike T, Kim SW (1998) A new non-viral DNA delivery vector: the terplex system, complexes. J Control Release 53:175–182

Konak C, Mrkvickova L, Nazarova O, Ulbrich K, Seymour LW (1999) Formation of DNA complexes with Diblock copolymers of poly[N-(2-hydroxypropyl)methacrylamide] and polycations. Supramol Sci 5:67–74

Kopecek J, Bazilova H (1973) Poly[N-(2-hydroxypropyl)methacrylamide]. 1. Radical polymerization and copolymerization, Eur Polym J 9:7–14

Lim YB, Han SO, Kong HU, Lee Y, Park JS, Jeong B, Kim SW (2000) Biodegradable polyester, poly[a-(4-aminobutyl)-L-glycolic acid], as a non-toxic gene carrier. Pharm Res 17: 811–816

Marshall E (1999) Clinical trials-Gene therapy death prompts review of adenovirus vector. Science 286:2244–2245

Marshall E (2002) Clinical research-Gene therapy a suspect in leukemia-like disease. Science 298:34–35

Maruyama A, Katoh M, Ishihara T, Akaike T (1997) Comb-type polycations effectively stabilize DNA triplex. Bioconjug Chem 8:3–6

Midoux P, Monsigny M (1999) Efficient gene transfer by histidylated polylysine/pDNA complexes. Bioconjug Chem 10:406–411

Miyata K, Kakizawa Y, Nishiyama N, Harada A, Yamasaki Y, Koyama H, Kataoka K (2004) Block catiomer polyplexes with regulated densities of charge and disulfide cross-linking directed to enhance gene expression. J Am Chem Soc 126:2355–2361

Murthy N, Campbell J, Fausto N, Hoffman AS, Stayton PS (2003) Bioinspired pH-responsive polymers for the intracellular delivery of biomolecular drugs. Bioconjug Chem 14:412–419

Nagasaki Y and Kataoka K (1998), The reactive polymeric micelle, convenient tool for targeting drug delivery system, In "Tailored polymeric materials for controlled delivery systems (ACS Symp. Ser.709)", Eds., McCulloch I and Shalaby SW, ACS, Washington, D. C., pp 105–116

Nah JW, Yu L, Han SO, Ann CH, Kim SW (2002) Artery wall binding peptide-poly(ethylene glycol)-grafted-poly(L-lysine)-based gene delivery to artery wall cells. J Control Release 78:273–284

Nishikawa M, Takemura S, Takakura Y, Hashida M (1998) Targeted delivery of plasmid DNA to hepatocytes *in vivo*: optimization of the pharmacokinetics of plasmid DNA/galactosylated poly(L-lysine) complexes by controlling their physicochemical properties. J Pharmcol Exp Ther 287:408–415

Osterberg E, Berrgstrom K, Holmberg K, Schuman TP, Riggs JA, Burns NL, Van Alstine JM, Harris JM (1995) Protein-rejecting ability of surface-bound dextran in end-on and side-on configurations: comparison to PEG. J Biomed Mater Res 29:741–747

Oupicky D, Parker AL, Seymour LW (2001) Laterally stabilized complexes of DNA with linear reducible polycations: strategy for triggered intracellular activation of DNA vectors. J Am Chem Soc 124:8–9

Oupicky D, Carlisle RC, Seymour LW (2001) Triggered intracellular activation of disulfide crosslinked polyelectrolyte gene delivery complexes with extended systemic circulation *in vivo*. Gene Ther 8:713–724

Oupicky D, Howard KA, Konak C, Dash PR, Ulbrich K, Seymour LW (2000) Steric stabilization of poly-L-lysine /DNA complexes by the covalent attachment of semitelechelic poly[*N*-(2-hydroxypropyl)methacrylamide], Bioconjug. Chem 11:492–501

Oupicky D, Konak C, Ulbrich K (1999) Preparation of DNA complexes with diblock copolymer of poly[N-(2-hydroxypropyl)methacrylamide] and polycations. Mater Sci Eng C 7:59–65

Petersen H, Fechner PM, Martin AL, Kunath K, Stolnik S, Roberts CJ, Fischer D, Davies MC, Kissel T (2002) Polyethyleneimine-*graft*-poly(ethylene glycol) copolymers: influence of copolymer block structure on DNA complexation and biological activities as gene delivery system. Bioconjug Chem 13:845–854

Plank CP, Oberhauser B, Mechtler K, Koch C, Wagner E (1994) The influence of endosome-disruptive peptides on gene transfer using synthetic virus-like gene transfer systems. J Biomol Chem 269:12918–12924

Roufai MB, Midoux R (2001) Histidylated polylysine as DNA vector: elevation of the imidazole protonation and reduced cellular uptake without change in the polyfection efficiency of serum stabilized negative polyplexes. Bioconjug Chem 12:92–99

Suh J, Paik HJ, Hwang, BK (1994) Inonization of poly(ethyleneimine) and poly(allylamine) at various pH's. BioorgChem 22:318–327

Toncheva V, Wolfert MA, Dash PR, Oupicky D, Ulbrich K, Seymour LW, Schacht EH, (1998) Novel vectors for gene delivery formed by self-assembly of DNA with poly(l-lysine) grafted with hydrophilic polymers. Biochim Biophys Acta 1380:354–368

Van de Wetering P, Moret EE, Schuurmans-Nieuwenbroek NME, Van Steenbergen M, Hennink MJ (1999) Structure-activity relationships of water-soluble cationic methacrylate/methacrylamide polymers for nonviral gene delivery. Bioconjug. Chem. 10:589–597

Vinogradov S, Batrakova E, Li S, Kabanov A (1999) Polyion complex micelles with protein-modified corona for receptor-mediated delivery of oliognucleotides into cells. Bioconjug Chem 10:851–860

Wakebayashi D, Nishiyama N, Yamasaki Y, Itaka K, Kanayama N, Harada A, Nagasaki Y, Kataoka K (2004) Lactose-conjugated polyion complex micelles incorporating plasmid DNA as a targetable gene vector system: their preparation and gene transfecting efficiency against cultured HepG2 cells. J Control Release 95:653–664

Wattiaux R, Laurent N, Wattiaux-De Coninck S, Jadot M (2000) Endosome, lysosome: their implification in gene transfer. Adv Drug Deliv Rev 41:201–208

Ward CM, Read ML, Seymour LW (2001) Systemic circulation of poly(L-lysine)/DNA vectors is influenced by polycation molecular weight and type of DNA: differential circulation in mice and rats and the implications for human gene therapy. Blood 97: 2221–2229

Wiethoff CM, Middaugh CR (2003) Barriers to nonviral gene delivery. J Pharm Sci 92: 203–217

Wu GY, Wu CH (1988) Receptor-mediated gene delivery and expression *in vivo*. J Biol Chem 263:14621–14624

Zauner W, Ogris M, Wagner E (1998) Polylysine-based transfection systems utilizing receptor-mediated delivery. Adv Drug Deliv Rev 30:97–113

Chitosan

Mayu Hashimoto, Zhihong Yang, Yohei Koya, and Toshinori Sato

1 Introduction

Chitosan is a β-1–4 linked copolymer of glucosamine and *N*-acetylglucosamine, and is obtained by the alkaline deacetylation of chitin, which is one of the most abundant natural polymers as the component of exoskeletons of crustaceans. Chitosan is a weak base with a pKa of around 6.4 and forms water-soluble salts with hydrochloride, glutamic acid and acetic acid. It has been employed as a drug carrier because of its low toxicity, low immunogenicity, biocompability, and biodegradability. In this chapter, we provide an overview of the utilization of chitosan and modified chitosan as gene carriers, their transfection mechanisms, and their intracellular trafficking.

2 DNA/Chitosan Complexes

In 1995, Mumper et al. proposed the usage of chitosan to deliver DNA. Since then, chitosan has been frequently investigated as a gene carrier. We have studied the formation of DNA complexes with cationic polymers, such as chitosan and polygalactosamine, and the interaction of these DNA complexes with tumor cells and white blood cells (Sato et al. 1996).

The physicochemical properties of DNA/polycation complexes, such as DNA compaction, size, and zeta potential, are important for efficient and versatile gene delivery. In the case of chitosan-mediated gene delivery systems, these features depend on the complex ratio (N/P ratio, the number of chitosan nitrogens per DNA phosphate), the preparation conditions (pH and ionic strength), the molecular weight, and the degree of deacetylation (DDA) of chitosan.

The self-assembly of DNA/chitosan complexes has been discussed in the literature. DNA compaction is dependent on the N/P ratio. DNA/chitosan (53 kDa, DDA = 93%) complexes were prepared by self-assembly and their morphologies observed by atomic force microscopy (AFM). Insufficient complex formation was observed at N/P

Department of Biosciences and Informatics, Keio University, 3-14-1 Hiyoshi, Kohoku-ku, Yokohama 223-8522, Japan

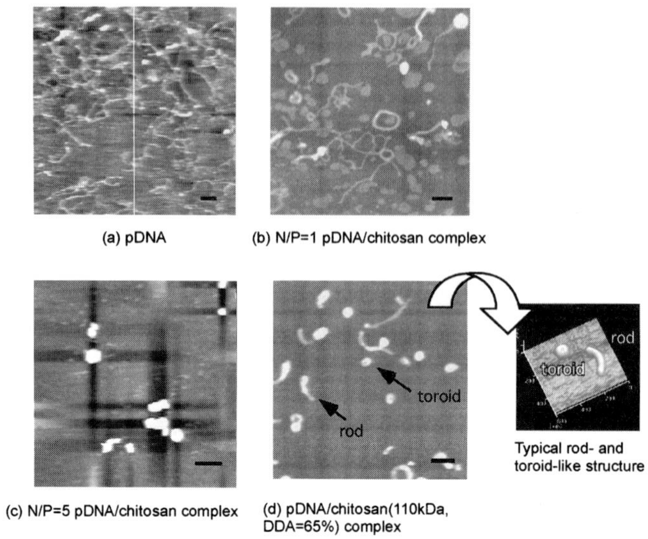

FIG. 1a–d. Atomic force microscopy images of plasmid DNA (pDNA) and pDNA/chitosan complexes. **a** pDNA; **b** pDNA/chitosan complex of N/P = 1; **c** pDNA/chitosan complex of N/P = 5; **d** pDNA/chitosan (DDA = 65%) complex of N/P = 5. *Bar* = 200 nm

= 1, while spherical particles with a diameter of 180 nm were formed at N/P = 5 (Fig. 1). Other authors observed donut- and toroid-like structures at N/P = 3 by transmission electron microscopy (TEM) (Erbacher et al. 1998; Köping-Höggård et al. 2001). The morphological changes associated with the N/P ratio were explained by the zeta potential of the complexes. The zeta potential of DNA/chitosan complex was close to 0 mV at N/P = 2 with a size range of 1–5 μm, and positive at N/P > 3 with a size range of 50–100 nm (Erbacher et al. 1998; Köping-Höggård et al. 2003) found that a uniformly dispersed globular structure, important for DNA protection against nuclease and an efficient cellular uptake, was formed by reducing the pH of the solvent.

The DDA values of chitosan affect complex formation with DNA because of the chain entanglement effect (Kiang et al. 2004). At DDAs of 90%, 70%, and 62%, the optimum N/P ratio was 3.3, 5, and 9, respectively. However, there was no significant difference in the size (150–200 nm) and morphology among these chitosans. Instead, the size of the complex (80–600 nm) is dependent on the molecular weight of chitosan (7–540 kDa) (MacLaughlin et al. 1998; Mumper et al. 1995).

Recently, other preparation methods, such as coacervation (Mao et al. 2001) and ionic gelation (Li et al. 2003) were developed in order to increase the structural stability of the DNA complexes. Using either method, DNA complexes of 100–250 nm diameter were obtained.

The molecular weight and DDA of chitosan are important factors that determine DNA compaction, the nuclease resistance of the complexes, and the dissociation of DNA from chitosan. However, since chitosan is a naturally occurring polymer prepared by the deacetylation of chitin, it is difficult to regulate its molecular weight and DDA. As shown in Tables 1 and 2, several kinds of chitosan having different molecu-

TABLE 1. Chitosan-based gene delivery in vitro

Chitosan	Complex size (nm)	Optimum N/P ratio	Cell lines	Result	Reference
70kDa	50–100	3	HeLa, HepG2, NBNLCL2	Gene expression increased over time, from 24 to 96h.	Erbacher et al. 1998
390kDa	260–750	2	HeLa, HEK293, IB-3	Lyophilized storage was possible for 4 weeks. Complexes were prepared by coacervation method.	Leong et al. 1998
102kDa (7–540kDa)	100–500	2	COS-1, NIH3T3, A549, C_2C_{12}	Complex was stable in the presence of salt and serum. Addition of endosomolytic peptide promoted gene expression.	Mac-Laughlin et al. 1998
162kDa, DDA = 83%	120	3	293, HT1080, Caco-2	Onset of gene expression with DNA/chitosan complex was slower than that with DNA/PEI complex. Low cytotoxicity.	Köping-Höggård et al. 2001
15, 52kDa	—	5	A549, B16, HeLa	Optimum condition for transfection was pH 6.9 and in the presence of serum.	Sato et al. 2001
150, 400, 600 kDa	100	—	MG63, HEK293, MSC	HEK293 cells were transfected more effectively than MG63 and MSC, but the efficiency was 1/100 that of lipofectamine. Low cytotoxicity.	Corsi et al. 2003
24-mer	100	10	293	Onset of gene expression with DNA/chitosan oligomer complex was earlier than that with DNA/high-molecular-weight chitosan complex	Köping-Höggård et al. 2003
Ultra pure and high-molecular-weight chitosan	300	2.4 or 3			
150kDa, DDA = 75%	40–80	5	293, CHO-K1	Gene expression continued for 10 days. Low cytotoxicity.	Li et al. 2003
380kDa (DDA = 60–90%)	150–200	3.3–9 (Depend on DDA)	HEK293, SW756, HeLa	DDA affected the transfection efficiency in vitro and in vivo.	Kiang et al. 2004

lar weights and DDAs have been investigated in transfection experiments in vitro and in vivo. In our studies, chitosan of 15–70kDa and 80%–90% DDA had a high transfection efficiency, as shown in Fig. 2 (Sato et al. 2001). Köping-Höggård et al. (2001) reported that chitosan of 162kDa and 83% DDA had the highest transfection efficiency. It was shown that the DDA value, rather than the molecular weight, of chitosan

TABLE 2. Chitosan-based gene delivery in vivo

Chitosan	Complex size (nm)	Administration	Gene	Result	Ref.
390 kDa	150–300	Oral	Dominant peanut allergen	A substantial reduction in allergen-induced anaphylaxis.	Roy et al. 1999
102 kDa, DDA = 80%	100–600	Transdermal	Luciferase, β-gal	Detectable gene expression by DNA in chitosan/carboxymethyle-cellulose nanoparticles.	Cui et al. 2001
—	—	Intranasal	RSV (Respiratory syncytial virus) antigen	Expression of RSV antigens in the lung. Reduction of viral titers.	Kumer et al. 2002
—	300–330	Intranasal	RSV antigen	Induction of CTL responses and reduction in the viral load.	Iqbal et al. 2003
—	—	Intranasal	INF-γ	Effective reduction of allergen-induced airway inflammation and hyperresponsiveness.	Kumer et al. 2003

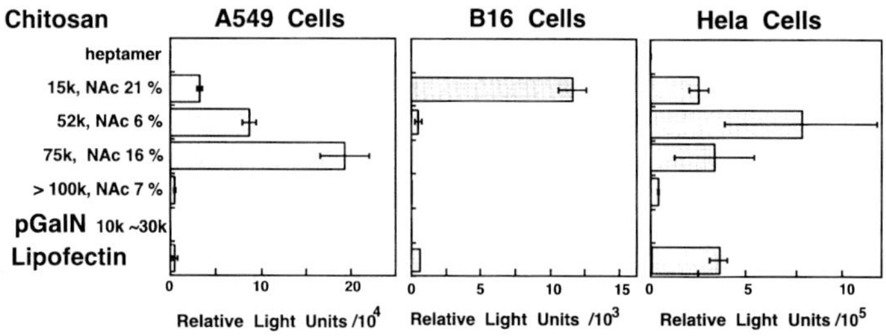

FIG. 2. Relative luciferase activity of pDNA (pGL) complexes with chitosan, polygalactosaine (pGal), and lipoplex

greatly affects the zeta potential, cellular uptake and cytotoxicity of DNA/chitosan complexes (Huang et al. 2004).

Depolymerized chitosan as a DNA carrier has also been studied (Köping-Höggård et al. 2003). DNA/chitosan oligomer (24-mer) complexes allowed the rapid onset of gene expression compared to chitosan polymer (100 kDa). Cross-linking of DNA/chitosan complex resulted in sustained gene expression for about 2 weeks, whereas the duration of expression of Lipofectamine and SuperFect was only 2–5 days (Li et al. 2003).

An advantage of chitosan-mediated gene delivery is high-level gene expression in the presence of serum and low cytotoxicity. The transfection efficiency of DNA/chitosan complexes in the presence of 10–20% serum was significantly higher than in serum-free medium (Sato et al. 2001). The complexes were less cytotoxic than DNA complexes containing Lipofectin, Lipofectamine, or PEI (Corsi et al. 2003; Köping-Höggård et al. 2001; Sato et al. 2001). For example, the IC_{50} of the DNA/PEI complex was 75 μg/ml for 293 cells, whereas that of the DNA/chitosan complex was 630 μg/ml (Köping-Höggård et al. 2001).

Transfection experiments have been carried out using several cell lines. The transfection efficiency of DNA/chitosan nanoparticles for HEK293 cells was superior to that seen for MG3 cells and mesenchymal stem cells (MSC) (Corsi et al. 2003). The cell-line dependence of chitosan-mediated gene delivery has also been reported by other authors (Erbacher et al. 1998; Kiang et al. 2004; Köping-Höggård et al. 2001), although the mechanism of cell dependence has not been clarified yet. In our recent studies, however, chitosan having the appropriate molecular weight had a higher transfection efficiency than Lipofectin in A549, HeLa, B16 (Sato et al. 2001) and HuH7 cells (unpublished data).

Chitosan has also been complexed with adenovirus, and adenovirus coated with chitosan resulted in increased transfection efficiency of mammalian cells lacking a receptor for the virus (Kawamata et al. 2002).

Chitosan is an attractive gene carrier in vivo, as shown in Table 2. It has an excellent targetability to mucus on intestinal epithelium and nasal-associated lymphoid tissue (NALT). Roy et al. (1999) investigated an immunoprophylactic strategy using oral allergen-gene immunization to moderate peanut-antigen-induced murine anaphylactic responses. The oral administration of DNA/chitosan nanoparticles resulted in the introduction of the transgene into the intestinal epithelium (Roy et al. 1999). In another study, the oral administration of chitosan nanoparticles containing the gene encoding blood coagulation factor IX resulted in sufficient production of factor IX for 2 weeks (Okoli et al. 2000). Chitosan has also been used for nasal administration. The delivery of DNA vaccine by chitosan induced a significant reduction of virus load in NALT (Kumar et al. 2002; Iqbal et al. 2003). The intranasal administration of chitosan nanoparticles containing IFN-γ-plasmid DNA effectively reduced allergen-induced airway inflammation (Kumar et al. 2003). In another experiment, topical administration of chitosan-based nanoparticles to skin induced variable but sufficient gene expression (Cui and Mumper 2001). Besides mammalian cells, chitosan-mediated gene delivery systems are useful for mass vaccination of fish (Romøren et al. 2002).

3 Chitosan Derivatives

For the practical use of DNA/chitosan complexes, several problems must be overcome, including physicochemical stability and cell targetability. Since the primary amine of chitosan is a reactive group for chemical modification, several kinds of chitosan derivatives have been synthesized (Fig. 3).

Trimethylated chitosan oligomers (TMO) result in the formation of DNA complexes that are smaller than those containing unmodified chitosan oligomers and showed good solubility and no cytotoxicity (Thanou et al. 2002). Furthermore, the transfec-

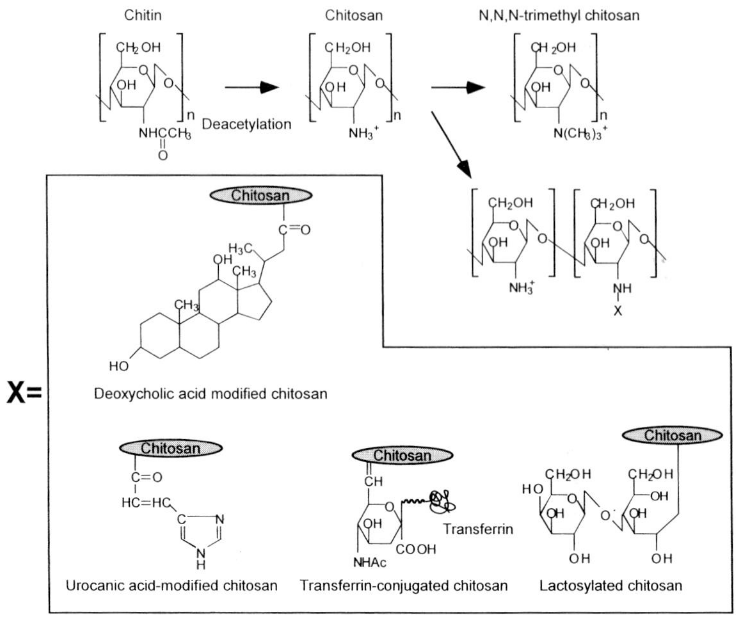

FIG. 3. Structure of chitosan derivatives

tion efficiency of DNA/TMO complexes was superior to that of DNA/DOTAP complexes and DNA/chitosan oligomer complexes in COS-1 cells, but not in Caco-2 cells.

The chitosan hydrophobilized with deoxycholic acid (DAMC) was prepared in order to control the size and structure of the complexes (Kim et al. 2001; Lee et al. 1998). The mixture of DAMC with DNA yielded polymer micelles of 160 nm, but their transfection efficiency was lower than that of Lipofectamine. Use of another hydrophobilized chitosan derivative, *N*-alkylated chitosan, enhanced the thermal stability of the DNA complexes and induced fusion with the cell membrane (Liu et al. 2003). The transfection efficiency was increased by elongating the alkyl chain from 4 to 16.

Uronic-acid-modified chitosan (UAC) was prepared by a condensation reaction (Kim et al. 2003). The imidazole ring of urocanic acid (UA) was expected to play a role in endosomal rupture through a proton sponge mechanism. The effect of UA-coupling is dependent on the particular cell line.

Viral fusion proteins have been employed to improve the transfection efficiency. Conjugation of the KNOB protein (the fiber protein on adenovirus capsid) to chitosan enhanced gene expression in HeLa cells by 130-fold (Mao et al. 2001). The attachment of an endosomolytic peptide (GM225.1 or GM227.3) to DNA/chitosan complexes also enhanced their transfection efficiency (MacLaughlin et al. 1998).

In order to deliver DNA to tumor cells, chitosan was first conjugated with transferrin (Mao et al. 2001). The transferrin-conjugated chitosan nanoparticle resulted in a four-fold increase in the transfection of HEK293 cells compared to using conventional DNA/chitosan nanoparticles.

Another approach to obtaining cell specificity is modification of the monosaccharide or disaccharide side chain of chitosan. Hepatocytes are known to express an asialoglycoprotein receptor that binds glycoprotein having a terminal galactose. Therefore, galactosylated or lactosylated chitosan is often employed. DNA complexes with galactosylated low-molecular-weight chitosan (21 kDa) were taken up via receptor-mediated endocytosis, and had a higher transfection efficiency of hepatocytes than of HeLa cells (Gao et al. 2003). Chitosan-graft-polyethylene-glycol (Park et al. 2003a) and chitosan-graft-poly (vinylpyarolidone) (Park et al. 2003b) were also galactosylated (GCP, GC-PVP). The resulting DNA/GCP complex allowed specific gene delivery to HepG2 cells. DNA/GC-PVP complexes were protected from enzymatic degradation by plasma proteins because the complex was covered with hydrophilic PVP. Galactosylation of trimethylated chitosan slightly enhanced the transfection efficiency of HepG2 cells (Murata et al. 1996). Erbacher et al. (1998) showed that lactosylated chitosan was not efficient in the transfection of hepatocytes. They discussed that the result was probably caused by the decrease in the zeta potential after lactosylation of the primary amine. In our recent research, however, lactosylated chitosan had a ten-fold higher transfection efficiency of HepG2 cells than obtained with conventional chitosan (unpublished result). Moreover, we found that lactosylation of chitosan enhanced the dispersion stability and suppressed albumin-induced aggregation.

Although several kinds of chitosan derivatives have been developed, one of the remaining problems is a decline in their DNA-binding ability following modification of the amino groups. Therefore, the method of chemical conjugation of the cell-specific ligand to chitosan needs to be improved. Moreover, the design of chitosan derivatives based on an understanding of the transfection mechanism is necessary for the development of an efficient gene delivery system.

4 Mechanism of Transcellular and Intracellular Transport of DNA/Chitosan Complexes

In order to understand the mechanism behind transgene expression of DNA/chitosan complexes, their physicochemical features and subcellular distribution have been investigated. In general, it is thought that the intracellular trafficking of DNA complexes occurs by endocytosis, followed by release from the endosome, nuclear transport and transcription (Fig. 4).

First, the DNA/chitosan complexes adsorb to the cell surface through electrostatic interactions and are internalized into the cells by endocytosis (Cho et al. 2003). TEM and conforcal laser scanning microscopy (CLMS) have revealed the distribution of DNA/chitosan complexes in endosomes (Ishii et al. 2001; Köping-Höggård et al. 2003).

In physicochemical studies, it was shown that chitosan has a unique interaction with the cell membrane. The interaction between chitosan and a dipalmitoyl-sn-glycero-3-phosphocholine (DPPC) bilayer was investigated by cross-polarization microscopy, differential scanning calorimetry (DSC) and Fourier transform Raman spectroscopy (Chan et al. 2001). It was demonstrated that chitosan induced the fusion of DPPC vesicles, and that the attractive inter-chain and intermolecular forces of the hydrophobic core (acyl chains) in the DPPC bilayer were significantly reduced by the chitosan-membrane interactions. This study provided valuable physical insights into

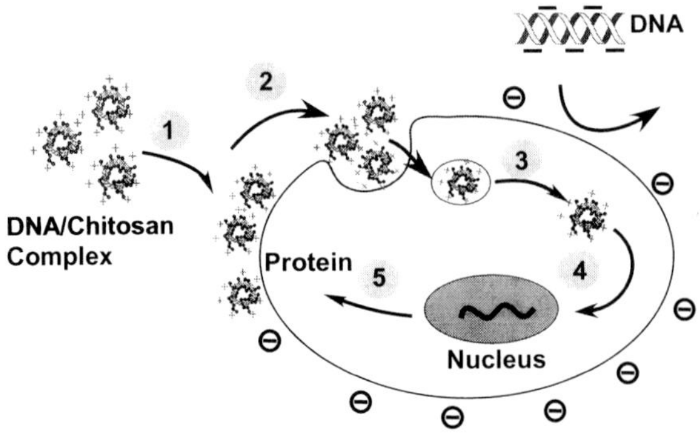

FIG. 4. Chitosan-based DNA delivery. *1* Adsorption to the cell membrane; *2* cell uptake by endocytosis; *3* release from the endosome; *4* nuclear import; *5* transcription

the mechanism of chitosan-induced perturbation of lipid membranes. The interaction of low-molecular-weight chitosan with lipid bilayer membranes has been investigated using electrochemical methods and surface plasmon resonance measurements (Yang et al. 2002). It was found that chitosan induced the formation of a mass-transfer channel in the lipid bilayer at neutral pH, and a perturbation effect of low-molecular-weight chitosan on cell membranes was shown. Furthermore, colloidal and polymer science studies have demonstrated that the molecular weight of chitosan and the pH of the medium are important factors influencing permeability and perturbation of the cell membrane (Fang et al. 2001). The interaction between chitosan and a mica-supported DPPC bilayer was investigated by AFM (Fang and Chan 2003). DPPC-chitosan binding and the intermolecular association of chitosan led to the aggregation of chitosan at the lipid-membrane surface. Furthermore, it was shown that a cell-surface protein might play an important role in chitosan-mediated cell transfection through non-ionic interactions with the carbohydrate backbone of chitosan (Venkatesh and Smith 1998).

After the DNA/chitosan complexes are internalized in cells by endocytosis, they must be released from endosomes into the cytoplasm for subsequent nuclear import. DNA would normally be attacked by nucleases in endosomes, lysosomes and the cytoplasm. However, complexation of DNA with chitosan results in protection of the DNA from enzymatic digestion (Ishii et al. 2001; Mao et al. 2001).

The buffering effect and the proton sponge hypothesis have been proposed to explain the release of DNA/chitosan complexes from endosomal compartments (Boussif et al. 1995; Cho et al. 2003). Since chitosan consists of glucosamine, the DNA/chitosan complexes in endosomes have the potential to absorb protons. In titration experiments, DNA/chitosan complexes showed high proton adsorption ability between pH 7.0 and 5.0 that was approximately 2.5-fold greater than adsorption by DNA/PLL complex (Ishii et al. 2001).

Chitosan at acidic pH is swelled by the protonation of amino group. The influx of H^+ and Cl^- increases the osmotic pressure and leads to the influx of water. Moreover, enzymatic degradation of chitosan also induces an increase in the osmolarity, thereby causing rupture of the endosome. Accordingly, we found that endocytosis inhibitors, such as bafilomycin and monenesin, significantly decrease transfection efficiency (unpublished data). While these results support the mechanism described above, release from endosomes is considered to be a barrier to chitosan-mediated gene delivery since DNA/chitosan complexes were observed in endosomes even at 72 h after cell uptake (Köping-Höggård et al. 2001).

The next step in the intracellular transport of DNA complexes is nuclear import. The subcellular distribution of DNA/chitosan complexes in the nucleus was observed by CLMS (Ishii et al. 2001). Recent studies showed that the upper diameter for transport through the nuclear pore complex (NPC) is 39 nm (Pante and Kann 2002). Since the size of DNA/chitosan complex is usually 80–500 nm, the mechanism of nuclear import of the DNA/chitosan complex needs to be further examined. In our recent study on nuclear import of DNA/chitosan complexes, using the microinjection technique, it was estimated that 0.1% of the DNA/chitosan complexes in the cytoplasm were transferred into the nucleus compared with 0.025% of naked DNA (unpublished data). Although complexation of DNA with chitosan enhanced nuclear import, addition of an active transport function to the DNA/chitosan complex is desired in order to further enhance transfer efficiency.

5 DNA/Chitosan/Sugar-PEG-C Ternary Complexes

In order to inhibit self-aggregation and improve cell targetability, ternary complexes prepared by coating the cationic pDNA/polycation complex with anionic polymers, such as poly(ethylene glycol) derivatives (PEG-C), have been developed (Carlisle et al. 2004; Maruyama et al. 2004; Trubetskoy et al. 1999, 2003; Koyama et al. 2003). In our study, DNA/chitosan complexes were coated with PEG derivatives bearing carboxylic acid and lactose residues (Fig. 5).

The zeta potential of DNA/chitosan complexes coated with Lac-PEG-C were negative. The size of the DNA/chitosan/Lac-PEG-C complexes was about 200 nm, which was almost the same as DNA/chitosan complexes. The ternary complexes were stable and did not self-aggregate. Transgene expression was evaluated in HuH-7 cells expressing the asialoglycoprotein receptor. The DNA/chitosan/PEG-C complexes

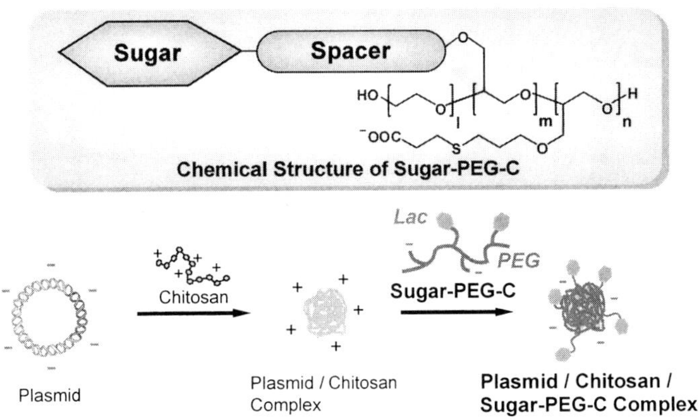

Fig. 5. Formation of plasmid DNA/chitosan/sugar-PEG-C ternary complex

without lactose had a slightly decreased transfection efficiency compared to DNA/chitosan complexes. This might due to electrostatic repulsion of the anionic complexes by the cell surface. By contrast, coating of the DNA/chitosan complexes with Lac-PEG-C resulted in a transfection efficiency that was three times higher than that of the DNA/chitosan complex, as measured by luciferase gene expression. The DNA/chitosan/Lac-PEG-C complexes were probably taken up by HuH-7 cells via a receptor-mediated interaction.

6 Conclusion

Chitosan possesses many favorable characteristics for use as a gene delivery system. However, further improvements in the structure and function of this system, such as optimization of chitosan (molecular weight or DDA), chemical modification using functional groups, and formation of ternary complexes, are still necessary. The results reviewed in this chapter indicate that such rational modifications to compensate for the limitations of chitosan as a gene carrier will lead to improved transfection efficiencies.

References

Boussif O, Lezounalc'h F, Zanta MA, Mergny MD, Scherman D, Demenex B, Behr JP (1995) A versatile vector for gene and oligonucleotide transfer into cells in culture and in vivo: Polyethylenimine. Proc Natl Acad Sci USA 92:7297–7310

Carlisle RC, Etrych T, Briggs SS, Preece JA, Ulbrich K, Seymour LW (2004) Polymer-coated polyethylenimine/DNA complexes designed for triggered activation by intracellular reduction. J Gene Med 6:337–344

Chan V, Mao HQ, Leong KW (2001) Chitosan-induced perturbation of dipalmitol-sn-glycero-3-phosphocholine membrane bilayer. Langmuir 17:3749–3756

Cho YW, Kim JD, Park K (2003) Polycation gene delivery system: escape from endosomes to cytosol. J Pharm Pharmacol 55:721–734

Corsi K, Chellat F, Yahia L, Fernandes JC (2003) Mesenchymal stem cells, MG63 and HEK293 transfection using chitosan-DNA nanoparticels. Biomaterials 24:1255–1264

Cui Z, Mumper RJ (2001) Chitosan-based nanoparticles for topical genetic immunization. J Control Rel 75:409–419

Erbacher P, Zou S, Bettinger T, Steffan AM, Remy JS (1998) Chitosan-Based Vector/DNA Complexes for Gene Delivery: Biophysical Characteristics and Transfection Ability. Pharmaceutical Res 15:1332–1339

Fang N, Chan V, Mao HQ, Leong KW (2001) Interactions of Phospholipid Bilayer with Chitosan: Effect of Molecular Weight and pH. Biomacromolecules 2:1161–1168

Fang N, Chan V (2003) Chitosan-induced restructuration of a mica-supported phospholipid bilayer: an atomic force microscopy study. Biomacromolecules 4:1596–1604

Gao S, Chen J, Su X, Ding A, Yang YH, Hua Z, Zhang J (2003) Galactosylated low molecular weight chitosan as DNA carrier for hepatocyte-targeting. Int J Pharm 255:57–68

Huang M, Khor E, Lim LY (2004) Uptake and cytotoxicity of chitosan molecules and nanoparticles: effects of molecular weight and degree of deacetylation. Pharm Res 21: 344–353

Iqbal M, Lin W, Jabbal-Gill I, Davis SS, Steward MW, Illum L (2003) Nasal delivery of chitosan-DNA plasmid expressing epitopes of respiratory syncytial virus (RSV) induces protective CTL responses in BALB/c mice. Vaccine 21:1478–1485

Ishii T, Okahata Y, Sato T (2001) Mechanism of cell transfection with plasmid/chitosan complexes. Biochim Biophys Acta 1514:51–64

Kawamata Y, Nagayama Y, Nakano K, Mizuguchi H, Hayakawa T, Sato T, Ishii N (2002) Receptor-independent augmentation of adenovirus-mediated gene tranfer with chitosan in vitro. Biomaterials 23:4573–4579

Kiang T, Wen J, Lim HW, Leong KW (2004) The effect of the degree of chitosan deacetylation on the efficiency of gene transfection. Biomaterials 25:5293–5301

Kim TH, Ihm JE, Choi YJ, Nah JW, Cho CS (2003) Efficient gene delivery by urocanic acid-modified chitosan. J Control Rel 93:389–402

Kim YH, Gihm SH, Park CR (2001) Structural Characteristics of Size-Controlled Self-Aggregates of Deoxycholic Acid-Modified Chitosan and Their Appplication as a DNA Delivery Carrier. Bioconjugate Chem 12:932–938

Köping-Höggård M, Tubulekas, Guan H, Edwards K, Nilsson M, Vårum KM, Artursson P (2001) Chitosan as a nonviral gene delivery system. Structure-property relationships and characteristics compared with polyethylenimine in vitro and after lung administration in vivo. Gene Ther 8:1108–1121

Köping-Höggård M, Mel'nikova YS, Vårum KM, Lindman B, Artusson P (2003) Relationship between the physical shape and the efficiency of oligomeric chitosan as a gene delivery system in vitro and in vivo. J Gene Med 5:130–141

Koyama Y, Ito T, Matsumoto H, Tanioka A, Okuda T, Yamamura N, Aoyagi H, Niidome T (2003) Novel poly (ethylene glycol) derivatives with carboxylic acid pendant groups: synthesis and their protection and enhancing effect on non-viral gene transfection systems. J Biomater Sci Polymer Edn 14:515–531

Kumar M, Behera AK, Lockey RF, Zhang J, Bhullar G, Cruz C, Chen L, Leong KW, Huang S, Mohapatra SS (2002) Intranasal Gene Transfer by Chitosan-DNA Nanoshperes Protects BALB/c Mice Against Acute Respiratory Syncytial Virus Infection Hum Gene Ther 13: 1415–1425

Kumar M, Kong X, Behera AK, Hellermann GR, Lockey RF, Mohapatra SS (2003) Chitosan IFN-g -pDNA Nanoparticle (CIN) Therapy for Allergic Asthma. Genet Vaccines Ther 1:3

Lee KY, Kwon IC, Kim YH, Jo WH, Jeong SY (1998) Preparation of chitosan self-aggregates as a gene delivery system. J Control Rel 51:213–220

Leong KW, Mao HQ, Trung-Le VL, Roy K (1998) DNA-polycation nanospheres as non-viral gene delivery vehicles. J Control Rel 53:183–193

Li XW, Lee DKL, Chan ASC, Alpar HO (2003) Sustained expression in mammalian cells with DNA complexed with chitosan nanoparticles. Biochim Biophy Acta 1630:7–18

Liu WG, Zhang X, Sun SJ, Sun GJ, Yao KD (2003) N-Alkylated Chitosan as a Potential Nonviral Vector for Gene Transfection. Bioconjugate Chem 14:782–789

MacLaughlin FC, Mumper RJ, Wang J, Tagliaferri JM, Gill I, Hinchcliffe, Rolland AP (1998) Chitosan and depolymerized chitosan oligomers as condensing carriers for in vivo plasmid delivery. J Control Rel 56:259–272

Mao HQ, Roy K, Troung-Le VL, Janes KA, Lin KY, Wang Y, August JT, Leong KW (2001) Chitosan-DNA nanoparticles as gene carriers: synthesis, characterization and transfection efficiency. J Control Rel 70:399–421

Maruyama K, Iwasaki F, Takizawa T, Yanagie H, Niidome T, Yamada E, Ito T, Koyama Y (2004) Novel receptor-mediated gene delivery system comprising plasmid/protamine/sugar-containing polyanion ternary complex. Biomaterials 25:3267–3272

Mumper RJ, Wang J, Claspell JM, Rolland AP (1995) Novel polymeric condensing carriers for gene delivery. Proc Intern Symp Control Rel Bioact Mater 22:178–179

Murata J, Ohya Y, Ouchi T (1996) Possibility of application of quaternary chitosan having pendant galactose residues as gene delivery tool. Carbohydrate Poymers 29:69–74

Okoli G, Hortelano G, Leong KW (2000) Oral delivery of plasmid DNA encoding the factor IX gene. Proc Intl Symp Control Rel Bioact Mater 27:74–75

Pante N, Kann M (2002) Nuclear pore complex is able to transport macromolecules with diameters of about 39 nm. Mol Bio Cell 13:425–434

Park IK, Kim TH, Kim SI, Park YH, Kim WJ, Akaike T, Cho CS (2003) Galactosylated chitosan (GC)-graft-poly (vinyl pyrrolidone)(PVP) as hepatocyte-targeting DNA carrier. Preparation and physicochemical characterization of GC-graft-PVP/DNA complex (1). J Control Rel 86:349–359

Park IK, Ihm JE, Park YK, Choi YJ, Kim SI, Kim WJ, Akaike T, Choi CS (2003) Visualization of transfection of hepatocytes by galactosylated chitosan-graft-poly (ethylene glycol)/DNA complexes by confocal laser scanning microscopy. Int J Pharm 257:103–110

Romøren K, Thu BJ, Evensen Ø (2002) Immersion delivery of plasmid DNA II. A study of the potentials of a chitosan based delivery system in rainbow trout (Onchorhynchus mykiss) fry. J Contl Rel 85:215–225

Roy K, Mao HQ, Huang SK, Leong KW (1999) Oral gene delivery with chitosan-DNA nanoparticles generates immunologic protection in a murine model of peanut allergy. Nature Medicine 5:387–391

Sato T, Shirakawa N, Nishi H, Okahata Y (1996) Formation of A DNA/polygalactosamine Complex and Its Interaction with Cells. Chem Lett 9:725–726

Sato T, Ishii T, Okahata Y (2001) In vitro gene delivery mediated by chitosan. Effect of pH, serum and molecular mass of chitosan on the transfection efficiency. Biomaterials 22: 2075–2080

Thanou M, Fleorea BI, Geldof M, Junginger HE, Borchard G (2002) Quaternized chitosan oligomers as novel gene delivery vectors in epithelial cell lines. Biomaterials 23:153–159

Trubetskoy VS, Loomis A, Hagstrom JE, Budker VG, WSOlff JA (1999) Layer-by-layer deposition of oppositely charged polyelectrolytes on the surface of condensed DNA particles. Nucleic Acids Res 27:3090–3095

Trubetskoy VS, Wong SC, Subbotin V, Budker VG, Loomis A, Hagstorm JE, Wolff JA (2003) Recharging cationic DNA complexes with highly charged polyanions for in vitro and in vivo gene delivery. Gene Therapy 10:261–271

Venkatesh S, Smith TJ (1998) Chitosan-membrane interactions and their probable role in chitosan-mediated transfection. Biotechnol Appl Biochem 27:256–267

Yang F, Cui X, Yang X (2002) Interaction of low-molecular-weight chitosan with mimic membrane studied by electrochemical methods and surface plasmon resonance. Biophys Chem 99:99–106

Dendrimers as DNA Carriers

Hidetoshi Arima

1 Introduction

Recently, numerous polycations have been used for formulating gene and oligonucleotides (ODN) into complexes now termed "polyplexes", which efficiently transfect cells (Wagner 2004). Polycations include histones, polylysine, polyethyleneimine (PEI), polypropyleneimine (PPI), cationic dendrimers, poly(2-(dimethyl-amino)ethyl methacrylate) and chitosan (De Smedt et al. 2000). The properties of these polymers and their uses in transfections have been reported previously (Abdallah et al. 1995; Garnett 1999; Ruponen et al. 2003).

Of the various polymers for gene transfer, dendrimers are highly branched three-dimensional macromolecules with well-defined structures constructed around a multifunctional central core (Tomalia et al. 1985). They have novel structural properties such as a single molecular weight, a large number of controllable peripheral functionalities and a tendency to adopt a globular shape (Tomalia et al. 1990). Indeed, they differ from classical monomers, oligomers and hyperbranched polymers. There are two synthetic approaches that have been used for the preparation of dendrimers: the divergent approach and the convergent approach (Grayson and Frechet 2001).

There are now more than fifty families of dendrimers, each with unique properties, since the surface, interior and core can be tailored to different sorts of applications (Klajnert and Bryszewska 2001). Table 1 summarizes the various types of cationic dendrimers and their conjugates for gene and ODN transfer. Polyamidoamine (PAMAM) starburst dendrimers (Bielinska et al. 1995; Haensler and Szoka 1993), partially hydrolyzed (degraded) PAMAM dendrimers (Tang et al. 1996), PPI dendrimers and phosphorous dendrimers (Loup et al. 1999) have been used for their delivery. The potential use of polylysine dendrimers for gene delivery has been reported as well (see the chapter by T. Niidome, this volume).

Of the various cationic dendrimers, PAMAM starburst dendrimers have steadily grown in popularity in the past decade in a variety of disciplines, ranging from materials science to biomedicine. Indeed, PAMAM starburst dendrimers are safe, nonim-

Graduate School of Pharmaceutical Sciences, Kumamoto University, 5-1 Oe-honmachi, Kumamoto 862-0973, Japan

TABLE 1. Various dendrimers and their conjugates for gene and oligonucleotide transfer

Dendrimers
Amphipathic asymmetric dendrimers
Cyclic core dendrimers
Degraded polyamidoamine (PAMAM) dendrimers (SuperFect)
Hydroxyl-terminated polyamidoamine dendrimer
Phosphorus dendrimers
PAMAM starburst dendrimers
Poly(ethylene glycol)-block-poly(L-lysine) dendrimers
Polypropyleneimine dendrimers
Polypropyleneimine (PPI)-DAB dendrimers
Dendrimer conjugates
PAMAM dendrimers conjugated with the fluorescent dye Oregon green 488.
PAMAM starburst dendrimer conjugate with cyclodextrins
Dendrimer conjugates with antibody

munogenic and can function as highly efficient cationic polymer vectors for delivering genetic material into cells (Esfand and Tomalia 2001). They have been shown to be more efficient and safer than either cationic liposomes or other cationic polymers for in vitro gene transfer (Gebhart and Kabanov 2001).

The major structural differences in PAMAM dendrimers relate to the core molecule, either ammonia (NH_3), as trivalent initiator core, or ethylenediamine (EDA), as a tetravalent initiator core, which dictates several structural characteristics of the molecule, including the overall shape, density, and surface charge (Tomalia et al. 1990). As shown in Fig. 1, the size of the PAMAM starburst dendrimer (EDA) is generally determined by the number of layers or generations present in the polymer (Klajnert and Bryszewska 2001). Each additional layer of a PAMAM dendrimer enlarges the size of the molecule by approximately 1 nm while doubling the number of surface amine groups. At the present, PAMAM dendrimers (Sigma-Aldrich (St. Louis, MO) and Dendritech Nano Technologies (Mount Pleasant, MZ)) and PPI dendrimer or Astramol (Sigma-Aldrich (St. Louis, MO) and DSM Fine Chemicals (Heerlen, Netherlands)) are commercially available (Cloninger 2002). In addition, the degraded PAMAM dendrimer Superfect can be obtained from Qiagen (Hilden, Germany).

2 Transfection Efficiency

Both the gene transfer activity and the cytotoxicity of PAMAM starburst dendrimers are significantly generation dependent. In general, the transfection activity of dendrimers with a high generation number is likely to be superior to those with a low generation number (Zhang and Smith 2000). For example, in the case of the PAMAM starburst dendrimers, optimum gene transfer is obtained with a molecular mass in excess of 20 kDa (Tang et al. 1996). In the presence of DEAE-dextran, in Rat2 cell line, a definitive relationship between the PAMAM dendrimer generation (G) and the transfection efficiency was observed, whereas an exponential increase in transfection efficiency was seen by increasing the generation number from G5 to G10, with a plateau in activity after G8 (Kukowska-Latallo et al. 1996). In addition, maximal trans-

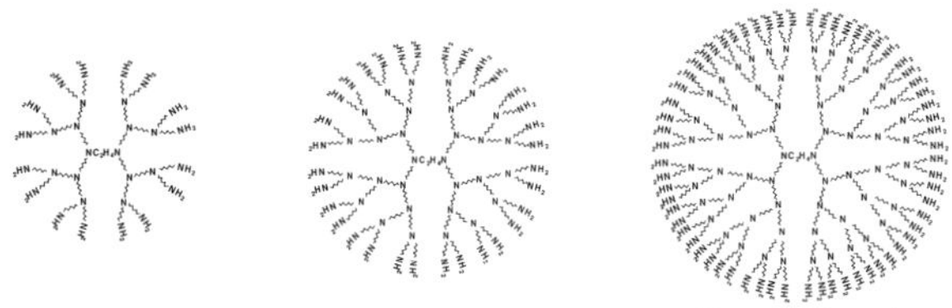

Generation 2 **Generation 3** **Generation 4**

〰 $=-C_2H_4CONHC_2H_4-$

Generation	Molecular weight	Measured diameter (nm)	Surface groups
0	517	1.5	4
1	1,430	2.2	8
2	3256	2.9	16
3	6909	3.6	32
4	14215	4.5	64
5	28826	5.4	128
6	58048	6.7	256
7	116433	8.1	512
8	233383	9.7	1024
9	467162	11.4	2048
10	934720	13.5	4096

Fig. 1. Physical characteristics of polyamidoamine (PAMAM) dendrimers (ethylenediamine core)

fection efficiency was obtained using G6 (NH_3) PAMAM starburst dendrimer rather than higher-generation dendrimers (Haensler and Szoka 1993), possibly due to the rigid structure and cytotoxicity of dendrimers with a generation number >G7. Thus, the use of dendrimers as gene delivery agents has largely been focused on high-generation (>G5) PAMAM dendrimers. However, there are a few reports on the use of PPI dendrimers containing 100% protonable nitrogens for gene delivery (Zinselmeyer et al. 2002). For example, lower-generation (G2) PPI dendrimers have been shown to be effective gene-transfer agents (Zinselmeyer et al. 2002). The potential use of PPI as a carrier of ODNs has been demonstrated as well (Santhakumaran et al. 2004).

Szoka et al. developed an activated PAMAM dendrimer, Superfect (Tang et al. 1996), which represents a new class of transfection reagents based on activated-dendrimer technology, in which some of the branches have been removed. Activated dendrimers assemble DNA into compact structures. Indeed, Superfect has been shown to enhance transfection activities due to the increased flexibility of the fractured dendrimers, which enable them to be compact when forming complexes with DNA and to swell when released from DNA (Hudde et al. 1999, Tang et al. 1996).

Recently, dendrimers have been reported to be able to transfer various types of DNA, including Epstein-Barr virus (EBV)-based plasmid vectors containing the

EBNA1 gene and oriP (Maruyama-Tabata et al. 2000) and a 60-Mb mammalian chromosome into murine and hamster cells (de Jong et al. 2001).

3 Factors Affecting Transfection Activity

There are various barriers to transfection by non-viral vectors, e.g. physicochemical properties, enzymatic stability, cellular association, endosomal release, cytoplasmic translocation, nuclear uptake, localization in the nucleus, transcription activity, epigenetic events and the cytotoxicity of the polyplexes (Nishikawa and Huang 2001; Pitkanen et al. 2003). In the following sections, each factor influencing the transfection activity of the polyplexes together with dendrimers and their conjugates is described.

3.1 Physicochemical Properties

Dendrimers offer many advantages for the in vitro transfection of cells. The DNA/dendrimer complexes are very soluble in most aqueous solutions and stable for many weeks in solution, mediating high-efficiency transfection in a wide variety of cell lines, including primary cells and non-adherent cell lines (Kukowska-Latallo et al. 1996).

In general, the size and surface charge of DNA complexes with cationic polymers determine the transfection efficiency (Bielinska et al. 1996; Kukowska-Latallo et al. 1996). For example, the size of the ODNs/DNA is an important factor in optimizing the efficiency of the dendrimer, and the great majority (>90%) of transfection activity is carried out by low-density and soluble complexes, which represent approximately only 10–20% of total complexed DNA (Bielinska et al. 1999). However, large polyplex aggregates have been found to be more active in transfection compared with small particles (Ogris et al. 1998). Thus, the relationship between size and transfection activity of DNA complexes with dendrimers is far from being resolved (Gebhart and Kabanov 2001), and size dependency cannot be generalized, but may instead be specific for each transfection reagent (Kiefer et al. 2004). Using atomic force microscopy, DNA nanoparticle formation in the presence of PAMAM dendrimers with some degree of uniformity was observed (Choi et al. 2004), although lipoplexes were reported to have a heterogeneous morphology. Therefore, polyplexes seem to have properties that are advantageous for gene transfer. Further elucidation of the relationship between the size and the morphology of the DNA complex with dendrimers, and of the effect on transfection efficiency is required.

Dendrimers protect DNA degradation by DNase through complexation, indicating that the therapeutic success of gene delivery depends on the availability and retention of intact DNA within the cell for prolonged periods of time (Santhakumaran et al. 2004). Taken together, the physicochemical properties of DNA complexes with dendrimers for DNA delivery are preferable to those of lipoplexes.

3.2 Cellular Uptake Mechanism

Cell transfection is a complicated multistep process that, in many cases, appears to be mediated by endocytosis, and there is evidence that exit from the endosome is the step that controls transfection efficiency.

The uptake mechanisms of DNA complexes with dendrimers are unlikely to be simple. While previously it was thought that polyplexes enter cells through adsorptive endocytosis (Thomas and Klibanov 2003; Lechardeur and Lukacs 2002; Hollins et al. 2004; Jevprasesphant et al. 2004), recent evidence suggests that they enter cells via a raft-dependent pathway (Manunta et al. 2004). Importantly, the different membrane lipid compositions, the pH and intracellular trafficking between endosomes and caveosomes have been reported to play important roles in transfection and may explain differences in transfection efficiencies (Bathori et al. 2004; Nichols 2003; Parton and Richards 2003). Since endocytotic mechanisms are cell-dependent in most cases, further investigation regarding the endocytotic mechanism of DNA complexes with dendrimers in a variety of cell types is needed.

3.3 Endosomal Release

Endosomal escape of DNA complexes with non-viral vectors is critical for gene transfer, and various membrane-active agents have been reported (Wagner 1998). The high transfection efficiency of dendrimers may not only be due to their well-defined shape but may also be caused by the low pK of the amines (3.9 and 6.9), which permits the dendrimer to buffer the pH change in the endosomal compartment (Klajnert and Bryszewska 2001). Thus, the enhanced transfection efficiency has been attributed to the dendrimer acting as a proton sponge similar to PEI in the acidic endosomes, leading to osmotic swelling and lysis of the endosomes/lysosomes (Tang et al. 1996; Haensler and Szoka 1993). In fact, chloroquine, a lysosomotropic drug, is ineffective in improving the transfection efficiency of DNA complexes with dendrimers (Haensler and Szoka 1993) and PEI (De Smedt et al. 2000). An alternative model for endosomal release has been proposed in which electrostatic interactions between the surface of dendrimers and charged groups on the lipid bilayers cause bending of the membrane (Zhang and Smith 2000).

3.4 Cytotoxicity

A major concern with the use of dendrimers as vectors for DNA delivery is their cytotoxicity, which may be due to the interaction between the positively charged dendrimer and the negatively charged cellular structure, especially glycosaminoglycans (heparan sulfate, hyaluronic acid and chondroitin sulfate). The toxicity of the dendrimers increases with increasing molecular size (Morgan et al. 1989), suggesting that the availability of multiple contact points between the dendrimer molecules and glycosaminoglycan is implicated in the toxicity of these molecules.

PPI dendrimers are likely to show molecular-weight-dependent cytotoxicity. For example, the cytotoxicity for B16F10 cells increased in the order of G2 < G3 = G4 dendrimers (Malik et al. 2000). The toxicity of PPI dendrimers against a human epidermoid carcinoma cell line has also been shown to be molecular weight-dependent (Zinselmeyer et al. 2002), but is lower than that of PEI in Vero cells (Lim et al. 2002a). Interestingly, SuperFect may be much better tolerated at longer exposure times of cells to the DNA complexes at any confluency (Gebhart and Kabanov 2001). It is important to note that the cytotoxicity of polyplexes appears to be lower than that of lipoplexes.

The mechanism of cytotoxicity resulting from polycations is not yet fully understood, although cationic lipids induce apoptosis (Aramaki et al. 1999). The cell death induced by PEI and its DNA complexes shows features of necrosis, as evidenced by an early membrane leakage without changes in nuclear morphology (Fischer et al. 2003).

3.5 *Ex Vivo and In Vivo Applications*

There are a few reports regarding the ex vivo and/or in vivo applications of DNA complexes with cationic polymers. DNA/dendrimer complexes can mediate gene transfer into murine cardiac transplants ex vivo (Qin et al. 1998). Billinska et al. (2000) demonstrated local application of the surface coating or incorporation of dendrimer/DNA complexes into poly(DL-lactide-co-glycolide) or the use of collagen-based biocompatible membranes as a possible means to facilitate transfection of dermal cells. Likewise, not only efficient adventitial gene delivery to rabbit carotid artery with plasmid complexes containing fractured PAMAM dendrimer (G6) (Turunen et al. 1999) but also intravascular and endobronchial DNA delivery to murine lung tissue using PAMAM starburst dendrimer (G9) were reported (Kukowska-Latallo et al. 2000). Moreover, applications of DNA complexes with dendrimers in tumor therapy have also been reported. The efficacy of dendrimer-mediated angiostatin and TIMP-2 gene delivery on the inhibition of tumor growth and angiogenesis after intratumor injection of these DNA complexes with dendrimers was demonstrated (Vincent et al. 2003). Intraperitoneal tumor targeting and imaging of intraperitoneal tumors by use of antisense oligo-DNA complexed with dendrimers and/or avidin in mice were reported (Sato et al. 2001).

From the in vivo safety point of view, attention should be paid to the occurrence of side effects. PAMAM dendrimers are not immunogenic or carcinogenic, enhancing their potential as vectors for an in vivo gene transfer system (Roberts et al. 1996), but dendrimers and PEI appear to activate the complement system after in vivo administration (Plank et al. 1996).

4 Combining Dendrimers with Other Strategies

Several strategies to enhance the gene expression of non-viral vectors are currently under development, e.g. the application of helper and pH-sensitive lipids, endosome-disruptive peptides, nuclear proteins, and nuclear localization signals (Davis 2002, De Smedt et al. 2000). For example, electroporation (Wang et al. 2001) and cyclodextrins (CDs) (Roessler et al. 2001) have been combined with DNA-dendrimer systems. Electroporation caused significant increases in the gene expression ability of a DNA: dendrimer-containing solution compared with direct injection plus electroporation, suggesting that the former allows plasmid vectors to directly enter a multitude of cell types (Wang et al. 2001). The addition of amphoteric or sulfonated β-CDs to PAMAM dendrimer (G5 EDA) caused the formation of smaller and more monodisperse particles, leading to an increase in the transfection efficiency of the dendrimer. By contrast, the potential use of a ternary complex of PPI-(1,4-diaminobutane) dendrimer, DNA, and cucurbituril as an example of a spontaneously assembled supermacromolecular gene delivery carrier was reported (Lim et al. 2002b). Other reports have shown that the combination of DNA/dendrimer complexes and surfactants improves gene transfer activity, e.g. the synthetic lung surfactant Exosurf and its component, tyloxapol, constitute a powerful enhancer for dendrimer-mediated gene transfer in vitro owing to improved complex internalization and intracellular release from endosomes rather than an increase in membrane permeability (Kukowska-Latallo et al. 1999).

The translocation of polyplexes into the nucleus is a critical step in gene expression, but gene delivery by the polyplex is limited by inefficient transfer of DNA from the cytoplasm to the nucleus. In order to improve transfer, Ritter et al. investigated the effect of combining a tetramer of the nuclear localization signal of the SV40 large-T-antigen peptide with DNA:dendrimer complexes. The combination resulted in a strong increase in transfection efficiency, indicating that the peptide mediates nuclear accumulation of transfected plasmid DNA (Ritter et al. 2003).

The potential uses of the combination (hybrid vectors) of viral vectors and dendrimers have been discussed (Schmidt-Wolf and Schmidt-Wolf 2003), e.g., reciprocal enhancement of gene transfers by combinatorial adenovirus transduction and plasmid DNA transfection in vitro and in vivo (Bonsted et al. 2004). Another example is efficient large-scale production and concentration of HIV-1-based lentiviral vectors for use in vivo (Coleman et al. 2003). Thus, the combination of dendrimers and other DNA transfer methods may be effective for the delivery of genes and ODNs.

5 Dendrimer Conjugates

Significant advances have been made in the synthesis and study of dendrimers with sugars (Cloninger 2002) and peptide (Sadler and Tam 2002) have recently been made. Table 1 summarizes the various types of dendrimer conjugates. Conjugating the fluorophore Oregon green 488 with a G5 dendrimer (G5) yielded a much better delivery agent for antisense compounds than unmodified dendrimer, possibly due to an increase in endosomal escape (Yoo and Juliano 2000). The use of CD-conjugated PAMAM dendrimers has also been described (Fig. 2). Arima et al. (2001) prepared PAMAM starburst dendrimers (G2, G3 and G4) conjugates (CDE conjugates) with CDs. As measured by luciferase gene expression using CDE conjugates with α-, β- or γ-CD, the transfection efficiency of PAMAM dendrimers (G2) functionalized with α-CD (α-CDE conjugate) was about 100 times higher than that of unfunctionalized PAMAM or non-covalent mixtures of PAMAM and α-CD (Arima et al. 2001). Of various α-CDE conjugates, an α-CDE conjugate (G3) with a degree of substitution of 2.4 had the best transfection efficiency as well as low cytotoxicity (Kihara et al. 2002, 2003). Moreover, the α-CDE conjugate was induced gene expression in the spleen after intravenous injection of the DNA-complex-containing suspension (Kihara et al. 2003). In addition, polyethylene glycol functionalization of PAMAM dendrimers (G5) produced a 20-fold increase in transfection efficiency using plasmid DNA relative to partially degraded PAMAM dendrimers (Luo et al. 2002). Dendrimer conjugates with antibody for cell-specific DNA delivery have also been demonstrated (Li et al. 2000). Thus, the conjugation of dendrimers with hydrophobic and hydrophilic compounds or antibodies is of interest.

6 Conclusion

This review has focused on the potential use of dendrimers, the combination of dendrimers with other transfection techniques, and the use of dendrimer conjugates for DNA and ODN delivery in vitro and in vivo.

(A)

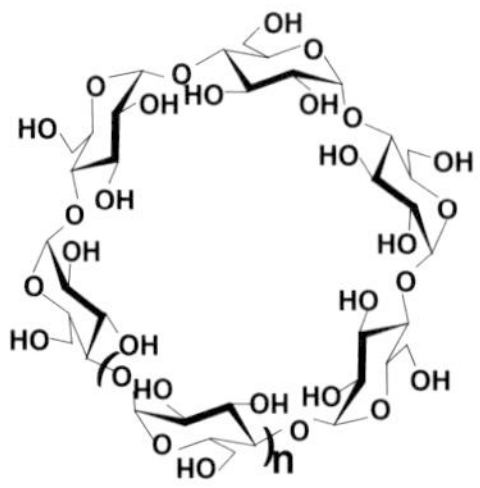

CyDs	Molecular weight	Cavity diameter (Å)	Inclusion affinity with membrane component
n = 1, α-CyD	972	4.7 ~ 5.3	Phospholipids
n = 2, β-CyD	1135	6.0 ~ 6.5	Cholesterol
n = 3, γ-CyD	1297	7.5 ~ 8.3	

(B)

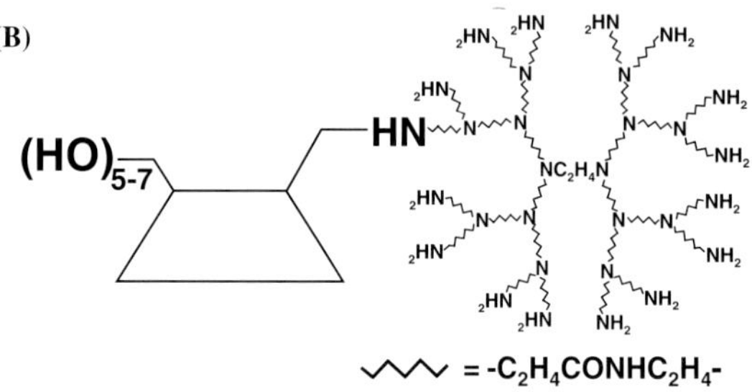

FIG. 2. Chemical structures **A** of α-, β- and γ-cyclodextrins and **B** PAMAM starburst dendrimer (G2) conjugate with α-, β- and γ-cyclodextrins

Recently, other biologically functional small nucleic acids, such as small interfering RNA and micro RNA, have been developed for clinical applications. Dendrimers may be useful as carriers for these small RNAs. In fact, we have shown the use of α-CDE conjugates as a carrier for siRNA in vitro.

Currently, there are no clinical trials using either dendrimers or other cationic polymers, despite the large number of in vitro and in vivo studies showing the potential applications of DNA complexes with dendrimers. The most important reason is the low efficiency of gene transfection, in vitro and in vivo, obtained with polyplexes. Thus, a more rational and intelligent molecular design of dendrimers and their conjugates is needed in order to improve transfection efficiency and to minimize cytotoxicity and side effects. In addition, the pharmaceutical technological aspects should be emphasized so that the carriers are safe for clinical use.

Acknowledgements. I thank Prof. Kaneto Uekama and Associate Prof. Fumitoshi Hirayama, Kumamoto University, for the many valuable discussions. I also thank Dr. Fumihiro Kihara, Dr. Koki Wada, Mr. Toshihito Tsutsumi and Ms. Yuko Chihara for their contributions of the results described in this article. The work was supported in part by Grant-in-Aid for Scientific Research (C) from the Japan Society for the Promotion of Science (14572158).

References

Abdallah B, Sachs L, Demeneix BA (1995) Non-viral gene transfer: applications in developmental biology and gene therapy. Biol Cell 85:1–7

Aramaki Y, Takano S, Tsuchiya S (1999) Induction of apoptosis in macrophages by cationic liposomes. FEBS Lett 460:472–476

Arima H, Kihara F, Hirayama F, Uekama K (2001) Enhancement of gene expression by polyamidoamine dendrimer conjugates with α-, β-, and γ-cyclodextrins. Bioconjug Chem 12:476–484

Bathori G, Cervenak L, Karadi I (2004) Caveolae-an alternative endocytotic pathway for targeted drug delivery. Crit Rev Ther Drug Carrier Syst 21:67–95

Bielinska AU, Chen C, Johnson J, Baker JR Jr (1999) DNA complexing with polyamidoamine dendrimers: implications for transfection. Bioconjug Chem 10:843–850

Bielinska A, Kukowska-Latallo J, Piehler LT, Tomalia DA, Spindler R, Yin R, Baker JR Jr (1995) Starburst PAMAM dendrimers: A novel synthetic vector for the transfection of DNA into mammalian cells. Polym Mater Sci Engineer 73:273–274

Bielinska A, Kukowska-Latallo JF, Johnson J, Tomalia DA, Baker JR Jr (1996) Regulation of in vitro gene expression using antisense oligonucleotides or antisense expression plasmids transfected using starburst PAMAM dendrimers. Nucleic Acids Res 24:2176–2182

Bielinska AU, Yen A, Wu HL, Zahos KM, Sun R, Weiner ND, Baker JR Jr, Roessler BJ (2000) Application of membrane-based dendrimer/DNA complexes for solid phase transfection in vitro and in vivo. Biomaterials 21:877–887

Bonsted A, Engesaeter BO, Hogset A, Maelandsmo GM, Prasmickaite L, Kaalhus O, Berg K (2004) Transgene expression is increased by photochemically mediated transduction of polycation-complexed adenoviruses. Gene Ther 11:152–160

Choi Y, Mecke A, Orr BG, Banaszak HMM, Baker JR Jr (2004) DNA-Directed synthesis of generation 7 and 5 PAMAM dendrimer nanoclusters. Nano Lett 4:391–397

Cloninger MJ (2002) Biological applications of dendrimers. Curr Opin Chem Biol 6: 742–748

Coleman JE, Huentelman MJ, Kasparov S, Metcalfe BL, Paton JF, Katovich MJ, Semple-Rowland SL, Raizada MK (2003) Efficient large-scale production and concentration of HIV-1-based lentiviral vectors for use in vivo. Physiol Genomics 12:221–228

Davis ME (2002) Nonviral gene delivery systems. Curr Opin Biotechnol 13:128–131

de Jong G, Telenius A, Vanderbyl S, Meitz A, Drayer J (2001) Efficient in-vitro transfer of a 60-Mb mammalian artificial chromosome into murine and hamster cells using cationic lipids and dendrimers. Chromosome Res 9:475–485

De Smedt SC, Demeester J, Hennink WE (2000) Cationic polymer based gene delivery systems. Pharm Res 17:113–126

Esfand R, Tomalia DA (2001) Poly(amidoamine) (PAMAM) dendrimers: from biomimicry to drug delivery and biomedical applications. Drug Discov Today 6:427–436

Fischer D, Li Y, Ahlemeyer B, Krieglstein J, Kissel T (2003) In vitro cytotoxicity testing of polycations: influence of polymer structure on cell viability and hemolysis. Biomaterials 24:1121–1131

Garnett MC (1999) Gene-delivery systems using cationic polymers. Crit Rev Ther Drug Carrier Syst 16:147–207

Gebhart CL, Kabanov AV (2001) Evaluation of polyplexes as gene transfer agents. J Control Release 73:401–416

Grayson SM, Frechet JMJ (2001) Convergent dendrons and dendrimers: from synthesis and applications. Chem Rev 101:3819–3867

Haensler J, Szoka FC Jr (1993) Polyamidoamine cascade polymers mediate efficient transfection of cells in culture. Bioconjug Chem 4:372–379

Hollins AJ, Benboubetra M, Omidi Y, Zinselmeyer BH, Schatzlein AG, Uchegbu IF, Akhtar S (2004) Evaluation of generation 2 and 3 poly(propylenimine) dendrimers for the potential cellular delivery of antisense oligonucleotides targeting the epidermal growth factor receptor. Pharm Res 21:458–466

Hudde T, Rayner SA, Comer RM, Weber M, Isaacs JD, Waldmann H, Larkin DF, George AJ (1999) Activated polyamidoamine dendrimers, a non-viral vector for gene transfer to the corneal endothelium. Gene Ther 6:939–943

Jevprasesphant R, Penny J, Attwood D, D'Emanuele A (2004) Transport of dendrimer nanocarriers through epithelial cells via the transcellular route. J Control Release 97: 259–267

Kiefer K, Clement J, Garidel P, Peschka-Suss R (2004) Transfection efficiency and cytotoxicity of non-viral gene transfer reagents in human smooth muscle and endothelial cells. Pharm Res 21:1009–1017

Kihara F, Arima H, Tsutsumi T, Hirayama F, Uekama K (2002) Effects of structure of polyamidoamine dendrimer on gene transfer efficiency of the dendrimer conjugate with α-cyclodextrin. Bioconjug Chem 13:1211–1219

Kihara F, Arima H, Tsutsumi T, Hirayama F, Uekama K (2003) In vitro and in vivo gene transfer by an optimized α-cyclodextrin conjugate with polyamidoamine dendrimer. Bioconjug Chem 14:342–350

Klajnert B, Bryszewska M (2001) Dendrimers: properties and applications. Acta Biochim Polon 48:199–208

Kukowska-Latallo JF, Bielinska AU, Johnson J, Spindler R, Tomalia DA, Baker JR Jr (1996) Efficient transfer of genetic material into mammalian cells using Starburst polyamidoamine dendrimers. Proc Natl Acad Sci USA 93:4897–4902

Kukowska-Latallo JF, Chen C, Eichman J, Bielinska AU, Baker JR Jr. (1999) Enhancement of dendrimer-mediated transfection using synthetic lung surfactant exosurf neonatal in vitro. Biochem Biophys Res Commun 264:253–261

Kukowska-Latallo JF, Raczka E, Quintana A, Chen C, Rymaszewski M, Baker JR Jr. (2000) Intravascular and endobronchial DNA delivery to murine lung tissue using a novel, nonviral vector. Hum Gene Ther 11:1385–1395

Lechardeur D, Lukacs GL (2002) Intracellular barriers to non-viral gene transfer. Curr Gene Ther 2:183–194

Lim YB, Kim SM, Suh H, Park JS (2002a) Biodegradable, endosome disruptive, and cationic network-type polymer as a highly efficient and nontoxic gene delivery carrier. Bioconjug Chem 13:952–957

Lim YB, Kim T, Lee JW, Kim SM, Kim HJ, Kim K, Park JS (2002b) Self-assembled ternary complex of cationic dendrimer, cucurbituril, and DNA: noncovalent strategy in developing a gene delivery carrier. Bioconjug Chem 13:1181–1185

Li S, Tan Y, Viroonchatapan E, Pitt BR, Huang L (2000) Targeted gene delivery to pulmonary endothelium by anti-PECAM antibody. Am J Physiol Lung Cell Mol Physiol 278:L504–511

Loup C, Zanta MA, Caminade AM, Majoral JP, Meunier B (1999) Preparation of water soluble cationic phosphorous containing dendrimers as DNA transfeting agents. Chem Eur J 5:3644–3650

Luo D, Haverstick K, Belcheva N, Han E, Saltaman WM (2002) Poly(ethylene glycol)-conjugated PAMAM dendrimer for biocompatible high-efficiency DNA delivery. Macromolecules 35:3456–3462

Malik N, Wiwattanapatapee R, Klopsch R, Lorenz K, Frey H, Weener JW, Meijer EW, Paulus W, Duncan R (2000) Dendrimers: relationship between structure and biocompatibility in vitro, and preliminary studies on the biodistribution of ^{125}I-labelled polyamidoamine dendrimers in vivo. J Control Release 65:133–148

Manunta M, Tan PH, Sagoo P, Kashefi K, George AJ (2004) Gene delivery by dendrimers operates via a cholesterol dependent pathway. Nucleic Acids Res 32:2730–2739

Maruyama-Tabata H, Harada Y, Matsumura T, Satoh E, Cui F, Iwai M, Kita M, Hibi S, Imanishi J, Sawada T, Mazda O (2000) Effective suicide gene therapy in vivo by EBV-based plasmid vector coupled with polyamidoamine dendrimer. Gene Ther 7:53–60

Morgan DM, Larvin VL, Pearson JD (1989) Biochemical characterisation of polycation-induced cytotoxicity to human vascular endothelial cells. J Cell Sci 94 (Pt 3):553–559

Nichols B (2003) Caveosomes and endocytosis of lipid rafts. J Cell Sci 116:4707–4714

Nishikawa M, Huang L (2001) Nonviral vectors in the new millennium: delivery barriers in gene transfer. Hum Gene Ther 12:861–870

Ogris M, Steinlein P, Kursa M, Mechtler K, Kircheis R, Wagner E (1998) The size of DNA/transferrin-PEI complexes is an important factor for gene expression in cultured cells. Gene Ther 5:1425–1433

Parton RG, Richards AA (2003) Lipid rafts and caveolae as portals for endocytosis: new insights and common mechanisms. Traffic 4:724–738

Pitkanen L, Ruponen M, Nieminen J, Urtti A (2003) Vitreous is a barrier in nonviral gene transfer by cationic lipids and polymers. Pharm Res 20:576–583

Plank C, Mechtler K, Szoka FC Jr, Wagner E (1996) Activation of the complement system by synthetic DNA complexes: a potential barrier for intravenous gene delivery. Hum Gene Ther 7:1437–1446

Qin L, Pahud DR, Ding Y, Bielinska AU, Kukowska-Latallo JF, Baker JR Jr, Bromberg JS (1998) Efficient transfer of genes into murine cardiac grafts by Starburst polyamidoamine dendrimers. Hum Gene Ther 9:553–560

Ritter W, Plank C, Lausier J, Rudolph C, Zink D, Reinhardt D, Rosenecker J (2003) A novel transfecting peptide comprising a tetrameric nuclear localization sequence. J Mol Med 81:708–717

Roberts JC, Bhalgat MK, Zara RT (1996) Preliminary biological evaluation of polyamidoamine (PAMAM) Stauburst dendrimers. J Biomed Mater Res 30:53–65

Roessler BJ, Bielinska AU, Janczak K, Lee I, Baker JR Jr (2001) Substituted β-cyclodextrins interact with PAMAM dendrimer-DNA complexes and modify transfection efficiency. Biochem Biophys Res Commun 283:124–129

Ruponen M, Honkakoski P, Ronkko S, Pelkonen J, Tammi M, Urtti A (2003) Extracellular and intracellular barriers in non-viral gene delivery. J Control Release 93:213–217

Sadler K, Tam JP (2002) Peptide dendrimers: applications and synthesis. J Biotechnol 90: 195–229

Santhakumaran LM, Thomas T, Thomas TJ (2004) Enhanced cellular uptake of a triplex-forming oligonucleotide by nanoparticle formation in the presence of polypropylenimine dendrimers. Nucleic Acids Res 32:2102–2112

Sato N, Kobayashi H, Saga T, Nakamoto Y, Ishimori T, Togashi K, Fujibayashi Y, Konishi J, Brechbiel MW (2001) Tumor targeting and imaging of intraperitoneal tumors by use of antisense oligo-DNA complexed with dendrimers and/or avidin in mice. Clin Cancer Res 7:3606–3612

Schmidt-Wolf GD, Schmidt-Wolf IG (2003) Non-viral and hybrid vectors in human gene therapy: an update. Trends Mol Med 9:67–72

Tang MX, Redemann CT, Szoka FC Jr (1996) In vitro gene delivery by degraded polyamidoamine dendrimers. Bioconjug Chem 7:703–714

Thomas M, Klibanov AM (2003) Non-viral gene therapy: polycation- mediated DNA delivery. Appl Microbiol Biotechnol 62:27–34

Tomalia DA, Baker H, Dewald J, Hall M, Kallos G, Martin S, Roeck J, Ryder J, Smith P (1985) A new class of polymers: starburst-dendritic macromolecules. Polymer J 17:117–132

Tomalia DA, Naylor AM, Goddard III WA (1990) Starburst dendrimers: Molecular-level control of size, shape, surface chemistry, topology and flexibility from atoms to macroscopic matter. Angrew Chem Int Edn 29:138–175

Turunen MP, Hiltunen MO, Ruponen M, Virkamaki L, Szoka FC Jr, Urtti A, Yla-Herttuala S (1999) Efficient adventitial gene delivery to rabbit carotid artery with cationic polymer-plasmid complexes. Gene Ther 6:6–11

Vincent L, Varet J, Pille JY, Bompais H, Opolon P, Maksimenko A, Malvy C, Mirshahi M, Lu H, Vannier JP, Soria C, Li H (2003) Efficacy of dendrimer-mediated angiostatin and

TIMP-2 gene delivery on inhibition of tumor growth and angiogenesis: in vitro and in vivo studies. Int J Cancer 105:419–429

Wagner E (1998) Effects of membrane-active agents in gene delivery. J Control Release 53: 155–158

Wagner E (2004) Strategies to improve DNA polyplexes for in vivo gene transfer: will "artificial viruses" be the answer? Pharm Res 21:8–14

Wang Y, Bai Y, Price C, Boros P, Qin L, Bielinska AU, Kukowska-Latallo JF, Baker JR Jr, Bromberg JS (2001) Combination of electroporation and DNA/dendrimer complexes enhances gene transfer into murine cardiac transplants. Am J Transplant 1:334–338

Yoo H, Juliano RL (2000) Enhanced delivery of antisense oligonucleotides with fluorophore-conjugated PAMAM dendrimers. Nucleic Acids Res 28:4225–4231

Zhang ZY, Smith BD (2000) High-generation polycationic dendrimers are unusually effective at disrupting anionic vesicles: membrane bending model. Bioconjug Chem 11: 805–814

Zinselmeyer BH, Mackay SP, Schatzlein AG, Uchegbu IF (2002) The lower-generation polypropylenimine dendrimers are effective gene-transfer agents. Pharm Res 19:960–967

Use of Synthetic Peptides for Non-viral Gene Delivery

Takuro Niidome and Yoshiki Katayama

1 Introduction

The development of non-viral gene delivery systems is an important key to solving several problems occurring in viral gene delivery, such as endogenous virus recombination, oncogenic effects and unexpected immune responses. As discussed in other chapters of this book, most of the basic techniques of non-viral gene delivery systems have relied on liposome and polymer chemistries; however, peptide chemistry has also contributed strongly to this field. Peptides can be synthesized automatically or manually and their chemistry allows the design and synthesis of complicated structures, e.g. ligand-modified peptides recognizable by specific cells, dye-modified peptides to trace their locations in cells and tissues, and other "intelligent" peptides to achieve functional gene delivery. It is expected that the relative ease of peptide construction will provide researchers with a wide range of molecules as well as important information about the structural requirements for functional gene delivery. Peptides play many roles in gene delivery, such as providing a simple cationic moiety to bind DNA, acting as a ligand or facilitating the release of DNA into the cytosol, carrying a signal for transport to the nucleus, or functioning as a sensor for the regulation of gene expression in cells. In this chapter, some examples of the use of peptides in gene delivery are described, and the prospects of utilizing such peptides in gene therapy are discussed.

2 Peptides as Cationic Moieties of Gene Carriers

2.1 Oligolysines

In the early 1990s, the most commonly used DN-condensing agent for gene transfection into cells was polylysine. Its amino groups were modified with several kinds of ligands, e.g., asialoorosomucoid, transferrin, and carbohydrates, in order to achieve cell recognition and receptor mediated uptake. However, the high molecular weight

---

Department of Applied Chemistry, Faculty of Engineering, Kyushu University, Fukuoka 812-8581, Japan and the polydispersity of polylysine complicated region-specific modification with ligands. To solve these problems, the use of synthesized oligolysines was reported by Gottschalk et al. (1996). They initially synthesized a peptide ($YKAK_4WK$), which had four clustered lysine moieties, as an analog of spermine, a tertamine. However, the peptide was not long enough to allow successful transfection. Having established a minimum length, $YKAK_8WK$ showed efficient gene transfection when an endosome-disruptive peptide (described below) was added. Wadhwa et al. (1997) also tested oligolysines with several chain lengths and found that 18 lysines followed by a tryptophan and a cysteine alkylated with iodoacetoamide ($AlkCWK_{18}$; $Cys(CH_2CONH_2)$-Trp-$(Lys)_{18}$) formed DNA complexes and showed efficient gene transfection into cultivated cells. In addition, a series of branched cationic oligopeptides that differed in the number and type of cationic amino acids was examined by Plank et al. (1999). They found that a minimal chain length of six cationic amino acids was required for the formation of DNA complexes that were incorporated into cells, and that a branched peptide consisting of lysine residues as cationic moiety was a weaker activator of the complement system in blood than arginine residues and high-molecular-weight polylysine.

Oligopeptides form small DNA complexes that show efficient gene transfection in cultivated cells; however, due to their lower affinity for DNA, they are not stable enough to be utilized in vivo. To stabilize the complex, peptides bound to DNA were cross-linked with glutaraldehyde (Adami and Rice 1999). After forming condensates between a 20-amino-acid peptide (CWK18) and plasmid DNA, glutaraldehyde was added. The improved stability imparted by glutaraldehyde cross-linking was demonstrated by the increased resistance of DNA condensates to serum endonucleases. Although a decrease in the magnitude of transient gene expression was determined for cross-linked DNA condensates, long lasting steady-state expression was observed. The advantage of interpeptide disulfide bonding was reported by McKenzie et al. (2000a). Disulfide cross-linking is attractive because of the release of DNA that is expected to be triggered by the reducing environment of the cytosol (Fig. 1). The authors used peptides containing multiple cysteines, such a Cys-Trp-$(Lys)_{18}$. Disulfide bond formation after condensing with plasmid DNA led a decrease in particle size of the condensates and prevented their dissociation. The cross-linked peptide DNA

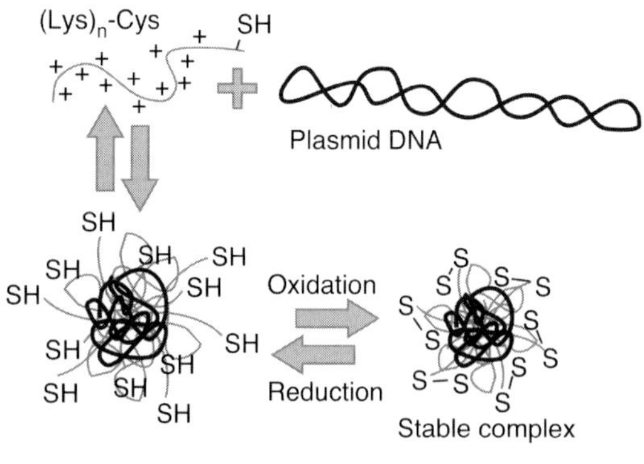

FIG. 1. Formation of cross-linked peptide-DNA complexes. The complex is formed through electrostatic interaction between cationic peptides and anionic DNA, followed by interpeptide oxidation to form disulfide bonds that stabilize the DNA complex

condensates had an higher transfection efficiency in vitro than uncross-linked ones. Furthermore, as the amount of incorporated DNA into cells was not affected by cross-linking of the peptides, it was suggested that the enhanced gene expression obtained with cross-linking was due to intracellular release of DNA triggered by disulfide bond reduction.

In order to improve cross-linked peptide DNA condensates with respect to in vivo gene delivery, PEG-modified and mannose-modified peptides were synthesized. After forming condensates with DNA, the peptides were cross-linked by glutaraldehyde (Yang et al. 2001) and disulfide bonds (Park et al. 2002; Kwok et al. 2003). The DNA of these cross-linked condensates containing PEG chain and branched mannose had a longer half-life in liver, and specifically accumulated in Kupffer cells after intravenous administration; long lasting gene expression was also shown. These are excellent examples demonstrating the function of PEG as a stealthy layer that blocks non-specific binding of blood components, and of mannose as a ligand for the mannose receptor, which is specifically expressed on Kupffer cells. The combination of several functional peptides in addition to the cross-linking technique based on the oligopeptide will enable a complicated, intelligent gene delivery system to be constructed with minimum effort.

2.2 Amphiphilic α-Helical Peptides

Gene transfection into cultivated cells mediated by a cationic peptide (KALA; WEAK-LAKA-LAKA-LAKH-LAKA-LAKA-LKAC-EA) was reported by Wyman et al. (1997). The peptide was derived from influenza hemagglutinin peptide and had an amphiphilic α-helical structure. In gel retardation and ethidium bromide displacement assays, KALA formed stable complexes when the C/A (cation of peptide/anion of DNA) ratio exceeded 1:1. The transfection ability of KALA was determined by the expression of luciferase from its cDNA in several kinds of cells. The complex composed of KALA and plasmid DNA at a C/A ratio of 10/1 showed the highest transfection efficiency in CV-1 cells. For this high efficiency, the amphiphilic structure was an important aspect, since, in general, amphiphilic peptides have membrane perturbation activity. After uptake of the complex into cells by endocytosis, KALA disrupted the endosomal membrane, resulting in transfer of the DNA to the cytosol. This peptide provided a starting point for the construction of a family of peptides incorporating other functions to improve DNA delivery systems.

Niidome et al. (1997) also reported on the DNA-binding and transfection abilities of several amphiphilic α-helical peptides. Peptides with a large hydrophobic region could strongly bind to DNA, and the formation of large twisted-fiber-like aggregates was observed. In transfection studies, these peptide showed high gene-transfer ability into COS-7 cells.

In order to clarify the significance of the hydrophobic region in amphiphilic peptides in relation to their transfection ability, Ohmori et al. (1998) employed five kinds of peptides with a systematically varied hydrophobic-hydrophilic balance in their amphiphilic structures (Hels 13–5, 11–7, 9–9, 7–11, and 5–13) (Fig. 2). The authors evaluated the relationship between the structure, the DNA-binding ability, and the gene transfer ability of the peptides into COS-7 cells (Table 1). This study also showed that the hydrophobic region plays an important role in disrupting the endosomal

A

Hel 13-5 : H-KLLK-LLLK-LWLK-LLKL-LL-OH
Hel 11-7 : H-KLLK-LLLK-LWKK-LLKL-LK-OH
Hel 9-9 : H-KLLK-KLLK-LWKK-LLKK-LK-OH
Hel 7-11 : H-KKLK-KLLK-KWKK-LLKK-LK-OH
Hel 5-13 : H-KKLK-KLKK-KWKK-LKKK-LK-OH

B

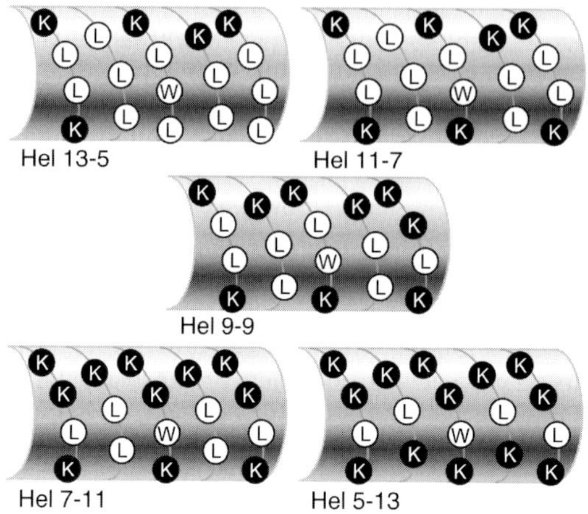

FIG. 2A, B. Structures of cationic α-helical peptides. Amino acid sequences of the designed peptides (**A**) and illustration of the amphiphilic structures of the α-helical peptides (**B**)

TABLE 1. Cationic α-helical peptides and their DNA-binding and transfection abilities

Peptide	Sequence	DNA binding	Transfection ability
Hel 13–5	KLLK-LLLK-LWLK-LLKL-LL	+++	+++
Hel 11–7	KLLK-LLLK-LWKK-LLKL-LK	+++	+++
Hel 9–9	KLLK-KLLK-LWKK-LLKK-LK	+	+
Hel 7–11	KKLK-KLLK-KWKK-LLKK-LK	–	–
Hel 5–13	KKLK-KLKK-KWKK-LKKK-LK	–	–

membrane, which can prevent the degradation of DNA in lysosomal vesicles. Furthermore, Niidome et al. (1999a) reported that, in a deletion series of the α-helical peptides, 16–17 amino acid residues were sufficient to form a complex with DNA and to transfer it into the cells. But even if the chain length was not enough for transfection, the efficiency could be recovered by modifying the peptide with a hydrophobic chain, such as a palmitoyl group (Niidome et al. 1999b).

As an application of α-helical peptides for functional gene delivery, a galactose-modified peptide was synthesized in order to achieve receptor-mediated gene delivery into hepatoma cells (Niidome et al. 2000). The galactose-modified peptides formed complexes with a plasmid DNA and showed gene transfer abilities into HuH-7 cells, a human hepatoma cell line. Considerable inhibition of the transfection efficiency by the addition of asialofetuin, which is a ligand for the asialoglycoprotein receptor, was observed in all galactose-modified peptides.

The use of amphiphilic α-helical peptides as gene carriers for transfection into cells has the following advantages. (1) The amphiphilic peptides, which have a smaller chain length than other gene carrier peptides such as a polylysine are still able to bind DNA and form stable complexes as the result of the strong intermolecular hydrophobic interaction. The aggregates can be internalized by endocytosis and remain resistant to digestion in lysosomal vesicles. (2) The membrane perturbation activity of amphiphilic peptides plays an important role in the escape from lysosomal degradation.

2.3 Dendritic Poly(L-Lysine)

The use of dendritic molecules as a gene carrier was reported by Haensler and Szoka (1993). Dendrimers are a new class of highly branched spherical polymers that are mono-dispersed and show high charge densities that are restricted to the surface of the molecule. These interesting features and the high flexibility of dendrimers, which allow defined molecules with functional groups to be constructed, are advantageous in the development of intelligent gene delivery systems. Previously, the synthesis of dendritic poly(L-lysine) using Boc-L-Lys(Boc)-OH as a branch unit was reported, and the properties were then described by Aharoni et al. (1982). Strictly speaking, the dendritic molecule cannot be classified in the original concept of the "dendrimer," described by Tomalia et al. in 1985, because it consists of asymmetrical L-lysine residues; nonetheless, dendritic molecules, consisting of amino acids, are expected to enable the construction of gene carriers using conventional peptide chemistry. For example, dendritic poly(L-lysine) modified by various ligands, such as a sugar chain, endosome-disrupting agents, and an intracellular targeting signal, should be readily synthesizable by selecting several protecting groups of the lysine, such as the Boc, Z, and Fmoc groups. As candidates for the gene carrier among several types of dendritic poly(L-lysine)s, block copolymers consisting of poly(ethylene glycol) and poly(L-lysine) dendrimer were tested by Choi et al. (1999, 2000). The copolymers were 50–150 nm and formed a spherical complex with DNA; they did not show any cytotoxity toward NIH3T3 cells even at higher concentrations. Shah et al. (2000) also reported the use of amphipathic asymmetric poly(L-lysine) dendrimers, in which the dendritic structure was attached to α-amino myristic acids as a hydrophobic tail. Amphipathic dendrimers bearing eight and 16 terminal amines formed stable complexes with plasmid DNA at a C/A ratio of 5/1 and allowed gene transfection into BHK-21 cells without significant cytotoxicity

The use of mono-dispersed dendritic poly(L-lysine) with a hexamethylenediamine core was reported by Ohsaki et al. (2002) (Fig. 3). In that study, the DNA-binding abilities and transfection efficiency of first- to sixth-generation dendritic poly(L-lysine)s were investigated. As shown by agarose gel shift and ethidium bromide titration assays, the dendritic poly(L-lysine)s of third generation and higher formed complexes with plasmid DNA, and the degree of compaction of the DNA was increased with increasing generation number. Fifth- and sixth-generation dendritic poly(L-lysine)s, which have 64 and 128 amine groups on the surface of the molecule, respectively, showed efficient gene transfection ability into several cultivated cell lines without significant cytotoxity.

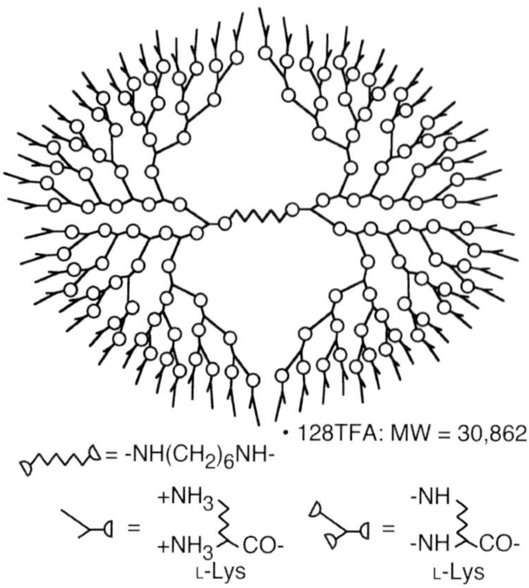

FIG. 3. Structure of sixth-generation dendritic poly(L-lysine) (KG6)

In order to understand the mechanism of complex formation of sixth-generation dendritic poly(L-lysine) (KG6) with plasmid DNA, the complex was analyzed using atomic force microscopy (Okuda et al. 2003a). After mixing for 15 min, 1- to 2-μm assemblies of complexes composed of several small particles (50–200 nm) had formed. At the same time, small, individual complexes of 50 to 500 nm were observed on a mica surface. After incubation for 2 h, only the large complexes were found on the mica surface. As the transfection efficiency of KG6 was correlated with the mean size of the DNA complexes, it was suggested that large complexes of more than 1 μm are the major species contributing to transfection in the case of in vitro transfection. With regard to the mechanism, it has been suggested that the dendrimer forms a complex with plasmid DNA, and the complex is incorporated via the endocytosis pathway. Subsequently, a part of the complex escapes from the endocytotic vesicle by the proton sponge effect of the dendrimer, similar to the case of polyethylenimine (Boussif et al. 1995), and the gene encoded in the plasmid DNA is expressed after transport into the nucleus.

Okuda et al. (2003b) also examined the effect of substituting terminal cationic groups on gene delivery into cells. KGR6 and KGH6, in which terminal amino acids were replaced by arginines and histidines, respectively, were synthesized. KGR6 bound to the plasmid DNA as strongly as KG6, whereas KGH6 showed decreased binding ability. The transfection efficiency of KGR6 into several types of cultivated cells was three- to 12-fold higher than that of KG6. The terminal guanidium groups of KGR6 are advantageous for interaction with the cell membrane or endosomal membrane and facilitate escape of the DNA into the cytosol. However, at present, more information about the efficiencies at each step of the transfection process is required to fully explain the enhancing effect of the arginines. By contrast, KGH6 showed no transfection efficiency ; however, once it was mixed with DNA under acidic conditions (pH 5.0), DNA-complexes were formed that showed high transfection efficiency, comparable to that obtained with KG6-mediated transfection. The unique character of KGH6 is a basic and valuable tool that will allow pH-dependent in vitro and in vivo gene transfection systems to be constructed.

Interestingly, the DNA-complex of this dendrimer had a low zeta potential (+3 mV), and showed high transfection efficiency even in the presence of 50% serum. The neutral surface of complexes with KG6 may contribute to its inert behavior in the presence of serum. From the viewpoint of in vivo gene delivery, this feature is advantageous, since non-specific interaction with serum components are thereby reduced, and the circulation half-life of the complexes in the blood will thus be extended. Moreover, it is not necessary to modify the carrier molecule with a PEG chain in order to confer stealth-like characteristics, as with the oligolysine, described above. To confirm the stealth ability of KG6-containing DNA complexes in the blood and the potential of KG6 as a functional gene carrier that can be applied in vivo, the biodistributions of plasmid DNA in normal and tumor-bearing mice after intravenous injection of DNA complexes with KG6 or other gene carrier molecules were evaluated (Kawano et al. 2004). Southern blotting analysis revealed that plasmid DNA complexes with KG6 at a C/A ratio of 8.0 circulated in the blood for 3 h after intravenous injection. In tumor-bearing mice, plasmid DNA injected with KG6 was observed in the tumor at 60 min after intravenous injection, while no DNA was present in the tumor using DOTAP/Chol liposomes. The stealth ability of KG6-containing DNA complexes in the blood would enhance their permeability and retention (EPR) effect in the tumor. This long-lasting circulation of DNA complexes in the blood and the permeability into the tumor achieved by simple cationic molecules hold great promise regarding the successful application of gene delivery systems. Furthermore, if KG6 is modified by a ligand without loss of the original character, highly controlled gene delivery will be achieved in vivo.

3 Peptides as Ligands

3.1 RGD Peptide

Studies of the bioactivity of RGD (arginyl-glycyl-aspartic acid tripeptide motif) peptide have a long history (reviewed by Ruoslahti 2003). Initially, Pierschbacher and Ruoslahti (1984) found that the RGDS (L-arginyl-glycyl-L-aspartyl-L-serine) peptide promoted cell adhesion. When included in extracellular matrix proteins, such as fibronectin, vitronectin, fibrinogen and laminin, this peptide motif was recognized by integrins, a family of cell-surface proteins (Pytela et al. 1985a, 1985b). The α5β1- and αV-containing integrins were particularly efficient in recognizing RGD peptide sequences (Hynes 1992). As the integrins act as receptors for cell adhesion on the substratum and in cell-cell interactions, RGD peptide and its analogues showed several bioactivities, such as the induction of apoptosis by endothelial cells (Brooks et al. 1994a) and the inhibition of angiogenesis (Brooks et al. 1994b; Buerkle et al. 2002). It is expected that RGD peptides will not only act as tools for studying the basic mechanisms of cell adhesion, cell-cell interaction, metastasis and angiogenesis, but that they will also function as therapeutic agents for the treatment of diseases, such as thrombosis and cancer. Recently, the imaging of tumors using RGD peptides modified with ^{18}F and ^{64}Cu was reported (Chen et al. 2004; Haubner et al. 2004).

The application of RGD peptide in gene delivery was reported by Hart et al. (1995) and Harbottle (1998). They used a peptide, $[K]_{16}$RGD peptide ($[K]_{16}$GGCRGDM-FGCA), containing the RGD motif and a DNA-binding domain of 16 lysine residues as a gene carrier. The peptide was oxidized to allow formation of intramolecular disulfide bridges that increase its affinity for integrins. Experiments carried out using several cell lines showed that the peptide bound to fibronectin and vitronectin integrin receptors and formed nuclease-resistant complexes with DNA. The transfection efficiency was enhanced by the addition of the RGD motif to $[K]_{16}$ peptide, whereas an RGE motif had no effect, indicating that the transfection was dependent on integrin (Colin et al. 1998).

Erbacher et al. (1999) modified polyethylenimine (PEI) with RGD peptide. In their study, thiol-derivatized PEI was conjugated to the integrin-binding peptide CYGGRGDTP via a disulfide bridge. This PEI-RGD conjugate formed 30- to 100-nm toroidal particles whose surface charge was close to neutral, as a consequence of the shielding effect of the prominent zwitterionic peptide residues. Following in vitro transfection, the expression of PEI-RGD was 10- to 100-fold higher in integrin-expressing epithelial (HeLa) and fibroblast (MRC5) cells than that of PEI without the RGD motif and with the RGE motif. The PEI-RGD system was further improved by the addition of a PEG chain between them. PEG-ylation of cationic polymers reduces non-specific binding to cell surfaces. A cyclic peptide, ACDCRGDCFC (Suh et al. 2002) and a tetra peptide, RGDC (Kunath et al. 2003) were modified into PEI via a hydrophilic poly(ethylene glycol) (PEG) spacer. Insertion of the PEG chain kept the DNA complexes of the conjugates neutral, whereas complexes consisting of RGD-PEI and PEI had a positively charged surface. In in-vitro transfection experiments using endothelial cell lines, RGD-PEG-PEI conjugates showed integrin dependent gene expression. Although an obvious effect of PEG-ylation was not observed in the in vitro system, the charge shielding effect of PEG would no doubt be advantageous in in-vivo gene therapy, especially for systemic injection.

RDG-modified lipid-protamine-DNA lipopolyplex (LPD) was prepared by Harvie et al. (2003). The incorporation of PEG-ylated lipid into LPD complex decreased their in vitro transfection activity. In order to restore particle binding and specifically target LPD formulations to tumor cells, a lipid-RGD peptide conjugate DSPE-PEG-ACD-CRGDCFCG was synthesized and then incorporated into LPD formulations. The resultant LPD-PEG-RGD showed significantly increased binding and uptake compared to that obtained with an LPD-PEG formulation. Moreover, transfection of LPD-PEG-RGD was specific for integrin-expressing cells. The combination of RGD with liposome-based systems could also be effective in targeted and systemic gene delivery.

3.2 Other Peptides

In addition to RGD peptides, the use of peptides bound to other types of integrin has been reported. A linear or cyclic PLAEIDGIEL, which binds to $\alpha9\beta1$-integrin, expressed in lung epithelia, hepatocyte and muscles, was connected to a DNA-binding moiety of 16 lysine residues (Schneider et al. 1999). In that case, specific gene delivery could only be achieved with the the cyclic form of the peptide. However, inclusion of a cationic liposome, lipofect-amine, into the peptide/DNA complexes the efficient gene transfer of both peptides with significant targeting specificity.

The serpin-enzyme complex receptor (SECR) was successfully targeted for gene delivery using peptide ligands covalently linked with oligolysine (Patel et al. 2001). The authors found that the peptide $[K]_{16}$CSIPPEVKFNKPFVFLI forms small complexes with DNA and showed high transfection efficiency in an Huh-7 human hepatocyte cell line expressing the receptor.

Recently, several peptides that can function as targeted ligands have been found in a phage display library. By repeating bio-panning on target tissues or cell lines, specific phage peptides can be screened and the peptide sequences subsequently identified. Peptide SIGYPLP was found to target vein endothelial cells and the phage was used as a vector for gene delivery (Nicklin et al. 2000). Other targeting peptides include: THALWHT, for targeting to human airway epithelia (Jost et al. 2001); CNGRC, for targeting to tumor (Colombo et al. 2002); CSRPRRSEC, CGKRK and CDTRL, for targeting to the neovasculature (Hoffman et al. 2003); and CGNKRTRGC, for targeting to tumor lymphatic and tumor cells (Laakkonen et al. 2004). Further accumulation of a wide variety of targeted peptides will allow the construction of tailor-made functional gene carriers that can be targeted to specific tissues.

4 Peptides as Tools for Endosome Disruption

4.1 Acidic Peptides

The endocytotic pathway is a major route of non-viral gene delivery into cells′, except for the method using DNA-encapsulated liposomes. Following uptake, most of the DNA complexes containing the gene carriers are degraded in acidic endocytotic vesicles or returned to the outside of the cells by exocytosis. Therefore, escape from the endosome to the cytosol is a critical step in efficient gene delivery and expression. To promote escape, endosome-disruptive peptides, derived from the amino-terminal sequence of influenza virus hemagglutinin HA-2, have been applied (Wagner et al. 1992; Midoux et al. 1993; Plank et al, 1994). The peptides were found as a membrane fusion domain of HA-2, which fused the virus and endosomal membrane in host cell, that is, the peptide domain had membrane perturbation activity that was triggered by an acidic environment. Furthermore, designer model peptides, which have anionic amino acids and take on an α-helical structure, have also been reported (Haensler et al. 1993; Plank et al. 1994; Ohmori et al. 1997, 1998). Such endosome-disruptive peptides introduce the DNA into the cytoplasm prior to fusion of the endosome with the lysosome. They contain several acidic amino acids, as shown in Table 2, and their α-helical structures, formed under acidic conditions, arise from protonation of the glutamic carboxylates at acidic pH, which decreases repulsion of the negatively charged side chains of glutamic acids. As a result, the peptides exhibit membrane-perturbation activities (Fig. 4). In the gene delivery protocol in vitro, anionic peptide was covalently conjugated to cationic carrier molecules, such as a polylysine, and the DNA complexes of the conjugates were then added into the cell culture medium (Wagner et al. 1992; Haensler et al. 1993). Alternatively, the anionic peptides were mixed with complexes of DNA and cationic carriers, and the resulting complexes were used in transfection (Plank et al. 1994; Midoux et al. 1993; Gottschalk et al. 1996; Ohmori et al. 1997, 1998). By the addition of anionic endosome-disruptive peptides, the transfection efficiencies of cationic gene carriers improved dramatically.

TABLE 2. Anionic endosome-disruptive peptides

Peptide	Structure
Peptide1, INF3	GLFE-AIAG-FIEN-GWEG-MIDG-GGC
Peptide2	GLFG-AIAG-FIEN-GWEG-MIDG-GGC
INF5	(GLFE-AIEG-FIEN-GWEG-nIDG)$_2$-K
INF7	GLFE-AIEG-FIEN-GWEG-WYG
GALAcys	WEAA-LAEA-LAEA-LAEH-LAEA-LAEA-LEAC-AA
GALA	WEAA-LAEA-LAEA-LAEH-LAEA-LAEA-LAAL-AAGG-SC
GALA-INF1	GLFG-AIAG-FIEN-GWEG-LAEA-LAEA-LAAL-AAGG-SC
Peptide I	GLFE-AIAE-FIEG-GWEG-LIEG-CA
JST-1	GLFE-ALLE-LLES-LWEL-LLEA
4$_3$E	LAEL-LAEL-LAEL

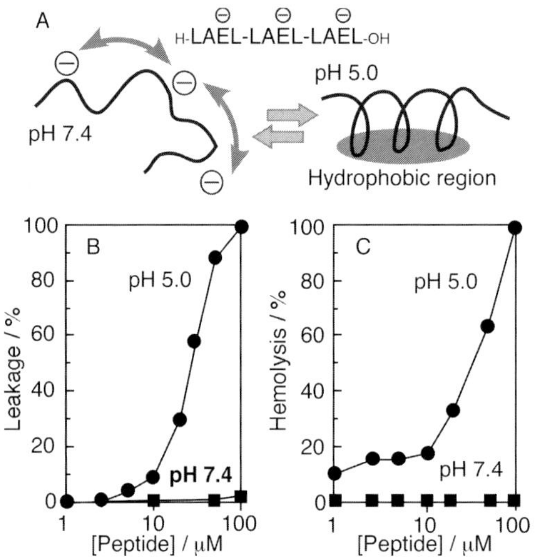

FIG. 4. The conformational change of an acidic peptide (LAEL-LAEL-LAEL) (**A**) and its pH-dependent liposome disruption (**B**) and hemolytic (**C**) activities

4.2 Peptides Containing Histidine Residues

Midoux et al. (1998) designed a peptide, H5WYG (GLFHAIAHFIHGGWHGLIHG-WYG), that undergoes a dramatic conformational change between pH 7.0 and 6.0 that correlates with the protonation of the histidyl residues. Thus, the peptide shows membrane perturbation activity at a slightly acidic pH but not at neutral pH. Addition of the peptide to a transfection system mediated by glycosylated polylysine led to a significant increase in gene expression. By contrast, McKenzie et al. (2000b) synthesized a peptide, Cys-His-(Lys)$_6$-His-Cys, by substituting histidine for some of the lysine residues in the peptide Cys-Trp-(Lys)$_{17}$-Cys. The new peptide provided buffering capacity that enhanced in vitro gene expression in the absence of chloroquine. As the pKa of the histidine residues is similar to the pH in endosomal vesicles, the histidines have a proton sponge effect (Boussif et al. 1995). From this viewpoint, the enhancing effect of the peptide, H5WYG may have originated from the buffering effect of the histidine residues in the peptide, in addition to its membrane-perturbing activity.

5 Peptides as Signals for Transport to the Nucleus

Gene expression from the transgene in plasmid DNA requires transcription of the DNA to mRNA, which occurs in the nucleus. Inefficient entry of DNA into the nucleus is a major limiting step in non-viral gene delivery systems. In the case of dividing cells, the DNA is internalized when the nuclear membrane is reconstituted. However, there are serious problems in the case of non-dividing cells, in which entry into the nucleus is thought to occur only through the nuclear pore complex. Therefore, in order to achieve active transport to the nucleus, nucleus localizing signal (NLS) peptides have been widely used. Recent efforts have been summarized in excellent reviews (Bremner et al. 2001; Tachibana et al. 2001; Cartier 2002). In most cases, NLS peptides are conjugated with a gene carrier, such as PEI or cationic liposome. In order to function as a NLS peptide, the carrier should not dissociate from the DNA in the cytosol. In the case of gene carriers that release DNA in the cytosol or whose time of endosome escape is mediated by membrane fusion, as in the case of lipoplex (Xu et al. 1996), NLS peptide modification is of no use. Moreover, it is hard to imagine that this type of DNA complex can transport through the nuclear pore. To clarify this problem, Zanta et al. (1999) directly modified DNA with a NLS peptide. An oligonucleotide cap modified with a NLS peptide (PKKKRKVEDPYC) was synthesized, and the cap was then ligated to a linearized luciferase gene. By modifying with the peptide, transfection was remarkably enhanced, whereas no enhancement of the control peptide (PKTKRKVEDPYC) was observed, suggesting that enhancement was due to importin-mediated translocation.

Although direct modification with NLS peptide is a straightforward strategy to achieve highly efficient gene expression, construction of the gene is complicated, costly, and time-consuming. Recently, a convenient method of DNA modification with NLS peptide was reported by Zelphati et al. (2000). They used an NLS peptide linked with a peptide nucleic acid (PNA), which can hybridize with plasmid DNA by forming a triplex strand invasion complex. The resulting DNA contained the NLS peptide via the PNA chain. This technique has been made commercially available as GeneGrip, from Gene Therapy Systems (San Diego, CA, USA). Bremner et al. (2004) prepared several NLS-modified DNAs using this system and compared their characteristics. Although the synthesis of PNA-peptide hybrid is relatively complicated, even for researchers in the field of gene therapy, the technique offers an easy approach to introducing not only the NLS onto plasmid DNA but also other functional peptides and groups without disturbing transcriptional activity.

6 Peptides as Sensors for Functional Gene Delivery

Peptides can be used as sensors that respond to cellular conditions. Each tissue expresses hundreds of genes, many of which are tissue specific, for example, gene expression in vascular smooth muscle cells is different from that of endothelial cells. Furthermore, there are large differences in gene expression between normal and abnormal tissues, such as inflammatory and tumor tissues. If a peptide was able respond to a specific protein in a target tissue and control gene delivery in the cells, then a tissue-specific gene expression system could be developed.

Katayama et al. (2002) focused on substrate peptides for cAMP-dependent protein kinase (PKA) and caspase-3 as sensors. The enzymes act as mediators of intracellular signals in the regulation of gene expression and apoptosis, respectively. Extraordinary activation of these enzymes is known to play a role in many diseases, such as melanoma, prostate tumor, and colon cancer for PKA, and hepatitis, Alzheimer's disease, Parkinson's disease, and other various nerve-denaturing diseases for caspase-3. Thus, these enzymatic activities are important determinants of the cellular condition. In the report of Katayama et al. (2002), the peptides were incorporated into polymers; specifically, graft-type copolymers were synthesized using methacryloyl-peptide monomer and acrylamide with radical copolymerization (Fig. 5A). In the case of the PKA system, polymer containing the substrate peptide (ALRRASLG) formed complexes with plasmid DNA via electrostatic interactions due to the cationic net charge of the peptides. Once PKA phosphorylates the peptide at the serine residues, the net charge of the peptide is reduced, and the polymer is then expected to release its DNA (Fig. 5B). PKA-dependent DNA release was observed by agarose gel retardation assay. In cell-free transcription and translation systems, the addition of PKA to the DNA complex of the polymer triggered gene expression from the plasmid DNA. After transfection of the complex into NIH 3T3 cells, the addition of forskolin, which actives PKA, significantly enhanced expression of the reporter gene. In the caspase-3

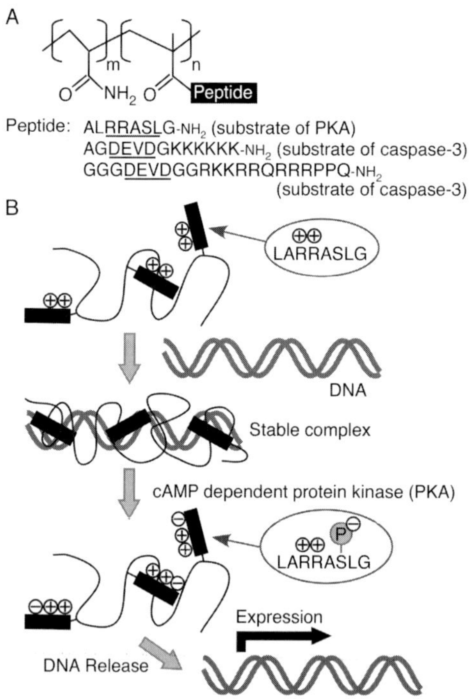

FIG. 5. **A** Structures of the polymer-peptide conjugates. **B** cAMP-dependent protein kinase(PKA)-dependent DNA release

system, a peptide containing a substrate for caspase-3 (AGDEVDG) and a cationic sequence (KKKKKK or GRKKRRQRRRPPQ; Tat peptide) for DNA binding was used as a pendant peptide on the polymer. When the DEVD portion of the peptide is recognized and cleaved by caspase-3 at the C-terminus of the second Asp (D), the cationic portion is released from the polymer. Therefore, polymer-covered DNA is released by the addition of caspase-3, allowing transcription of the gene. In a cell-free in vitro transfection system stimulated by staurosporin, controlled DNA release and gene expression mediated by caspase-3 were achieved. These systems are the first examples of an artificial gene regulation system controlled by a cationic polymer containing a peptide sensor.

7 Prospects

As described in this chapter, peptides have several applications in gene delivery. Advanced techniques of peptide synthesis make it possible to design and precisely synthesize complicated peptides, such as a multifunctional peptide containing an additional PNA chain and a ligand. In addition, the molecular science of bioactive peptides has progressed. Basic research on peptide, such as their physical characteristics, their detailed structure-function relationships, the mechanism of their biological activity, and further screening of functional peptides using a peptide library, will provide a large number of tools for use in gene delivery. Thus, peptide science will connecting a wide variety of research fields and will strongly contribute to the development of functional gene delivery systems.

References

Adami RC, Rice KG (1999) Metabolic stability of glutaraldehyde cross-linked peptide DNA condensates. J Pharm Sci 88:739–746

Aharoni SM, Crosby CR. III, Walsh EK (1982) Size and solution sroperties of globular *tert*-butyloxycarbonyl-poly(α,ε-L-lysine). Macromolecules 15:1093–1098

Boussif O, Lezoualc'h F, Zanta MA, Mergny MD, Scherman D, Demeneix B, Behr JP (1995) A versatile vector for gene and oligonucleotide transfer into cells in culture and *in vivo*: polyethylenimine. Proc Natl Acad Sci USA 92:7297–7301

Bremner KH, Seymour LW, Logan A, Read ML (2004) Factors influencing the ability of nuclear localization sequence peptides to enhance nonviral gene delivery. Bioconjug Chem 15:152–161

Bremner KH, Seymour LW, Pouton CW (2001) Harnessing nuclear localization pathways for transgene delivery. Curr Opin Mol Ther 3:170–177

Brooks PC, Montgomery AM, Rosenfeld M, Reisfeld RA, Hu T, Klier G, Cheresh DA (1994a) Integrin αvβ3 antagonists promote tumor regression by inducing apoptosis of angiogenic blood vessels. Cell 79:1157–1164

Brooks PC, Clark RA, Cheresh DA (1994b) Requirement of vascular integrin αvβ3 for angiogenesis. Science 264:569–571

Buerkle MA, Pahernik SA, Sutter A, Jonczyk A, Messmer K, Dellian M (2002) Inhibition of the αlpha-nu integrins with a cyclic RGD peptide impairs angiogenesis, growth and metastasis of solid tumours in vivo. Br J Cancer 86:788–795

Cartier R, Reszka R (2002) Utilization of synthetic peptides containing nuclear localization signals for nonviral gene transfer systems. Gene Ther 9:157–167

Chen X, Park R, Tohme M, Shahinian AH, Bading JR, Conti PS (2002) MicroPET and autoradiographic imaging of breast cancer alpha v-integrin expression using ^{18}F- and ^{64}Cu-labeled RGD peptide. Bioconjug Chem 15:41–49

Choi JS, Joo D, Kim CH, Kim K, Park JS (2000) Synthesis of a barbell-like triblock copolymer, poly(l-lysine) dendrimer-block-poly(ethylene glycol)-block-poly(l-lysine) dendrimer and its self-assembly with plasmid DNA. J Am Chem Soc 122:474–480

Choi JS, Lee EJ, Choi YH, Park JS (1999) Poly(ethylene glycol)-block-poly(l-lysine) dendrimer: novel linear polymer/dendrimer block copolymer forming spherical water-soluble polyionic complex with DNA. Bioconjugate Chem 10, 62–65

Colin M, Harbottle RP, Knight A, Kornprobst M, Cooper RG, Miller AD, Trugnan G, Capeau J, Coutelle C, Brahimi-Horn MC (1998) Liposomes enhance delivery and expression of an RGD-oligolysine gene transfer vector in human tracheal cells. Gene Ther 5:1488–1498

Colombo G, Curnis F, De Mori GM, Gasparri A, Longoni C, Sacchi A, Longhi R, Corti A (2002) Structure-activity relationships of linear and cyclic peptides containing the NGR tumor-homing motif. J Biol Chem 277:47891–47897

Erbacher P, Remy JS, Behr JP (1999) Gene transfer with synthetic virus-like particles via the integrin-mediated endocytosis pathway. Gene Ther 6:138–145

Gottschalk S, Sparrow JT, Hauer J, Mims MP, Leland FE, Woo SL, Smith LC (1996) A novel DNA-peptide complex for efficient gene transfer and expression in mammalian cells. Gene Ther 3:48–57

Haensler J, Szoka FC (1993) Polyamidoamine cascade polymers mediate efficient transfection of cells in culture. Bioconjugate Chem 4:372–379

Harbottle RP, Cooper RG, Hart SL, Ladhoff A, McKay T, Knight AM, Wagner E, Miller AD, Coutelle C. (1998) An RGD-oligolysine peptide: a prototype construct for integrin-mediated gene delivery. Hum Gene Ther 9:1037–1047

Hart SL, Harbottle RP, Cooper R, Miller A, Williamson R, Coutelle C (1995) Gene delivery and expression mediated by an integrin-binding peptide. Gene Ther 2:552–554

Harvie P, Dutzar B, Galbraith T, Cudmore S, O'Mahony D, Anklesaria P, Paul R (2003) Targeting of lipid-protamine-DNA (LPD) lipopolyplexes using RGD motifs. J Liposome Res 13:231–247

Haubner R, Kuhnast B, Mang C, Weber WA, Kessler H, Wester HJ, Schwaiger M (2004) [^{18}F]Galacto-RGD: synthesis, radiolabeling, metabolic stability, and radiation dose estimates. Bioconjug Chem 15:61–69

Hoffman JA, Giraudo E, Singh M, Zhang L, Inoue M, Porkka K, Hanahan D, Ruoslahti E (2003) Progressive vascular changes in a transgenic mouse model of squamous cell carcinoma. Cancer Cell 4:383–391

Hynes RO (1992) Integrins: versatility, modulation, and signaling in cell adhesion. Cell 69:11–25

Jost PJ, Harbottle RP, Knight A, Miller AD, Coutelle C, Schneider H (2001) A novel peptide, THALWHT, for the targeting of human airway epithelia. FEBS Lett 489:263–269

Katayama Y, Fujii K, Ito E, Sakakihara S, Sonoda T, Murata M, Maeda M (2002) Intracellular signal-responsive artificial gene regulation for novel gene delivery. Biomacromolecules 3:905–909

Kawano T, Okuda T, Aoyagi H, Niidome T (2004) Long circulation of intravenously administered plasmid DNA delivered with dendritic poly(L-lysine) in the blood flow. J Control Release in press

Kunath K, Merdan T, Hegener O, Haberlein H, Kissel T (2003) Integrin targeting using RGD-PEI conjugates for in vitro gene transfer. J Gene Med 5:588–599

Kwok KY, Park Y, Yang Y, McKenzie DL, Liu Y, Rice KG (2003) In vivo gene transfer using sulfhydryl cross-linked PEG-peptide/glycopeptide DNA co-condensates. J Pharm Sci 92: 1174–1185

Laakkonen P, Akerman ME, Biliran H, Yang M, Ferrer F, Karpanen T, Hoffman RM, Ruoslahti E (2004) Antitumor activity of a homing peptide that targets tumor lymphatics and tumor cells. Proc Natl Acad Sci USA 101:9381–9386

McKenzie DL, Kwok KY, Rice KG (2000a) A potent new class of reductively activated peptide gene delivery agents. J Biol Chem 275:9970–9977

McKenzie DL, Smiley E, Kwok KY, Rice KG (2000b) Low molecular weight disulfide cross-linking peptides as nonviral gene delivery carriers. Bioconjug Chem 11:901–909

Midoux P, Kichler A, Boutin V, Maurizot JC, Monsigny M (1998) Membrane permeabilization and efficient gene transfer by a peptide containing several histidines. Bioconjug Chem 9:260–267

Midoux P, Mendes C, Legrand A, Raimond J, Mayer R, Monsigny M, Roche AC(1993) Specific gene transfer mediated by lactosylated poly-L-lysine into hepatoma cells. Nucleic Acids Res 21:871–878

Nicklin SA, White SJ, Watkins SJ, Hawkins RE, Baker AH (2000) Selective targeting of gene transfer to vascular endothelial cells by use of peptides isolated by phage display. Circulation 102:231–237

Niidome T, Ohmori N, Ichinose A, Wada A, Mihara H, Hirayama T, Aoyagi H (1997) J Biol Chem 272:15307

Niidome T, Takaji K, Urakawa M, Ohmori N, Wada A, Hirayma T, Aoyagi H (1999a) Chain length of cationic α-helical peptide sufficient for gene delivery into cells. Bioconjugate Chem 10:773–780

Niidome T, Urakawa M, Takaji K, Matsuo Y, Ohmori N, Wada A, Hirayama T, Aoyagi H (1999b) Influence of lipophilic groups in cationic α-helical peptides on their abilities to bind with DNA and deliver gene into cells. J Peptide Res 54:361–367

Niidome T, Urakawa M, Sato H, Takahara Y, Anai T, Hatakeyama T, Wada A, Hirayama T, Aoyagi H (2000) Gene transfer into hepatoma cells mediated by galactose-modified α-helical peptides. Biomaterials 21:1811–1819

Ohmori N, Niidome T, Wada A, Hirayama T, Hatakeyama T, Aoyagi H (1997) The enhancing effect of anionic alpha-helical peptide on cationic peptide-mediating transfection systems. Biochem Biophys Res Commun 235:726–729

Ohmori N, Niidome T, Kiyota T, Lee S, Sugihara G, Wada A, Hirayama T, Aoyagi H (1998) Biochem Biophys Res Commun 245:259

Ohsaki M, Okuda T, Wada A, Hirayama T, Niidome T, Aoyagi T (2002) In vitro gene tansfection using dendritic poly(L-lysine), Bioconjugate Chem 13:510–517

Okuda T, Kidoaki S, Ohsaki M, Koyama Y, Yoshikawa K, Niidome T, Aoyagi H (2003a) Time-dependent complex formation of dendritic poly(L-lysine) with plasmid DNA and correlation with in vitro transfection efficinecies. Org Biomon Chem 1:1270–1273

Okuda T, Sugiyama A, Niidome T, Aoyagi H (2003b) Characters of dendritic poly(L-lysine) analogues with the terminal lysines replaced with arginines and histidines as gene carriers in vitro. Biomaterials 25:537–544

Patel S, Zhang X, Collins L, Fabre JW (2001) A small, synthetic peptide for gene delivery via the serpin-enzyme complex receptor. J Gene Med 3:271–279

Pierschbacher MD, Ruoslahti E (1984) Cell attachment activity of fibronectin can be duplicated by small synthetic fragments of the molecule. Nature 309:30–33

Plank C, Oberhauser B, Mechtler K, Koch C, Wagner E (1994) The influence of endosome-disruptive peptides on gene transfer using synthetic virus-like gene transfer systems. J Biol Chem 269:12918–12924

Plank C, Tang MX, Wolfe AR, Szoka FC (1999) Branched cationic peptides for gene delivery: role of type and number of cationic residues in formation and in vitro activity of DNA polyplex. Hum Gene Ther 10:319–332

Pytela R, Pierschbacher MD, Ruoslahti E (1985a) A 125/115-kDa cell surface receptor specific for vitronectin interacts with the arginine-glycine-aspartic acid adhesion sequence derived from fibronectin. Proc Natl Acad Sci USA 82:5766–5770

Pytela R, Pierschbacher MD, Ruoslahti E (1985b) Identification and isolation of a 140 kd cell surface glycoprotein with properties expected of a fibronectin receptor. Cell 40: 191–198

Ruoslahti E (2003) The RGD story: a personal account. Matrix Biology 22:459–465

Schneider H, Harbottle RP, Yokosaki Y, Jost P, Coutelle C (1999) Targeted gene delivery into alpha9beta1-integrin-displaying cells by a synthetic peptide. FEBS Lett 458:329–332

Shah DS, Sakthivel T, Toth I, Florence AT, Wilderspin AF (2000) DNA transfection and transfected cell viability using amphipathic asymmetric dendrimers. Int J Pharm 208: 41–48

Suh W, Han SO, Yu L, Kim SW (2002) An angiogenic, endothelial-cell-targeted polymeric gene carrier. Mol Ther 6:664–672

Tachibana R, Harashima H, Shinohara Y, Kiwada H (2001) Quantitative studies on the nuclear transport of plasmid DNA and gene expression employing nonviral vectors. Adv Drug Deliv Rev 52:219–226

Tomalia DA, Baker H, Dewald J, Hall M, Kallos G, Martin S, Roeck J, Ryder J, Smith, P (1985) A new class of polymers: starburst dendritic macromolecules. Polym J 17:117–132

Wadhwa MS, Collard WT, Adami RC, McKenzie DL, Rice KG (1997) Peptide-mediated gene delivery: influence of peptides structure on gene expression. Bioconjugate Chem 8:81–88

Wagner E, Plank C, Zatloukal K, Cotten M, Birnstiel ML (1992) Influenza virus hemagglutinin HA-2 N-terminal fusogenic peptides augment gene transfer by transferrin-polylysine-DNA complexes: toward a synthetic virus-like gene-transfer vehicle. Proc Natl Acad Sci USA 89:7934–7938

Wyman TB, Nicol F, Zelphati O, Scaria PV, Plank C, Szoka FC (1997) Design, synthesis, and characterization of a cationic peptide that binds to nucleic acids and permeabilizes bilayers. Biochemistry 36:3008–3017

Xu Y, Szoka FC Jr (1996) Mechanism of DNA release from cationic liposome/DNA complexes used in cell transfection. Biochemistry 35:5616–5623

Yang Y, Park Y, Man S, Liu Y, Rice KG (2001) Cross-linked low molecular weight glycopeptides-mediated gene delivery: relationship between DNA metabolic stability and the level of transient gene expression in vivo. J Pharm Sci 90:2010–2022

Zanta MA, Belguise-Valladier P, Behr JP(1999) Gene delivery: a single nuclear localization signal peptide is sufficient to carry DNA to the cell nucleus. Proc Natl Acad Sci USA 96: 91–96

Zelphati O, Liang X, Nguyen C, Barlow S, Sheng S, Shao Z, Felgner PL (2000) PNA-dependent gene chemistry: stable coupling of peptides and oligonucleotides to plasmid DNA. Biotechniques 28:304–316

An Oligonucleotide Carrier Based on β-1,3-Glucans

Kazuo Sakurai[1] and Seiji Shinkai[2]

1 Oligonucleotide/β-1,3-Glucan Complexes

Schizophyllan (SPG) is a member of the β-1,3-glucans (Tabata et al. 1981) that is produced as a cell-wall polysaccharide by some fungi (Fig. 1). The compound has been shown to cure some gynecological cancers (Okamura et al. 1986; Tabata et al. 1990). Norisuye and coworkers (Yanaki et al. 1980; Norisuye et al. 1980; Kashiwagi et al. 1981) extensively studied the properties of schizophyllan in dilute solution and determined that it dissolves in water as a triple helix and in DMSO as a single chain (s-SPG). When water is added to the SPG/DMSO solution, s-SPG collapses owing to the hydrophobic interaction, and both intra- and inter-molecular hydrogen bonds are formed (Sato et al. 1983; Stokke et al. 1993; McIntire and Brant 1998). Since the local structure of the s-SPG in water is considered to retrieve the triple helix, this process can be regarded as "renaturing" of the schizophyllan triple helix. Sakurai and Shinkai (2000) found that when a single-stranded polynucleotide is present in this renaturing process, the nucleotide and s-SPG form a macromolecular complex (Sakurai et al. 2001; Mizu et al. 2001). Their spectroscopic data showed that the hydrogen bond and hydrophobic interactions are essential for complexation, and the stoichiometric number indicated that the complex consists of two glucan and one polynucleotide chains. Recently, X-ray crystallography (Sakurai et al. 2004), electron microscopy and (Bae et al. 2004) computational chemistry showed that the complex takes on a triple helix and the helicity is similar to the original SPG triple helix. Some of the other novel features for this complex are: it is remarkably stable (large binding constant) and considerably water-soluble under physiological conditions; two main chain glucose units and one base unit bind with each other to form hydrogen bonds; when a s-SPG/DNA complex meets the corresponding complementary sequence, the complex is dissociated immediately and hybridization takes place (Koumoto et al. 2004); the complex automatically dissociates at a pH < 6.0, because protonation of the nucleotide base induces

[1] Department of Chemical Process & Environments, The University of Kitakyushu, 1-1 Hibikino, Wakamatu-ku, Kitakyushu, Fukuoka 808-0135, Japan
[2] Faculty of Engineering, Department of Chemistry & Biochemistry, Graduate School of Engineering, Kyushu University, 6-10-1 Hakozaki, Higashi-ku, Fukuoka 812-8581, Japan

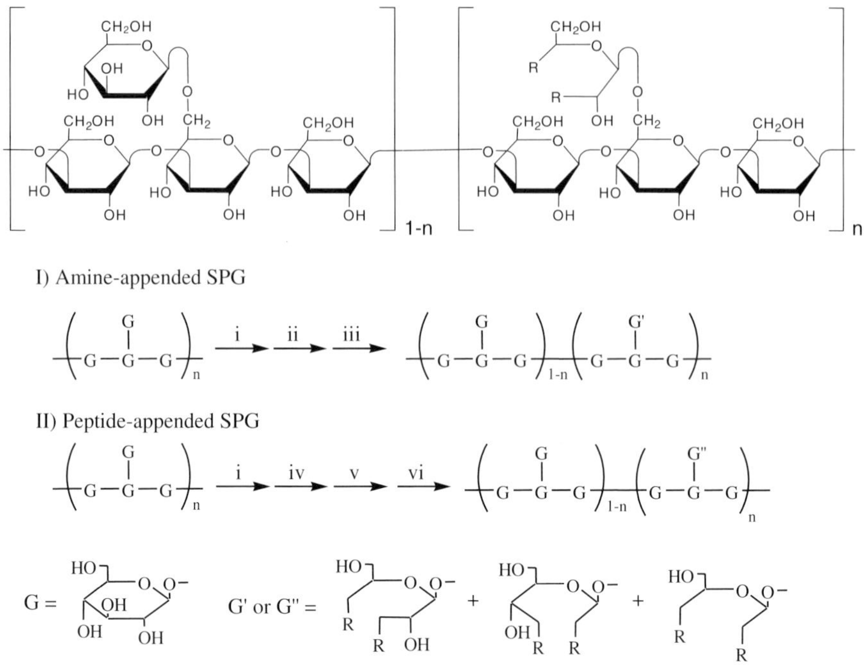

FIG. 1. Chemical structure of chemically modified schizophyllan (SPG). Intact structure of SPG (*left*), modified structure (*right*). The modification level in Table 1 is defined by "*n*" in the figure. The modification was made by the selective cleavage of the 1,2-diol group of the glycosyl side chain with periodate, leading to formyl group formation, and subsequently introducing a functional group. *i* $NaIO_4$, H2O; *ii* amine or animo acid, DMSO; *iii* $NaBH_4$, DMSO; *iv* NH_3 aq. $NaBH_3$, CN, DMSO, *N*-succinimidyl-3-maleinidopropionate; *vi* peptide, DMSO

conformational changes, which cause dissociation of the complex (Sakurai et al. 2003). Figure 2 schematically presents the above novel features of this complex.

2 Biofunctional Oligonucleotides, Antisense and CpG Oligonucleotides

Oligodeoxynucleotides (ODNs) containing unmethylated CpG sequence (CpG motif) have been showed to stimulate a cell-mediated immune response in mammals (Krieg 1999, 2003). This immune response is a defense system that mammals have evolved based on genetic differences between species; that is, unmethylated CpG sequences occur more frequently in bacterial DNA than in mammalian DNA. During the first event of the response, the CpG motif activates antigen-presenting cells. For example, activated macrophages secrete a wide variety of cytokines such as interleukin (IL)-6, IL-12, IL-18, and tumor necrosis factor (TNF)-α. Considerable attention is currently being devoted to this response because the CpG motif can be an extraordinarily

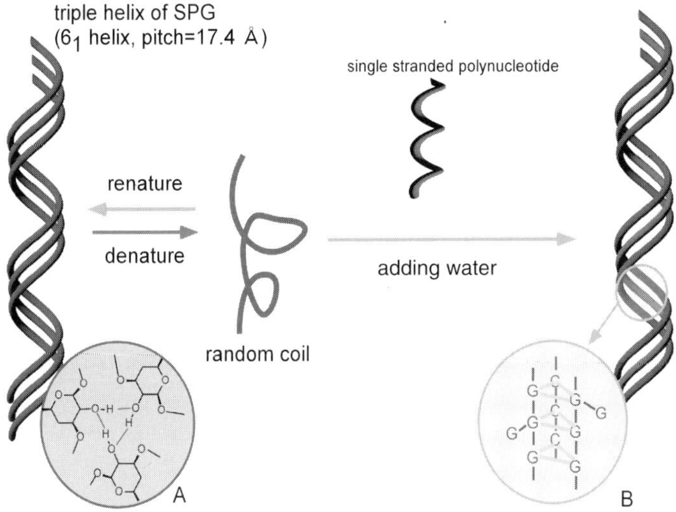

FIG. 2. Complex formation from polynucleotide and s-SPG. *Black and gray lines* represent the s-SPG and polynucleotide molecules, respectively. In the *right inset* G and C represent glucose and base, and the *gray lines* show the hydrogen bonds. The *left inset* shows the hydrogen bonds forming the SPG triple helix

effective adjuvant for many vaccines against infectious agents, cancer antigens, and allergens.

Hemmi et al. (2000) demonstrated that CpG motifs are recognized by a pattern-recognition receptor called Toll-like receptor 9 (TLR-9), which is localized in endosome and lysosome compartments (Ahmad et al. 2002). This finding suggests that, if the CpG motifs can be delivered to these organelles with an appropriate carrier that can protect them from enzymatic degradation and release them after transport to these organelles, the immune response can be artificially and more accurately controlled compared with naked administration. As mentioned above, the SPG/ODN complex is automatically dissociated when the pH is less than 6.0 (Sakurai et al. 2003). This pH response seems ideal for releasing the complexed CpG DNA in late endosomes, in which the pH is maintained at about 5.0.

Antisense oligonucleotides (AS ODNs), designed to suppress expression of a particular gene, have considerable potential as therapeutic agents. However, a device or innovation to protect AS ODNs from enzymatic degradations is also necessary to translate this strategy into practical applications. The degradation problem can be overcome by use of ODN analogues, such as phosphorothioates, phosphoramidates, and peptide nucleic acids. Phosphorothioates are the leading candidate among the first-generation antisense compounds, and several of them are in phase I/II clinical trials (Mani et al. 2002). However, those analogues are easily absorbed by serum proteins, causing an unfavorable "non-antisense effect" (Stein and Cheng 1993; Stein and

Krieg 1994). The materials that can form a complex with phosphorothioates AS ODNs can be used as carriers, preventing the bound AS ODN from interacting with hydrolytic enzymes and other serum proteins (Chirila et al. 2002). Synthetic polycations, such as poly(L-lysine) and polyethyleneimine (PEI), have been studied as AS ODN carriers, because they can form polyion complexes with polynucleotides. Although polycations have great advantages with respect to improving cellular uptake, serious drawbacks, such as the toxicity of the polycations and the poor solubility of the resultant polyion complexes, have been pointed out. In addition, efficient endosomal escape remains a challenging task. The targets of AS ODNs are mRNAs and they are normally present in the cytosol, so that AS ODNs have to escape the endosomal pathway in order to bind mRNA.

Material design for AS ODN and CpG motif carriers has several aspects in common, including that the carrier has to protect the bound nucleotides from enzymatic degradation and unfavorable absorption with proteins. However, the final destinations of these compounds differ from each other.

3 Strategy and Chemistry

Figures 3 and 4 schematically present our strategy to deliver oligonucleotides to the targets and the target organelles for the functional oligonucleotides, respectively. Schizophyllan is unable to bind to the plasma membrane to induce endocytosis or other uptake processes. Therefore, we needed to modify the schizophyllan chain with

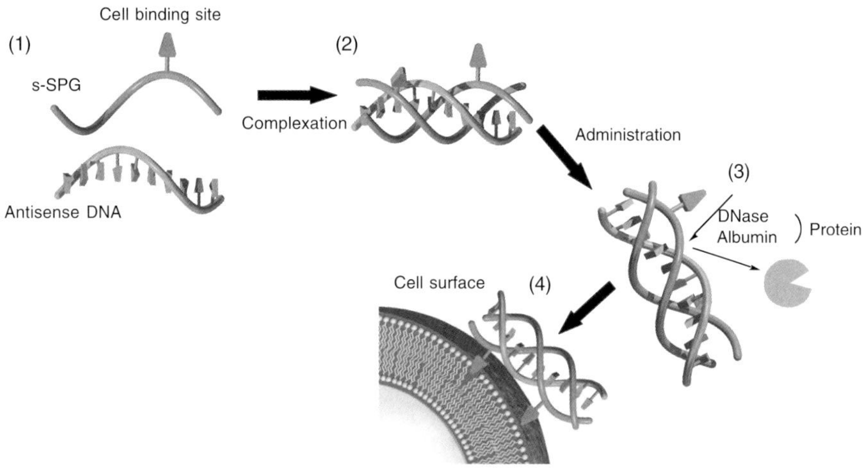

FIG. 3. Our strategy to deliver oligonucleotides to cells. A functional group that can induce the cellular uptake of the complex is attached to the s-SPG side chain, where s-SPG stands for the single chain of schizophyllan (*1*). When the modified s-SPG and oligonucleotide are mixed, they form a complex consisting of two modified s-SPG chains and one nucleotide chains (*2*). The complex can prevent the bound oligonucleotide from forming unfavorable interactions with proteins (*3*). Once the functional group binds to the plasma membrane, cellular uptake can be induced to ingest the complex (*4*)

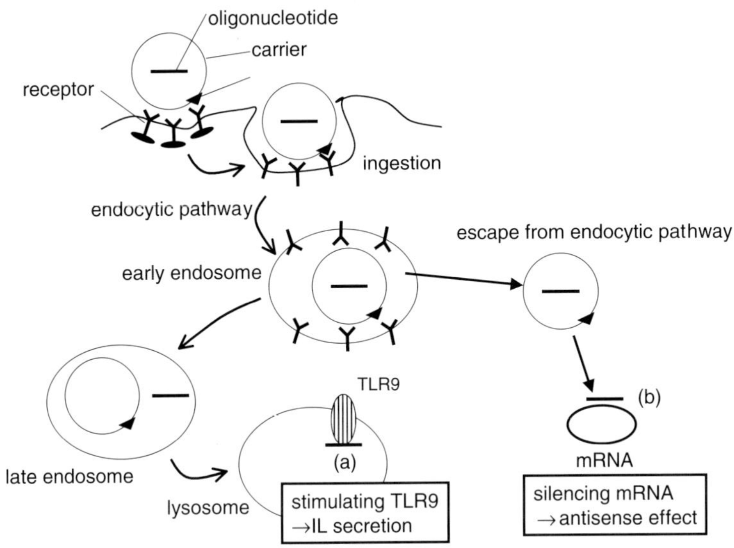

FIG. 4. Final target organs for antisense DNA and CpG motif. The target of antisense is mRNA, located in the cytosol, and that of CpG motif is TLR-9, located in the lysosome

a functional group that induces cellular uptake (Matsumoto et al. 2004). Such functional groups include cationic groups, which can bind to the anionic cellular surface; or a specific peptide that can induce receptor-mediated endocytosis or other cellular up-take mechanisms. Only the side chain of schizophyllan needs to be modified, because the glucan main chain is critical to forming the complex. As presented in the lower part of Fig. 1, this site-specific modification can be carried out by oxidizing schizophyllan with sodium periodate anion to yield the formyl terminate, and subsequent reactions to attach a functional group to the formyl group (Koumoto et al. 2001). The resultant modified s-SPG and oligonucleotides should form a complex. Since we already know that the complex can protect the bound oligonucleotides against nuclease-mediated hydrolysis or unfavorable binding to the serum proteins (Mizu et al. 2004a, b), the population of oligonucleotides surviving before reaching the cells should be increased. Once the complex has reached the cell, the cell-binding site interacts with the cell surface to induce cellular uptake.

As presented in Fig. 4, after ingestion by cells, the target organelles differ between AS ODNs and CpG motifs. The AS ODNs should be released from the endocytic pathway to bind the target mRNA, which is present in the cytosol. This event should preferably happen before the AS ODNs start to suffer from enzymatic degradation in the lysosomes. By contrast, CpG motifs have to be released from the carrier either in the lysosome or late endosome because the receptor for CpG motifs is present in these organelles.

4 Delivery of CpG Motifs

4.1 Experiments

We introduced spermine (SP), arginine-glycine-aspartic acid tripeptide (RGD), octaarginine (R8), or cholesterol (Chol) (see Table 1) to s-SPGs with a weight average molecular weight (M_w) of 15000. The modification levels were about 5 mol% for SP and Chol samples, and 1 mol% for the peptide-appended samples. As CpG DNA, 5′-TCCATGACGTTCCTGATG-(dA)40-3′ (the immunostimulatory sequence; PuPuCG-PyPy is underlined) was used (Aramaki et al. 2002). As the negative control (denoted non-CpG DNA), the sequence 5′-TCCATGAGCTTCCTGAGT-(dA)40-3′, in which only CG sequence is reversed (underlined), was used. In both sequences, a poly(dA)40 tail was attached at the 3′ end to increase complex stability. Complexation was carried out by the established method and the molar ratio ($M_{s\text{-}SPG}/M_{ODN}$) was fixed at 1.5, where $M_{s\text{-}SPG}$ and M_{ODN} are the molar concentration of the repeating unit of SPG and CpG DNA, respectively.

4.2 Results and Discussion

In Fig. 5, a comparison of cytokine secretion between the naked dose and the five carrier-mediated doses is shown. Exposure of the cells to naked CpG DNA yielded secretions of IL-6, IL-12, and TNF-α of 9.0 ± 1, 2.6 ± 1, and 0.75 ± 0.1 ng/ml, respectively. All of these secretion levels were much higher than in the controls. When CpG DNA was added as a complex with non-modified SPG (CpG DNA/s-SPG), secretion is increased by about 20–40% compared to the naked assay. This difference can be ascribed to the fact that complexed CpG DNA does not bind to serum proteins and does not undergo hydrolysis by deoxyribonucleases. However, SPG is unable to enter cells, thus the secretions are only slightly increased. Modified SPG increases the secre-

TABLE 1. Sample codes and the introduced chemical groups

Sample code	R	Modification level[a]	N/P ratio[b]
SP-SPG	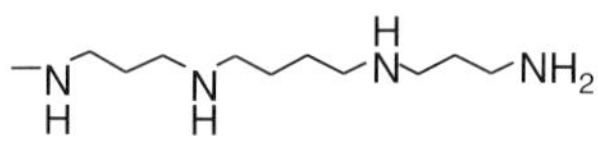	4.6 ± 0.3 mol%	0.27
Chol-SPG	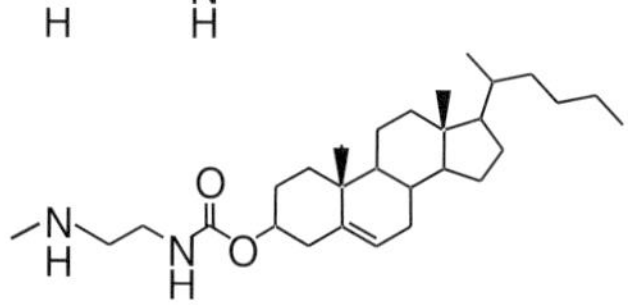	6.9 ± 1.0 mol%	0.21
R8-SPG	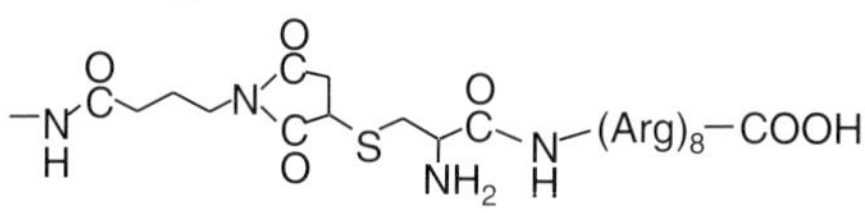	0.5 ± 0.1 mol%	~0
RGD-SPG	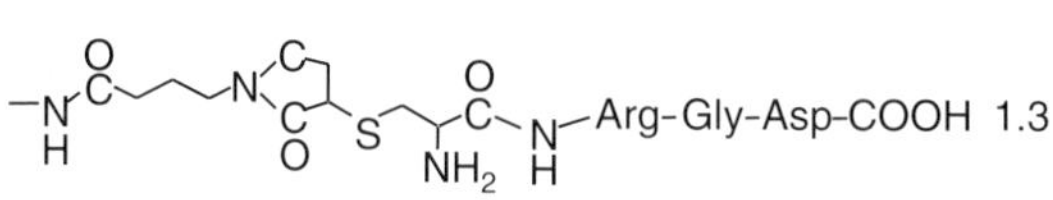	1.3 ± 0.3 mol%	~0

[a] Determined by N elemental analysis. [b] Cation (N)/anion (P) ratio.

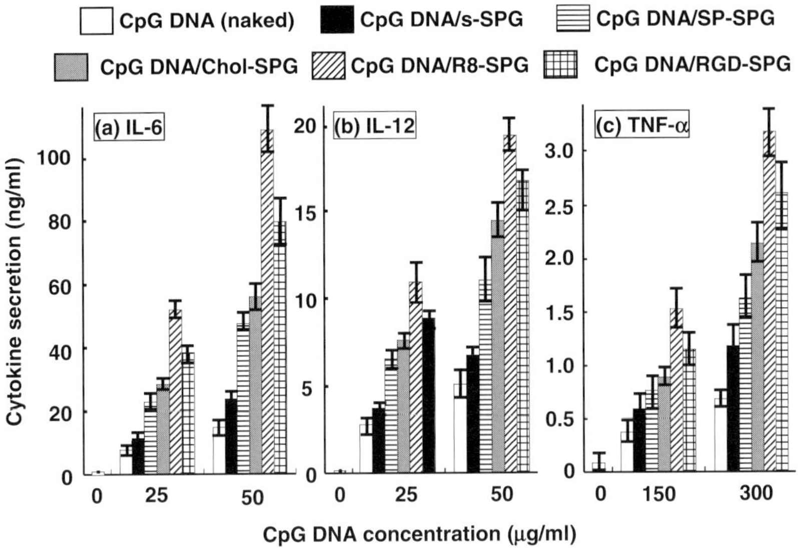

FIG. 5. Effect of chemical modification of SPG on CpG-motif-mediated cytokine secretion. The murine macrophage-like cell J774.A1 (1 × 106 cells/ml, 100 μl/well) was stimulated with IL-6 and IL-12 at 25 and 50 μg/ml, and with TNF-α. at 150 and 300 μg/ml

tion dramatically, i.e., five- to ten-fold compared with the naked assays. Among the modified SPG compounds, R8-SPG had the highest performance, followed by RGD-SPG and then Chol-SPG. Difference between RGD-SPG and Chol-SPG were prominent for IL-6 but relatively small for IL-12 and TNF-α.

The cellular ingestion mechanisms differ depending on the chemical moieties of the carrier. The cellular membrane is negatively charged, so that cations such as spermine can bind to the surface via coulombic forces and should be ingested by pinocytosis. While this should be the case for SP-SPG; ingestion via an electrostatic interaction did not occur because the complex is negatively charged in total (see Table 1), and thus it interferes with interaction of spermine with the cell membrane. Generally, cholesterol-appended carriers are ingested through the low-density lipoprotein (LDL) receptor and RGD-appended ones by the integrin receptor. Cellular ingestion for these compounds is considered to proceed by receptor-mediated endocytosis. Therefore, after ingestion, the RGD-SPG or Chol-SPG/CpG DNA complex is eventually transported to the endosome and finally to the lysosome, where the compartment pH is kept at about 5 and digestive enzymes are highly activated. The complex releases the CpG DNA owing to the low pH, the naked CpG DNA is easily hydrolyzed and the fragments are recognized by TLR-9 on the vesicle membrane. This may be the reason for the relatively high secretion levels induced by RGD-SPG and Chol-SPG complexes. In contrast, with RGD, arginine-rich peptides such as R8 and Tat are ingested by a pathway different from the one that interacts with cations, RGD and cholesterol (Fischer et al. 2004; Futaki et al. 2002; Richard et al. 2003). Although there is still controversy and little agreement on the uptake mechanism, some data suggest that the pathway induced by R8 avoids capture by lysosomes and endosomes. If this were the only pathway to ingest the R8-SPG complex, cytokine secretion would be less than obtained with the other complexes. However, the R8-SPG complex induced the highest level of secretion—almost ten times larger than obtained with naked CpG DNA. As far as we know, this is the highest enhancement of cytokine secretion by a carrier. The precise mechanism of R8-SPG complex uptake and the reason for the enhancement remain to be clarified.

Although the data are not shown, when cells were exposed to non-CpG DNA and to only carriers under exactly the same conditions as shown in Fig. 5, essentially no secretion was measured. These negative-control experiments confirmed that the cytokine secretion observed in Fig. 5 is strongly related to the CpG sequence. It is particularly noteworthy that only two sequence difference between CpG and non-CpG DNA creates such a difference and the presence of the complex dramatically enlarges the difference. For the naked assay, non-CpG and CpG provided 0.2 and 5.1 ng IL-12/ml, respectively, when 50 mg oligonucleotides/ml were administered. When the cells were exposed to the complex, non-CpG and CpG provided 0.2 and 20 ng/ml, respectively. The secretion level is four-fold over that of the background. The cytotoxicities of unmodified-SPG, R8-SPG, RGD-SPG, and Chol-SPG were negligibly small, whereas SP-SPG and commercially available cationic liposome had the same degree of cytotoxicity as polyethyleneimine. The lower cytotoxicity of modified SPG is especially important for the CpG DNA carrier, because inflammatory immune response can thus be avoided.

Figure 6 shows confocal microscopic images comparing the distribution of fluorescein-labeled phosphorothioate dA50 in J774.A1 cells. The cells were exposed to naked dA40 or to three complexes, Chol-SPG, R8-SPG, or RGD-SPG. It can be seen that dA40 does not enter the cells, whereas chemically modified SPG is ingested. Careful examination of the images reveals that labeled dA40 seems to localize in the cytosol and is excluded from the nucleus. Furthermore, the green dots are distributed in a spotty fashion, which may correspond to the endosome or lysosome.

5 Delivery of Antisense Oligonucleotides to Avoid Lysosomal Degradation

The same chemical modifications as described in the above section were used to design a carrier for AS ODNs. Several SPG complexes with antisense oligonucleotides were constructed and an antisense assay was carried out in which phosphorothioate AS ODN was administered to melanoma A375 or leukemia HL-60 cell lines (Matsumoto et al. 2004). Similar to the CpG carrier, the R8- or RGD-modified SPGs had the strongest antisense effect, most likely due to the enhancement of endocytosis by these functional peptides (Matsumoto et al. 2004). However, most AS ODNs were presumably transported to lysosomes and eventually degraded. Furthermore, the phosphorothioate caused cytotoxicity.

Polyethylene glycol (PEG) has been shown to be the most effective polymer in inducing aggregation and fusion of vesicles and cells. Sunamoto and coworkers (Higashi et al. 1996) attached PEG to a lipid and found that the resultant liposome had enhanced capability to fuse with plasma membrane, depending on the chain length of PEG. Kuhl et al. (1996) established that PEG of molecular weight 8000–10000 was

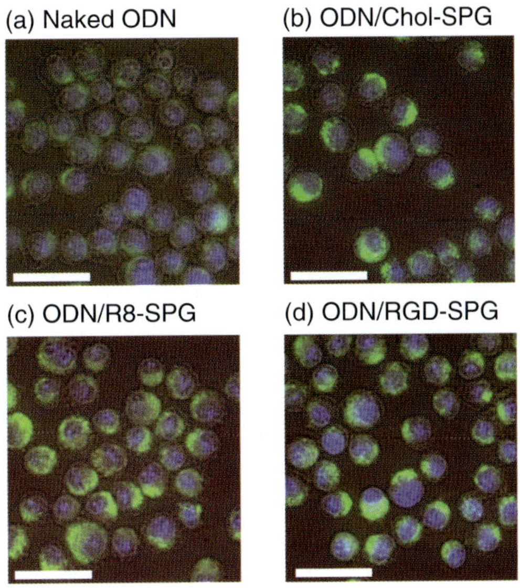

FIG. 6. Confocal microscopy projection images of the cellular distribution of fluorescein-labeled dA40 (as a model oligonucleotide) in J774.A1, when the dA40 was administrated in the states **(a)** naked, **(b)** complexed with Chol-SPG, **(c)** with R8-SPG, and **(d)** with RGD-SPG. The nucleus was stained with DAPI. *Bars* 50 μm

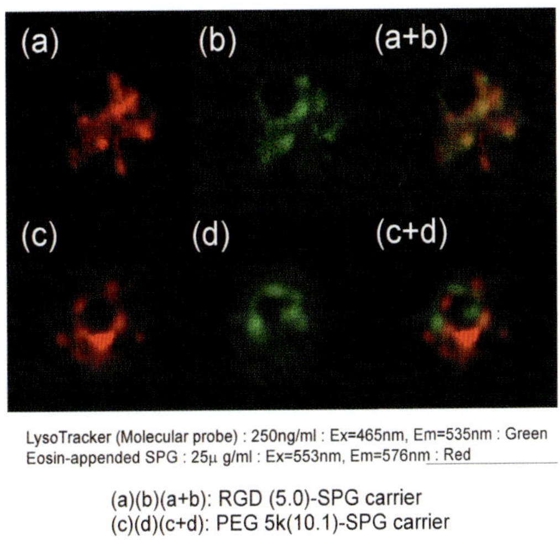

FIG. 10a–d. Confocal laser scanning microscopy images comparing the PEG (*lower images*) and RGD (*upper images*) systems. **a, c** Distribution of eosin-labeled SPG; **b, d** LysoTracker

effective in causing the aggregation of vesicles, and showed the existence of an attractive osmotic force due to a polymer-depleted layer near the bilayer surface. These pioneering works led us to attach PEG to SPG in order to provide endosomal escape ability to SPG/AS ODN complexes.

5.1 Experiments

Using a similar method to that presented in Fig. 1, methoxypolyethylene glycol amine ($M_w = 5000$) was attached to 10 mol% of the side chain in s-SPG ($M_w = 15000$), denoted

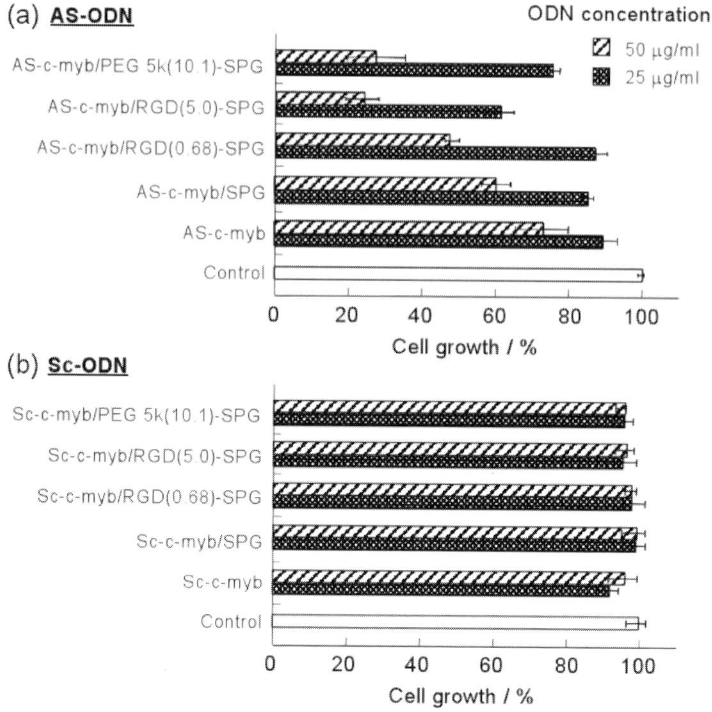

FIG. 7a, b. Comparison of cell growth when A375 cells were exposed to AS-c-myb and the complexes with s-SPG or PEG 5K(10.1)-SPG, RGD(5.0)-SPG and RGD(0.68)-SPG. **a** AS-c-myb assays, **b** Sc-c-myb assays

as PEG 5K(10.1)-SPG. The ODN sequence of 5′-GTGCCGGGGTCTTCGGGC-3′ is well-known to bind to c-myb mRNA and thus cause depression of c-myb. For the same reason as mentioned in Sect. 4.1, a poly(dA)$_{40}$ was attached. The resulting 5′-GTGCCGGGGTCTTCGGGC-(dA)40-3′ (Gryaznov et al. 1996; Gewirtz and Calabretta 1988) phosphorothioate and phosphodiester were used as an AS ODN and denoted AS-c-myb. As a negative control, a scrambled sequence of AS-c-myb, 5′-TGCTGCGCGTGGTCGGCG-(dA)40-3′, denoted Sc-c-myb, was used. The cell number was evaluated using the Cell Counting Kit-8, WST-8 assay (Dojindo, Japan).

5.2 Results and Discussion

The upper panel in Fig. 7 shows the cell growth when AS-c-myb was dosed at 25 or 50 μg/ml, and comparisons with naked AS-c-myb, s-SPG/AS-c-myb, RGD(0.68)-SPG/AS-c-myb, RGD(5.0)-SPG/AS-c-myb, and PEG 5K(10.1)-SPG/AS-c-myb are shown. The lower panel shows the results when the cells were exposed to Sc-c-myb and its complexes. Naked AS-c-myb decreased growth by 10% at 25 μg/ml and 30% at 50 μg/ml. However, in the case of Sc-c-myb, growth was 92–96%, slightly less than the control. This small decrement can be ascribed to anti-proliferation activity (or cyto-

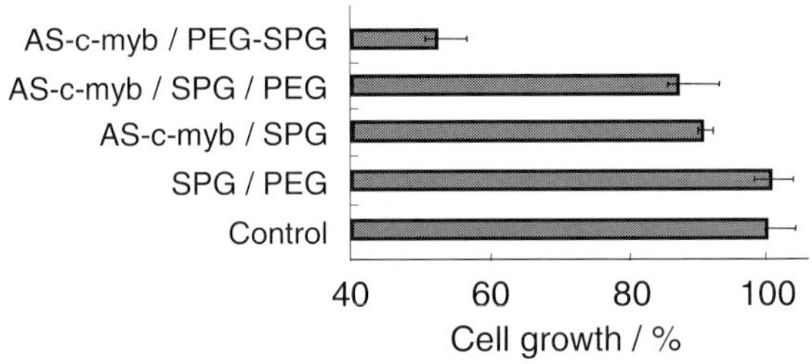

Fig. 8. Effect of mixing PEG with SPG and appended PEG with SPG on the antisense assay

toxicity) caused by the non-antisense effect of the contiguous G residues and/or the cytotoxicity of phosphorothioate itself. Comparison of the AS-c-myb and Sc-c-myb assays clarifies that the antisense effect overwhelmed the other effects in the assay.

In the upper panel in Fig. 7, s-SPG/AS-c-myb decreased growth more efficiently than achieved with the naked doses. When AS-c-myb was added as the s-SPG complex, growth was reduced to 83% at 25 μg/ml and to 58% at 50 μg/ml. By comparing the corresponding Sc-c-myb assays, it is clear that the complex enhances the antisense effect. When RGD was attached to the s-SPG chain, growth was dramatically reduced because RGD enhances contact of the complex with the cell-surface, thus the ingestion probability is increased. It is interesting that PEG 5K(10.1)-SPG/AS-c-myb yielded a similar reduction in the cell growth. Since there is no receptor to recognize PEG on the cell surface, the up-take of the PEG/AS-c-myb complex should occur by a different mechanism. Figure 8 compares the cell growth between the PEG-appended SPG and PEG-mixed SPG. When PEG was simply mixed into the complex (the same amount of PEG as in the appended samples), there was no decrease in the growth. This negative control experiment clarified that appended PEG is critical to enhancing the antisense effect.

There are many agents whose ability to specifically inhibit biological reactions in endocytosis is well known. For example, chlorpromazine (Wang et al. 1993) inhibits receptor recycling, wortmannin (Jones and Clague 1995; Jones et al. 1998) inhibits formation of early endosomes, nigericin (Uherek et al. 1998) inhibits vesicular transport from the endosome to the lysosome, and chloroquine (Fredericksen et al. 2002) interferes with pH lowering in the lysosome.

Figure 9 compares cell growth between the RGD and PEG systems when endocytosis is inhibited with each of the above four compounds. Chlorpromazine and wortmannin increased cell growth from 60% (non-inhibition) to 70–80% and 90%, respectively. These increments indicate that both PEG and RGD complexes are processed by receptor-mediated cellular uptake, and subsequently the early endosome pathway. There is a possibility that the PEG complex can be ingested directly into the cytosol by a cell-fusion like mechanism, without experiencing endocytosis. However, the chlorpromazine and wortmannin results do not support a direct ingestion mechanism for our PEG complex. Inhibition with nigericin was considerably different in the PEG and RGD systems; the antisense effect was almost eliminated in the RGD system, while the PEG system was not inhibited at all. Since nigericin interrupts the transporting processes from endosome to lysosome, the difference between the PEG and RGD systems indicates that, in the PEG system, AS ODNs were released into the cytosol before they could be degraded in late endosomes or lysosomes.

To compare the cellular distribution of AS ODN in the PEG and RGD systems, A375 cells were exposed to complexes made of eosin-appended RGD(5.0)-SPG and dA40 or complexes made of eosin-appended PEG 5k(10.1)-SPG and dA40. Here, dA40 was used as a model compound for AS ODNs. Figure 10 shows the typical confocal microscopy images. Figures 10a and 10c show many red spots in the cell, corresponding to the SPG distribution. The green spots in Fig. 10b and Fig. 10d correspond to lysosomes. When Fig. 10a and Fig. 10b are superimposed for the RGD system, the resultant image (Fig. 10a+b) shows many yellow spots, indicating that most SPG molecules are located in lysosome. The superimposed image made from Fig. 10c and Fig. 10d, however, shows few yellow spots, i.e., the green and red spots are distributed independently. This result clearly shows that RGD-appended SPG but not PEG-appended SPG is localized in the lysosome.

Figures 9 and 10 indicate that PEG-appended SPG is able to release the bonded AS ODN before being transported to the lysosome, suggesting that phosphodiester AS ODNs can be delivered to the cytosol while avoiding lysosomal degradation. Figure 11 compares the cell growth between the phosphodiester (PO) and phosphorothioate (PS) assays. PEG-appended SPG increased the antisense effect to some extent; however, the PS sequence reduced growth more efficiently than the PS sequence did. The PS scrambled sequence decreased growth because of the cytotoxicity of phos-

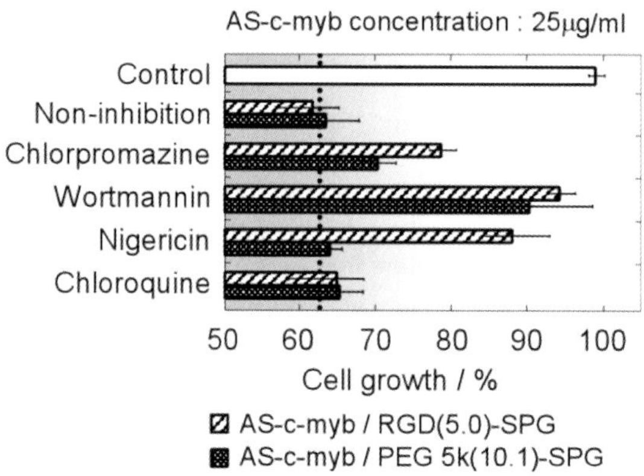

FIG. 9. Endocytosis inhibition assays using chlorpromazine, wortmannin, nigericin, and chloroquine. The PEG and RGD systems are compared

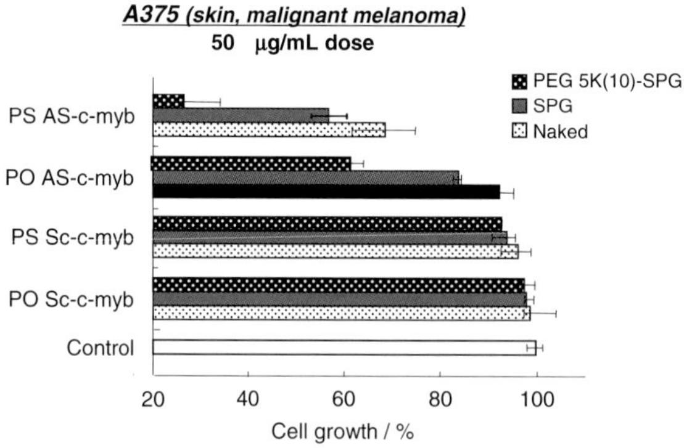

FIG. 11. Comparison of the phosphodiester (*PO*) and phosphorothioate (*PS*) antisense sequences and enhancement by SPG carriers

phorothioates, while the corresponding PO sequence showed little toxicity. This experiment shows that the SPG carrier must be improved in order to attain the same level achieved with the PS dose.

6 Conclusions

This study demonstrates the potential of SPG as an oligonucleotide carrier, for example, in antisense and CpG DNAs. The most distinguishing feature of this carrier compared to other cation types is that the driving force of the complexation is hydrogen bonds instead of electrostatic forces. Therefore, the total charge of the SPG/nucleotide complex is negative and there is no DNA-compaction problem in this system. Using chemically modified SPG, the antisense effect and CpG immunostimulatory response can be enhanced.

References

Ahmad-Nejad P, Hacker H, Rutz M, Bauer S, Vabulas RM, Wagner H (2002) Bacterial CpG-DNA and lipopolysaccharides activate Toll-like receptors at distinct cellular compartments. Eur J Immunol 32:1958–1968

Aramaki Y, Yotsumoto S, Watanabe H, Tsuchiya S (2002) NADPH-oxidase may contribute to IL-12 production in macrophages stimulated with CpG phosphorothioate oligodeoxynucleotides. Biol Pharm Bull 25:351–355

Bae A, Lee S, Ikeda M, Sano M, Shinkai S, Sakurai K (2004) Rod-like architecture and helicity of the poly(C)/schizophyllan complex observed by AFM and SEM. Carbohydr Res 339:251–258

Chirila TV, Rakoczy PE, Garrett KL, Lou X, Constable IJ (2002) The use of synthetic polymers for delivery of therapeutic antisense oligodeoxynucleotides. Biomaterials 23: 321–342

Fischer R, Kohler K, Fotin-Mleczek M, Brock R (2004) A stepwise dissection of the intracellular fate of cationic cell-penetrating peptides. J Biol Chem 279:12625–12635

Fredericksen BL, Wei BL, Yao J, Luo T, Garcia JV (2002) Inhibition of endosomal/lysosomal degradation increases the infectivity of human immunodeficiency virus. J Virol 76:11440–11446

Futaki S, Nakase I, Suzuki T, Youjun Z, Sugiura Y (2002) Translocation of branched-chain arginine peptides through cell membranes: flexibility in the spatial disposition of positive charges in membrane-permeable peptides. Biochemistry 41:7925–7930

Gewirtz AM, Calabretta B (1988) A c-myb antisense oligodeoxynucleotide inhibits normal human hematopoiesis in vitro. Science 242:1303–1306

Gryaznov S, Skorski T, Cucco C, Nieborowska-Skorska M, Chiu CY, Lloyd D, et al. (1996) Oligonucleotide N3′→P5′ phosphoramidates as antisense agents. Nucleic Acids Res 24: 1508–1514

Hemmi H, Takeuchi O, Kawai T, Kaisho T, Sato S, Sanjo H, et al. (2000) A Toll-like receptor recognizes bacterial DNA. Nature 408:740–745

Higashi N, Yamauchi M, Okumura Y, Nakanishi M, Sunamoto J (1996) Fusion between Jurkat cell and PEO-lipid modified liposome. Biochim Biophys Acta 1285: 183–191

Jones AT, Clague MJ (1995) Phosphatidylinositol 3-kinase activity is required for early endosome fusion. Biochem J 311(Pt 1):31–34

Jones AT, Mills IG, Scheidig AJ, Alexandrov K, Clague MJ (1998) Inhibition of endosome fusion by wortmannin persists in the presence of activated Rab5. Mol Biol Cell 9:323–332

Kashiwagi Y, Norisuye T, Fujita H (1981) Triple helix of Schizophyllum commune polysaccharide in dilute solution. 4. Light scattering and viscosity in dilute aqueous sodium hydroxide. Macromolecules 14:1220–1225

Koumoto K, Kimura T, Mizu M, Sakurai K, Shinkai S (2001) Chemical modification of schizophyllan by introduction of a cationic charge into the side chain which enhances the thermal stability of schizophyllan-poly(C) complexes. Chem Commun (Camb) 1962–1963

Koumoto K, Mizu M, Sakurai K, Kunitake T, Shinkai S (2004) Complementary-strand-induced release of single-stranded DNA bound in the schizophyllan complex. Chemistry & Biodiversity 1:520–529

Krieg AM (1999) Mechanism and application of immune stimulatory CpG oligonucleotides. Biochimica et Biophysica Acta 1489:107–116

Krieg AM (2003) CpG motifs: the active ingredient in bacterial extracts? Nature Medicine 9:831–835

Kuhl T, Guo Y, Alderfer JL, Berman AD, Leckband D, Israelachvili J, et al. (1996) Direct Measurement of Polyethylene Glycol Induced Depletion Attraction between Lipid Bilayers. Langmuir 12:3003–3014

Mani S, Rudin CM, Kunkel K, Holmlund JT, Geary RS, Kindler HL, et al. (2002) Phase I clinical and pharmacokinetic study of protein kinase C-alpha antisense oligonucleotide ISIS 3521 administered in combination with 5-fluorouracil and leucovorin in patients with advanced cancer. Clin Cancer Res 8:1042–1048

Matsumoto T, Numata M, Anada T, Mizu M, Koumoto K, Sakurai K, et al. (2004) Chemically modified polysaccharide schizophyllan for antisense oligonucleotides delivery to enhance the cellular uptake efficiency. Biochim Biophys Acta 1670:91–104

McIntire TM, Brant DA (1998) Observations of the (13)–D-Glucan Linear Triple Helix to Macrocycle Interconversion Using Noncontact Atomic Force Microscopy. J Am Chem Soc 120:6909–6919

Mizu M, Kimura T, Koumoto K, Sakurai K, Shinkai S (2001) Thermally induced conformational transition of polydeoxyadenosine in the complex with schizophyllan and the base-length dependence of its stability. Chem Commun ••:429–430

Mizu M, Koumoto K, Kimura T, Sakurai K, Shinkai S (2004a) Protection of polynucleotides against nuclease-mediated hydrolysis by complexation with schizophyllan. Biomaterials 25:3109–3116

Mizu M, Koumoto K, Anada T, Karinaga R, Kimura T, Nagasaki T, et al (2004b) Enhancement of the Antisense Effect of Polysaccharide? Polynucleotide Complexes by Preventing the Antisense Oligonucleotide from Binding to Proteins in the Culture Medium. Bulletin of the Chemical Society of Japan 77:1101–1110

Mizu M, Koumoto K, Anada T, Karinaga R, Nagasaki T, Sakurai K, et al. (2004c) A Polysaccharide Carrier for Immunostimulatory CpG DNAs to Enhance Cytokine Secretion. J Am Chem Soc 126:8372–8373

Norisuye T, Yanaki T, Fujita H (1980) Triple helix of a schizophyllum commune polysaccharide in aqueous solution. J Polym Sci Polym Phys Ed 18:547–558

Okamura K, Suzuki M, Chihara T, Fujiwara A, Fukuda T, Goto S, et al. (1986) Clinical evaluation of schizophyllan combined with irradiation in patients with cervical cancer. A randomized controlled study. Cancer 58:865–872

Richard JP, Melikov K, Vives E, Ramos C, Verbeure B, Gait MJ, et al. (2003) Cell-penetrating peptides. A reevaluation of the mechanism of cellular uptake. J Biol Chem 278:585–590

Sakurai K, Fuji M (2000) Optical Properties of a Low Birefringence Polyester Containing Fluorene Side Chain (I). Polym J 32:676–682

Sakurai K, Shinkai S (2000) Molecular Recognition of Adenine, Cytosine, and Uracil in a Single-Stranded RNA by a Natural Polysaccharide: Schizophyllan. J Am Chem Soc 122: 4520–4521

Sakurai K, Mizu M, Shinkai S (2001) Polysaccharide–polynucleotide complexes. 2. Complementary polynucleotide mimic behavior of the natural polysaccharide schizophyllan in the macromolecular complex with single-stranded RNA and DNA. Biomacromolecules 2:641–650

Sakurai K, Iguchi R, Mizu M, Koumoto K, Shinkai S (2003) Polysaccharide-polynucleotide complexes. Part 7. Hydrogen-ion and salt concentration dependence of complexation between schizophyllan and single-stranded homo RNAs. Bioorg Chem 31:216–226

Sakurai K, Koumoto K, Shinkai S (2004) Macromolecular Nanostructured Materials. Yokyo: Koudan-sha

Sato S, Sakurai K, Norisuye T, Fujita H (1983) Collapse of Randomly Coiled Schizophyllan in Mixture of Water and Dimethylsulfoxide. Polym J 15:87–92

Stein CA, Cheng YC (1993) Antisense oligonucleotides as therapeutic agents is the bullet really magical. Science 261:1004–1012

Stein CA, Krieg AM (1994) Problems in interpretation of data derived from in vitro and in vivo use of antisense oligodeoxynucleotides. Antisense Res Dev 4:67–69

Tabata K, Ito W, Kojima T, Kawabata S, Misaki A (1981) Ultrasonic degradation of schizophyllan, an antitumor polysaccharide produced by Schizophyllum commune Fries. Carbohydr Res 89:121–135

Tabata K, Itoh W, Hirata A, Sugawara I, Mori S (1990) Preparation of polyclonal antibodies to an anti-tumor (1–3)-beta-D-glucan, schizophyllan. Agric Biol Chem 54:1953–1959

Uherek C, Fominaya J, Wels W (1998) A modular DNA carrier protein based on the structure of diphtheria toxin mediates target cell-specific gene delivery. J Biol Chem 273: 8835–8841

Wang LH, Rothberg KG, Anderson RG (1993) Mis-assembly of clathrin lattices on endosomes reveals a regulatory switch for coated pit formation. J Cell Biol 123:1107–1117

Yanaki T, Norisuye T, Fujita H (1980) Triple helix of schizophyllum commune polysaccharide in dilute solution. 3. Hydrodynamic Properties in Water. Macromolecules 13: 1462–1466

Biological and Chemical Hybrid Vectors

Mahito Nakanishi

1 Introduction: Concept of the Hybrid Vector

It is generally accepted that current gene delivery vectors, either viral or non-viral, have a number of drawbacks as ideal tools for gene therapy. Viral vectors have intrinsic efficient machineries for delivering their genomes into target cells, but it is not feasible to maintain the delivered genes stably beyond the characteristics of the parent virus genome in the cells. By contrast, non-viral (or synthetic) vectors can accommodate various genetic materials more flexibly than viral vectors, but their delivery machineries are still primitive and much less efficient than those of viral vectors. In order to overcome these difficult problems, Cotton and Wagner (1999) proposed the development of a novel non-viral vector equipped with the function of a viral vector, but the ideal vector still remains conceptual.

In 1998, I proposed the development of a more concrete gene delivery system, the "hybrid vector" (Nakanishi et al. 1998). Hybrid vector is an ambiguous term, sometimes referring to non-viral vectors containing some biological moieties and sometimes to viral vectors combined with various synthetic materials. In this chapter, however, the term is used to describe the vector system defined as follows: Firstly, the hybrid vectors have a virus-like structure and function both for preserving their genetic materials and for efficiently delivering them into the target cells. Secondly, the hybrid vectors contain either DNA produced in a microorganism defective in DNA recombination or RNA, which in itself lacks the recombination mechanism. These genetic materials may also have the function to stabilize themselves in the cells.

The most critical point in my proposal is to direct impartial attention to these two different aspects of the vector system (the gene delivery machineries and the genetic platforms to be installed in these machineries). Both of these aspects are equally important and relate with each other, as two sides of a coin. Development of a novel genetic platform is essential for curing patients suffering from congenital metabolic diseases, which is a difficult final goal of gene therapy. In this chapter, basic informa-

Gene Function Research Center, National Institute of Advanced Industrial Science and Technology (AIST), 1-1 Higashi, Central 4, Tsukuba Science City 305-8562, Japan tion related to these topics, and recent achievements in realizing the hybrid vector are discussed.

1.1 Structural Requirements of Hybrid Vectors

All intact virus particles share one general and important characteristic: the genetic material is encapsulated in nano-sized particles for effective protection from environmental nucleases. Viral genome DNA (or RNA) has unique sequences, called "packaging signals," that are responsible for active encapsulation of the genome into the virus particles through the interaction with viral structural proteins. By contrast, most current non-viral vectors encapsulate DNA in a sequence-independent passive mechanism, either through electrostatic interaction with cationic molecules or by mechanical packaging into small lipid vesicles. Hence, complete and efficient encapsulation of nucleic acids into nano-sized particles remains a challenging goal for conventional non-viral vectors.

The structure of the vectors is closely related to their delivery function, especially to active nuclear transport of DNA (Nakanishi et al. 2001). As most DNA viruses (and some RNA viruses) transcribe and replicate their genetic information in the host cell nucleus, these karyophilic viruses should have intrinsic active mechanisms for transporting their genome from the cytoplasm to the nucleus. Although the detailed mechanism is still controversial, it is almost certain that these viruses transport their genome exclusively through the nuclear pore complex (NPC) (Whittaker et al. 2000). Transport of cytoplasmic cargo through the NPC is primarily regulated by the peptidic nuclear localization signal (NLS), which functions as a binding ligand to cytoplasmic carrier proteins (Chook and Blobel 2001) but is also restricted by the physical size of the cargo. The upper size limit of the particles transportable through the NPC has been estimated as 25 nm~39 nm in diameter (Dworetzky et al. 1988; Pante and Kann 2002), depending on the experimental design. However, we recently demonstrated that, under optimal conditions, lambda phage particles with a diameter of 55 nm could accumulate into the nucleus efficiently through the NPC in an NLS-dependent manner (Akuta et al. 2002; Eguchi et al., 2005). Our findings are consistent with recent observations using cryoelectron tomography that the inner diameter of the NPC is about 60 nm (Beck et al. 2004). Considering the ultrastructure of the NPC, hybrid vectors with active nuclear delivery activity should pack their genetic materials tightly in nano-sized particles with diameters of 55 nm or below. Our observation also leads to a second important conclusion, that the maximum DNA size transportable to the nucleus is about 50 kbp, equal to the size of lambda phage genome DNA, which is packed concisely to a density of 0.49 g/ml (Earnshaw and Casjens 1980) (see also Sect. 5).

1.2 Genetic Materials Used in Hybrid Vectors

One of the most significant differences distinguishing viral vectors from non-viral vectors is the mode of replication of their genetic materials. Most of the current viral vectors replicate their genome DNA (or RNA) in mammalian cells, using double-stranded (ds) DNA as a replication intermediate. However, recombination between the ds DNA molecules in mammalian cells resulted in the emergence of viruses with undesired characteristics. For example, recombination between vector genome DNA and helper DNA encoding the viral structural proteins regenerates replication-competent viruses (RCV) at a low but significant frequency. The Center for Biologics Evaluation and Research (USA) recommends that vector preparation for human gene therapy should contain no more than one RCV in 3×10^{10} virus particles in order to avoid unexpected side effects. As this strict regulation makes vector production expensive, various approaches for suppressing the generation of RCV have been reported (Yee 1999). Nevertheless, it is still difficult to suppress DNA recombination to a satisfactory level in mammalian cells: homologous regions as short as ten base pairs long are sufficient to induce recombination (Otto et al. 1994).

As large-scale chemical synthesis of long ds DNA is still unrealistic, the production of DNA for use in non-viral vectors relies on the bacterial system. In this case, the emergence of undesired DNA structures due to recombination is negligible if the DNA is propagated carefully in recombination-defective *E. coli*. Even if bacterial genome DNA is inserted accidentally by recombination, it may not be harmful by itself because the bacterial genes have no physiological function in mammalian cells. Mutant *E. coli* strains defective in DNA recombination has been used widely in research as well as in commercial DNA production, while no mammalian cell line with a defect in recombination is available for vector production.

The bacterial system has another advantage in that it can support the production of DNA that is difficult to insert into viral vectors. DNA fragments required for stable maintenance of functional chromosome are one of the examples. Human artificial chromosomes are regarded as ideal platforms for maintaining therapeutic genes in cells, because the random chromosomal integration of vector DNA occasionally induces malignant transformation of somatic cells, as observed in recent clinical gene therapy trials (Hacein-Bey-Abina et al. 2003). However, functional chromosomal components (centromere, telomere and the replication origin) consist of long repetitive DNA fragments, which are difficult to insert into the viral genome due to its limited capacity.

Tumor-specific cytotoxic suicide genes are another example of DNA that is difficult to insert into viral vectors. These suicide genes, in which the expression of toxic proteins is driven by promoters activated specifically in cancer cells, could serve as useful tools for cancer gene therapy if they are inserted into efficient delivery vehicles. However, they cannot be inserted into viral vectors as they kill the packaging cells before vector production can take place. Suicide genes are readily propagated in *E. coli* for large-scale production, and we demonstrated their effectiveness in an animal model by delivering them into disseminated gastric cancer cells with the aid of fusogenic liposomes (FL) (Horimoto et al. 2000) (see Sect. 2.2).

2 Fusogenic Liposome: A Delivery Vehicle Mimicking Envelope Viruses

Animal viruses are classified into two major groups, envelope viruses and non-envelope viruses, based on their structures. In the former, the genomes is encapsulated in a lipid envelope and delivered by fusion of the envelope with the cell membrane. In the latter, the genome is encapsulated in a proteinaceous capsid and delivered by disrupting the membrane of cell organelles (endosome, caveollae and microphagosome). Fusogenic liposomes (FL) mimic the function of envelope viruses (Nakanishi et al. 1998).

Fusogenic liposome is a generic term referring to lipid vesicles (liposomes) capable of fusing with the cell membrane. Previously, simple liposomes, consisting of natural or synthetic lipids, were thought to readily fuse with the cell membrane and deliver their contents into the cells directly. However, it now appears that most liposome-mediated macromolecule delivery occurs through the passive endocytotic pathway, associated with membrane disruption rather than with membrane fusion, regardless of the constituents of the liposomes (Barron and Szoka 1999). This is true even when the liposomes are modified with fusogenic chemicals. For example, liposomes modified with polyethylene glycol (PEG), a popular agent frequently used for generating hybrid cells, fused with the cell membrane 10,000 times less efficiently than virus-based FL (Higashi et al. 1996). In the following, the characteristics of virus-based FLs and their possible application are discussed.

2.1 Basic Science of Fusogenic Liposomes

Fusogenic liposomes are prepared from Sendai virus, a prototype virus of the paramyxoviruses (see Sect. 4.2.). Although any envelope animal virus may possibly become a source for preparing FLs, Sendai virus is the only practical choice for large-scale production, because it is much more productive than other attenuated virus strains used for vaccine production: about 10 mg (protein) of purified Sendai virus can be recovered from a single fertilized chicken egg. Moreover, Sendai virus is not a human pathogen, and infectivity is readily inactivated by ultraviolet light or alkylation. Sendai virus Z_{YO} strain, established by Dr. Yoshio Okada from the original Z strain through selection for more than 20 years, is particularly suitable for FL production. The Z_{YO} strain is much more fusogenic than the parental Z strain, possibly due to alteration of the structure of envelope glycoproteins (Segawa et al., unpublished result).

Another important characteristic of Sendai virus is its exceptionally wide host range. Although natural target of infection is the lung epithelium, this virus can deliver the genome into a wide range of animal cells derived from various tissues and species at almost equal efficiency (Nakanishi et al. 1998). Sendai virus uses sialic acid, a ubiquitous molecule on the animal cell surface, as a primary receptor. However, we recently found that the cell specificity of Sendai virus infection was determined not by the primary receptor but by an unidentified host cell factor(s) (Eguchi et al. 2000). These findings indicate that the mechanism of virus-mediated membrane fusion is more complex than supposed previously. Uncovering this mechanism will provide insight into the development of synthetic vehicles with fusogenic activity.

Transplantation of the fusogenic activity of Sendai virus into a simple liposome was first achieved by reconstituting the viral-envelope-like structure from detergent-solubilized envelope glycoproteins (Uchida et al. 1979). However, this procedure only partially restored fusogenic activity and is thus unsuitable for encapsulating very large molecules such as DNA. Currently, FL preparation is based on our finding that Sendai virus particles efficiently fuse with simple liposomes at 37°C in a receptor-independent manner (Kato K. et al. 1991; Nakanishi et al. 1985). The FLs generated by this primary fusion encapsulate the contents of the liposome and deliver them efficiently into cells by secondary fusion with the cell membrane (Nakanishi et al. 1998). Therefore, this type of FL satisfies the structural criteria of a hybrid vector. As these processes are independent from viral gene expression and from the method of liposome preparation, inactivated Sendai virus and liposome encapsulating plasmid DNA can be used to generate FL capable of delivering the encapsulated DNA into cells efficiently and harmlessly. If unilamellar liposomes with a diameter of 200–250 nm are used, the resultant FLs have a structure and function almost indistinguishable from those of intact Sendai virus particles (Fujita et al. submitted).

2.2 Application of Fusogenic Liposomes

Sendai-virus-based FL can deliver their contents within a few minutes after exposure to cells, both in vitro and in vivo, as the parent virus does (Mizuguchi et al. 1996). However, Sendai-virus-based delivery vehicles are not suitable for direct intravenous administration because they bind to erythrocytes and lyse them immediately after injection. Moreover, when we delivered regular DNA-based transcription units by FL, gene expression was both weak and transient due to the absence of nuclear transport machinery (Nakanishi et al. 2001). Therefore, these characteristics must be taken into account when designing applications of Sendai-virus-based FLs.

We have observed that FL functions most effectively as a vector for novel cancer gene therapy when used in combination with tumor-specific cytotoxic suicide genes (Horimoto et al. 2000). Conventional suicide genes used in current clinical trials are not toxic to cancer cells by themselves: they encode enzymes that metabolize non-toxic substrates (prodrugs) to toxic compounds (Morris et al. 1999). Therefore, following systemic administration of their substrates, local expression of these suicide genes in cancer cells can induce potent anti-cancer activity in situ. However, this two-step approach has resulted in only limited success in clinical trials, partly because retrovirus-mediated gene delivery to cancer cells is inefficient in vivo.

We constructed novel suicide genes by driving cDNA encoding potent toxic proteins, consisting of a mutant sodium channel (degenerin) and fragment A of diphtheria toxin, with a promoter of tumor-specific genes (α-fetoprotein and the catalytic subunit of telomerase) (Horimoto et al. 2000; Tsukinoki, et al. in preparation). Weak and transient expression of these suicide genes is sufficient for obtaining strong cytotoxicity in cancer cells. However, a meaningful anti-tumor effect requires that the genes be delivered into the majority of the target cancer cells in situ. This goal cannot achieved using current synthetic vector systems, and it is not feasible to insert these genes into viral vectors. Nonetheless, this deadlock could be overcome by using FLs as the delivery vehicle. For example, we succeeded in curing mice with gastric cancer disseminated in the peritoneum by delivering the suicide genes directly into the abdominal cavity with the aid of FLs (Horimoto et al. 2000). Unresectable peritoneal dissemination frequently appears in patients with end-stage cancer, and anti-cancer drugs are usually ineffective. Therefore, our approach may provide a means of treating patients with intractable cancers.

Other applications of FLs include cancer therapy by local expression of anti-tumor cytokines (Mizuguchi et al. 1998). Tumor-controlling arteries have unique characteristics compared to normal arteries, such as the higher sensitivity to TNF-α. This cytokine is cleared rapidly from the blood stream ($t_{1/2} = 3{\sim}5$ min), but systemic administration of high doses of TNF-α leads to severe side effects (Mizuguchi et al. 1998). Quick and efficient delivery of TNF-α expression vector by FLs resulted in the production of small amounts (40–600 pg) of this cytokine locally in the tumor-controlling arteries, which was sufficient for destroying them without side effects (Mizuguchi et al. 1998).

3 Peptide-Displaying Phage Particles: Delivery Vehicles Mimicking Non-envelope Viruses

Polyplex is the generic name of synthetic gene delivery systems consisting of non-lipidic polymer and nucleic acid (Felgner et al. 1999). Polyplex made of block copolymer has a compact structure resembling that of non-envelope viruses (see the chapter by K. Kataoka in this volume). However, the activity of these synthetic vectors depends mostly on their physical nature, and attempts to facilitate delivery by the addition of various transport-related peptides (e.g., the protein transduction domain and the nuclear localization signal) have yet to succeed.

One of the obstacles hampering the application of these peptides is that they are generally rich in basic amino acid residues (lysine and arginine) and thus may interfere with complex formation through electrostatic interaction with nucleic acid. Furthermore, this interaction prevents the peptides from being displayed on the surface of the complex. Therefore, a model system is needed that allows these peptides to be evaluated without affecting the structure of the complex.

This goal was achieved by using the peptide display system based on a recombinant bacteriophage lambda (Sternberg and Hoess 1995). In the following sections, basic information about phage-based peptide display systems is provided, and our recent results, obtained by displaying the protein transduction domain peptide and the NLS peptide, are described.

3.1 *Basic Science of Peptide Display Systems*

Phage-based peptide display systems have been used widely for screening peptides with affinity for various biological molecules. We applied this system for evaluating the role of transport-related peptides in gene delivery (Akuta et al. 2002; Eguchi et al. 2001). Among various display systems, the lambda phage-based one was chosen for several reasons. Firstly, lambda phage has a nucleocapsid with a diameter of 55 nm, resembling the structure of non-envelope animal viruses and of compact polyplex. Double-stranded, linear genomic DNA up to 50 kbp is packed tightly in an outer proteinaceous shell consisting of two major capsid proteins (E and D), and is protected completely from environmental nucleases. Secondly, lambda phage can display 405 copies of random peptides of up to 400 amino acid residues flexibly as N- or C-terminal chimera with D protein (Yang et al. 2000), without altering the size and structure of the phage particles. Thirdly, lambda phage can encapsulate expression cassettes of mammalian marker genes, such as the luciferase gene and the gene encoding green fluorescent protein (GFP), as a part of their genomic DNA. This allows measurement of the efficiency of gene transfer by using gene expression as an index (Akuta et al. 2002; Eguchi et al. 2001). Fourthly, large amount of information has been accumulated for genetic engineering and for large-scale production during decades of research experiences. Therefore, lambda phage particles satisfy the structural requirements for hybrid vectors and are a good model of polyplexes.

3.2 Role of the Protein Transduction Domain Peptide in Gene Delivery

The cell membrane is a tight barrier against various ions and hydrophilic molecules. Bacterial and plant protein toxins have hydrophobic domains in their primary structure, and these are essential for efficient delivery of their catalytic subunits into the cytoplasm. However, the utility of these hydrophobic peptides in assisting the membrane penetration of heterologous molecules has not yet been established.

Recently, several hydrophilic proteins were found to penetrate the cell membrane efficiently and to carry out their specific functions within the cell. The domains essential for membrane penetration (protein transduction domains, PTDs) were identified in the primary structures of these proteins (Derossi et al. 1998). Hydrophilic PTD peptides with abundant arginine residues facilitated the delivery of heterologous molecules across the cell membrane. However, their strong basic nature interferes with precise evaluation of their activity in polyplex-mediated gene delivery.

Using the lambda phage system, the effectiveness of the PTD peptide as an active component of the gene delivery vehicles were demonstrated for the first time (Eguchi et al. 2001). An 18-mer peptide derived from the Tat protein of human immunodeficiency virus (Tat peptide) was chosen as a candidate PTD peptide, because this peptide has been characterized in detail. Recombinant lambda phage particles displaying Tat peptide and encapsulating marker genes (Tat Phage) were prepared, and were found to induce expression of the marker genes following exposure to cultured mammalian cells in vitro (Eguchi et al. 2001) and to the hepatocytes in vivo (Nakanishi et al. 2003). The DNA did not associate with the Tat peptide directly, indicating that the PTD facilitates intracellular delivery of whole phage particles. The efficiency of gene delivery into sensitive COS cells was estimated to be about 1% of that of the animal virus simian virus 40 (SV40) (Eguchi et al. 2001). Since peptides with high affinity for the cell surface could not promote gene expression, and since endosomotropic reagents (chloroquine and monensin) did not affect Tat-peptide-mediated gene delivery, we concluded that the Tat peptide did not simply facilitate the passive binding and uptake of the phage particles. This notion was further supported by the observation that Tat-peptide-dependent gene delivery had the same characteristics (temperature dependency, effect of inhibitors, etc.) (Eguchi et al. 2001) as Tat-peptide-dependent protein delivery through macropinocytosis (Wadia et al. 2004).

As practical gene transfer vectors, however, the Tat-peptide-based system has several drawbacks. For example, Tat phages have to be exposed to the target cells for several hours in order to achieve sufficient gene expression; exposure shorter than 10 min resulted in gene expression below the detection limit (Eguchi et al. 2001). This is in contrast to virus- and FL-mediated gene delivery, which is accomplished within a few minutes, and is disadvantageous for in vivo applications. Moreover, cellular factor(s) determining the efficiency of Tat-mediated gene delivery, which varies significantly among different cell lines, have not yet been identified. Recently, we found that other PTD peptide derived from VP22 of herpes simplex virus also assisted the delivery of phage particles (Eguchi et al., in preparation). Interestingly, VP22 requires a specific orientation on the phage particle for efficient gene delivery, indicating that the role of PTD is not as a simple binding ligand through electrostatic adhesion (Eguchi et al., in preparation). Further investigation will reveal the optimal PTD peptide structure for gene delivery with the aid of the peptide display system.

3.3 Role of the Nuclear Localization Signal Peptide in Gene Delivery

As described in Sect. 1.1, the nuclear membrane is the second major barrier to gene delivery in mammalian cells. Overcoming this barrier is critical for gene delivery in vivo because breakdown of the nuclear membrane cannot be expected in arrested tissue cells. Most of the information about nuclear-cytoplasmic exchange has been obtained by using natural and recombinant proteins, and it is now established that karyophilic proteins are transported through the nuclear pore complex (NPC) by a signal-dependent active mechanism.

The NLS is a signal peptide responsible for the nuclear transport of proteins. It was initially identified as a domain essential for transporting karyophilic proteins, and was then found to also assist in the transport of heterologous molecules (cytoplasmic proteins, oligonucleotides, and gold particles) (Chook and Blobel 2001). The NLS is a domain bound to the cytoplasmic carrier protein importin, a molecular engine that drives the cargo into the nucleus. As DNA has no intrinsic ability to localize in the nucleus, several groups have reported the covalent- and non-covalent modification of plasmid DNA with NLS in order to achieve active nuclear targeting. However, the utility of this approach for vector construction remains unclear (Nakanishi et al. 2001; Sebestyen and Wolff 1999).

There are several problems in modifying plasmid DNA directly with NLS peptide (Nakanishi et al. 2001). Firstly, the size of the modified plasmid DNA is much larger than the inner diameter of the NPC. According to physical estimates, the inner diameter of the NPC is about 60 nm (Beck et al. 2004), while the size of plasmid DNA is 200 nm or larger even in the compact closed circular form. Secondly, modification of template DNA interferes with its transcription and replication. Thirdly, modified DNA is sensitive to environmental nucleases and is thus degraded rapidly in the cytoplasm. Moreover, there is no direct evidence that the copy number and structure of the NLS peptides used in these studies had been optimized for transporting large DNA molecules.

We succeeded in resolving these problems by displaying NLS peptide on lambda phage particles. Most importantly, the size and structure of the phage head (icosahedral capsid, 50–55 nm in diameter) is almost identical to that of some karyophilic non-envelope viruses (polyomavirus and papillomavirus), which might pass through the NPC as an intact particle. We prepared phage particles displaying SVLT32 (NLS peptide derived from SV40 T antigen) (NLS-phage) and demonstrated that intact phage particles accumulated in the nucleus within 30 min after injection into the cytoplasm, both by biochemical assay (Akuta et al. 2002) and by electron microscopy (Eguchi et al., 2005). This accumulation is sensitive to wheat germ agglutinin (WGA), an inhibitor of the NPC-mediated transport of karyophilic proteins, indicating that the phage particles are transported through the NPC (Akuta et al. 2002). In these experiments, it was estimated that about 2% of the phage particles were transported into the nucleus. This is the first demonstration of the delivery of unmodified, intact DNA (40 kbp) into the nucleus with the aid of NLS peptide. More recently, we found that further purification of NLS-phage particles dramatically enhances nuclear transport (Eguchi et al., in preparation).

Based on the results of experiments using NLS-phage, it was also found that optimization of the NLS is critical for assisting the transport of nano-sized particles (Eguchi et al., 2005). There is some evidence that various NLS peptides have different activities with respect to assisting the transport of heterologous macromolecules. For example, minimum NLS (PKKKRKV) and SVLT32 (32-mer peptide comprising minimum NLS and additional amino acid residues) have different capacities to assist in the transport of IgM (molecular mass = 900 kDa): SVLT32 transported IgM into the nucleus whereas minimum NLS did not (Yoneda et al. 1992). We recently identified the structures of SVLT32 responsible for assisting in the transport of phage particles by displaying variant NLS peptides, and found that the minimum NLS was necessary but not sufficient for the transport of nano-sized particles (Eguchi et al., 2005). Interestingly, SVLT32 and minimum NLS are indistinguishable in the binding affinity for importin-α and in their ability to assist in the transport of smaller proteins (Eguchi et al., 2005). Therefore, NLS peptides to be inserted into polyplexes should be designed by displaying them on model particles.

4 Construction of Stable Replicons

Using the information in the previous sections, a rough sketch of the ideal delivery vehicle can be drawn. The FL is one of the most efficient tools for delivering the across the cell membrane in vivo. Nano-sized particles encapsulating DNA can move through the nuclear membrane when an optimized NLS is displayed on the surface. Now we are ready to design the genetic platform to be inserted into these vehicles.

In this section, the current status of the development of genetic platforms is discussed with the aim of providing a model for use in this field. In contrast to disposable delivery vehicles, a stable genetic platform is needed, one that is based on natural models of stable genetic platforms in mammalian cells: chromosomes and viral genomes. Current researches focus on uncovering the principles determining the stability of these model platforms, and on reconstructing the novel platform based upon these principles. In the following section, I briefly introduce two different types of genetic platforms: DNA-based and RNA-based replicons. The replicon is a term used to represent a single replication unit of genetic materials, and I use it as an independent genetic platform into which therapeutic genes are installed.

4.1 *Stable DNA Replicon*

The chromosome is a representative DNA replicon in mammalian cells. Human chromosomes contain linear ds DNA molecules (300–30 mega base-pairs), and consist of several functional units: the telomere (terminal structure of chromosome), centromere (structure required for even separation of the replicated chromosomes to the daughter cells), and replication origin (autonomous replication sequence, ARS). The DNA fragments corresponding to these functional units have been cloned by using microorganisms (*E. coli* and yeast), while the structure was defined precisely in relation to its function only in the telomere: repetitive DNA $(TTAGGG)_n$ is required and sufficient for reconstructing functional telomeres in mammalian cells (Hanish et al. 1994; Okabe et al. 2000).

By jointing these cloned DNA fragments, successful reconstruction of human artificial chromosomes was reported (Harrington et al. 1997; Ikeno et al. 1998; Katoh et al. 2004). However, the resultant artificial chromosomes contained the DNA of approximately 2,000 kbp to 320 kbp, and no shorter artificial chromosome has been generated yet. These results suggest that there is a lower limit to the size of linear chromosomal DNA required for stable maintenance, as was found in linear yeast artificial chromosomes (Wada et al. 1994). The size of the smallest linear artificial chromosome (320 kbp) is still much larger than the estimated upper limit (50 kbp) for encapsulation into nano-sized particles (see Sect. 3). Furthermore, our recent results showed that the mechanism to stabilize the telomere (and hence the chromosomes) was suppressed in normal tissue cells with limited proliferating capacity (Okabe et al. 2004). Therefore, the linear DNA molecule may not be a good candidate for a stable genetic platform that can be inserted into the hybrid vector.

Circular genome DNA of the Epstein-Barr virus (EBV) and of human papillomavirus (HPV) is maintained stably in the nucleus during latent infection. These circular genomes have their own replication origins, and replicate themselves dependent on host DNA polymerase with the aid of virus-encoded proteins. However, several unique repetitive sequences essential for chromosome stability are not found in the circular viral genomes. Therefore, these viruses should have unique mechanisms for maintaining their genome stably in the nucleus.

EBV requires a single virus protein, EBNA1, both for DNA replication and for stable genome maintenance (Sun et al. 1994; Yates et al. 1985). As the majority of adults are infected with EBV latently, without any symptoms, the EBV replicon has been considered as a candidate platform for human gene therapy (Sun et al. 1994). However, EBNA1 may have oncogenic activity (Tsimbouri et al. 2002), and this safety concern has limited the utility of the EBV-based replicon as a practical vector component. HPV is another platform candidate but the mechanism of genome maintenance is obscure due to the absence of appropriate experimental systems in cultured cells. Recently, bovine papillomavirus has been found to use a unique strategy for stabilizing the genome: it sticks to the genome on the host chromosome through interaction between viral E2 protein and some host protein (You et al. 2004). Considering the size limit required for the DNA platform, this parasitic strategy using a circular DNA replicon will be more practical than a human linear artificial chromosome based on natural chromosome function.

4.2 Stable RNA Replicon

A number of animal viruses use RNA as a platform for carrying their genetic information. However, only a few (retrovirus, lentivirus and paramyxovirus) can establish persistent infection, a mode of virus infection by which viruses coexist with host cells while maintaining their genome stably and harmlessly in the cells. Retrovirus and lentivirus convert their RNA genome into DNA by reverse transcriptase, and stabilize this cDNA by integrating into the host chromosome. Paramyxovirus, by contrast, lyses the host cells in the normal replication cycle, but infects the cells persistently under special circumstances: either by mutations in the viral genomes or through the coexistence of a defective interfering viral genome (Eguchi et al. 2000). Paramyxoviruses use no DNA intermediate in their replication; instead, their genome RNA is copied directly by virus-encoded RNA-dependent RNA polymerase. No recombination was detected between genomic RNA molecules or between genomic RNA and host chromosomal DNA (Pringle 1991). Cells infected with paramyxovirus persistently carry the viral RNA genome stably in the cytoplasm.

Paramyxovirus is a large virus family that includes human and animal pathogens (mumps virus, measles virus, human parainfluenzavirus and Newcastle disease virus). Sendai virus, classified as mouse parainfluenza virus type 1, is considered to be a proto-type of this family. The genome of Sendai virus is a single negative-strand RNA with a length of 15,384 nucleotides, encoding eight proteins. Since the finding that recombinant Sendai viruses could be reconstructed from cDNA (Kato et al. 1996), this virus has provided a useful tool for gene expression in vitro and in vivo (Li et al. 2000). However, current Sendai virus vectors based on the wild-type Z strain can induce foreign gene expression only transiently (Li et al. 2000).

There are several mutant Sendai virus strains capable of infecting cells persistently, but the genetic background underlying this phenotype has not been examined in detail. We succeeded in identifying the genetic background leading to persistent infection caused by mutant Sendai virus cl.151 strain (Kondo et al. 1993; Nishimura and Nakanishi, in preparation), originally isolated as a temperature-sensitive mutant (Yoshida et al. 1979). Furthermore, based on these results, we succeeded in expressing marker genes stably in cultured cells from the RNA replicon (Nishimura and Nakanishi, in preparation). Importantly, this stable RNA replicon was purified from the cytoplasm as a functional nucleocapsid (RNA-protein complex), and was delivered to uninfected cells with the aid of commercial delivery agents (cationic liposomes and peptides) (Nishimura and Nakanishi, in preparation). Therefore, apart from its application as a recombinant RNA vector, this stable RNA replicon may be a good candidate genetic platform for use in hybrid vectors.

5 Future Perspectives: Towards the Development of Ideal Hybrid Vectors

In this section, the current status of the development of hybrid vectors is described. As mentioned in the Introduction, the goal is to combine ideal delivery vehicles with stable genetic platform. We have developed FLs and peptide-displaying phage particles as model systems mimicking the function and structure of animal viruses. The next challenge is to reconstruct these model vehicles from synthetic materials. This requires a more precise understanding of the molecular mechanisms underlying the delivery phenomenon.

Development of a stable genetic platform still remains an enormous challenge. We developed a stable RNA replicon, but its utility in tissue cells of living animals remains to be examined. Furthermore, gene expression from the RNA replicon cannot be controlled in response to environmental stimuli, in contrast to DNA-based transcription units. Thus, the development of a stable DNA replicon is an absolute prerequisite for the sophisticated control of gene expression. Our investigation of delivery vehicles has provided important information on this issue, namely, that the size of DNA which can be inserted into active delivery vehicles should be 50 kbp or less. Therefore, a stable DNA replicon must be developed within this restriction.

Regarding the DNA replicon, the lambda phage particle may not be merely a model of the polyplex but a candidate practical delivery vehicle. In addition to the functions revealed by displaying various peptides, as described in Sect. 3, this phage can encapsulate DNA actively by using a short packaging signal named *Cos* (cohesive end site) (Feiss et al. 1983), as in animal viruses. Moreover, by utilizing the *Cos* sequence, the nano-sized particles can be prepared in which most of the phage genomic DNA is replaced with human DNA. This may be a quite important characteristic for fully utilizing the limited capacity of nano-sized particles for a genetic platform.

The development of a hybrid vector is a challenging project and may require a long period of time to be completed. Successful application of these vectors will benefit patients with various incurable diseases.

References

Akuta T, Eguchi A, Okuyama H, Senda T, Inokuchi H, Suzuki Y, Nagoshi E, Mizuguchi H, Hayakawa T, Takeda K, Hasegawa M, Nakanishi M (2002) Enhancement of phage-mediated gene transfer by nuclear localization signal. Biochem Biophys Res Commun 297:779–786

Barron LG, Szoka FC (1999) The perplexing delivery mechanism of lipoplexes. In: Huang L, Hung M-C, Wagner E, eds. Nonviral vectors for gene therapy, Academic Press, San Diego CA, pp 229–266

Beck M, Forster F, Ecke M, Plitzko JM, Melchior F, Gerisch G, Baumeister W, Medalia O (2004) Nuclear pore complex structure and dynamics revealed by cryoelectron tomography. Science 306:1387–1390

Chook YM, Blobel G (2001) Karyopherins and nuclear import. Curr Opin Struct Biol 11: 703–715

Derossi D, Chassaing G, Prochiantz A (1998) Trojan peptides: The penetratin system for intracellular delivery. Trends Cell Biol 8:84–87

Dworetzky SI, Lanford RE, Feldherr CM (1988) The effects of variations in the number and sequence of targeting signals on nuclear uptake. J Cell Biol 107:1279–1287

Earnshaw WC, Casjens SR (1980) DNA packaging by the double-stranded DNA bacteriophages. Cell 21:319–331

Eguchi A, Akuta T, Okuyama H, Senda T, Yokoi H, Inokuchi H, Fujita S, Hayakawa T, Takeda K, Hasegawa M, Nakanishi M (2001) Protein transduction domain of HIV-1 Tat protein promotes efficient delivery of DNA into mammalian cells. J Biol Chem 276:26204–26210

Eguchi A, Furusawa H, Yamamoto A, Akuta T, Hasegawa M, Okahata Y, Nakanishi M (2005) Optimization of Nuclear Localization Signal for Nuclear Transport of DNA-encapsulating Particles. J Cont Rel, in press.

Eguchi A, Kondoh T, Kosaka H, Suzuki T, Momota H, Masago A, Yoshida T, Taira H, Ishii-Watabe A, Okabe J, Hu J, Miura N, Ueda S, Suzuki Y, Taki T, Hayakawa T, Nakanishi M (2000) Identification and characterization of cell lines with a defect in a post-adsorption stage of Sendai virus-mediated membrane fusion. J Biol Chem 275:17549–17555

Feiss M, Widner W, Miller G, Johnson G, Christiansen S (1983) Structure of the bacteriophage lambda cohesive end site: Location of the sites of terminase binding (*cos*b) and nicking (*cos*n). Gene 24:207–218

Felgner PL, Zelphati O, Liang X (1999) Advances in synthetic gene-delivery system technology. In: Friedman T, ed. The development of human gene therapy, Cold Spring Harbor Laboratory Press, Cold Spring Harbor NY, pp 241–260

Hacein-Bey-Abina S, Von Kalle C, Schmidt M, McCormack MP, Wulffraat N, Leboulch P, Lim A, Osborne CS, Pawliuk R, Morillon E, Sorensen R, Forster A, Fraser P, Cohen JI, de Saint Basile G, Alexander I, Wintergerst U, Frebourg T, Aurias A, Stoppa-Lyonnet D, Romana S, Radford-Weiss I, Gross F, Valensi F, Delabesse E, Macintyre E, Sigaux F, Soulier J, Leiva LE, Wissler M, Prinz C, Rabbitts TH, Le Deist F, Fischer A, Cavazzana-Calvo M (2003) *LMO2*-associated clonal T cell proliferation in two patients after gene therapy for SCID-X1. Science 302:415–419

Hanish JP, Yanowitz JL, de Lange T (1994) Stringent sequence requirements for the formation of human telomeres. Proc Natl Acad Sci U S A 91:8861–8865

Harrington JJ, Van Bokkelen G, Mays RW, Gustashaw K, Willard HF (1997) Formation of de novo centromeres and construction of first-generation human artificial microchromosomes. Nat Genet 15:345–355

Higashi N, Yamauchi M, Okumura Y, Nakanishi M, Sunamoto J (1996) Fusion between jurkat cell and PEO-lipid modified liposome. Biochim Biophys Acta 1285:183–191

Horimoto M, Sasaki Y, Ugawa S, Wada S, Toyama T, Iyoda K, Yakushijin T, Minami Y, Ito T, Hijioka T, Eguchi A, Nakanishi M, Shimada S, Tohyama M, Hayashi N, Hori M (2000) A novel strategy for cancer therapy by mutated mammalian degenerin gene transfer. Cancer Gene Ther 7:1341–1347

Ikeno M, Grimes B, Okazaki T, Nakano M, Saitoh K, Hoshino H, McGill NI, Cooke H, Masumoto H (1998) Construction of YAC-based mammalian artificial chromosomes. Nat Biotechnol 16:431–439

Kato A, Sakai Y, Shioda T, Kondo T, Nakanishi M, Nagai Y (1996) Initiation of Sendai virus multiplication from transfected cDNA or RNA with negative or positive sense. Genes Cells 1:569–579

Kato K, Nakanishi M, Kaneda Y, Uchida T, Okada Y (1991) Expression of hepatitis B virus surface antigen in adult rat liver. Co-introduction of DNA and nuclear protein by a simplified liposome method. J Biol Chem 266:3361–3364

Katoh M, Ayabe F, Norikane S, Okada T, Masumoto H, Horike S, Shirayoshi Y, Oshimura M (2004) Construction of a novel human artificial chromosome vector for gene delivery. Biochem Biophys Res Commun 321:280–290

Kondo T, Yoshida T, Miura N, Nakanishi M (1993) Temperature-sensitive phenotype of a mutant Sendai virus strain is caused by its insufficient accumulation of the M protein. J Biol Chem 268:21924–21930

Li HO, Zhu YF, Asakawa M, Kuma H, Hirata T, Ueda Y, Lee YS, Fukumura M, Iida A, Kato A, Nagai Y, Hasegawa M (2000) A cytoplasmic RNA vector derived from nontransmissible Sendai virus with efficient gene transfer and expression. J Virol 74:6564–6569

Mizuguchi H, Nakagawa T, Nakanishi M, Imazu S, Nakagawa S, Mayumi T (1996) Efficient gene transfer into mammalian cells using fusogenic liposome. Biochem Biophys Res Commun 218:402–407

Mizuguchi H, Nakagawa T, Toyosawa S, Nakanishi M, Imazu S, Nakanishi T, Tsutsumi Y, Nakagawa S, Hayakawa T, Ijuhin N, Mayumi T (1998) Tumor necrosis factor alpha-mediated tumor regression by the in vivo transfer of genes into the artery that leads to tumors. Cancer Res 58:5725–5730

Morris JC, Touraine R, Wildner O, Blease RM (1999) Suicide genes: Gene therapy applications using enzyme/prodrug strategies. In: Friedman T, ed. The development of human gene therapy, Cold Spring Harbor Laboratory Press, Cold Spring Harbor NY, pp 477–526

Nakanishi M, Akuta T, Nagoshi E, Eguchi A, Mizuguchi H, Senda T (2001) Nuclear targeting of DNA. Eur J Pharm Sci 13:17–24

Nakanishi M, Eguchi A, Akuta T, Nagoshi E, Fujita S, Okabe J, Senda T, Hasegawa M (2003) Basic peptides as functional components of non-viral gene transfer vehicles. Curr Protein Pept Sci 4:141–150

Nakanishi M, Mizuguchia H, Ashihara K, Senda T, Akuta T, Okabe J, Nagoshi E, Masago A, Eguchi A, Suzuki Y, Inokuchi H, Watabe A, Ueda S, Hayakawa T, Mayumi T (1998) Gene transfer vectors based on Sendai virus. J Controlled Release 54:61–68

Nakanishi M, Uchida T, Sugawa H, Ishiura M, Okada Y (1985) Efficient introduction of contents of liposomes into cells using HVJ (Sendai virus). Exp Cell Res 159:399–409

Okabe J, Eguchi A, Masago A, Hayakawa T, Nakanishi M (2000) TRF1 is a critical trans-acting factor required for de novo telomere formation in human cells. Hum Mol Genet 9:2639–2650

Okabe J, Eguchi A, Wadhwa R, Rakwal R, Tsukinoki R, Hayakawa T, Nakanishi M (2004) Limited capacity of the nuclear matrix to bind telomere repeat binding factor TRF1 may restrict the proliferation of mortal human fibroblasts. Hum Mol Genet 13:285–293

Otto E, Jones-Trower A, Vanin EF, Stambaugh K, Mueller SN, Anderson WF, McGarrity GJ (1994) Characterization of a replication-competent retrovirus resulting from recombination of packaging and vector sequences. Hum Gene Ther 5:567–575

Pante N, Kann M (2002) Nuclear pore complex is able to transport macromolecules with diameters of about 39 nm. Mol Biol Cell 13:425–434

Pringle CR (1991) The genetics of paramyxoviruses. In: Kingsbury DW, ed. The paramyxovirus, Plenum Press, New York NY, pp 1–39

Sebestyen MG, Wolff JA (1999) Nuclear transport of exogenous DNA. In: Huang L, Hung M-C, Wagner E (Eds) Nonviral vectors for gene therapy, Academic, San Diego CA, pp 139–169

Sternberg N, Hoess RH (1995) Display of peptides and proteins on the surface of bacteriophage lambda. Proc Natl Acad Sci U S A 92:1609–1613

Sun TQ, Fernstermacher DA, Vos JM (1994) Human artificial episomal chromosomes for cloning large DNA fragments in human cells. Nat Genet 8:33–41

Tsimbouri P, Drotar ME, Coy JL, Wilson JB (2002) Bcl-Xl and RAG genes are induced and the response to IL-2 enhanced in emuEBNA-1 transgenic mouse lymphocytes. Oncogene 21:5182–5187

Uchida T, Kim J, Yamaizumi M, Miyake Y, Okada Y (1979) Reconstitution of lipid vesicles associated with HVJ (Sendai virus) spikes. Purification and some properties of vesicles containing nontoxic fragment a of diphtheria toxin. J Cell Biol 80:10–20

Wada M, Kohno K, Kano Y, Mitsui H, Kuwano M, Schlessinger D, Imamoto F (1994) Increased retention of small chromosomes in a novel yeast mutant. J Biochem (Tokyo) 116:694–696

Wadia JS, Stan RV, Dowdy SF (2004) Transducible TAT-HA fusogenic peptide enhances escape of TAT-fusion proteins after lipid raft macropinocytosis. Nat Med 10:310–315

Whittaker GR, Kann M, Helenius A (2000) Viral entry into the nucleus. Annu Rev Cell Dev Biol 16:627–651

Yang F, Forrer P, Dauter Z, Conway JF, Cheng N, Cerritelli ME, Steven AC, Pluckthun A, Wlodawer A (2000) Novel fold and capsid-binding properties of the lambda-phage display platform protein gpD. Nat Struct Biol 7:230–237

Yates JL, Warren N, Sugden B (1985) Stable replication of plasmids derived from Epstein-Barr virus in various mammalian cells. Nature 313:812–815

Yee J-K (1999) Retrovirus vectors. In: Friedman T, ed. The development of human gene therapy, Cold Spring Harbor Laboratory Press, Cold Spring Harbor NY, pp 21–45

Yoneda Y, Semba T, Kaneda Y, Noble RL, Matsuoka Y, Kurihara T, Okada Y, Imamoto N (1992) A long synthetic peptide containing a nuclear localization signal and its flanking

sequences of SV40 T-antigen directs the transport of IgM into the nucleus efficiently. Exp Cell Res 201:313–320
Yoshida T, Nagai Y, Maeno K, Iinuma M, Hamaguchi M, Matsumoto T, Nagayoshi S, Hoshino M (1979) Studies on the role of M protein in virus assembly using a ts mutant of HVJ (Sendai virus). Virology 92:139–154
You J, Croyle JL, Nishimura A, Ozato K, Howley PM (2004) Interaction of the bovine papillomavirus E2 protein with brd4 tethers the viral DNA to host mitotic chromosomes. Cell 117:349–360

2. Controlled Gene Delivery

Pharmacokinetics of Gene Delivery in Cells

Hidetaka Akita[1,2], Ikramy A. Khalil[1,2], Kentaro Kogure[2], and Hideyoshi Harashima[1,2]

1 Introduction

1.1 Intracellular Barriers to Gene Delivery

Since the first report, in the 1980s, of gene delivery with a cationic liposome, numerous attempts have been made to improve non-viral gene vectors. It has gradually become clear that transgene expression can be interrupted by several intracellular barriers (Fig. 1). The cellular association of naked DNA molecules is very poor, since negative charges on both the cell surface and the DNA molecules interrupt contact with each other via electrostatic interactions. Thus, in order to enhance cellular association, DNA was condensed with cationic polymers (Oupicky et al. 2000; Brown et al. 2001) and cationic liposomes (Li et al. 1999; Tranchant et al. 2004) that neutralize the effect of the negative charge. A cationic vector enhances cell-surface binding through interactions with the negative constituents of the cell surface (e.g. heparan sulfate proteoglycans) or through selective binding to specific receptors, resulting in a strong transgene expression. This method of condensation also enables targeting of the cells by modulating different ligands to the surfaces of the gene vectors (Table 1).

Currently, several ligands are used for recognition by their specific receptors and uptake via cellular receptor-mediated endocytosis. In this pathway, however, lysosomal degradation presents a second barrier for gene vectors. Before the endosomal membrane fuses with the lysosome, the gene vectors must be released into the cytosol. The importance of endosomal escape is clearly evidenced by the drastically enhanced transfection efficiency in the presence of lysosomotropic reagents, such as chloroquine (Erbacher et al. 1996), which accumulate in the acidic lysosome and destabilize the membrane by causing it to swell. Various devices, such as pH-sensitive fusiogenic lipids (Farhood et al. 1995; Hui et al. 1996; Harashima et al. 2001), polycations that have proton sponge ability (Boussif et al. 1995; Kichler et al. 2001), and pH-sensitive

[1] Graduate School of Pharmaceutical Sciences, Hokkaido University, Kita 12, Nishi 6, Kita-ku, Sapporo 060-0812, Japan

[2] CREST, Japan Science and Technology Agency (JST), Laboratory for Molecular Design of Pharmaceutics, Graduate School of Pharmaceutical Sciences, Hokkaido University, Kita 12 Nishi 6, Kita-ku, Sapporo 060-0812, Japan

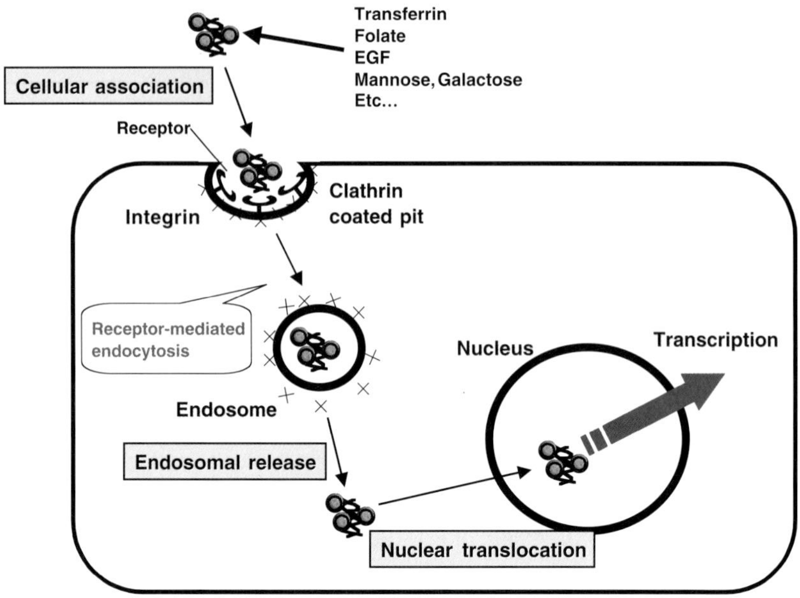

FIG. 1. The commonly accepted intracellular trafficking pathways of non-viral gene vectors. It is generally accepted that endocytosis is a major pathway for cellular entry. After internalization, the gene vector must be released to the cytosol to escape lysosomal degradation. Finally, the vector must be translocated through the nuclear membrane for transcription. In order to overcome these barriers, it is necessary to develop functional devices and combine them in such a way so as to exhibit all of the functions at the appropriate site and with realistic timing

TABLE 1. Various functional devices for overcoming cellular barriers

Barriers	Plasma membrane	Endosomal membrane	Nuclear membrane
Functional devices	Specific ligands (transferrin, insulin, EGF, etc.)	pH-sensitive peptides (GALA, HA2, etc.)	NLS peptides (mu, protamine, etc.)
	PTD peptides (octaarginine, Tat, etc.)	pH sensitive lipids (CHEMS, DOPE, etc).	PTD peptides (octaarginine)
	Cationic lipids (DOTMA, DOTAP, DOSPA, etc.)	Proton sponge polymers (PEI)	
		PTD peptides (octaarginine)	

membrane lytic peptides (Wagner et al. 1992; Plank et al. 1994; Wagner 1998) can also aid in endosomal escape (Table 1), as described below.

The final barrier to the intracellular trafficking of DNA is the nuclear membrane. Since the size threshold for freely passing through the nuclear pore complex is 50 kDa, commonly used plasmid DNA (pDNA) is too large to pass, unassisted, through the nuclear-pore complexes (Lycke et al. 1988; Hagstrom et al. 1997; Sodeik et al. 1997;

Sebestyen et al. 1998; Branden et al. 1999; Ludtke et al. 1999; Wilson et al. 1999). Therefore, plasmid DNA is thought to enter into the nucleus when the nuclear membrane structure disappears during the M-phase of mitosis (Wilke et al. 1996; Marenzi et al. 1999; Mortimer et al. 1999; Tseng et al. 1999). In fact, the percent of plasmid DNA, reaching the nucleus has been reported to be less than 1% of the total cytoplasmically microinjected DNA (Pollard et al. 1998). Therefore, in order to improve entry into the nucleus, nuclear localization signal peptides (NLS) are attached either to the plasmid DNA itself (Sebestyen et al. 1998; Zanta et al. 1999; Nagasaki et al. 2003; Tanimoto et al. 2003) or to counter-polycations (Chan and Jans 1999; Chan et al. 2000). Various reports on the use of NLS peptides as polycations have recently been published (Li and Huang 1997; Li et al. 1998; Murray et al. 2001; Rittner et al. 2002; Tagawa et al. 2002; Keller et al. 2003; Rudolph et al. 2003). Although these approaches succeed in enhancing transgene expression, the technology for delivering plasmid DNA to the nucleus in non-dividing cells needs further development.

1.2 Topics in This Chapter

In order to develop efficient gene vectors, new devices to overcome the intracellular barriers discussed above must be designed. Furthermore, these devices should be assembled into one vector, taking the topology of these devices into consideration so that their functions are utilized at the appropriate time and position. In this chapter, we first introduce the concept of Programmed Packaging; a novel packaging concept for a non-viral gene delivery system. We then describe our recent development of a Multifunctional Envelope-type Nano-Device (MEND) for use in either the endocytic or non-endocytic pathways.

In order to optimize gene delivery, it is necessary to quantitatively evaluate the function of the devices in cells. However, quantitative information regarding, e.g., endosomal release and nuclear translocation, are lacking. We recently succeeded in addressing this problem by developing a three-dimensional quantitative analysis for the intracellular trafficking of plasmid DNA delivered with non-viral vectors. The importance of the pharmacokinetic considerations of intracellular trafficking of DNA based on this methodology are discussed below.

2 A New Packaging Concept for Non-viral Vectors for Efficient Gene Delivery: Programmed Packaging

Various non-viral systems have been developed to date and can be classified as follows. The first generation consists of DNA/cation complexes, in which DNA is simply mixed with positively charged molecules, such as cationic polymers (Kawai et al. 1984; Boussif et al. 1995; Tang et al. 1997), cationic detergents (Blessing et al. 1998) and cationic liposomes (Felgner et al. 1987; Felgner et al. 1994; Sorgi et al. 1997; Gao et al. 1996), to form complexes via electrostatic interactions. For example, a complex consisting of DNA and polyethylenimine (PEI), which is known to function as a proton sponge, had significantly high transgene activity in vitro (Boussif et al. 1995). However, such complexes easily aggregate with serum proteins and/or red blood cells in blood and accumulate in the lung capillary (Lee et al. 1996). This problem was overcome by a second generation of gene delivery systems, in which plasmid DNA encapsulated in lipid vesicles and/or a PEI/DNA complex were covered with polyethylene glycol (PEG) in order to avoid interactions with blood components (Lee et al. 1996; Ogris et al. 1999; Wheeler et al. 1999). Furthermore, these second-generation gene delivery systems can be equipped with specific targeting ligands, such as transferrin and folate at the PEG terminal (Lee et al. 1996; Ogris et al. 1999; Kircheis et al. 2001), which prolongs the circulation time in the blood (Wheeler et al. 1999) and allows selective tumor targeting (Ogris et al. 1999; Kircheis et al. 2001; Tam et al. 2000). However, while these vectors can reach the target cells in vivo, they cannot efficiently escape from the endosomes to the nucleus (Wheeler et al. 1999; Kircheis et al. 2001; Tam et al. 2000).

For efficient gene delivery, the advantages of both generations need to be integrated into a packaged delivery system. Moreover, intracellular trafficking of the delivery system should be controllable with the devices introduced. However, it is not easy to assemble such devices in a manner in which each device can function appropriately. Therefore, we have proposed a new concept, Programmed Packaging, to represent the next (third) generation of gene delivery systems. In Programmed Packaging, functional devices are able to overcome the barriers posed by cells at the correct time and at the appropriate place in order to control the pharmacokinetics (PK) and pharmacodynamics (PD) of a gene delivery system. Programmed Packaging consists of three features: (1) a strategy to overcome the barriers; (2) the design of functional devices to overcome the barriers and their assignments; and (3) the assembly of these components into a nano structure. We recently developed a novel non-viral gene delivery system, which we named a "Multifunctional Envelope-type Nano Device (MEND)" (Kamiya et al. 2003; Kogure et al. 2004). The ideal scheme of MEND is shown in Fig. 2. The complete system consists of a condensed DNA core and a lipid envelope structure equipped with various functional devices. The condensed core has several advantages, such as the protection of DNA from DNase, size control and enhanced packaging efficiency. Furthermore, the core-envelope structure confers the ability to control the topology of the functional devices. The NLS should be located on the surface of the core particles for targeting the nucleus and the protein transduction domain (PTD) peptide should be located on the envelope in order to control the entrance pathway. In addition, a pH-sensitive fusiogenic peptide should be incorporated on the envelope for enhanced endosomal escape (Kakudo et al. 2004). Fourth-generation gene delivery systems might thus be represented by an "artificial-intelligent nano-device (AIND)." The AIND would be equipped with arms to kill the virus and a system to convert chemical energy to kinetic energy; which can be denoted a "nano-machine."

3 Control of the Intracellular Trafficking of Genes

3.1 Receptor-mediated Endocytosis: Enhanced Endosomal Escape with Cholesterol GALA Peptide

It is important to select the appropriate entrance route in order to optimize intracellular trafficking, since the intracellular fate of internalized DNA with the carrier system is determined principally by a constitutive map of intracellular trafficking of

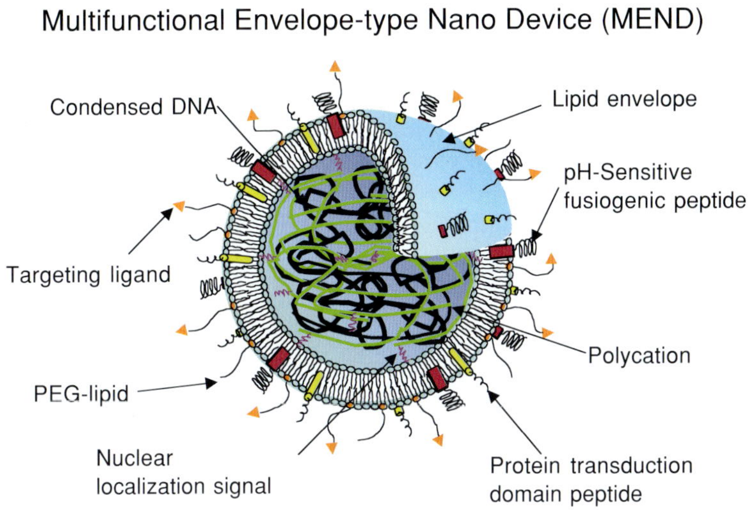

FIG. 2. Ideal multifunctional envelope-type nano device (MEND). MEND was designed to have multifunctional devices, such as specific ligands, for selective cellular delivery, PEG for prolonged circulation in the blood, a pH-sensitive fusiogenic peptide for endosomal escape after receptor-mediated endoscytosis, protein transduction domains (PTDs) for efficient cellular uptake, and a nuclear localization signal for efficient nuclear translocation. MEND also has a specific mechanism for enhancing intranuclear transcriptional efficiency, thus maximizing the efficacy of gene expression

the endogenous substance. In general, conventional endosomes are able to fuse with lysosomes, where DNA/proteins are degraded. Therefore, it is essential for the device to escape from endosomes/lysosomes prior to degradation. The utilization of receptor-mediated endocytosis would be expected to function as a pathway for the intracellular delivery of anticancer drugs in cancer chemotherapy as well as nucleic acids in gene therapy (Harashima et al. 2001). However, once endocytosed, the control of intracellular trafficking is difficult, since it is under the regulation of the cell. Therefore, in order to optimize intracellular trafficking after receptor-mediated endocytosis to a target organelle, such as the cytosol and the nucleus, artificial sorting devices are required (Kamiya et al. 2001). Liposomes are one of the most promising systems for selective cellular targeting by introducing specific ligands for cell-surface receptors.

We have succeeded in elucidating the intracellular fate of transferrin-modified liposomes and in altering it by introducing the pH-sensitive fusiogenic peptide, GALA (WEAALAEALAEALAEHLAEALAEALEALAA). Transferrins that are chemically attached to the liposomal surface (Tf-L) are internalized via receptor-mediated endocytosis more slowly than unmodified transferrins (Kakudo et al. 2004). In contrast to the recyclable nature of transferrin, liposome-attached transferrins, together with encapsulated rhodamines, were retained in the vesicular compartments.

The intracellular fate of Tf-containing Chol-GALA was analyzed by confocal laser scanning microscopy (CLSM) using K562 cells. As shown in Fig. 3c, a dramatic cytosolic release of encapsulated sulfo-rhodamine (S-Rh) was observed, while S-Rh in Tf-L with encapsulated GALA was not released (Fig. 3b). These results suggest that the endosomal release of S-Rh proceeded quite efficiently with the aid of the GALA present on the liposome surface. It is noteworthy that, in the case of GALA-encapsulated Tf-L, Tf was entrapped in the endosomes/lysosomes, while in the case of Chol-GALA, Tf was recycled and remained on the plasma membranes (Fig. 3c, Fig. 4). Therefore,

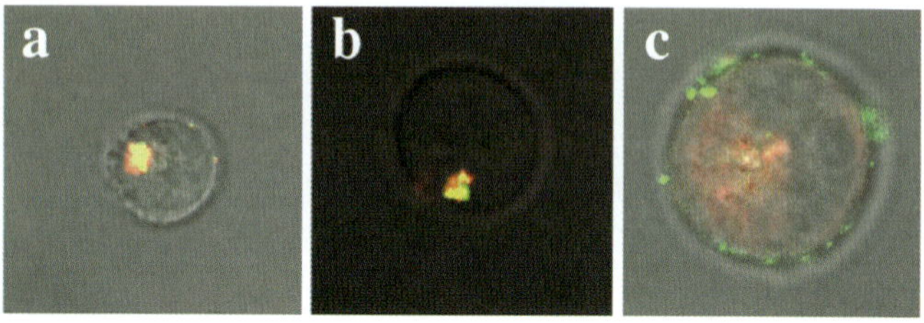

FIG. 3. **a** Confocal laser scanning microscopy (CLSM) of Tf-L, **b** Tf-L with GALA, and **c** Tf-L with Chol-GALA. FITC-labeled Tf-modified liposomes encapsulating sulforhodamine as an aqueous-phase marker were administered to K562 cells and analyzed by CLSM after 18 h

FIG. 4. Scheme for the intracellular trafficking of Tf-L with Chol-GALA. Tf-L were internalized by Tf-receptor-mediated endocytosis. In an endosome, the pH decreases from 7.4 to ~6.0, which induces a conformational change in Chol-GALA that enhances membrane fusion between the endosomal liposomal membranes. The aqueous marker is released to the cytosol and liposomally attached Tf is distributed to the endosomal membrane. After recycling of the endosome, the Tf remains on the plasma membrane, since the Tf originally introduced to the liposomal membrane is transferred to the endosomal membrane by fusion via Chol-GALA, then fused with the plasma membrane

Chol-GALA significantly affected the intracellular fate of Tf-L. These results suggest that the topology of GALA, namely, its surface disposition on the liposomal membrane, is a critical factor in these effects.

GALA was designed to preferentially interact with neutral bilayers at a low pH, considering factors such as the hydrophobicity of the residues, the conformational preference of the amino acids, the length of the peptides, and the topology of the residues on the peptide (Subbarao et al. 1987; Nir and Neiva 2000). The peptide contains 30 amino acids, with a repeated sequence of Glu-Ala-Leu-Ala (GALA). When the pH of the solution is decreased, protonation of the Glu residues weakens the repulsive forces, thus inducing the formation of a hydrophobic helical structure. In the presence of lipid membranes, the peptide, readily fuses with the lipid membranes (Goormaghtigh et al. 1991; Parente et al. 1988; Parente et al. 1990; Duzgunes and Nir 1999). In vitro energy transfer and dynamic light scattering experiments indicated that endosomal escape of the encapsulates in Tf-L equipped with Chol-GALA could be attributed to the pH-dependent membrane fusion (Kakudo et al. 2004).

This system has great potential for gene delivery applications by delivering encapsulated, condensed plasmid DNA to tissue tumors, where transferrin liposomes would be expected to be internalized by receptor-mediated endocytosis with the aid of Chol-GALA by membrane fusion between liposomes and the endosome (Fig. 4). The condensed plasmid DNA could then be effectively released into the cytosol.

3.2 Non-endocytic Routes for Gene Delivery

3.2.1 Protein Transduction Domains

Since endocytosis acts as a significant intracellular barrier, as explained above, it would be more advantageous to bypass the endocytic pathway and somehow achieve the cytosolic delivery of genes using other pathways. The non-endocytic delivery of genes can be achieved by microinjection, electroporation and cell permeabilization; however, these techniques are highly invasive and cannot be used for in vivo gene delivery. One interesting, non-invasive technique is the use of different peptides and proteins with membrane-translocating properties (Schwarze et al. 2000; Lindgren et al. 2000). Several proteins that possess the ability to enter cells have been identified, including the HIV-1 TAT protein and the *Drosophila* Antennapedia (Antp) transcription factor. Surprisingly, internalization of these PTDs is largely due to relatively short peptide sequences, termed membrane-translocationable signals (MTSs). For example, the minimal signal peptide required for intracellular translocation consists of the basic residues 48-60 of the TAT protein (Futaki 2002). The mechanism of uptake of these peptides is currently controversial; there are arguments for and against endocytic uptake (Polyakov et al. 2000; Torchilin et al. 2001; Terrone et al. 2003; Richard et al. 2003; Ferrari at al. 2003; Wadia et al. 2004). However, uptake appears to be different from classical clathrin-dependent endocytosis. The sequence of TAT-(48-60), which is critical for the translocation, is highly basic, containing six arginines and two lysines in 13 amino acid residues. The arginine-substituted analog of TAT was found to be internalized into cells as efficiently as the TAT-(48-60) peptide, suggesting the importance of arginine residues in the translocation process (Futaki et al. 2001a). Peptides consisting of only arginine residues were also internalized and the optimum number of arginine residues for efficient internalization was shown to be approximately eight (Futaki et al. 2001a). Therefore, we became interested in the R8 peptide for use in the non-endocytic delivery of genes.

First, the mechanism by which the R8 peptide is internalized was investigated. CLSM of fixed cells showed that the R8 peptide was internalized equally well at 37°C and 4°C and was transported to the cytosol and nucleus (Fig. 5a). Similarly, different endocytosis inhibitors, including hypertonic treatment, *N*-ethyl maleimide and nystatin, had no effect on internalization of the peptide (Khalil et al. 2004). These results were confirmed using flow cytometry of live cells. R8 was then used to condense plasmid DNA, and a positively charged complex of ~100 nm in diameter was obtained. Again, using CLSM, the R8/DNA complexes were shown to be internalized into the cytosol of cells (Khalil et al. 2004). However, CLSM of fixed cells under the same conditions used in the case of R8 peptide showed that internalization of the R8/DNA complexes was significantly inhibited (~90%) at low temperature (Fig. 5b) or by treatment with a hypertonic agent that specifically inhibits clathrin-mediated endocytosis. Furthermore, it was observed that the complexes were trapped in endosomes and lysosomes, thus decreasing the effectiveness of the vector (Akita et al. 2004). Therefore, the uptake of R8 peptide and R8/DNA complexes under similar conditions was compared, and significant differences in the uptake process in the presence of endocytosis inhibitors were found, indicating that the mechanism of internalization of the R8 peptide is not exactly the same as that for its complexes with DNA. While the R8/DNA complexes were taken up by clathrin-mediated endocytosis, the R8 peptide seemed to use different uptake pathways. Therefore, by direct complexation with DNA, the peptide fails to internalize the DNA away from endocytosis, where the endosomal

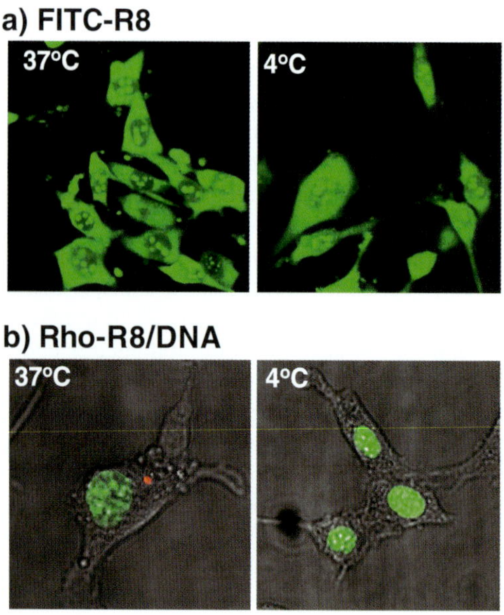

Fig. 5a,b. Internalization mechanisms of R8 peptide and R8/DNA complexes. NIH3T3 cells were incubated with **a** FITC-labeled R8 (2 μM) or **b** rhodamine-labeled plasmid DNA complexed with R8 (2.5 μg/ml) for 1 h at 37°C or at 4°C followed by fixation and observation by CFLM. In the case of R8/DNA complexes, the nucleus was stained with Syto-24 green fluorescence barrier still exits. Whether the peptide is taken up through penetration or through endocytic pathways other than classical endocytosis is not clear.

3.2.2 Stearylated Octaarginine

One possibility for explaining the endocytic delivery of R8/DNA complexes is that the peptide is in contact with the DNA, and as a result, is not free to interact with cellular membranes. Therefore, R8-modified liposomes (R8-Lip); lipid vesicles modified with free R8 peptide that can carry genes or proteins in their cores, were prepared. In order to present the R8 peptide on the surface of the liposomes, it was necessary to add an anchor to the peptide to be embedded in the lipid of the vesicle, thus leaving the free peptide on the surface. Therefore, a stearylated-R8 peptide (STR-R8) was used in which the stearyl moiety would be expected to be anchored in the lipid membrane of the liposomes. Surprisingly, complexes formed by simply mixing the STR-R8 peptide and plasmid DNA showed improved transfection activities compared to R8/DNA complexes (Futaki et al. 2001b). This result was confirmed by demonstrating that STR-R8/DNA complexes showed transfection activities at least three orders of magnitudes higher than those of the R8/DNA complexes (Fig. 6a). The mechanism of the improved transfection activity of the STR-R8 peptide was subsequently investigated. Cellular uptake of DNA complexed with STR-R8 was significantly higher than that of DNA complexed with R8 (Fig. 6b). Furthermore, the internalized DNA was partially localized in the nucleus in the case of STR-R8, but not in case of R8 (Fig. 6b). The internalization of both complexes was inhibited in the presence of endocytosis inhibitors, indicating endocytic delivery (Khalil et al. 2004). In order to explain the improved nuclear delivery in the case of STR-R8, cells were incubated with complexes formed from differently labeled DNA and peptides. In the case of the R8/DNA complexes, a high degree of co-localization was found between the peptide and the DNA, indicating that the DNA is present in cells in the complexed state (Fig. 6c). In contrast, free DNA was found intracellularly in the case of STR-R8, indicating that the DNA is released more easily from the complexes. Furthermore, using our novel quantitative intracellular trafficking technique, described below, it was observed that the DNA was able to escape from endosomes in the case of STR-R8 only (Akita et al. 2004). These collective findings suggest that STR-R8 has the ability to disrupt endosomes by itself, releasing free DNA to the cytosol, which may be more easily delivered to the nucleus. Using flow cytometry to differentiate between internalized and surface-bound DNA, surface association of DNA in the case of STR-R8 was found to be much higher than that in the case of R8 (Khalil et al. 2004). This improvement was confirmed by atomic force microscopy (AFM), which showed that STR-R8, but not R8, had the ability to condense the DNA completely into stable cores (Fig. 6d). Aided by both electrostatic and hydrophobic interactions, the STR-R8/DNA complexes would be highly adsorbed to cell membranes and subsequently highly internalized.

3.2.3 Octaarginine-Modified Liposomes

Since our data indicated that direct complexation between R8 and DNA allowed endocytic delivery of the complexes, we focused on our main objective, i.e., effective cellular internalization of genes in the absence of typical endocytosis. R8-Lip were prepared by mixing the lipids of the liposomes with STR-R8. R8-Lip, containing 5

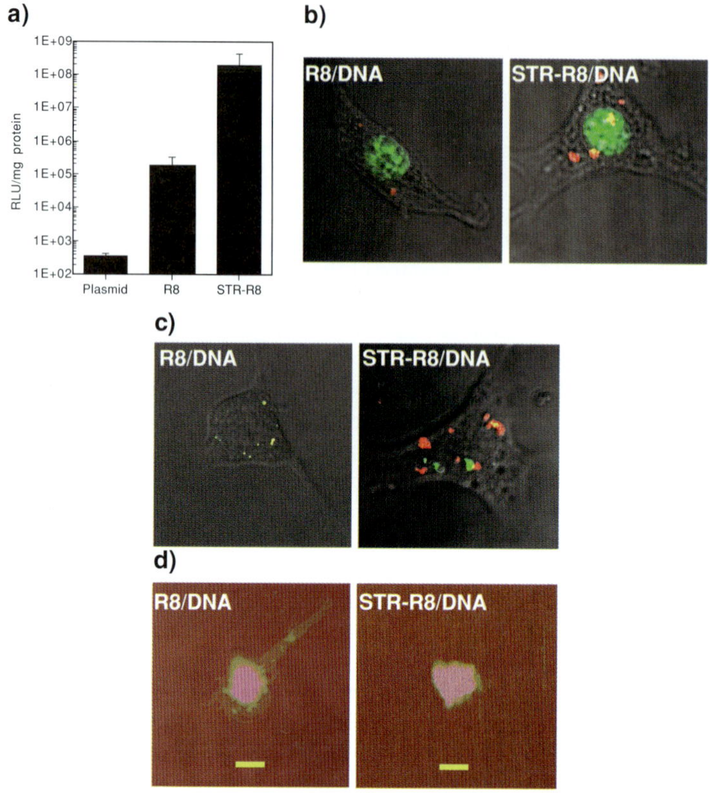

FIG. 6. **a** Transfection activities of R8 and STR-R8 peptides. NIH3T3 cells were transfected with plasmid DNA coding for the luciferase reporter gene (2.5 μg/ml) complexed with R8 or STR-R8 (cation-anion charge ratio 2:1) for a total of 48 h. The *vertical axis* represents the luciferase activities in cell lysates, expressed as relative light units (RLU) per mg protein. **b** Confocal images of cells transfected with rhodamine-labeled plasmid DNA complexed with R8 or STR-R8 for 3 h. **c** Confocal images of cells transfected with rhodamine-labeled plasmid DNA complexed with FITC-labeled R8 or STR-R8 for 3 h. **d** Atomic force microscopic (AFM) analysis of R8/DNA and STR-R8/DNA complexes. *Bar*, 100 nm

mol% STR-R8, were positively charged and ~100 nm in diameter. CLSM of living cells incubated with R8-Lip containing a rhodamine aqueous phase for 1 h showed that the liposomes were efficiently internalized and located in the cytosol. Surprisingly, internalization of the liposomes was only slightly inhibited at low temperature, indicating that the major uptake pathway is not typical endocytosis (Fig. 7a). Liposomes were internalized as intact vesicles since a high degree of co-localization between the lipid and the aqueous phase markers in the cytosol was found. Furthermore, inhibitors of clathrin-mediated endocytosis slightly inhibited uptake of the liposomes; however, the liposomes were partially co-localized with acidic compartments in the cytosol (Fig. 7b), indicating that a certain fraction of the liposomes are internalized through endocytic pathways that may be different from clathrin-mediated endocytosis. It has

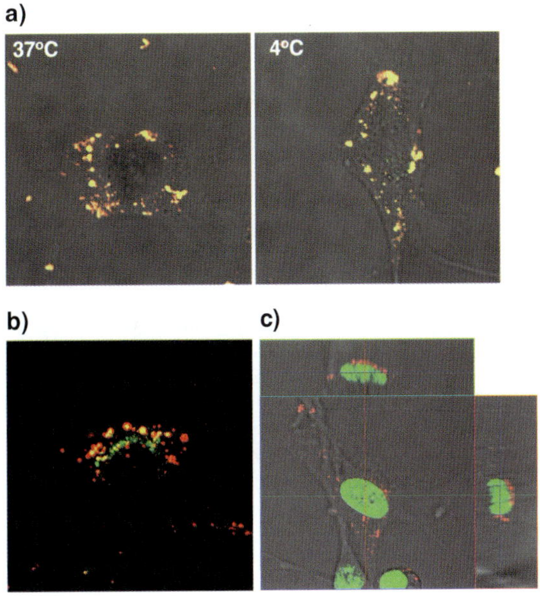

FIG. 7. **a** Images of live cells showing the uptake of R8-Lip containing rhodamine-labeled lipids and FITC-dextran in their aqueous phase at different temperatures. **b** Co-localization between R8-Lip containing rhodamine aqueous phase and FITC-lysosensor after 3 h. **c** Three-dimensional analysis of cells incubated for 9 h with R8-Lip containing a rhodamine aqueous phase. The nucleus was stained with Syto-24 green fluorescence

recently been shown that uptake of the TAT-Cre fusion protein occurs through macropinocytosis (Wadia et al. 2004). Since R8 and TAT peptides would be expected to share a similar internalization pathway (Suzuki et al. 2002), a possible contribution of macropinocytosis in the uptake of R8-Lip is highly likely. It has been suggested that macropinosomes are leaky and do not fuse with lysosomes (Swanson and Watts 1995). Although the exact mechanism is currently unclear, uptake through non-endocytic pathways or through macropinocytosis is still more advantageous than through classical endocytosis, because lysosomal degradation is highly avoided in both cases. Another interesting observation was that, after a 9-h incubation, R8-Lip were located largely in the perinuclear region, bound to the nuclear membrane (Fig. 7c). Therefore, using R8-Lip, the nucleus can be accessed using pathways different from typical endocytosis. This demonstrates the importance of the topology of the peptide in achieving a more preferred entrance pathway, since the interaction between free peptides and plasma membrane seems to be important for the peptide to exert its function.

3.2.4 Transgene Ability of MEND Modified with STR-R8

Based on the mechanistic study of R8-Lip, it was expected that the gene vector MEND modified with STR-R8 would show a high transgene ability. A MEND prototype was constructed using a three-step lipid-film hydration method consisting of: (1) DNA condensation; (2) lipid film hydration; and (3) sonication for lipid coating (Kogure et al. 2004). Packaging of the condensed DNA with the fusiogenic lipid (DOPE/CHEMS) improved transfection activity, possibly due to enhancing the endosomal escape of the complexes (Fig. 8). Surprisingly, as shown in Fig. 8, the localization of STR-R8 on

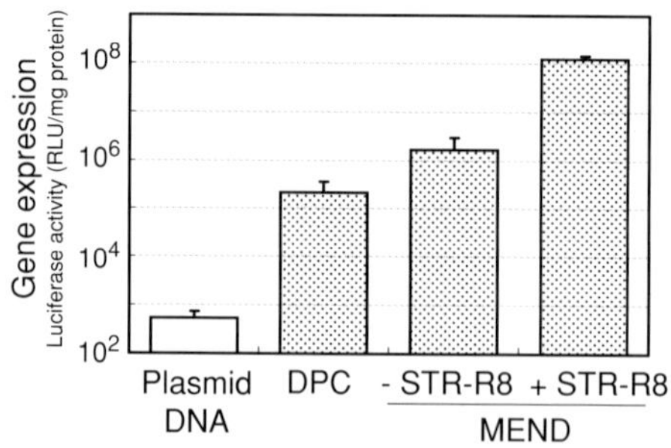

FIG. 8. Transfection assay for plasmid DNA, DNA/polycation complex (DPC), MEND and MEND(+STR-R8) in NIH3T3 cells. Samples containing 0.4 μg DNA suspended in 0.25 ml of serum and antibiotic-free DMEM were added to 4×10^4 NIH3T3 cells and incubated for 3 h at 37°C. One ml of DMEM containing 10% fetal calf serum was added to the cells, followed by a further incubation for 45 h. The cells were then washed, solubilized, and luciferase activity was measured using a luminometer. Luciferase activities are expressed as relative light units (RLU) per mg protein. Data represent the mean±S.D. ($n = 3$)

the surface of the MEND enhanced transfection activity by two orders of magnitude compared to MEND without STR-R8 (Kogure et al. 2004). Furthermore, MEND (+STR-R8) produced transfection activities as high as that of adenovirus, which is one of the most potent viral vectors (unpublished result). A negligible cytotoxicity was observed for the MEND(+STR-R8). Thus, the high transfection ability of MEND(+STR-R8) results from its entrance via a pathway that is different from classical endocytosis.

4 Optimization of Intracellular Trafficking: Pharmacokinetic Considerations

Although transgene expression is governed by various rate-limiting intracellular barriers, most of the current research efforts have focused on individual barriers to enhance transgene expression. However, the increased efficiency of one process may reduce that of others. For example, it is generally accepted that the tight condensation of pDNA, so as to produce a small complex, enhances cellular uptake by endocytosis, but excess condensation inhibits transcription. In order to optimize intracellular trafficking it is necessary to balance all of these processes. A computer-assisted intracellular kinetic model (Fig. 9) with kinetic parameters (i.e. first-order rate constant: $time^{-1}$) determined using quantitative data is a useful tool to analyze, simulate, and optimize transgene expression (Ledley and Ledley 1994; Varga et al. 2001; Banks et al. 2003). Varga et al. (2001) integrated a kinetic model for cellular uptake, endosomal release, nuclear binding, nuclear translocation, dissociation, and protein synthesis with first-order mass action kinetics and demonstrated the utility of kinetic modeling for optimization.

In spite of the great advantages of kinetic modeling, most vector development studies measure final output, e.g., transfection activity, while intracellular events remain a black box, because of the lack of a quantitative assay system for pDNA in

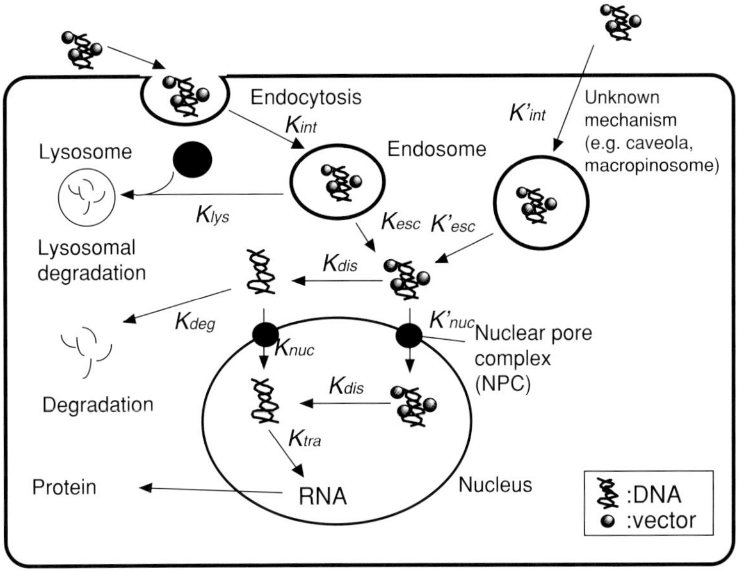

FIG. 9. The intracellular pharmacokinetic model of exogenous DNA. The intracellular disposition of the gene is represented. A complex of DNA and gene vector is internalized via endocytosis (*kint*). Endosomal plasmid DNA fused with lysosome is degraded (*klys*). Alternatively, the complex can enter the cells via a non-classical pathway (*k'int*). A certain fraction of the endosomal DNA is released into the cytosol (*kesc*) where some of the DNA dissociates from the cationic lipids/polycations (*kdis*). Free plasmid DNA is subject to degradation by nucleases (*kdeg*) or partially translocated to the nucleus (*knuc*). Alternatively, DNA enters the nucleus in a complexed form with cationic lipids/polycations (*k'nuc*) followed by dissociation in the nucleus (*kdis*). Finally, free DNA in the nucleus is transcribed (*ktra*)

each organelle. In fact, the kinetic parameters used in the simulation demonstrated by Varga et al. were mainly collected from existing data in the literature, which were evaluated using different cell cultures and different gene vectors.

Recently, we and other researchers established a methodology for quantifying the amount of plasmid DNA in the nucleus by subcellular fractionation of the nucleus followed by the polymerase chain reaction (PCR) for quantification (Leopold et al. 1998; Tachibana et al. 2001; Oh et al. 2002; Tachibana et al. 2002). PCR is a very convenient technique for measuring the amount of pDNA. However, it cannot distinguish intact plasmid DNA from a degraded sample. To resolve this problem, Southern blotting can be applied (Tachibana et al. 2001; Tachibana et al. 2002) to estimate the stability of the plasmid DNA from its length. Using these methods, we were able to clarify the relationship between the applied dose of plasmid DNA, the amount of nuclear DNA, and transgene expression. These data indicated that the nuclear plasmid DNA increased linearly depending on the dose, however, a dramatic saturation of transgene expression was observed. These results indicate that it is necessary to enhance not only the nuclear delivery of plasmid DNA, but also the transcription efficiency of the plasmid itself in the nucleus. In contrast to the nucleus, very few reports are available concerning the amount of plasmid DNA in the endosome/lysosome compartment, and, therefore, it is very difficult to quantitatively evaluate the efficiency of endosomal release. Although the subcellular fractionation of endosomes/lysosomes may solve this problem, several factors, such as the complexity of the steps in the protocol, uncertainties associated with the recovery of the endosomal fraction and mutual contamination of each organelle may preclude the use of this methodology in practical applications.

For these reasons, we have proposed a novel strategy for quantifying the distribution of pDNA in the cytosol, endosomes/lysosomes, and the nucleus simultaneously, using sequential Z-series images captured by CLSM, which is called confocal image-assisted 3-dimensionally integrated quantification (CIDIQ) (Akita et al. 2004). A schematic diagram of this technique and the numerical formulas are provided in Fig. 10. After transfection of rhodamine-labeled pDNA, acidic compartments (endosomes/lysosomes) and the nucleus are stained with LysoSensor DND-189 (green) and Hoechst 33258 (blue), respectively, to determine the subcellular localization of the pDNA. At an early time after transfection, plasmid DNA is detected as clusters. When plasmid DNA is localized in the endosomes/lysosomes, the cytosol, and the nucleus, the molecules are detected as yellow, red, and pink clusters, respectively. For quantification, the pixel areas of the clusters are used as an index of the amount of pDNA. The total pixel area for the clusters of plasmid DNA in each subcellular compartment are first determined in each xy-plane. These values are then further integrated and denoted as S(cyt), S(endosome/lysosome) and S(nucleus), which represent the amount of plasmid DNA in each organelle, in one cell. S(tot), determined by the integration of these values, represents the total cellular uptake of plasmid DNA in one cell. Finally, the fraction of plasmid DNA in each compartment in the total cell is calculated. As shown above, the transgene expression level of STR-R8 was drastically higher than that of R8, while it was much less than that of the commercially available lipoplex, LipofectAMINE PLUS (Fig. 11). In order to clarify which process is rate-determining, this approach was applied to an analysis of the intracellular trafficking of pDNA, transfected by LipofectAMINE PLUS, stearylated octaarginine (STR-R8), and R8. The results showed that most of the pDNA was trapped by endosomes/lysosomes in the case of R8, whereas STR-R8 underwent endosomal escape followed by nuclear translocation in a time-dependent manner. These data suggest that a stearyl moiety enhances the endosomal escape process. Furthermore, LipofectAMINE PLUS was the most effective for rapidly delivering DNA to the nucleus as well as the cytosol. Surprisingly, nuclear localization was observed within 1 h, which is as rapid as that for an adenovirus. This phenomenon is consistent with the fact that transgene expression began to be exhibited within 3 h. Collectively, the differences in transgene expression can be readily explained by intracellular trafficking, as assessed by CIDIQ. In addition, combining the inhibition study of the internalizing process, we succeeded in assessing the contribution of the endocytotic pathway to the total cellular uptake of Lipoplex. Such quantitative data on the contribution of various pathways to overall cellular uptake are essential for establishing intracellular pharmacokinetic models in the future. Collectively, this method can be applied to the intracellular pharmacokinetic analysis of various gene vectors and would be useful in the development of new gene delivery systems. In the future, an application that enables all of these quantification procedures to be automated is needed, since it is necessary to analyze a large number of

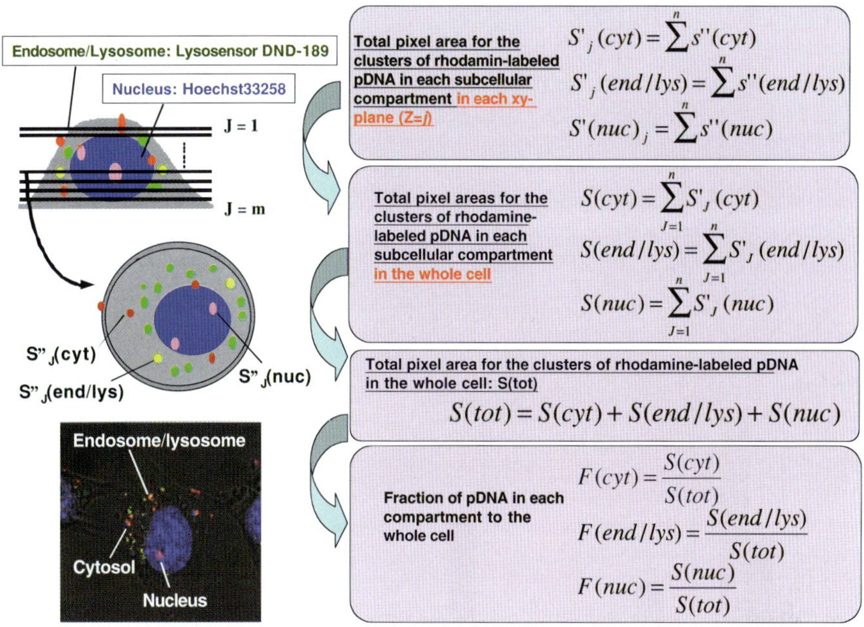

$$S'_j(cyt) = \sum^{n} s''(cyt)$$
$$S'_j(end/lys) = \sum^{n} s''(end/lys)$$
$$S'(nuc)_j = \sum^{n} s''(nuc)$$

$$S(cyt) = \sum_{J=1}^{n} S'_J(cyt)$$
$$S(end/lys) = \sum_{J=1}^{n} S'_J(end/lys)$$
$$S(nuc) = \sum_{J=1}^{n} S'_J(nuc)$$

$$S(tot) = S(cyt) + S(end/lys) + S(nuc)$$

$$F(cyt) = \frac{S(cyt)}{S(tot)}$$
$$F(end/lys) = \frac{S(end/lys)}{S(tot)}$$
$$F(nuc) = \frac{S(nuc)}{S(tot)}$$

FIG. 10. Methods for quantifying the subcellular distribution of pDNA. After transfection of rhodamine-labeled pDNA, the endosome/lysosome and nuclear fractions were stained with LysoSenser DND-189 and Hoechat33342, respectively, to identify the subcellular localization of pDNA. Each 8-bit TIFF image was transferred to an Image-Pro Plus ver 4.0 (Media Cybernetic, Silver Spring, Md., USA) in order to quantify the total brightness and pixel area of each region of interest. For the data analysis, endosomes/lysosomes, s_i(end/lys), cytosol, s_i (cyt) and nucleus; s_i (nuc), were separately summed up in each xy-plane, and are denoted as $S'_{Z=j}$(mem), $S'_{Z=j}$(end/lys), $S'_{Z=j}$(cyt), and $S'_{Z=j}$(nuc), respectively. The values of $S'_{Z=j}$(mem), $S'_{Z=j}$(end/lys), $S'_{Z=j}$(cyt), and $S'_{Z=j}$(nuc) in each xy-plane were further summed up and are denoted as S(end/lys), S(cyt), and S(nuc), respectively. These parameters represent the total amount of pDNA in each compartment in the whole cell. Furthermore, the total area of the pDNA, denoted as S(tot), was calculated by integrating S(end/lys), S(cyt), and S(nuc). This value represents the total cellular uptake of pDNA. The fractions of pDNA present on, endosomes/lysosomes, cytosol and nucleus relative to the whole cell are denoted as F(end/lys), F(cyt), and F(nuc), which are calculated as S(end/lys), S(cyt), and S(nuc) divided by S(tot), respectively

confocal images in order to obtain an accurate time course for the determination of kinetic parameters by fitting.

5 Conclusions

We have introduced a new concept, Programmed Packaging, for the rational development of a non-viral gene delivery system. Based on this concept, a Multifunctional Envelope-type Nano Device (MEND) was developed that overcomes cellular and intracellular barriers. R8-modified MEND yields transfection activities that are as

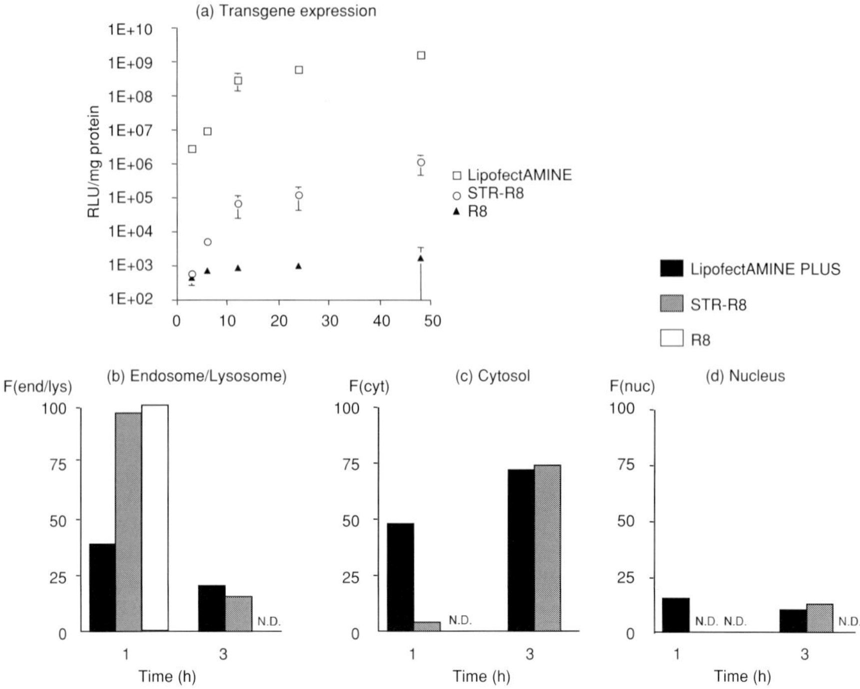

FIG. 11. Transgene expression (**a**) and intracellular disposition (**b–d**) of pDNA transfected by the LipofectAMINE, STR-R8 and R8. **a** pDNA encoding luciferase was condensed with LipofectAMINE PLUS, STR-R8 or R8 and incubated for 3 h at 37°C. One ml of DMEM containing 10% fetal calf serum was added to the cells, followed by a further incubation. At the indicated times, cells were harvested and the luciferase expression level was evaluated. **b–d** At 1 and 3 h post-transfection with LipofectAMINE PLUS, STR-R8 or R8, subcellular distributions of pDNA in the endosome/lysosome (**b**), cytosol (**c**), and nucleus (**d**) were quantified by CIDIC as described in the text

high as those of adenovirus under in vitro conditions. A quantitative assay of the intracellular trafficking of the gene delivery system (CIDIQ) will clarify the rate-limiting step of non-viral vectors, which will be useful for the development of nano-machines.

References

Akita H, Ito R, Khalil IA, Futaki S, Harashima H (2004) Quantitative three-dimensional analysis of the intracellular trafficking of plasmid DNA transfected by a nonviral gene delivery system using confocal laser scanning microscopy. Mol Ther 9:443–451

Banks GA, Roselli RJ, Chen R, Giorgio TD (2003) A model for the analysis of nonviral gene therapy. Gene Ther 10:1766–1775

Blessing T, Remy JS, Behr JP (1998) Monomolecular collapse of plasmid DNA into stable virus-like particles. Proc Natl Acad Sci USA 95:1427–1431

Boussif O, Lezoualc'h F, Zanta MA, Mergny MD, Scherman D, Demeneix B, Behr JP (1995) A versatile vector for gene and oligonucleotide transfer into cells in culture and in vivo: polyethylenimine. Proc Natl Acad Sci U S A 92:7297–7301
Branden LJ, Mohamed AJ, Smith CI (1999) A peptide nucleic acid-nuclear localization signal fusion that mediates nuclear transport of DNA. Nat Biotechnol 17:784–787
Brown MD, Schatzlein AG, Uchegbu IF (2001) Gene delivery with synthetic (non viral) carriers. Int J Pharm 229:1–21
Chan CK, Jans DA (1999) Enhancement of polylysine-mediated transferrinfection by nuclear localization sequences; polylysine does not function as a nuclear localization sequence. Hum Gene Ther 10:1695–1702
Chan CK, Senden T, Jans DA (2000) Supramolecular structure and nuclear targeting efficiency determine the enhancement of transfection by modified polylysine. Gene Ther 7:1690–1697
Duzgunes N, Nir S (1999) Mechanisms and kinetics of liposome cell interaction. Adv. Drug Delivery Rev. 40:3–18
Erbacher P, Roche AC, Monsigny M, Midoux P (1996) Putative role of chloroquine in gene transfer into a human hepatoma cell line by DNA/lactosylated polylysine complexes. Exp Cell Res 225:186–194
Farhood H, Serbina N, Huang L (1995) The role of dioleoyl phosphatidylethanolamine in cationic liposome mediated gene transfer. Biochim Biophys Acta 1235:289–295
Felgner PL, Gadek TR, Holm M, Roman R, Chan HW, Wenz M, Northrop JP, Ringold GM, Danielsen M (1987) Lipofection: a highly efficient, lipid-mediated DNA-transfection procedure. Proc Natl Acad Sci USA 84:7413–7417
Felgner JH, Kumar R, Sridhar CN, Wheeler CJ, Tsai YJ, Border R, Ramsey P, Martin M, Felgner PL (1994) Enhanced gene delivery and mechanism studies with a novel series of cationic lipid formulations. J Biol Chem 269:2550–2561
Ferrari A, Pellegrini V, Arcangeli C, Fittipaldi A, Giacca M, Beltram F (2003) Caveolae-mediated internalization of extracellular HIV-1 Tat fusion proteins visualized in real time. Mol. Ther. 8:284–294
Futaki S (2002) Arginine-rich peptides: potential for intracellular delivery of macromolecules and the mystery of the translocation mechanisms. Int. J. Pharm. 245:1–7
Futaki S, Suzuki T, Ohashi W, Yagami T, Tanaka S, Ueda K, Sugiura Y (2001a) Arginine-rich peptides: An abundant source of membrane-permeable peptides having potential as carriers for intracellular protein delivery. J. Biol. Chem. 276:5836–5840
Futaki S, Ohashi W, Suzuki T, Niwa M, Tanaka S, Ueda K, Harashima H, Sugiura Y (2001b) Stearylated arginine-rich peptides: a new class of transfection systems. Bioconjug. Chem. 12:1005–1011
Gao X, Hunag L (1996) Potentiation of cationic liposome-mediated gene delivery by polycations. Biochemistry 35:1027–1036
Goormaghtigh E, De Meutter J, Szoka F, Cabiaux V, Parente RA, Ruysschaert JM (1991) Secondry structure and orientation of amphipathic peptide GALA in lipid structures. Eur. J. Biochem. 195:421–429
Hagstrom JE, Ludtke JJ, Bassik MC, Sebestyen MG, Adam SA, Wolff JA (1997) Nuclear import of DNA in digitonin-permeabilized cells. J Cell Sci 110 (Pt 18):2323–2331
Harashima H, Shinohara Y, Kiwada H (2001) Intracellular control of gene trafficking using liposomes as drug carriers. Eur J Pharm Sci 13:85–89
Hui SW, Langner M, Zhao YL, Ross P, Hurley E, Chan K (1996) The role of helper lipids in cationic liposome-mediated gene transfer. Biophys J 71:590–599
Kakudo T, Chaki S, Futaki S, Nakase I, Akaji K, Kawakami T, Maruyama K, Kamiya H, Harashima H (2004) Transferrin-modified liposomes equipped with a pH-sensitive fusogenic peptide: an artificial viral-like delivery system. Biochemistry 43:5618–5628
Kamiya H, Tsuchiya H, Yamazaki J, Harashima H (2001) Intracellular trafficking and transgene expression of viral and non-viral gene vectors. Adv. Drug Delivery Rev. 52:153–164

Kamiya H, Akita H, Harashima H (2003) Pharmacokinetic and pharmacodynamic considerations in gene therapy. Drug Discov Today 8:990–996

Kawai S, Nishizawa M (1984) New procedure for DNA transfection with polycation and dimethyl sulfoxide. Mol Cell Biol 4:1172–1174

Keller M, Harbottle RP, Perouzel E, Colin M, Shah I, Rahim A, Vaysse L, Bergau A, Moritz S, Brahimi-Horn C, Coutelle C, Miller AD (2003) Nuclear localisation sequence templated nonviral gene delivery vectors: investigation of intracellular trafficking events of LMD and LD vector systems. Chembiochem 4:286–298

Khalil IA, Futaki S, Niwa M, Baba Y, Kaji N, Kamiya H, Harashima H (2004) Mechanism of improved gene transfer by the N-terminal stearylation of octaarginine: enhanced cellular association by hydrophobic core formation. Gene Ther 11:636–644

Kichler A, Leborgne C, Coeytaux E, Danos O (2001) Polyethylenimine-mediated gene delivery: a mechanistic study. J Gene Med 3:135–144

Kircheis R, Wightman L, Schreiber A, Robitza B, Rössler V, Kursa M, Wagner E (2001) Polyethylenimine/DNA complexes shielded by transferrin target gene expression to tumors after systemic application. Gene Ther 8:28–40

Kogure K, Moriguchi R, Sasaki K, Ueno M, Futaki S, Harashima H (2004) Development of a nonviral multifunctional envelope-type nano device by a novel lipid film hydration method. J Cont Release 98:317–323

Ledley TS, Ledley FD (1994) Multicompartment, numerical model of cellular events in the pharmacokinetics of gene therapies. Hum Gene Ther 5:679–691

Lee RJ, Huang L (1996) Folate-targeted, anionic liposome-entrapped polylysine-condensed DNA for tumor cell-specific gene transfer. J Biol Chem 271:8481–8487

Leopold PL, Ferris B, Grinberg I, Worgall S, Hackett NR, Crystal RG (1998) Fluorescent virions: dynamic tracking of the pathway of adenoviral gene transfer vectors in living cells. Hum Gene Ther 9:367–378

Li S, Huang L (1997) In vivo gene transfer via intravenous administration of cationic lipid-protamine-DNA (LPD) complexes. Gene Ther 4:891–900

Li S, Rizzo MA, Bhattacharya S, Huang L (1998) Characterization of cationic lipid-protamine-DNA (LPD) complexes for intravenous gene delivery. Gene Ther 5:930–937

Li S, Tseng WC, Stolz DB, Wu SP, Watkins SC, Huang L (1999) Dynamic changes in the characteristics of cationic lipidic vectors after exposure to mouse serum: implications for intravenous lipofection. Gene Ther 6:585–594

Lindgren M, Hallbrink M, Prochiantz A, Langel U (2000) Cell-penetrating peptides. Tips 21:99–103

Ludtke JJ, Zhang G, Sebestyen MG, Wolff JA (1999) A nuclear localization signal can enhance both the nuclear transport and expression of 1 kb DNA. J Cell Sci 112 (Pt 12): 2033–2041

Lycke E, Hamark B, Johansson M, Krotochwil A, Lycke J, Svennerholm B (1988) Herpes simplex virus infection of the human sensory neuron. An electron microscopy study. Arch Virol 101:87–104

Marenzi S, Adams RL, Zardo G, Lenti L, Reale A, Caiafa P (1999) Efficiency of expression of transfected genes depends on the cell cycle. Mol Biol Rep 26:261–267

Mortimer I, Tam P, MacLachlan I, Graham RW, Saravolac EG, Joshi PB (1999) Cationic lipid-mediated transfection of cells in culture requires mitotic activity. Gene Ther 6: 403–411

Murray KD, Etheridge CJ, Shah SI, Matthews DA, Russell W, Gurling HM, Miller AD (2001) Enhanced cationic liposome-mediated transfection using the DNA-binding peptide mu (mu) from the adenovirus core. Gene Ther 8:453–460

Nagasaki T, Myohoji T, Tachibana T, Futaki S, Tamagaki S (2003) Can nuclear localization signals enhance nuclear localization of plasmid DNA? Bioconjug Chem 14:282–286

Nir S, Nieva JL (2000) Interactions of peptides with liposomes: pore formation and fusion. Prog. Lipid Res. 39:181–206

Ogris M, Brunner S, Schüller S, Kircheis R, Wagner E (1999) PEGylated DNA/transferrin-PEI complexes: reduced interaction with blood componets, extended circulation in blood and potential for systemic gene delivery. Gene Ther 6:595–605

Oh YK, Suh D, Kim JM, Choi HG, Shin K, Ko JJ (2002) Polyethylenimine-mediated cellular uptake, nucleus trafficking and expression of cytokine plasmid DNA. Gene Ther 9: 1627–1632

Oupicky D, Konak C, Ulbrich K, Wolfert MA, Seymour LW (2000) DNA delivery systems based on complexes of DNA with synthetic polycations and their copolymers. J Control Release 65:149–171

Parente RA, Nir S, Szoka FC (1988) pH-dependent fusion of phosphatidyl choline small vesicles. J. Biol. Chem. 263:4724–4730

Parente RA, Nadasdi L, Subbarao NK, Szoka FC (1990) Association of a pH-sensitive peptide with membrane vesicles: role of amino acid sequence. Biochemistry 29: 8713–8719

Plank C, Oberhauser B, Mechtler K, Koch C, Wagner E (1994) The influence of endosome-disruptive peptides on gene transfer using synthetic virus-like gene transfer systems. J Biol Chem 269:12918–12924

Pollard H, Remy JS, Loussouarn G, Demolombe S, Behr JP, Escande D (1998) Polyethylenimine but not cationic lipids promotes transgene delivery to the nucleus in mammalian cells. J Biol Chem 273:7507–7511

Polyakov V, Sharma V, Dahlheimer JL, Pica CM, Luker GD, Piwnica-Worms D (2000) Novel Tat-peptide chelates for direct transduction of Technetium-99 nm and Rhenium into human cells for imaging and radio therapy. Bioconjugate Chem 11:762–771

Richard JP, Melikov K, Vives E, Ramos C, Verbeure B, Gait MJ, Chernomordik LV, Lebleu B (2003) Cell-penetrating peptides: A reevaluation of the mechanism of cellular uptake. J. Biol. Chem. 278:585–590

Rittner K, Benavente A, Bompard-Sorlet A, Heitz F, Divita G, Brasseur R, Jacobs E (2002) New basic membrane-destabilizing peptides for plasmid-based gene delivery in vitro and in vivo. Mol Ther 5:104–114

Rudolph C, Plank C, Lausier J, Schillinger U, Muller RH, Rosenecker J (2003) Oligomers of the arginine-rich motif of the HIV-1 TAT protein are capable of transferring plasmid DNA into cells. J Biol Chem 278:11411–11418

Schwarze SR, Hruska KA, Dowdy SF (2000) Protein transduction: unrestricted delivery into all cells? Trends Cell Biol. 10, 290–295

Sebestyen MG, Ludtke JJ, Bassik MC, Zhang G, Budker V, Lukhtanov EA, Hagstrom JE, Wolff JA (1998) DNA vector chemistry: the covalent attachment of signal peptides to plasmid DNA. Nat Biotechnol 16:80–85

Sodeik B, Ebersold MW, Helenius A (1997) Microtubule-mediated transport of incoming herpes simplex virus 1 capsids to the nucleus. J Cell Biol 136:1007–1021

Sorgi FL, Bhattacharya S, Huang L (1997) Protamine sulfate enhances lipid-mediated gene transfer. Gene Ther 4:961–968

Subbarao NK, Parente RA, Szoka FC, Nadasdi L, Pongracz K (1987) pH-dependent bilayer destabilization by an amphipathic peptide. Biochemistry 26:2964–2972

Suzuki T, Futaki S, Niwa M, Tanaka S, Ueda K, Sugiura Y (2002) Possible existence of common internalisation mechanisms among arginine-rich peptides. J. Biol. Chem. 277: 2437–2443

Swanson JA, Watts C (1995) Macopinocytosis. Trends Cell Biol. 5:424–428

Tachibana R, Harashima H, Ide N, Ukitsu S, Ohta Y, Suzuki N, Kikuchi H, Shinohara Y, Kiwada H (2002) Quantitative analysis of correlation between number of nuclear plasmids and gene expression activity after transfection with cationic liposomes. Pharm Res 19:377–381

Tachibana R, Harashima H, Shinohara Y, Kiwada H (2001) Quantitative studies on the nuclear transport of plasmid DNA and gene expression employing nonviral vectors. Adv Drug Deliv Rev 52:219–226

Tagawa T, Manvell M, Brown N, Keller M, Perouzel E, Murray KD, Harbottle RP, Tecle M, Booy F, Brahimi-Horn MC, Coutelle C, Lemoine NR, Alton EW, Miller AD (2002) Characterisation of LMD virus-like nanoparticles self-assembled from cationic liposomes, adenovirus core peptide mu and plasmid DNA. Gene Ther 9:564–576

Tam P, Monck M, Lee D, Ludkovski O, Leng EC, Clow K, Stark H, Scherrer P, Graham RW, Cullis PR (2000) Stabilized plasmid-lipid particles for systemic gene therapy. Gene Ther 7:1867–1874

Tang MX, Szoka FC (1997) The influence of polymer structure on the interactions of cationic polymers with DNA and morphology of the resulting complexes. Gene Ther 4:823–832

Tanimoto M, Kamiya H, Minakawa N, Matsuda A, Harashima H (2003) No enhancement of nuclear entry by direct conjugation of a nuclear localization signal peptide to linearized DNA. Bioconjug Chem 14:1197–1202

Terrone D, Sang SLW, Roudaia L, Silvius JR (2003) Penetratin and related cell-penetrating cationic peptides can translocate across lipid bilayers in the presence of a transbilayer potential. Biochemistry 42:13787–13799

Torchilin VP, Rammohan R, Weissig V, Levchenko TS (2001) Tat peptide on the surface of liposomes affords their efficient intracellular delivery even at low temperature and in presence of metabolic inhibitors. Proc. Natl. Acad. Sci. USA 98:8786–8791

Tranchant I, Thompson B, Nicolazzi C, Mignet N, Scherman D (2004) Physicochemical optimisation of plasmid delivery by cationic lipids. J Gene Med 6 Suppl 1:S24–35

Tseng WC, Haselton FR, Giorgio TD (1999) Mitosis enhances transgene expression of plasmid delivered by cationic liposomes. Biochim Biophys Acta 1445:53–64

Varga CM, Hong K, Lauffenburger DA (2001) Quantitative analysis of synthetic gene delivery vector design properties. Mol Ther 4:438–446

Wadia JS, Stan RV, Dowdy SF (2004) Transducible TAT-HA fusogenic peptide enhances escape of TAT-fusion proteins after lipid raft macropinocytosis. Nat. Med. 10:310–315

Wagner E (1998) Effects of membrane-active agents in gene delivery. J Control Release 53: 155–158

Wagner E, Plank C, Zatloukal K, Cotten M, Birnstiel ML (1992) Influenza virus hemagglutinin HA-2 N-terminal fusogenic peptides augment gene transfer by transferrin-polylysine-DNA complexes: toward a synthetic virus-like gene-transfer vehicle. Proc Natl Acad Sci U S A 89:7934–7938

Wheeler JJ, Pakmer L, Ossanlou M, MacLanchlan I, Graham RW, Zhang YP, Hope MJ, Scherrer P, Cullis PR (1999) Stabilized plasmid-lipid particles: construction and characterization. Gene Ther 6:271–281

Wilke M, Fortunati E, van den Broek M, Hoogeveen AT, Scholte BJ (1996) Efficacy of a peptide-based gene delivery system depends on mitotic activity. Gene Ther 3:1133–1142

Wilson GL, Dean BS, Wang G, Dean DA (1999) Nuclear import of plasmid DNA in digitonin-permeabilized cells requires both cytoplasmic factors and specific DNA sequences. J Biol Chem 274:22025–22032

Zanta MA, Belguise-Valladier P, Behr JP (1999) Gene delivery: a single nuclear localization signal peptide is sufficient to carry DNA to the cell nucleus. Proc Natl Acad Sci USA 96: 91–96

Active DNA Release from Complexes

Takeshi Nagasaki

1 Introduction

As pharmaceuticals have progressed from small-molecule drugs (e.g., antibiotics) to peptides and proteins (e.g., growth hormone and antibodies) and to genes, delivery of the drug to the appropriate site in the body and the cells, in the appropriate amount, at the appropriate time, and without damage to the drug has become increasingly important. Since plasmid DNA is generally considered to be safe and does not stimulate an immune response, it has been widely used in non-viral gene delivery. Plasmid DNA delivery, however, is limited by the inability of cells to efficiently internalize and express the delivered gene. Although cationic compound-mediated delivery, such as cationic lipid/DNA complexes (lipoplexes) and polymer/DNA complexes (polyplexes) (Fig. 1), can enhance gene transfer and produce physiological responses even in vivo, at present there is no quantitative or mechanistic understanding of plasmid DNA uptake and expression by a DNA release mechanism. Therefore, active gene release systems can hypothesize about where dissociation would be most beneficial. For example, dissociation on the membrane of endocytic vesicles may aid plasmid release into the cytoplasm, whereas keeping the plasmid in the complex until the vehicle enters the nucleus may protect the DNA from enzymatic degradation. Alternatively, release of the plasmid DNA from the complex may facilitate transport through the nuclear pores into the nucleus, or not. In non-viral gene delivery systems, transfection efficiency is directly influenced by gene release from the carrier/gene complex, because interaction between anionic DNA and the cationic carrier compound interferes with transcription in the nucleus as the final destination.

Indeed, for nuclear microinjections of DNA complexes with cationic lipid or polymer, it is possible to estimate the accessibility of the complexed DNA for mRNA transcription and expression. For the cationic lipids DMRIE/DOPE (a 1 : 1 mixture of *N*-[1-(2,3-dimyristyloxy)propyl]-*N,N*-dimethyl-*N*-(2-hydroxyethyl)ammonium bromide (DMRIE) and dioleoyl phosphatidylethanolamine (DOPE)), DOTAP (N-[1-(2,3-dioleolyloxy)propyl]-*N,N,N*-trimethylammonium methanesulfonate), or DOGS

Department of Applied and Bioapplied Chemistry, Graduate School of Engineering, Osaka City University, 3-3-138 Sugimoto, Sumiyoshi-ku, Osaka 558-8585, Japan

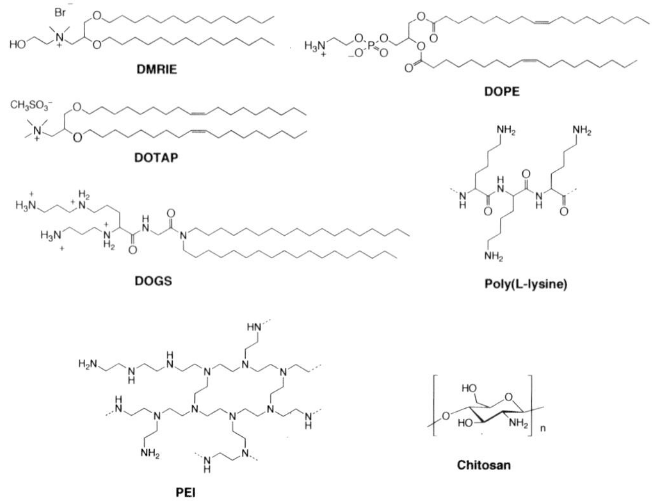

Fig. 1. Chemical structure of compounds used as gene delivery reagents

(dioctadecylamidoglycyl spermine), no reporter-gene expression was observed when these lipoplexes, with an excess N/P ratio, were microinjected into nuclei (Zabner et al. 1995; Pollard et al. 1998). The mechanism of DNA release from cationic liposome/DNA complexes is thought to be as follows. After the lipoplex is internalized into cells by endocytosis, cationic lipids interact with anionic vesicle lipids and destabilize the endocytic vesicle membrane. This, in turn, causes the fusion to form a neutral ion pair in the membrane, which results in displacement of the DNA from the cationic lipid and release of DNA into the cytoplasm (Xu et al. 1996; Bhattacharya et al. 1998). Nonetheless, vector unpacking as a potential barrier for receptor-mediated polyplex gene delivery has been demonstrated using a series of poly(L-lysine) (PLL) derivatives with various molecular weights and DNA affinities. In this model system, PLL derivatives were linked to epidermal growth factor (EGF) as a ligand to mouse fibroblasts bearing the EGF receptor. Maximal gene expression was observed using a polyplex with intermediate molecular weight and DNA affinity, because stable complexes suppressed mRNA transcription and labile complexes allowed rapid DNA degradation (Schaffer et al. 2000). In another work, the dissociation of plasmid DNA complexed with polycations was explored by confocal microscopy using fluoresce resonance energy transfer (FRET) between a pair of donor-acceptor fluorescent dyes tagged on a single plasmid DNA molecule. The observation of FRET revealed that the easy dissociation of DNA complex with linear polyethylenimine (PEI) contributes to its higher transfection efficiency compared to branched PEI (Ikata et al. 2004). Additionally, polyplexes based on chitosan oligomers (10- to 50-mers) dissociate more easily than those of a high-molecular-weight ultrapure chitosan (UPC, approximately a 1,000-mer) and release plasmid DNA in the presence of anionic heparin. The more easily dissociated polyplexes mediated a faster onset of action and led to higher gene expression both in 293 cells in vitro and after lung administration in vivo as compared to the more stable UPC polyplexes (Köping-Höggård et al. 2004).

These results indicate that DNA release should be added to the list of barriers to non-viral gene delivery. However, until non-viral vectors reach their destination, the high stability of the DNA complex should be retained. At the DNA release point, a decrease in the stability of the DNA complex is required. As DNA association and dissociation are opposite phenomena, ideal non-viral vectors should possess differential biophysical and biological properties. In order to overcome this problem, many efforts have been made to develop non-viral vectors capable of responding to various endogenous and exogenous stimuli. In this chapter, recent advances in active or controlled DNA release using non-viral gene delivery systems containing stimuli-responsive mechanisms, such as pH-, redox-, and photo-responsive carriers, are discussed. It should be noted, however, that thermo-responsive gene carriers are discussed in another chapter.

2 pH-Responsive Vectors

A decrease in the pH is a common physiological and pathological response that occurs as part of several intracellular functions, such as endocytic processing, tumor growth, and inflammation (Gerweck et al. 1996). When plasmid DNA is incorporated into the cell via a non-viral vector, irrespective whether by non-specific endocytosis, phagocytosis, pinocytosis, or receptor-mediated endocytosis, it is spatially separated from the cytoplasm by the vesicle membrane. The normal intracellular trafficking of non-viral vectors usually directs endocytosed vesicles to lysosomes for enzymatic degradation. As the endosomes mature from "early" to "late" stage, the pH of the vesicles decreases to less than that of the cytoplasm by 1–2 pH units, and late endosomes fuse with lysosomes (Luzio et al. 2001). Since lysosomal enzymatic degradation of DNA strongly limits the transfection efficiency, endosomal escape of exogenous DNA internalized via a non-viral vector is one of the important factors that determine transfection efficiency (Cho et al. 2003).

To date, a great deal of effort has been invested into methods to promote endosomal disruption using pH-responsive fusogenic vectors (Fig. 2). Viral envelopes have been noted to fuse and destabilize the endosomal and/or lysosomal membrane. For example, influenzavirus hemagglutinin (HA) has been extensively investigated and utilized as a membrane-destabilizing agent. In addition, enhanced transfection efficiency with transferrin-conjugated PLL/DNA complex has been achieved by using N-terminal fusogenic peptides of HA-2. Activation of membrane disruption in the presence of influenza peptide conjugates was demonstrated in a liposome leakage assay, either by electrostatic interaction (Wagner et al. 1992) or by biotin-streptavidin cross-linking (Plank et al. 1994). The enhancement of transfection efficiency with HA peptide was also observed during transferrin-mediated lipofection (Simoes et al. 1998). Furthermore, gene transfer using cationic lipid vesicles (dioleyldimethylammonium chloride: DODAC) could be mediated by the fusogenic protein hemagglutinin. Transfection of BHK-21 cells with hemagglutinin/DODAC/DNA ternary complexes resulted in high-level gene expression, and transfection activity was shown to be dependent on the functional activity of hemmagglutinin (Schoen et al. 1999). Similar to applications with viral peptides, rationally synthesized and pH-dependent fusogenic peptides, such as GALA (Li et al. 2004), have also been widely used to

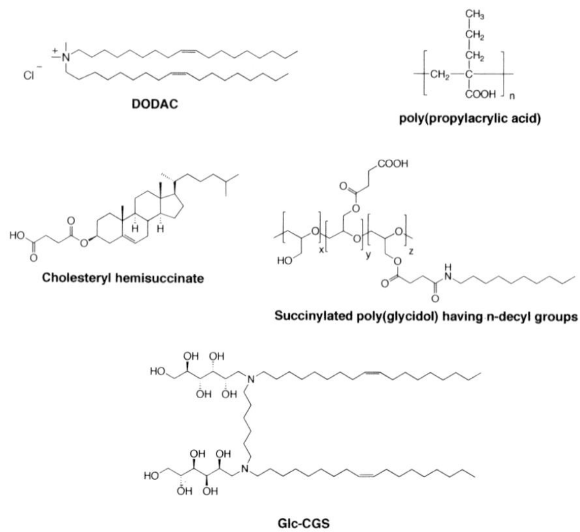

FIG. 2. Chemical structure of compounds used in pH-responsive gene delivery systems

promote DNA escape from endocytic vesicles. The covalent attachment of the amphipathic peptide GALA to polyamidoamine dendrimers significantly enhances transfection efficiency in both primary cells and cell lines (Haensler et al. 1993). Increased expression of an exogenous gene using the GALA system was observed not only in such polyplex carriers but also in lipoplex systems (Simoes et al. 1998 1999).

Synthetic, pH-sensitive, and anionic polymers have also been designed to aid in endosomal escape and improve transfection efficiencies of non-viral vectors. It was proposed that poly(2-alkylacrylic acid)s are endocytosed by cells into acidified vesicles where they are triggered by low pH to disrupt vesicle membranes and release DNA into the cytoplasm (Jones et al. 2003). Indeed, poly(propylacrylic acid) can enhance cationic-lipid-mediated gene transfer both in vitro (Cheung et al. 2001) and in vivo (Kyriakides et al. 2002). Succinylated-poly(glycidol)-modified liposome also exhibited high transfection efficiency against HeLa and K562 cells in transferrin-mediated gene delivery (Kono et al. 2002). An efficient gene delivery system using pH-sensitive lipid has been reported with cholesteryl hemisuccinate (Shi et al. 2002). Additionally, the pH-dependent morphological change in assembly using cationic gemini surfactants (Glc-CGS) causes destabilization of endosomal membranes and release of DNA into the cytoplasm (Bell et al. 2003).

3 Redox-Responsive Vectors

An increasing number of attractive drug delivery systems that exploit differences in the reduction potential at different locations within cells and at their surfaces have been studied during the last decade (Fig. 3). A disulfide bond (–S–S–) is a covalent

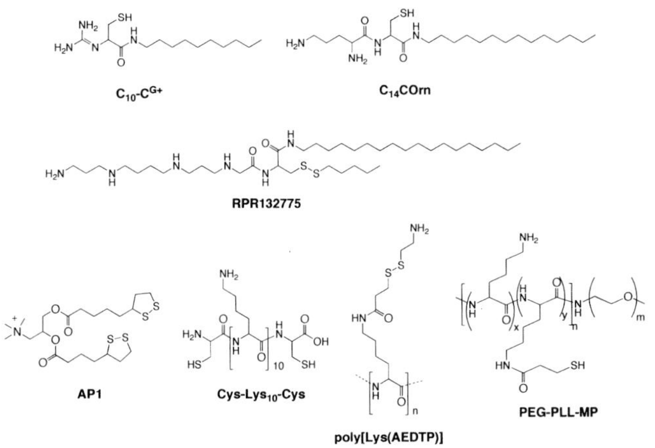

FIG. 3. Chemical structure of compounds used in redox-responsive gene delivery systems

linkage that is obtained from the oxidation of two sulfhydryl (–SH) groups. Two advantages that render disulfide-bonds attractive in drug delivery systems are their reversibility and their relative stability in plasma. In contrast to the cytoplasm, in the extracellular space, the concentrations of glutathione (L-γ-glutamyl-L-cysteinyl-glycine), the most abundant non-protein thiol source in mammalian cells, are much lower, with typical concentrations in plasma being ~10 μM (Meister 1991). As disulfide bonds appear to be important in assembly and unpacking of virus (MacCarthy et al. 1998), this strategy has wide application for the controlled release of DNA from non-viral DNA complexes.

Cysteine-containing cationic lipids, such as $C_{10}C^{G+}$ and C_{14}COrn, were prepared and used to regulate DNA affinity and transfection efficiency. Dimerized lipid with a low cmc condensed plasmid DNA into 30- to 40-nm particles and efficiently transfected into cells in vitro. In these studies, the chemical structures of lipids were set to a cmc (critical micelle concentration) high enough for monomers to perform monomolecular DNA condensation but low enough for the dimers to form stable mesophases capable of cell transfection (Blessing et al. 1998; Dauty et al. 2001).

Another cysteine-containing cationic lipid (RPR132775), bearing spermine as hydrophilic head group, had a 1,000-fold higher transfection efficiency than an analogue without a disulfide bond (Wetzer et al. 2001). In order to reduce the cytotoxicity, natural non-toxic and redox-responsive residues have been incorporated into a cationic lipid. This cationic amphiphile (AP1), made from the natural provitamin lipoic acid, reversibly binds and releases plasmid DNA, depending on the redox state. The increase of transfection efficiency is considered due to intracellular DNA complex reduction by glutathione and/or NAD(P)H and consequent DNA release (Balakirev et al. 2000).

Strategies involving reversible disulfide cross-linking have also been applied to cationic-polymer-mediated gene delivery. For example, many cysteine-containing cationic peptides have been investigated as redox-responsive gene carriers. Multiple cysteinyl residues were inserted into 20-amino-acid cationic peptides to afford reductively activated gene delivery systems. In addition, very stable peptide/DNA condensates were prepared by oxidation of reducible polyplexes, so that intracellular DNA release triggered by disulfide bond reduction resulted in high transfection efficiency (McKenzie et al. 2000b). The structure–activity relationship with respect to transfection efficiency was systematically estimated using a combination of Lys, His, and Cys residues, resulting in low-molecular-weight peptides (McKenzie et al. 2000a). Cell targeting of these disulfide-cross-linked cationic peptide/DNA complexes was obtained by incorporation of a saccharide residue, such as lactose (Park et al. 2002). Another cysteine-containing cationic peptide (Cys-Lys$_{10}$-Cys) was also demonstrated to enhance lipid-mediated gene delivery. A 187-fold higher level of reporter gene expression indicated that intracellular delivery of plasmid DNA using Cys-Lys$_{10}$-Cys/DOTAP was more efficient than that obtained with vectors based on non-reducible PLL (Read et al. 2003).

Furthermore, PLL was modified with 3-(2-aminoethyldithio)propinoyl residues so that each amino group interacted with a phosphate DNA linked to the polymer backbone via a disulfide bond. The cationic polymer (poly[Lys-(AEDTP)]) allowed dissociation of plasmid DNA from complexes in reductive medium and mediated gene delivery into 293T7 and HepG2 cells that was 10- and 50-fold more efficient, respectively, than that with normal PLL (Pichon et al. 2002). Poly(ethylene glycol)-PLL block copolymers bearing sulfhydryl groups (PEG-PLL-MP) were prepared and evaluated as non-viral vectors. The regulation of both the cationic charge and disulfide-cross-linking densities of the backbone polycations allowed a high stability in extracellular medium and an efficient release of plasmid DNA in the intracellular compartment (Miyata et al. 2004). Moreover, low-molecular-weight PEI (800 Da) was also cross-linked with disulfide bonds in order to construct high-molecular-weight PEI having sufficient DNA affinity. Although control of the molecular weight and DNA affinity was confirmed by redox change, the transfection efficiency using disulfide cross-linked PEI was only slightly enhanced (Gosselin et al. 2001).

4 Photo-Responsive Vectors

As light is a very controllable stimulus, both temporally and spatially, photo-irradiation has been used to improve endosomal escape of exogenous DNA and DNA release from complexes (Fig. 4). Photochemical transfections were actively investigated using reactions initiated by photosensitizing compounds localized in endocytic vesicles and inducing the rupture of these vesicles upon light exposure (Høgest et al. 2002; Høgest et al. 2004). In photochemical transfection, photosensitizers, such as porphyrins and phthalocyanines, are initially excited to a state that changes the triplet ground-state of molecular oxygen to highly reactive singlet oxygen. The latter is a powerful oxidant that can oxidize various biomolecules, such as unsaturated fatty acid, certain amino acids, and nucleic acid. Thus, activated oxygen damages and disrupts neighboring endocytic membranes. The release of plasmid DNA from early endosomes seemed to be of importance because photochemical transfection is significantly inhibited by bafilomycinA1, which increases vesicular pH and interferes with endocytic transport during the early stage (Prasmickaite et al. 2000). Furthermore, photochemical transfection by the photosensitizer aluminum phthalocyanine

FIG. 4. Chemical structure of compounds used in photo-responsive gene delivery systems

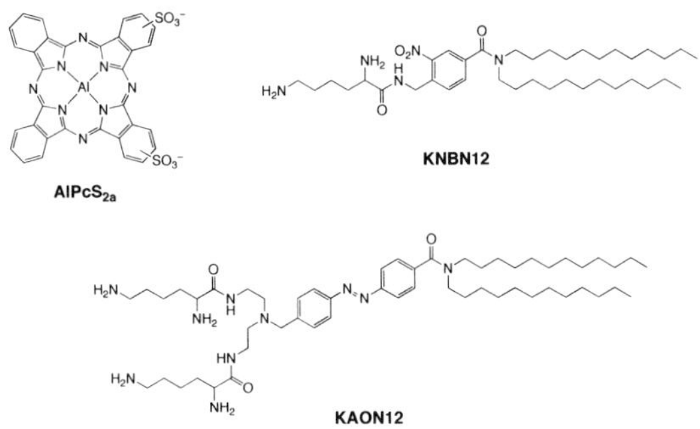

($AlPcS_{2a}$) strongly improves transfection efficiency of cationic polymers (e.g., PLL and PEI), while the effect on the efficiency of cationic lipids is more variable.

The direct introduction of photo-sensitive moieties into cationic lipids has been also reported in order to construct photo-responsive gene delivery systems. The transfection efficiency of a novel photo-responsive cationic lipid (KAON12) having an azobenzene structure between hydrophilic and hydrophobic groups was enhanced with UV irradiation, which causes *trans* to *cis* isomerization and subsequent destabilization of endocytic membranes (Nagasaki et al. 2003b). The photo-cleavage of the backbone of a cationic lipid also resulted in enhanced gene delivery due to the promotion of dissociation of the cationic lipoplex (Nagasaki et al. 2003a). UV irradiation to a cationic lipoplex consisting of plasmid DNA and a cationic lipid (KNBN12) possessing an *o*-nitrobenzyl moiety as photo-cleavable spacer led not only to fusion with endocytic vesicles but also to release of DNA into the cytosol. Photo-induced membrane fusion has also been confirmed with photosensitive liposomes made from KNBN12 and natural phospholipid liposomes.

5 Conclusions

Since most non-viral gene delivery systems involve complex formation with other compounds, disassembly of the complexes must occur before mRNA transcription can proceed. Both the stability of complexation and the rate of dissociation influence transfection efficiency. In other words, the disassembly of DNA complexes diametrically opposes the previous stages of gene delivery; premature dissociation could lead to DNA degradation and low efficacy of gene delivery, while delayed or incomplete dissociation would interfere with efficient gene expression. In order to overcome this dilemma, stimuli-responsive gene delivery is of obvious benefit to achieve temporal and spatial gene expression, both in vitro and in vivo. It is still unclear whether the DNA complex must dissociate before reaching the nuclear envelope or after. The answer might depend on the carrier structure and/or the kind of lipids or polymer. Recently, an intracellular signal-responsive artificial gene-regulation system was developed (Katayama et al 2002). This type of system, which responds to endogenous enzymes and hormones, can be an important step in the development of non-viral gene delivery systems capable of active DNA catch and release.

References

Balakirev M, Schoehn G, Chrobozek J (2000) Lipoic acid-derived amphiphiles for redox-controlled DNA delivery. Chem Biol 7:813–819

Bhattacharya S, Mandal SS (1998) Evidence of interlipidic ion-pairing in anion-induced DNA release from cationic amphiphile-DNA complexes. Mechanistic implications in transfection. Biochem 37:7764–7777

Bell PC, Bergsma M, Dolbnya IP, Bras W, Stuart MCA, Rowan AE, Feiters MC, Engberts JBFN (2003) Transfection mediated by Gemini surfactants: Engineered escape from the endosomal compartment. J Am Chem Soc 125:1551–1558

Blessing T, Remy JS, Behr JP (1998) Monomolecular collapse of plasmid DNA into stable virus-like particles. Proc Natl Acad Sci USA 95:1427–1431

Cheung CY, Murthy N, Stayton PS, Hoffman AS (2001) A pH-sensitive polymer that enhances cationic lipid-mediated gene transfer. Bioconjug Chem 12:906–910

Cho YW, Kim J-D, Park K (2003) Polycation gene delivery systems: escape from endosomes to cytosol. J Pharm Pharmacol 55:721–734

Dauty E, Remy JS, Blessing T, Behr JP (2001) Dimerizable cationic detergents with a low cmc condense plasmid DNA into nanopmetric particles and transfect cells in culture. J Am Chem Soc 123:9227–9234

Gerweck LE, Seetharaman K (1996) Cellular pH gradient in tumor versus normal tissue: Potential exploitation for the treatment of cancer. Cancer Res 56:1194–1198

Gosselin MA, Guo W, Lee RJ (2001) Efficient gene transfer using reversibly cross-linked low molecular weight polyethylenimine. Bioconjug Chem 12:989–994

Haensler J, Szoka FC Jr. (1993) Polyamidoamine cascade polymers mediate efficient transfection of cells in culture. Bioconjug Chem 4:372–379

Høgset A, Prasmickaite L, Hellum M, Engesæter BØ, Olsen VM, Tjelle TE, Wheeler CJ, Berg K (2002) Photochemical transfection: a technology for efficient light-directed gene delivery. Somat Cell Mol Genet 27:97–113

Høgset A, Prasmickaite L, Selbo PK, Hellum M, Engesæter BØ, Bonsted A, Berg K (2004) Photochemical internalization in drug and gene delivery. Adv Drug Deliv 56:95–115

Itaka K, Harada A, Yamasaki Y, Kawaguchi H, Kataoka K (2004) In situ single cell observation by fluorescence energy transfer reveals fast intra-cytoplasmic delivery and easy release of plasmid DNA complexed with linear polyethylenimine. J Gene Med 6:76–84

Jones RA, Cheung CY, Black FE, Zia JK, Stayton PS, Hoffman AS, Wilson MR (2003) Poly(2-alkylacrylic acid) polymers deliver molecules to the cytosol by pH-sensitive disruption of endosomal vesicles. Biochem J 372:65–75

Katayama Y, Fujii K, Ito E, Sakakihara S, Sonoda T, Murata M, Maeda M (2002) Intracellular signal-responsive artificial gene regulation for novel gene delivery. Biomacromol 3: 905–909

Kono K, Torikoshi Y, Mitsutomi M, Itoh T, Emi N, Yanagie H, Takagishi T (2002) Novel gene delivery systems: complexes of fusogenic polymer-modified liposomes and lipoplexes. Gene Ther 8:5–12

Köping-Höggård M, Vårum KM, Issa M, Danielsen S, Christensen BE, Stokke BT, Artursson P (2004) Improved chotosan-mediated gene delivery based on easily dissociated chitosan polyplexes of highly defined chitosan oligomers. Gene Ther 10:in press

Kyriakides TR, Cheung CY, Murthy N, Bornst6ein P, Stayton PS, Hoffman AS (2002) pH-sensitive polymer6s that enhance intracellular drug delivery in vivo. J Control Release 78:295–303

Li W, Nicol F, Szoka FC Jr (2004) GALA: a designed synthetic pH-responsive amphipathic peptide with application in drug and gene delivery. Adv Drug Deliv 56:967–985

Luzio JP, Mullock BM, Pryor PR, Lindsay MR, James DE, Piper RC (2001) Relationship between endosomes and lysosomes. Biochem Soc Trans 29:476–480

MacCarthy MP, White WI, Palmer-Hill F, Koenig S, Suzich JA (1998) Quantitative disassembly and reassembly of human papilomavirus type 11 viruslike particles in vitro. J Virol 72:32–41

McKenzie DL, Smiley E, Kwok KY, Rice KG (2000) Low molecular weight disulfide cross-linking peptide as nonviral gene delivery carriers. Bioconjug Chem 11:901–909

McKenzie DL, Kwok KY, Rice KG (2000) A potent new class of reductively activated peptide gene delivery agents. J Biol Chem 275:9970–9977

Meister A (1991) Gluththione deficiency produced by inhibition of its synthesis, and its reversal; applications in research and therapy. Pharmacol Ther 51:155–194

Miyata K, Kakizawa Y, Nishiyama N, Harada A, Yamasaki Y, Koyama H, Kataoka K (2004) Block catiomer polyplexes with regulated densities of charge and disulfide cross-linking directed to enhance gene expression. J Am Chem Soc 126:2355–2361

Nagasaki T, Taniguchi A, Tamagaki S (2003) Photoenhancement of transfection efficiency using novel cationic lipids having a photocleavable spacer. Bioconjug Chem 14:513–516

Nagasaki T, Wada K, Tamagaki S (2003) Photo-enhancement of transfection efficiency with a novel azobenzene-based cationic lipid. Chem Lett 32:88–89

Park Y, Kwok KY, Boukarim C, Rice KG (2002) Synthesis of sulfhydryl cross-linking poly(ethylene glycol)-peptides and glycopeptides as carriers for gene delivery. Bioconjug Chem 13:232–239

Pichon C, LeCam E, Guérin B, Coulaud D, Delain E, Midoux P (2002) Poly[Lys-(AEDTP)]: A cationic polymer that allowes dissociation of pDNA/cationic polymer complexes in reductive medium and enhance polyfection. Bioconjug Chem 13:76–82

Plank C, Oberhauser B, Mechtler K, Koch C, Wagner E (1994) The influence of endosome-disruptive peptides on gene transfer using synthetic virus-like gene transfer systems. J Biol Chem 269:12918–12924

Pollard H, Remy JS, Loussouarn G, Demolombe S, Behr JP (1998) Polyethylenimine but not cationic lipids promotes transgene delivery to the nucleus in mammalian cells. J Biol Chem 273:7507–7511

Prasmickaite L, Høgset A, Tjelle TE, Olsen VM, Berg K (2000) Role of endosomes in gene transfection mediated by photochemical internalization (PCI). J Gene Med 2:477–488

Read ML, Bremner KH, Oupicky D, Green NK, Searle PF, Seymour LW (2003) Vector based on reducible polycations facilitate intracellular release of nucleic acids. J Gene Med 5:232–245

Schaffer DV, Fidelman NA, Dan N, Lauffenburger DA (2000) Vector unpacking as a potential barrier for receptor-mediated polyplex gene delivery. Biotech Bioeng 67:598–606

Schoen P, Chonn A, Cullis PR, Wilschut J, Scherrer P (1999) Gene transfer mediated by fusion protein hemagglutinin reconstituted in cationic lipid vesicles. Gene Ther 6: 823–832

Shi G, Guo W, Stephenson SM, Lee RJ (2002) Efficient intracellular drug and gene delivery using folate receptor-targeted pH-sensitive liposomes composed of cationic/anionic lipid combinations. J. Control Release 80:309–319

Simoes S, Slepushkin V, Gaspar R, de Lima MC, Duzgunes N (1998) Gene delivery by negatively charged ternary complexes of DNA, cationic liposomes and transferrin or fusigenic peptides. Gene Ther 5:955–964

Simoes S, Slepushkin V, Pires P, Gaspar R, de Lima MP, Duzgunes N (1999) Mechanisms of gene transfer mediated by lipoplexes associated with targeting ligands or pH-sensitive peptides. Gene Ther 6:1798–1807

Wagner E, Plank C, Zatloukal K, Cotton M, Birnstiel ML (1992) Influenza virus hemagglutinin HA-2 N-terminal fusogenic peptides augment gene transfer by transferrin-polylysine-DNA complexes: Toward a synthetic virus-like gene-transfer vehicle. 89:7934–7938

Wetzer B, Byk G, Frederic M, Airiau M, Blanche F, Pitard B, Scherman D (2001) Reducible cationic lipids for gene transfer. Biochem J 356:747–756

Xu Y, Szoka FC Jr (1996) Mechanism of DNA release from cationic liposome/DNA complexes used in cell transfection. Biochem 7:5616–5623

Zabner J, Fasbender AJ, Moninger T, Poellinger KA, Welsh MJ (1995) Cellular and molecular barriers to gene transfer by a cationic lipid. J Biol Chem 270:18997–19007

Controlled Release of DNA Using Thermoresponsive Polymers

Masayuki Yokoyama

1 Significance of DNA Release from Synthetic Gene Carriers

The primary role of synthetic DNA vectors is their translocation through cellular membranes, since only a small amount of naked DNA goes through these membranes due to its large molecular weight and negative charge at very high density. Thus, in order to obtain high rates of uptake, the DNA is commonly complexed with positively charged lipids or polymers, yielding ion complexes packaged in synthetic vectors. The complexes usually have a net positive charge and thus interact with negatively charged cellular membranes, resulting in cellular uptake by endocytosis. Another advantage of packaging DNA within complexes is that the resistance to DNA hydrolytic enzymes is enhanced.

A large number of studies have focused on optimizing complex formation using various chemical structures of lipids and polymers to maximize the transfection efficiency of synthetic vectors. In contrast, only a few studies have examined the dissociation of the DNA-carrier complexes. In order for the transfected DNA to be recognized by transcription enzymes or form base pairs, thereby allowing, respectively, transcription or the formation of duplexes or triplexes via antisense DNA, it must be in the naked form. Therefore, both the formation and the dissociation of the complex are important to achieving highly effective synthetic DNA carrier systems.

As illustrated in Fig. 1, the optimization of DNA-carrier complex formation is important for the efficient entry of the complexes into cells and the protection of DNA from degradation. However, subsequent quick release of the DNA from the complex is necessary in order to accelerate the expression of DNA functions. In addition to the enhancement of transfection efficiency, controlled DNA release allows selective expression of DNA functions, such as gene expression and antisense action, which are important for medical applications. In many types of gene therapy, controlled gene expression is not strictly required for some genes. Adenosine deaminase is a typical example in that therapeutic effects are obtained over a wide range of adenosine deam-

Kanagawa Academy of Science and Technology, KSP East 404, 3-2-1 Sakado, Takatsu-ku, Kawasaki 213-0012, Japan

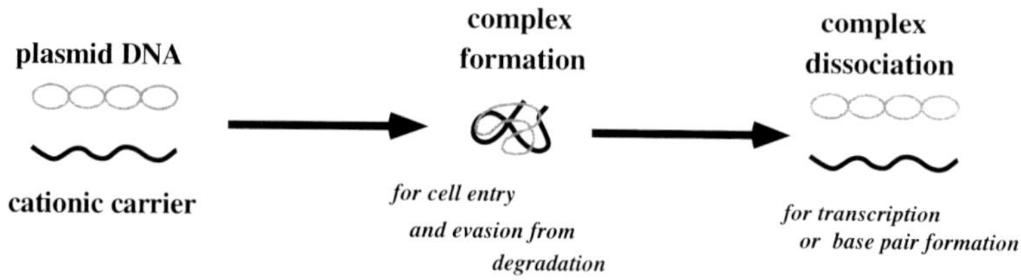

FIG. 1. Significance of complex formation and dissociation

TABLE 1. Methodologies for specific gene expression.

1. Specific promoters
* signal-inducible (e.g., heat shock, lactose, tertacycline)
* organ-, tissue-specific
2. Selective delivery to target with carriers
3. Stimuli-responsive gene carriers

inase concentrations in the blood. Additionally, the cell type and the location of protein production are not restricted as they are in nature if the produced protein is to be present in the blood. In contrast, the quantity, site, and timing of protein production must be strictly controlled in order to obtain the therapeutic effects of many other kinds of proteins, for example, cytokines for gene therapies of cancer. Naturally, the amount of cytokines, as well as the site, timing, and duration of production are strictly controlled by the immune system. Therefore, cytokine expression in an excess amount, at unwanted sites, or for too long causes severe toxic side effects.

If DNA release from the DNA-carrier complex is controlled by external stimuli or internal signals that are specific to the therapeutic sites, the desired amount of selective gene expression at a specific site, with the correct timing, and for the appropriate duration could be accomplished. This is analogous to drug targeting systems (for conventional drugs) that achieve selective drug action at the therapeutic sites without toxic side effects. In most gene therapy protocols that have been carried out to date, gene expression was not precisely controlled in terms of the parameters mentioned above. The specific regulation of gene expression, particularly in vivo, would substantially improve progress in gene therapy.

2 Stimuli-Responsive Synthetic DNA Carriers

Table 1 summarizes three approaches to obtaining site-, timing-, and duration period-specific gene expression,. Firstly, the use of specific promoters (Russel 1999) can initiate transcription of the gene of interest selectively in a specific environment. Physical signals, such as γ-ray (Stackhouse and Buchsbaum 2000; Nuyts et al. 2001) and temperature (Schweinfest et al. 1988), and chemical signals, such as lactose (Jacobsen and Willumsen 1995) and tertacycline (Gossen et al. 1995; Furth et al. 1994), induce transcription of genes placed downstream of the respective signal-inducible promoter. Alternatively, tissue- or organ-specific promoters have been used to obtain the corresponding tissue- or organ-specific gene expression, even if the exact nature of the signal is not always known. Using these specific promoters, induction ratios, defined as the ratio of gene expression in the presence and absence of the signal, of up to several thousand-fold have been obtained. The utility of many of these promoters, however, is considerably limited, since their specificity is based on the the binding properties of naturally occurring transcription factors. For example, the heat-shock-protein promoter responds only to a temperature (42°C) that is dangerous to living bodies. Artificial protein engineering to change the signal temperature is very difficult using current technologies. Furthermore, the specific delivery of chemicals to therapeutic sites poses another problem for chemical-signal-inducible promoters, since, in general, the specific concentrations and duration of action of these chemicals at the target sites are not easily achieved in vivo.

The second approach to selective gene expression is delivery by carriers. Similar to the delivery of pharmaceutical drugs, several types of carriers, such as antibodies, liposomes, and synthetic polymers, have been studied with the goal of achieving the selective delivery of genes. Even without specific promoters, selective gene expression is possible by the selective delivery of genes. In nature, viral vectors work as selective carriers by utilizing their infection tropism for specific cell types.

The third approach to obtaining specific gene expression is by using stimuli-responsive gene carriers. Synthetic vectors induce transcription by releasing the DNA component of the DNA-carrier complex at specific environments. Thus, stimuli-responsive gene carriers not only allow site-specific gene expression, but they also contributes to greatly enhancing the efficiency of gene expression. As stated above, it is very difficult to optimize complex formation and dissociation using ion-complex-type synthetic gene carriers, since these two processes are in opposition to each other. In order to overcome this difficulty, intelligent systems to control complex formation/dissociation by external stimuli are being studied. By applying physical stimuli. such as light and temperature, complex formation/dissociation or tight/loose complex formation may be achieved.

One of the benefits of stimuli-responsive vector systems is high gene expression efficiency. As shown in Fig. 2, in conventional cationic vector systems, the complex must be formed with an intermediate tightness in order to simultaneously fulfill the two opposing requirements of protection from enzymatic degradation and free access to RNA polymerase. However, an intermediate tightness prohibits maximum efficiency for either function. Similarly, for stimuli-responsive vector systems, the avoidance of DNA degradation can be optimized by forming a very tight complex, while transcription efficiency is maximized by complex dissociation or loose complex formation. Gene expression efficiency (transfection efficiency) is a multi-step process that is dependent upon the efficiency of various cellular processes, including attachment to the cell surface, endocytosis, translocation from endosome to cytosol and then from cytosol to nucleus, and, finally, transcription. In contrast to cationic vector systems, efficiency may be maximized in stimuli-responsive gene carrier systems, since the efficiency of each cellular step can be increased without decreasing the efficiencies of the other steps.

The second benefit of stimuli-responsive vectors is that site-, timing-, and duration-specific gene expression may be achieved by releasing the DNA from the DNA-carrier complex in a stimuli-responsive manner. Compared with the other methodologies

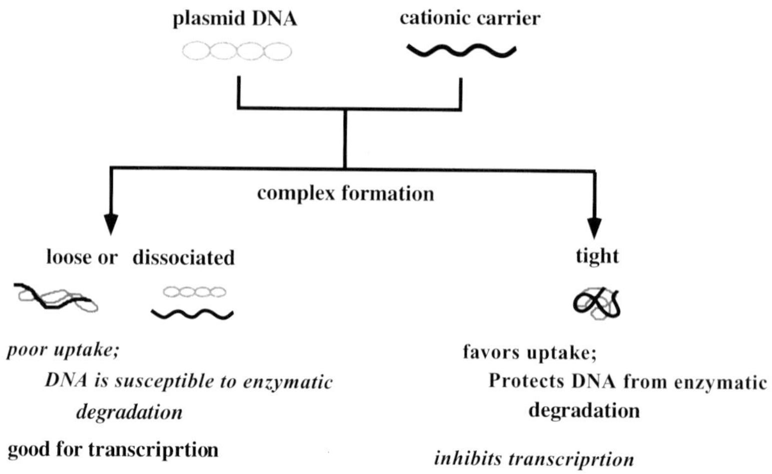

Fig. 2. The properties of tight vs. dissociated or loose DNA-cationic carrier complexes

listed in Table 1, stimuli-responsive synthetic vectors possess several advantages. First, fine adjustments in stimuli applied to living bodies are possible; for example, while in the heat-shock-protein promoter system, the inducing temperature is fixed for, in the stimuli-responsive synthetic systems it can be easily changed by designing appropriate chemical structures. Second, site specificity is easily obtained by the application of light and temperature. While light possesses advantages over temperature with respect to site precision, temperature can also be applied with considerable specificity, such as by using a 5-mm ultrasound device (Umemura et al. 1996; Kawabata and Umemura 1996). In addition, a temperature stimulus has advantages over light in terms of available depth from the surface. In particular, such site-specificity can be readily obtained in a uniform tissue (e.g. dermis and muscle) or organ.

Nagasaki et al. developed novel synthetic light-inducible gene expression systems using dendrimers and lipids. These studies are described elsewhere in this book.

3 Temperature-Responsive Polymers and Their Applications

Temperature is one of the safest external signals that can be applied to a variety of sites of living bodies. Site-specific heating is now routinely carried out in hyperthermia therapy against solid tumors (Falk and Issels 2001). The hyperthermia, which is localized, is usually applied by microwave, which selectively heats the target site with a precision of approximately 1 cm. Furthermore, since microwave heating can reach a considerable depth from the body surface, this therapy can be used to treat most internal organs and tissues, even though the microwave radiation is applied from outside the body.

In a separate approach, thermoresponsive synthetic polymers have been actively studied due to strong interest in their unique physico-chemical characteristics as well

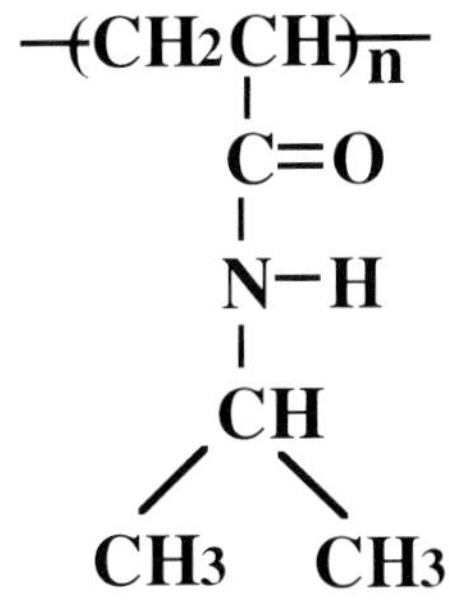

Fig. 3. Chemical structure of poly(*N*-isopropylacrylamide) (P(IPAAm))

as to their potential applications in medicine and biotechnology. Among these thermoresponsive polymers, poly(*N*-isopropylacrylamide) [P(IPAAm)] (Fig. 3) has received the most attention, mainly owing to the fact that its phase-transition of temperature 32°C is near body temperature (Heskins and Guillet 1968). P(IPAAm) has been applied extensively to biomedical applications, such as hydrogels (Yoshida et al. 1995; Kaneko et al. 1995), bioconjugates (Matsukata et al. 1996), and polymeric micelles as drug carriers (Chung et al. 1998, 1999; Kohori et al. 1998). The phase-transition temperature of P(IPAAm) is a lower critical solution temperature (LCST). Below the LCST, P(IPAAm) is water-soluble, hydrophilic, and assumes an extended chain form, while above the LCST, it undergoes a reversible phase transition to form an insoluble and hydrophobic aggregate. The other benefits of P(IPAAm) for biomedical applications are its simple chemical structure and sharp phase-transition behavior. The phase transition temperature of P(IPAAm) can be altered by copolymerizing IPAAm with another monomer unit. Copolymerization with hydrophilic and hydrophobic monomers results in an increase or decrease of the LCST, respectively. For example, in order to obtain a thermoresponsive drug releasing system, the phase-transition temperature was adjusted to 39°C by introducing a hydrophilic comonomer, *N*,*N*-dimethylacrylamide.

4 DNA Carrier Systems Using Temperature-Responsive Polymers: Temperature-Responsive Polymeric Vector Systems

A temperature-responsive polymeric vector system based on poly(*N*-isopropylacrylamide) was developed by Kurisawa, Takeda, and Okano (Takeda et al. 2004; Kurisawa et al. 2000; Yokoyama et al. 2001). Using this polymer as carrier, a tight complex with DNA formed above the transition temperature by hydrophobic aggregation of poly(*N*-

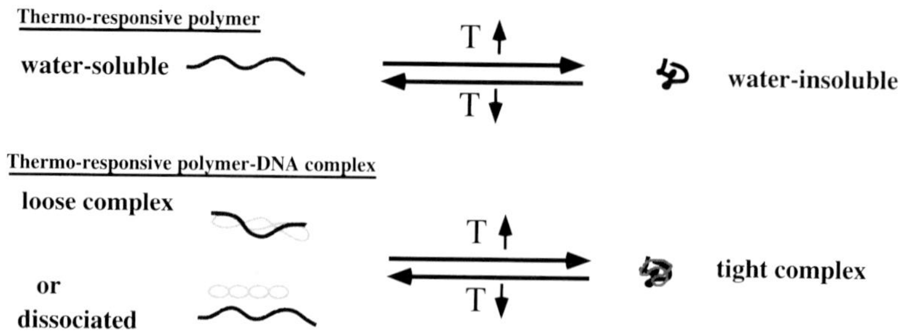

FIG. 4. Concept of thermoresponsive gene carriers. *T* Phase-transition temperature

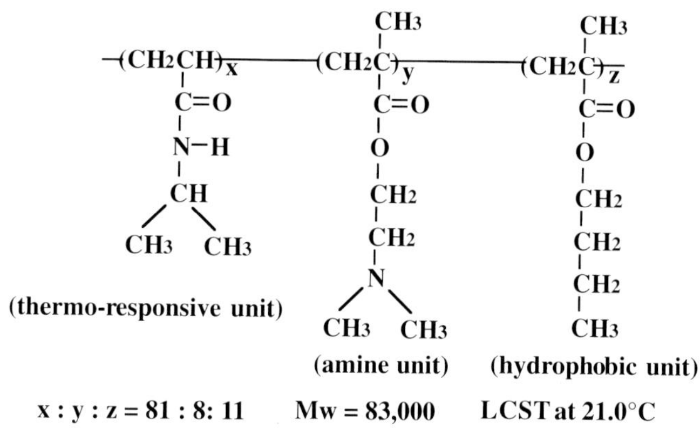

FIG. 5. Chemical structures of thermoresponsive gene carriers. *LCST* Lower critical solution temperature

isopropylacrylamide), as shown in Fig. 4. Below the transition temperature, the complex dissociated or loosened owing to the dissolution of the hydrophobic aggregate. A tightly formed complex is favorable for efficient cellular uptake and avoidance of enzymatic degradation of DNA, while free DNA released from the dissociated or loosely packed complex is favorable for transcription. Therefore, by switching from a tight complex and to a dissociated (or loosely formed) one contribute to optimizing the efficiency of those cellular processes needed for selective gene expression such as binding to cells, entry into cells, and transcription.

A temperature-responsive polymeric gene carrier was designed as shown in Fig. 5. This polymeric carrier was composed of three components; a temperature-responsive unit, a cationic unit, and a hydrophobic unit. The cationic unit interacts with anionic DNA to give high DNA yields in the polymer-DNA complex. The hydrophobic unit adjusts the phase-transition temperature. The poly(*N*-isopropylacrylamide) homopolymer possesses a LCST of 32°C in water. Introduction of an amine unit to homopolymer chain raises the LCST, whereas the addition of hydrophobic units lowers the LCST. The LCST of this copolymer was set below physiological body temperature in order to test temperature enhancement of gene expression. If the LCST is

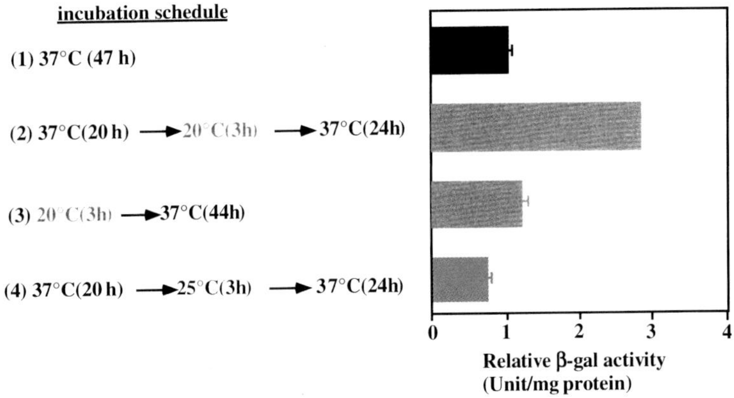

FIG. 6. Increase in β-galactosidase expression by lowering incubation temperature

set above 37°C, then enhanced gene expression following an increase in temperature can be considered to be due to both cell activation and formation/dissociation control of the complex. When the temperature is decreased to less than 37°C, then only enhanced gene expression occurs, since cell activities such as protein synthesis are reduced at low (<37°C) temperature.

The copolymer shown in Fig. 5 was mixed with plasmid DNA encoding the β-galactosidase gene at 37°C to form a polymer-DNA complex. The phase-transition temperature of this copolymer was 21°C, which was also the transition temperature after complex formation with DNA. This complex (charge ratio of polymer: DNA = 3:1) was added to COS-1 cells, and incubated for 1 h at 37°C. After removal of the complex solution, fresh DMEM medium containing 10% FBS was added to the cells, which were then cultured at various times and temperature schedules, as shown in Fig. 6. Compared with cells cultured according to incubation schedule (1), which lacked any cooling period at 20°C, β-galactosidase activity of cells cultured according to schedule (2) was significantly higher. Under this condition, the number of β-galactosidase-expressing cells also increased, as shown in Fig. 7. Generally, temperatures lower than 37°C are disadvantageous for transfection. In fact, the transfection efficiency of a homopolymer of the amine unit was found to decrease by half when the incubation temperature was lowered (Kurisawa et al. 2000). Interestingly, gene expression of the temperature-responsive vector system was controlled by the precise conditions of the incubation schedules. For example, when the cells were cooled immediately after the incubation period with the complex [schedule (3)], the transfection efficiency was almost the same in for schedule (1) cells. When the cells were cooled at 25°C, which is higher than the phase-transition temperature of this vector system, no increase in transfection efficiency was observed [schedule (4)]. These results indicate that the enhanced transfection efficiency obtained by lowering the temperature was most likely mediated by preferential DNA release during late cellular processes in the cytoplasm and nucleus.

In order to achieve an even greater enhancement of gene expression, the incubation period of the DNA-carrier complex with cells was varied (Yokoyama et al. 2001).

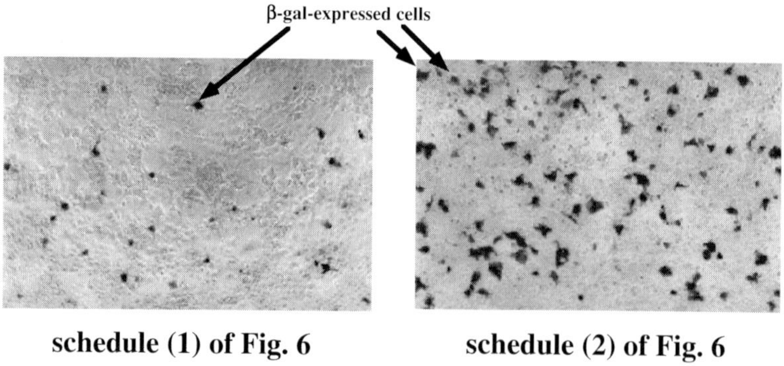

Fig. 7. Increase in the number of β-galactosidase-expressing cells by lowering incubation temperature

The enhancement ratio (activity with the cooling procedure/activity without the cooling procedure) reached 8.6 following a 3-h incubation period. The ability to selectively enhance expression may be very useful in obtaining site and time-specific expression of foreign genes in biological and medical applications.

While the above-mentioned copolymer showed thermoresponsive gene expression, the transfection efficiency was the same as that obtained with a homopolymer of the cationic unit used as the thermoresponsive copolymer, even when the appropriate temperature change was applied. Therefore, in order to raise the transfection efficiency, the amounts of both the amine and hydrophobic units were increased (Takeda et al. 2004). Figure 8 compares the transfection efficiency of two thermoresponsive copolymers and the amine homopolymer in the presence of a very small amount of DNA in order to evaluate transfection in the background of a high level of cell viability. Copolymer A indicates the copolymer whose composition is shown in Fig. 5, and copolymer B is the modified copolymer possessing more amine and hydrophobic units. Using a small amount of DNA per cell, only copolymer B exhibited prominent transfection activity (Fig. 8a) and maintained thermo-responsive transfection behavior (Fig. 8b). Consequently, a synthetic gene carrier possessing both thermoresponsive gene expression and greatly enhanced transfection efficiency was successfully synthesized by optimizing the content of both amines and hydrophobic units.

The above mentioned copolymer system was the first example of temperature-mediated control of gene expression using a synthetic vector. In this system, the phase-transition temperature was below 37°C, and expression of the target gene was enhanced by DNA release inside the cells. Alternatively, enhanced gene expression can be obtained by applying heat in the following way. If the phase-transition temperature of the DNA-carrier complex is set above that of body temperature, selective cellular uptake of the complex may be obtained by localized heating at a specific site and time. Subsequently, the complex in the cytoplasm or nucleus is dissociated at body temperature, which is below the phase-transition temperature, and gene expression occurs via DNA released from the complex. Thus, gene expression is selectively obtained at the heated site. The feasibility of this approach

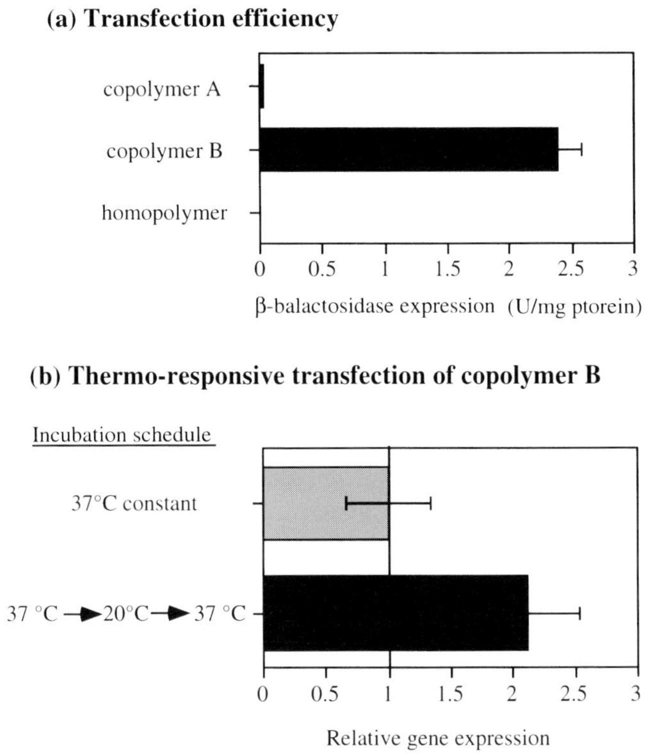

FIG. 8. A copolymer showing high transfection efficiency (**a**) while maintaining its thermoresponsive properties (**b**)

has been reported (Yokoyama et al. 2001). Taken together, the results show that temperature-responsive gene carrier systems may be widely applicable to in vivo gene therapies, both by heating and cooling the therapeutic sites to control gene expression.

Lastly, the possible importance of DNA release outside cells should also be pointed out. This chapter has focused on intracellular delivery of DNA to the nucleus, since in vivo applications (e.g. intravenous injection) using naked DNA are not possible, due to its negligible transfection efficiency. Intramuscular injection of naked DNA, however, is an exception. Additionally, high in vivo transfection efficiency of naked DNA using a hydrodynamics-based procedure was recently reported (Liu et al. 1999; Zhang et al. 2000). In this process, a large volume of DNA solution (1.5 ml to 3.0 ml per mouse) is injected over a short time period (several seconds). The naked DNA, without any vector, transfects cells at high efficiency. Gene expression efficiencies obtained by intramuscular injection or the hydrodynamic method are considerably lower than obtained with vector-based systems due to enzymatic degradation of naked DNA in muscular tissue fluid and the bloodstream (Miyao et al. 1995; Mahato et al. 1995). Therefore, DNA release both inside and outside the target cells using temperature-responsive vector systems deserves further study.

References

Chung JE, Yokoyama M, Aoyagi T, Sakurai Y, Okano T (1998) Effect of molecular architecture of hydrophobically modified poly(*N*-isopropylacrylamide) on the formation of thermo-responsive core-shell micellar drug carriers. J Contr Rel 53:119–130

Chung JE, Yokoyama M, Yamato M, Aoyagi T, Sakurai Y, Okano T (1999) Thermo-responsive drug delivery from polymeric micelles constructed using block copolymers of poly(*N*-isopropylacrylamide) and poly(butylmethacrylate). J Contr Rel 62:115–127

Falk MH, Issels RD (2001) Hyperthermia in oncology. Int J Hyperthermia 17:1–18

Furth PA, St Onge L, Boger H, Gruss P, Gossen M, Kistner A, Bujard H, Hennighausen L (1994) Temporal control of gene expression in transgenic mice by a tetracycline-responsive promoter. Proc Natl Acad Sci USA 91:9902–9306

Gossen M, Freundlieb S, Bender G, Muller G, Hillen W, Bujard H (1995) Transcriptional activation by terracyclines in mamalian cells. Science 268:1766–1769

Heskins M, Guillet JE (1968) Solution properties of poly(*N*-isopropylacrylamide). J Macromol Sci Chem A2:1441–1455

Jacobsen KD, Willumsen BM (1995) Kinetics of expression from vectors containing beta-galactosidase in murine fibroblasts: High initial rate compared to steady-state expression. J Mol Biol 252:289–295

Kaneko Y, Sakai K, Kikuchi A, Yoshida R, Sakurai Y, Okano T (1995) Influence of freely mobile grafted chain length on dynamic properties of comb-type grafted poly(*N*-isopropylacrylamide) hydrogels. Macromolecules 28:7717–7723

Kawabata K, Umemura S (1996) Use of second-harmonic superimposition to induce chemical effects on ultrasound2. J Phys Chem 100:18784–89

Kohori F, Sakai K, Aoyagi T, Yokoyama M, Sakurai Y, Okano T (1998) Preparation and characterization of thermally responsive block copolymer micells comprising poly(*N*-isopropylacrylamide-*b*-DL-lactide). J Contr Rel 55:87–98

Kurisawa M, Yokoyama M, Okano T (2000) Gene expression control by temperature with thermo-responsive polymeric gene carriers. J. Controlled Release 69:127–137

Liu F, Song Y, Liu D (1999) Hydrodynamics-based transfection in animals by systemic administration of plasmid DNA. Gene Therapy 6:1258–1266

Mahato RI, Kawabata K, Takakura Y, Hashida M (1995) In vivo disposition characteristics of plasmid DNA complexed with cationic liposomes. J. Drug Targeting 3:149–157

Matsukata M, Aoki T, Sanui K, Ogata N, Kikuchi A, Sakurai Y, Okano T (1996) Effect of molecular architecture of poly(N-isopropylacrylamide)-trypsin conjugates on their solution and enzymatic properties. Bioconjugate Chem 7:96–101

Miyao T, Takakura Y, Akiyama T, Yoneda F, Sezaki H, Hashida M (1995) Stability and pharmcokinetic characteristics of oligonucleotides modified at terminal linkages in mice. Antisense Research and Development 5:115–121

Nuyts S, Van Mellaert L, Theys J, Landuyt W, Bosmans E, Anne J, Lambin P (2001) Radio-responsive recA promoter significantly increases TNFα production in reconbinant clostridia after 2 Gy irradiation. Gene Therapy 8:1197–1201

Russell M (1999) Inducible mamalian expression systems. In: Fernandez JM and Hoeffler JP (Eds) Gene Expression Systems, Academic Press, San Diego, pp 235–257

Schweinfest CW, Jorcyk CL, Fujiwara S, Papas TS (1988) A heat shock-inducible eukaryotic epxression vector. Gene 71:207–210

Stackhouse MA, Buchsbaum DJ (2000) Radiation to control gene expression. Gene Therapy 7:1085–1086

Takeda N, Nakamura E, Yokoyama M, Okano T (2004) Temperature-responsive polymeric carriers incorporating hydrophobic monomers for effective transfection in small doses. J Controlled Release 95:343–355

Umemura S, Kawabata K, Sasaki K, Yumita N, Umemura K, Nishigaki R (1996) Recent advances in sonodynamic approach to cancer therapy. Ultrasonics Sonochemistry 3: S187–191

Yokoyama M, Kurisawa M, Okano T (2001) Influential factors on temperature-controlled gene expression using thermo-responsive polymeric gene carriers. J Artificial Organs 4: 138–145

Yoshida R, Uchida K, Kaneko Y, Sakai K, Kikuchi A, Sakurai Y, Okano T (1995) Comb-type grafted hydrogels with rapid de-swelling response to temperature changes. Nature 374: 240–242

Zhang G, Song YK, Liu D (2000) Long-term expression of human alpha1-antitrypsin gene in mouse liver achieved by intravenous administration of plasmid DNA using a hydrodynamics-based procedure. Gene Therapy 7:1344–1349

Active Transport of Exogenous Genes into the Nucleus

Takeshi Kawazu and Takeshi Nagasaki

1 Introduction

Gene delivery into mammalian cells requires the nuclear import of exogenous DNA. As non-viral vectors lack the ability to migrate into the nucleus by themselves, the efficiency of their nuclear import is very low. Indeed, in non-dividing cells, transgene expression is very difficult. When the herpes simplex virus thymidine-kinase gene was injected by glass micropipettes into the nucleus of a mouse cell line deficient in thymidine kinase activity, it was expressed by 50–100% of the injected cells. But when the gene was injected into the cytoplasm, enzymatic activity was not detected (Capecchi 1980). Therefore, expression of DNA introduced using a non-viral vector was expected to improve if migration into the nucleus occurred during mitosis, in which the nuclear membrane disappears. Indeed, the expression efficiency of a transgene is strongly dependent on the cell-cycle phase at transfection. When cells were incubated with lipoplexes or polyplexes during or just before mitosis, expression efficiency of the transgene was increased, suggesting that transfection close to M phase is facilitated by nuclear membrane breakdown (Mortimer et al. 1999; Tseng et al. 1999; Brunner et al. 2000). In contrast, this effect was not observed with recombinant adenovirus, which has an efficient nuclear entry machinery (Brunner et al. 2000). Accordingly, nuclear import of a transgene is currently one of the most difficult issues in non-viral gene delivery.

In non-viral gene delivery, the mechanism of nuclear import of a transgene is poorly understood. When the transgene migrates from the cytoplasm to the nucleus, except for during mitosis, it must pass through the nuclear pore complexes (NPC) that are present in the nuclear envelope. The NPC acts as a water-filled channel and has a diameter of about 10 nm (Paine et al. 1975). The size of freely diffusible molecules through the NPC is exactly limited. Molecules smaller than 40~60 kDa can migrate into the nucleus by diffusion, but molecules larger than 60 kDa must utilize active-transport-system-mediated importins. As for the entry of DNA molecules, the limit of the size of DNA that can enter the nucleus via passive diffusion was found to be

Department of Applied and Bioapplied Chemistry, Graduate School of Engineering, Osaka City University, 3-3-138 Sugimoto, Sumiyoshi-ku, Osaka 558-8585, Japan

between 200 and 310 bp (Ludtke et al. 1999). Microinjection experiments have also shown that plasmid DNA introduced in the cytosol diffuse only a short distance from the site of injection (Dowty et al. 1995). When biotin- or gold-labeled plasmids were microinjected into the cytoplasm of primary rat myotubes in culture, the majority of the injected plasmid DNA was sequestered by cytoplasmic elements. Therefore, in order to import DNA larger than 310 bp or plasmid DNA into the nucleus, it is essential to utilize the active transport system.

To date, a great deal of effort has been made to achieve the active transport of exogenous genes from the cytoplasm into the nucleus. In order to enhance the nuclear import of DNA, endogenous cellular mechanisms of nucleo-cytoplasmic traffic have been utilized. Nuclear localization signals (NLS) were conjugated to DNA by various means, including covalent or noncovalent binding, and association with karyophilic proteins that have NLSs as part of their amino acid sequences. With this approach, the nuclear import of DNA/NLS or DNA/karyophilic protein was observed following direct microinjection of these constructs into the cells or using digitonin-treated cells. Also, the expression efficiencies of the constructs in the presence of various polymers or lipids were evaluated. In this chapter, we will discuss the various strategies to actively transport DNA into the nucleus. For additional articles on the nuclear import of exogenous genes, the reader is referred to recent reviews (Cartier et al. 2002; Escriou et al. 2003; Hébert 2003; Munkoge et al. 2003; Tachibana et al 2001).

2 Nuclear-Pore Complex and Nuclear Import of Karyophilic Proteins

The transport of molecules between the cytoplasm and the nucleus occurs through the NPCin the nuclear envelope. The NPC is a huge protein complex of about 125 MDa and is composed of 50–100 different proteins. Molecules smaller than 40–60 kDa (up to 9 nm in diameter) can freely diffuse through the central channel of the NPC, while molecules larger than 40–60 kDa cannot and are instead transported across the NPC by an active process (Yoneda et al. 1999). The size of molecules that can pass through the NPC by active transport is also limited. A recent study showed that cargo-receptor-gold complexes with diameters close to 39 nm were translocated by the NPC (Pante et al. 2002).

Active nuclear transport is mediated by a family of proteins called importins. Transport occurs either directly or indirectly via nuclear localization signals (NLSs), which are contained in the amino acid sequences of karyophilic proteins. NLSs are roughly classified into classical and non-classical types. The most wellcharacterized NLS is that of SV40 large T-antigen, called the classical NLS, which contains basic amino acids, like lysine or arginine. Importin-α primarily recognizes and binds to the classical NLS. Subsequently importin-β, which can interact with NPC, recognizes and binds to importin-α. In the other words, importin-α acts as an adaptor between a cargo molecule and a carrier, importin-β. This ternary complex then migrates into the nucleus through the NPC (Görlich et al. 1996). By contrast, the NLS M9, which exists in heterogeneous nuclear ribonucleoprotein A1 (hnRNP A1), is not highly cationic. Transportin, which is a member of the importin-β family and can interact with NPC, directly recognizes and binds to M9, allowing transport of this complex into the nucleus (Nakielny et al. 1996) (Fig.1).

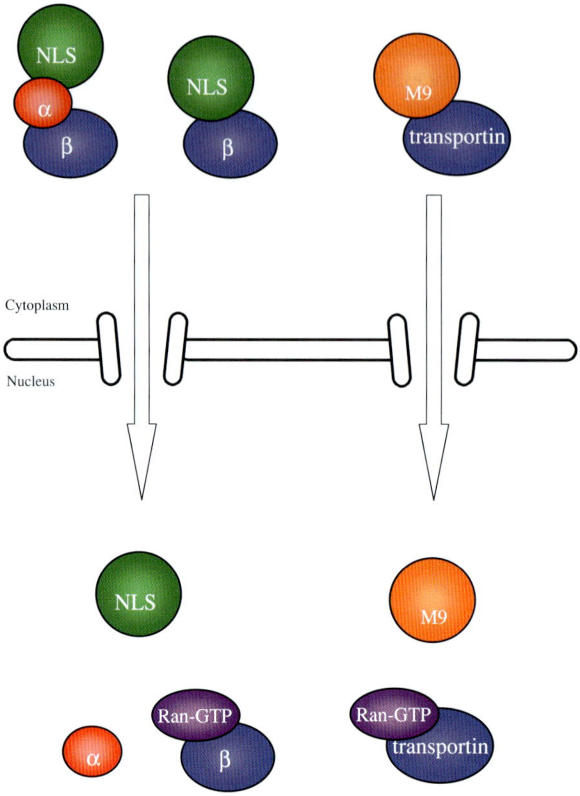

FIG. 1. Nucleocytoplasmic transport system mediated by a nuclear localization signal (*NLS*). *Ran*, Ras related nuclear protein

Active nuclear transport mediated by the NLS enables non-karyophilic proteins to migrate into the nucleus (Goldfarb et al. 1986; Lanford et al. 1986; Yoneda et al. 1992). A 13-mer synthetic peptide containing seven amino acids homologous to SV40 T antigen was capable of inducing nuclear transport, but no transport was observed when proteins were coupled with a synthetic peptide homologous to a nuclear-transport-defective T antigen. The largest protein-peptide conjugate efficiently transported was ferritin (465 kDa) (Lanford et al. 1986). From these studies, the active nuclear import of a transgene coupled to a NLS would be expected.

Nuclear import of glycosylated proteins in a sugar-dependent manner has also been shown (Duverger et al. 1995; Monsigny et al. 2004) and was found to be energy dependent, since transport does not occur when cells are maintained at 4°C or when they are ATP-depleted by treatment with apyrase. The nuclear import of neoglycoproteins occurs through the nuclear pore. It is inhibited by preincubation of the cells with wheat-germ agglutinin, a lectin that binds to nuclear-pore glycoproteins and blocks the translocation step of NLS-bearing proteins through the NPC. Moreover, the nuclear import of neoglycoproteins is distinct from transport via the classical NLS pathway, since the latter depends on cytosolic factors and is inhibited by treatment of cells with *N*-ethylmaleimide. By contrast, the nuclear import of neoglycoproteins neither requires added cytoplasmic factors nor is it sensitive to alkylation by *N*-ethylmaleimide. In addition, upon incubation in the presence of a large excess of NLS-bearing protein, the nuclear import of neoglycoproteins is not inhibited (Duverger et al. 1995). This study suggests that nuclear import of a transgene is enhanced by the use of glycosylated carriers.

3 Nuclear Import of Virus Genomes

Whereas the efficiency of nuclear import of a transgene in an artificial gene delivery system is very low, viruses can efficiently move their genomes into the nucleus of host cells for replication. Viral genomes are transported into the nucleus by three different mechanisms, according to the size of the capsid enclosing the genome (Greber et al. 2003). The first mechanism involves disassembly of the capsid in the cytosol or in a docked state at the NPC, so that a subviral genomic complex is transported through the pore. Adenovirus particles directly bind to the NPC protein CAN/Nup214, located at the cytoplasmic filaments. The viral affinity for the NPC is independent of cytoplasmic factors but virus particles are not imported without importins (Trotman et al. 2001; Wisnivesky et al. 1999). In the second mechanism, the genome is injected from a capsid that is bound to the NPC. Capsids for purified herpes simplex virus-1 were found to dock to the cytoplasmic face of NPCs on rat-liver nuclear envelopes in the absence of cytoplasmic factors (Ojala et al. 2000). DNA ejection from the docked capsid required cytoplasmic factors and energy. The third mechanism consists of the recruitment of the import factors importins into cytosolic capsids, thus increasing the affinity of the capsid for the NPC and mediating translocation, followed by disassembly in the nucleus.

Thus, the mechanisms of nuclear import of viral genomes obviously imply that efficient nuclear import of a transgene requires the utilization of inherent cellular machinery for nucleo-cytoplasmic traffic.

4 Nuclear Import of DNA Conjugated with NLS Peptides

To date, a great deal of effort has focused on enhancing transfection efficiencies with non-viral vectors by conjugating the various NLS peptides to exogenous DNA.

Several reports showed that NLS non-covalently bound to exogenous DNA enhanced the nuclear import of the latter (Malecki 1996; Aronsohn et al. 1998, Schwartz et al. 1999). Additionally, plasmid DNA coupled to the NLS derived from SV40 large T-antigen was efficiently targeted to zebrafish embryo nuclei (Collas et al. 1996).

However, different results were obtained regarding the sort or size of DNA that can be used. In vitro studies have shown that the nuclear import of DNA (PCR products) utilized an active nuclear transport process through the NPC and was energy dependent, but it did not require the addition of soluble cytoplasmic extract (Hagstrom et al. 1997; Ludtke et al. 1999). Plasmid DNA, however, was imported into the nucleus in the presence of both cytoplasmic and nuclear extracts, suggesting that nuclear import of plasmid DNA occurs by a protein-mediated mechanism (Wilson et al. 1999).

DNA/NLS conjugates prepared by non-covalent bonding run the risk of dissociating in the cytoplasm after internalization by the cells. Therefore, covalently modified DNA/NLS conjugates have been developed. For example, the covalent modification of plasmid DNA with the SV40 large T antigen NLS peptide led to nuclear accumulation of the conjugate DNA in digitonin-permeabilized cells through the NPC in a manner similar to that of the nuclear import of karyophilic proteins. After microinjection into the cytoplasm of live HeLa cells, significant nuclear import of plasmid DNA, however, required more than 100 NLSs per 1 kb of DNA (Sebestyén et al. 1998). The modification of cationic polymers with NLS-like peptides has also been reported. Melittin, which contains the NLS KRKR, was conjugated with polyethylenimine (PEI) and subsequently enhanced the nuclear access of plasmid DNA (Orgis et al. 2001).

The classical NLS-DNA conjugate must be recognized by importin-α, the adapter for importin-β. However, in the covalent or non-covalent conjugate with DNA, the classical NLS may be buried in the DNA due to electrostatic interactions, and thus cannot interact with importin-α. It has been reported that accumulation in the nucleus needed more than 40 classical NLSs per 1 kb of DNA (Sebestyén et al. 1998; Ciolina et al. 1999). Therefore, when a classical NLS like that of SV40 large T antigen is attached to plasmid DNA, the DNA/NLS has to include a spacer that allows NLS recognition by importin-α. Indeed, compared to NLS-plasmid DNA conjugates containing a short methylene spacer, the conjugate that was covalently bonded by diazo coupling through a PEG chain (MW 3400) efficiently stimulated complexation with the nuclear-transport proteins importin-α and importin-β (Nagasaki et al. 2003).

The spacer problem in the DNA/classical NLS system has been resolved by using a non-classical NLS, that contains the M9 sequence of heterogeneous nuclear ribonucleoprotein. The M9 sequence was conjugated to a cationic peptide derived from a scrambled sequence of the SV40 T-antigen consensus NLS (ScT) in order to improve DNA binding ability of the M9 sequence. Lipofection of confluent bovine aortic endothelial cells with plasmid complexed with the M9-ScT conjugate resulted in 83% transfection and a 63-fold increase in reporter gene expression compared to the control (Subramanian et al. 1999). Moreover, M9-assistant lipofection resulted in a 20- to 100-fold enhancement of transfection, compared to lipofection alone, of primary neurons and neuronal cells, such as embryonic-derived retinal ganglion cells, rat pheochromocytoma (PC12) cells, embryonic rat ventral mesencephalon neurons, and the clinically relevant human NT2 cells or retinoic-acid-differentiated NT2 neurons (Ma et al. 2002).

The random and abundant attachment of an NLS to DNA could inhibit transcription of the transgene. Therefore, site-specific modification and control of introduction number is necessary. Dumb-bell DNAs have been selectively modified and evaluated. For example, a capped 3.3-kbp CMV-luciferase gene containing SV-40 T antigen NLS peptide was synthesized. Transfection improvement (10- to 1000-fold) as a result of the signal peptide was observed irrespective of the cationic vector or the cell type used. Increased expression with only one NLS linked to a ca. 4-kbp linear DNA has also been obtained (Zanta et al. 1999). Nonetheless, other results reported that the conjugation of one or two NLS peptides did not enhance the nuclear import of linearized DNA (Tanimoto et al. 2003).

The peptide nucleic acid (PNA)-DNA interaction is very tight and specific. Thus, the insertion of a PNA-binding site into DNA enables PNA to specifically hybridize with the complementary DNA site . In this respect, the PNA molecules linked to an SV40 NLS peptide could work as a nuclear targeting tool of DNA, and the increase in transfection efficiency of PNA-NLS hybridized plasmid DNA has been observed (Branden et al. 1999; Vaysse et al. 2004).

5 Nuclear Import of DNA Conjugated with DNA-Binding Proteins

DNA-binding proteins and NLS-fused DNA-binding proteins have also been utilized to enhance the nuclear import of DNA. A recombinant histone (NLS-N1), which contains both the SV40 large T antigen NLS and the carboxy-terminal domain of human histone H1, was complexed with plasmid DNA. NIH3T3 or COS-7 cells transfected with NLS-H1-plasmid DNA-lipofectin complexes expressed at least 20 times more luciferase or had at least 2.5 times more β-galactosidase activity than cells transfected with plasmid DNA-lipofectin complexes (Fritz et al. 1996). A complex consisting of the non-histone nuclear protein high-mobility group-1 (HMG-1) protein and plasmid DNA was successfully transported into the nucleus. In addition, efficient transfection experiments that involved cationic liposomes and Sendai virus HVJ were performed (Kaneda et al. 1989; Isaka et al. 1998). In vivo experiments carried out with liposomes /HVJ/HMG-1 complexes have revealed efficient gene transfer in the liver (Kato et al. 1991; Namiki et al. 1998).

The insertion of an enhancer sequence or a transcription-factor-binding sequence in DNA could lead to the binding of a transcriptional activator protein or transcriptional factor that functions in the nucleus. This, in turn, would facilitate DNA transport into the nucleus by an active transport system. A 100-fold increase in luciferase activity was observed in cells stably expressing EBNA 1 that had been transiently transfected with plasmids containing EBV oriP sequences. Cytoplasmic injection of these plasmids also led to a 100-fold enhancement of luciferase activity. This difference in the fold activation after either nuclear or cytoplasmic injection can be ascribed to the increased import of plasmids containing oriP from the cytoplasm to the nucleus in the presence of EBNA1 (Kaneda et al. 2000; Langel-Rouault et al. 1998; Tu et al. 2000). The high affinity and specificity of tetracycline repressor protein (TetR) for the tetracycline operator sequence *tetO* was utilized to bind the SV40 T antigen NLS to DNA. TetR-NLS, but not the control protein TetR, specifically enhanced expression efficiency from lipofected *tetO*-containing DNA between 4- and 16-fold. Intracellular trafficking studies demonstrated an increased accumulation of fluorescently labeled DNA in the nucleus after TetR-NLS binding (Vaysse et al. 2004). DNA vectors that possess repetitive binding sites for the inducible transcription factor NFκB, which is imported into the nucleus through the NPC, were imported into the nucleus with a 12-fold efficiency. This enhancement was associated with a corresponding increase in gene expression (Mesika et al. 2001).

Targeting to the nucleus in a cell-specific manner was carried out using DNA elements that contained binding sites for transcription factors expressed in unique cell types. A series of reporter plasmids were expressed selectively in smooth muscle cells by using the promoter from the smooth-muscle gamma actin gene promoter localized to the nucleus of the cells (Vacik et al 1999). The SV40 enhancer is known to have dual functions, promoting nuclear targeting of DNA and activating transcription. A single 72-bp element of the SV40 enhancer caused as much as a 20- to 40-fold increase in CMV-promoter-driven gene expression in murine tibialis muscle (Li et al. 2001) or rat mesenteric vessels (Young et al. 2003). In a strategy called steroid-mediated gene delivery (SMGD), steroid receptors, which are DNA-binding transcription regulators, were used as shuttles to facilitate the uptake of transfected DNA into the nucleus. The nuclear accumulation of transfected steroid-derivatives-decorated plasmid DNA was confirmed in glucocorticoid-receptor-positive cells (Rebuffat et al. 2001, 2002).

6 Nuclear Import of DNA Conjugated with Viral Proteins

The NLSs derived from viral proteins have been used to enhance nuclear import of transgenes in non-viral gene delivery systems. The C-terminal half of the human immunodeficiency virus type-1 viral protein R (Vpr) mediates DNA transfection in a variety of cell lines with efficiencies comparable to those of the best-known transfection agents. The import of Vpr/DNA complexes into the nucleus was enhanced through a pH-independent mechanism (Kichler et al. 2000). Adenovirus hexon protein was covalently linked to PEI. Hexon-containing complexes showed ten-fold greater trransgene expression in HepG2 cells than PEI/DNA. These results are consistent with the hexon protein mediating nuclear transport of plasmid complexes via the NPC, using mechanisms that are only partially dependent on the classical NLS import pathway (Carlisle et al. 2001).

7 Nuclear Import of DNA Directly Conjugated with Importins

The NLS-mediated strategy for nuclear import of a transgene often requires the contribution of the importin-α adapter, which reduces the efficiency of plasmid DNA nuclear import. Therefore, the direct conjugation of importins to plasmid DNA was expected to stimulate its nuclear import. First, a 62-amino-acid peptide consisting of the importin-β-binding domain of importin-α was coupled to plasmid DNA covalently. However, this peptide did not enhance either the nuclear import of plasmid DNA or the efficiency of cationic-lipid-mediated transfection (Carrière et al. 2003). In addition, a plasmid DNA/importin-β conjugate, consisting of biotinylated plasmid DNA and a recombinant streptavidin-fused importin-β was prepared, since importin-β plays an essential role in nuclear import thorough the NPC. The results suggested that plasmid DNA/importin-β conjugates are an important tool to enhance the nuclear localization of exogenous DNA in non-viral gene delivery systems (Nagasaki et al. 2005).

8 Nuclear Import of DNA with a Glycosylated Carrier

Mammalian cells express several types of lectins involved in intracellular trafficking, including endocytosis, interorganelle routing and putatively nuclear import. Glycosylated proteins lacking an NLS are able to enter the nucleus through sugar-dependent pathways that are distinct from the classical NLS pathway (Duverger et al. 1995; Monsigny et al. 2004). Therefore, in order to enhance gene transfer efficiency, glycosylated carriers were utilized for the nuclear import of DNA (Roche et al. 2003; Fajac et al 2002). The intracellular fate of the lactosylated poly-L-lysine/cDNA complex was studied using confocal microscopy. In the presence of chloroquine, the complex remained intact during internalization, intracellular transport, and, most importantly, transport into the nucleus (Klink et al. 2001). Furthermore, lactosylated PEI has been used to enhance the nuclear import of plasmid DNA. Greater gene transfection efficiency into airway epithelial cells was observed, although less than 5% of the intracellular polyplex was imported into the nucleus (Grosse et al. 2004).

9 Conclusions

In order to enhance transgene expression, it is crucial to promote nuclear import of exogenous DNA contained in non-viral gene delivery systems. In mammalian cells, factors concerning nucleocytoplasmic transport have steadily been identified. As a result, significant progress has been made in the development of non-viral gene delivery systems capable of nuclear import. Currently, two approaches to active nuclear import of DNA are available. One is the importin-mediated pathway and the other is the lectin-mediated pathway. Both pathways have confirmed the import of plasmid DNA from the cytoplasm to the nucleus of eukaryotic cells via an active nuclear import system. These studies have also shown that the control of the complex size is an important factor in the nuclear import of DNA.

For use in gene therapies aimed at treating genetic diseases, not only the safety of the complex but also the enhancement of nuclear import of the transgene are essential considerations. Along with further elucidation of nucleocytoplasmic transport systems of mammalian cells, progress in utilizing intracellular active transport systems for shuttling exogenous DNA into the nucleus can be expected in the near future.

References

Aronsohn AI, Hughes JA (1998) Nuclear localization signal peptides enhance cationic liposome-mediated gene therapy. J Drug Target 5:163–169

Branden LJ, Mohamed AJ, Smith CI (1999) A peptide nucleic acid-nuclear localization signal fusion that mediates nuclear transport of DNA. Nat Biotechnol 17:784–787

Brunner S, Sauer T, Carotta S, Cotten M, Saltik M, Wagner E (2000) Cell cycle dependence of gene transfer by lipoplex, polyplex and recombinant adenovirus. Gene Ther 5:401–407

Capecchi MR (1980) High effciency transformation by direct microinjection of DNA into cultured mammalian cells. Cell 22:479–488

Carlisle RC, Bettinger T, Ogris M, Hale S, Mautner V, Seymour LW (2001) Adenovirus hexon protein enhances nuclear delivery and increases transgene expression of polyethylenimine/plasmid DNA vectors. Mol Ther 4:473–483

Carrière M, Escriou V, Savarin A, Scherman D (2003) Coupling of importin beta binding peptide on plasmid DNA: transfection efficiency is increased by modification of lipoplex's physico-chemical properties. BMC Biotechnol 3:1–14

Cartier R, Reszka R (2002) Utilization of synthetic peptides containing nuclear localization signals for nonviral gene transfer systems. Gene Ther 9:157–167

Ciolina C, Byk G, Blanche F, Thuillier V, Scherman D, Wils P (1999) Coupling of nuclear localization signals to plasmid DNA and specific interaction of the conjugates with importin alpha. Bioconjug Chem 10:49–55
Collas P, Husebye H, Alestrom P (1996) The nuclear localization sequence of the SV40 T antigen promotes transgene uptake and expression in zebrafish embryo nuclei. Transgenic Res 5:451–458
Dowty ME, Williams P, Zhang G, Hagstrom JE, Wolff JA (1995) Plasmid DNA entry into postmitotic nuclei of primary rat myotubes. Proc Natl Acad Sci USA 92:4572–4576
Duverger E, Pellerin-Mendes C, Mayer R, Roche AC, Monsigny M (1995) Nuclear import of glycoconjugates is distinct from the classical NLS pathway. J Cell Sci 108:1325–1332
Escriou V, Carrière M, Scherman D, Wils P (2003) NLS bioconjugates for targeting therapeutic genes to the nucleus. Adv Drug Deliv Rev 55:295–306
Fajac I, Grosse S, Briand P, Monsigny M (2002) Targeting of cell receptors and gene transfer efficiency: a balancing act. Gene Ther 9:740–742
Fritz JD, Herweijer H, Zhang G, Wolff JA (1996) Gene transfer into mammalian cells using histone-condensed plasmid DNA. Hum Gene Ther 7:1395–1404
Goldfarb DS, Gariepy J, Schoolnik G, Kornberg RD (1986) Synthetic peptides as nuclear localization signals. Nature 322:641–644
Görlich D, Henklein P, Laskey RA, Hartmann E (1996) A 41 amino acid motif in importin-alpha confers binding to importin-beta and hence transit into the nucleus. EMBO J 15: 1810–1817
Greber UF, Fassati A (2003) Nuclear import of viral DNA genomes. Traffic 4:136–143
Grosse S, Aron Y, Honore I, Thevenot G, DanelC, Roche AC, Monsigny M, Fajac I (2004) Lactosylated polyethylenimine for gene transfer into airway epithelial cells: role of the sugar moiety in cell delivery and intracellular trafficking of the complexes. J Gene Med 6:345–356
Hagstrom JE, Ludtke JJ, Bassik MC, Sebestyen MG, Adam SA, Wolff JA (1997) Nuclear import of DNA in digitonin-permeabilized cells. J Cell Sci 110:2323–2331
Hébert E (2003) Improvement of exogenous DNA nuclear importation by nuclear localization signal-bearing vectors: a promising way for nonviral gene therapy? Biol Cell 95: 59–69
Isaka Y, Akagi Y, Kaneda Y, Imai E (1998) The HVJ liposome method. Exp Nephrol 6: 144–147
Kaneda Y, Iwai K, Uchida T (1989) Increased expression of DNA cointroduced with nuclear protein in adult rat liver. Science 243:375–378
Kaneda Y, Saeki Y, Nakabayashi M, Zhou WZ, Kaneda MW, Morishita R (2000) Enhancement of transgene expression by cotransfection of oriP plasmid with EBNA-1 expression vector. Hum Gene Ther 11:471–479
Kato K, Nakanishi M, Kaneda Y, Uchida T, Okada Y (1991) Expression of hepatitis B virus surface antigen in adult rat liver. Co-introduction of DNA and nuclear protein by a simplified liposome method. J Biol Chem 266:3361–3364
Kichler A, Pages JC, Leborgne C, Druillennec S, Lenoir C, Coulaud D, Delain E, Le Cam E, Roques BP, Danos O (2000) Efficient DNA transfection mediated by the C-terminal domain of human immunondeficiency virus type 1 viral protein R. J Virol 74:5424–31
Klink DT, Chao S, Glick MC, Scanlin TF (2001) Nuclear translocation of lactosylated poly-L-lysine/cDNA complex in cystic fibrosis airway epithelial cells. Mol Ther 3:831–841
Lanford RE, Kanda P, Kennedy RC (1986) Induction of nuclear transport with a synthetic peptide homologous to the SV40 T antigen transport signal. Cell 46:575–582
Li S, MacLaughlin FC, Fewell JG, Gondo M, Wang J, Nicol F, Dean DA, Smith LC (2001) Muscle-specific enhancement of gene expression by incorporation of SV40 enhancer in the expression plasmid. Gene Ther 8:494–497
Ludtke JJ, Zhang G, Sebestyen MG, Wolff JA (1999) A nuclear localization signal can enhance both the nuclear transport and expression of 1 kb DNA. J Cell Sci 112:2033–2041

Malecki M (1996) Preparation of plasmid DNA in transfection complexes for fluorescence and electron spectroscopic imaging. Scanning Microsc Suppl 10:1–16

Ma H, Zhu J, Maronski M, Kotzbauer PT, Lee VM, Dichter MA, Diamond SL (2002) Non-classical nuclear localization signal peptides for high efficiency lipofection of primary neurons and neuronal cell lines. Neuroscience 112:1–5

Mesika A, Grigoreva I, Zohar M, Reich Z (2001) A regulated, NFkappaB-assisted import of plasmid DNA into mammalian cell nuclei. Mol Ther 3:653–657

Monsigny M, Rondamino C, Duverger E, Fajac I, Roche AC (2004) Glyco-dependent nuclear import of glycoproteins, glycoplexes and glycosylated plasmids. Biochim Biophys Acta 1673:94–103

Mortimer I, Tam P, MacLachlan I, Graham RW, Saravolac EG, Joshi PB (1999) Cationic lipid-mediated transfection of cells in culture requires mitotic activity. Gene Ther 6:403–411

Munkonge FM, Dean DA, Hillery E, Griesenbach U, Alton E (2003) Emerging significance of plasmid DNA nuclear import in gene therapy. Adv Drug Deliv Rev 55:749–760

Nagasaki T, Myohoji T, Tachibana T, Futaki S, Tamagaki S (2003) Can nuclear localization signals enhance nuclear localization of plasmid DNA? Bioconjug Chem 14:282–286

Nagasaki T, Kawazu T, Tachibana T, Tamagaki S, Shinkai S (2005) Enhanced nuclear import and transfection efficiency of plasmid DNA using streptavidin-fused importin-β. J Control Release 103:199–207

Nakielny S, Siomi MC, Siomi H, Michael WM, Pollard V, Dreyfuss G (1996) Transportin: nuclear transport receptor of a novel nuclear protein import pathway. Exp Cell Res 229: 261–266.

Namiki Y, Takahashi T, Ohno T (1998) Gene transduction for disseminated intraperitoneal tumor using cationic liposomes containing non-histone chromatin proteins: cationic liposomal gene therapy of carcinomatosa. Gene Ther 5:240–246

Ojala PM, Sodeik B, Ebersold MW, Kutay U, Helenius A (2000) Herpes simplex virus type 1 entry into host cells: reconstitution of capsid binding and uncoating at the nuclear pore complex in vitro. Mol Cell Biol 20:4922–4931

Paine PL, Moore LC, Horowitz SB (1975) Nuclear envelope permeability. Nature 254: 109–114

Pante N, Kann M (2002) Nuclear pore complex is able to transport macromolecules with diameters of about 39 nm. Mol Biol Cell 13:425–434

Rebuffat A, Bernasconi A, Ceppi M, Wehrli H, Verca SB, Ibrahim M, Frey BM, Frey FJ, Rusconi S (2001) Selective enhancement of gene transfer by steroid-mediated gene delivery. Nat Biotechnol 19:1155–1161

Rebuffat AG, Nawrocki AR, Nielsen PE, Bernasconi AG, Bernal-Mendez, E, Frey BM, Frey FJ (2002) Gene delivery by a steroid-peptide nucleic acid conjugate. FASEB J 16: 1426–1428

Roche AC, Fajacb I, Grosseb S, Frisona N, Rondaninoa N, Mayera R and Monsignya M (2003) Glycofection: facilitated gene transfer by cationic glycopolymers. Cell Mol Life Sci 60:288–297

Schwartz B, Ivanov MA, Pitard B, Escriou V, Rangara R, Byk G, Wils P, Crouzet J, Scherman D (1999) Synthetic DNA-compacting peptides derived from human sequence enhance cationic lipid-mediated gene transfer in vitro and in vivo. Gene Ther 6:282–292

Sebestyén MG, Ludtke JJ, Bassik MC, Zhang G, Budker V, Lukhtanov EA, Hagstrom JE, Wolff JA (1998) DNA vector chemistry: the covalent attachment of signal peptides to plasmid DNA. Nat Biotechnol 16:80–85

Subramanian A, Ranganathan P, Diamond SL (1999) Nuclear targeting peptide scaffolds for lipofection of nondividing mammalian cells. Nat Biotechnol 17:873–877

Tanimoto M, Kamiya H, Minakawa N, Matsuda A, Harashima H (2003) No enhancement of nuclear entry by direct conjugation of a nuclear localization signal peptide to linearized DNA. Bioconjug Chem 14:1197–1202

Tachibana R, Harashima H, Shinohara Y, KiwadaH (2001) Quantitative studies on the nuclear transport of plasmid DNA and gene expression employing nonviral vectors. Adv Drug Deliv Rev 52:219–226

Trotman LC, Mosberger N, Fornerod M, Stidwill RP, Greber UF (2001) Import of adenovirus DNA involves the nuclear pore complex receptor CAN/Nup214 and histone H1. Nat Cell Biol 3:1092–1100

Tseng WC, Haselton FR, Giorgio TD (1999) Mitosis enhances transgene expression of plasmid delivered by cationic liposomes. Biochim Biophys Acta 1445:53–64

Tu G, Kirchmaier AL, Liggitt D, Liu Y, Liu S, Yu WH, Heath TD, Thor A, Debs RJ (2000) Non-replicating Epstein-Barr virus-based plasmids extend gene expression and can improve gene therapy in vivo. J Biol Chem 275:30408–30416

Vacik J, Dean BS, Zimmer WE, Dean DA (1999) Cell-specific nuclear import of plasmid DNA. Gene Ther 6:1006–1014

Vaysse L, Harbottle R, Bigger B, Bergau A, Tolmachov O, Coutelle C (2004) Development of a self-assembling nuclear targeting vector system based on the tetracycline repressor protein. J Biol Chem 279:5555–5564

Wilson GL, Dean BS, Wang G, Dean DA (1999) Nuclear import of plasmid DNA in digitonin-permeabilized cells requires both cytoplasmic factors and specific DNA sequences. J Biol Chem 274:22025–22032

Wisnivesky JP, Leopold PL, Crystal RG (1999) Specific binding of the adenovirus capsid to the nuclear envelope. Hum Gene Ther 10:2187–2195

Yoneda Y, Semba T, Kaneda Y, Noble RL, Matsuoka Y, Kurihara T, Okada Y, Imamoto N (1992) A long synthetic peptide containing a nuclear localization signal and its flanking sequences of SV40 T antigen directs the transport of IgM into the nucleus efficiently. Exp Cell Res 201:313–320

Yoneda Y, Hieda M, Nagoshi E, Miyamoto Y (1999) Nucleocytoplasmic protein transport and recycling of Ran. Cell Struct Funct 24:425–433

Young JL, Benoit JN, Dean DA (2003) Effect of a DNA nuclear targeting sequence on gene transfer and expression of plasmids in the intact vasculature. Gene Ther 10:1465–1470

Zanta MA, Belguise-Valladier P, Behr JP (1999) Gene delivery: a single nuclear localization signal peptide is sufficient to carry DNA to the cell nucleus. Proc Natl Acad Sci USA 96:91–96

Controlled Intracellular Localization of Oligonucleotides by Chemical Conjugation

TAKANORI KUBO[1], RUMIANA BAKALOVA[2], ZHIVKO ZHELEV[2], HIDEKI OHBA[2], and MASAYUKI FUJII[1,3]

1 Introduction

Artificial control of genetic expression by oligonucleotides is a powerful tool for biological studies and medical therapies. Nucleic-acid drugs, such as antisense oligonucleotides, ribozymes, decoys and siRNAs, have attracted much attention and have been intensively studied for the past two decades (Crooke et al. 2004; Tung et al. 2000; Fischer et al. 2001; Eisele et al. 1999; Zubin et al. 2002; Antopolsky et al. 1999). Difficulties in using oligonucleotides as therapeutic agents involve their transport through the cell membrane, delivery and localization in the targeted cellular structure, and targeting of the specific mRNA or DNA sequence with sufficient affinity and specificity. For these reasons, DNA-peptide conjugates have been attracting intensive attention as alternative and advanced materials for the technology of genetic medicines and novel functional nucleic acids (Stetsenko et al. 2000, 2002; Soukchareum et al. 1995; Haralambidis et al. 1987; Antopolsky et al. 2002). In this chapter, the cellular uptake and controlled localization of oligonucleotide-peptide conjugates are discussed.

2 Synthesis of Oligonucleotide-Peptide Conjugates by Solid-Phase Fragment Condensation

Synthetic methods of oligonucleotide-peptide conjugates so far studied can be classified into two categories: solution-phase synthesis and solid-phase synthesis. The former mostly involves coupling procedures of oligonucleotide and peptide fragments using small linker molecules having two different functionalities (Antopolsky et al. 1999; Stetsenko et al. 2000).

[1] Department of Biological and Environmental Chemistry, Kyushu School of Engineering, Kinki University, 11-6 Kayanomori, Iizuka, Fukuoka 820-8555, Japan
[2] Kyushu National Industrial Research Institute, Agency of Industrial Science and Technology (AIST-Kyushu), Ministry of International Trade and Industry, 807-1 Shuku-machi, Tosu, Saga 841-0052, Japan
[3] Molecular Engineering Institute, Kinki University, 11-6 Kayanomori, Iizuka, Fukuoka 820-8555, Japan

Tandem synthesis of oligonucleotide-peptide conjugates on solid phase has been intensively studied (Haralambidis et al. 1987; Soukchareum et al. 1995; Antopolsky et al. 2002; Stetsenko et al. 2002; Zubin et al. 1999). One of the major problems is how to prepare fluorenylmethoxycarbonyl (fmoc) derivatives of all amino acids having protective groups on side chains labile under treatment with ammonia. Another problem of solid-phase synthesis of oligonucleotide-peptide conjugates, in which first peptides are usually prepared first and then oligonucleotides are prepared, is the low coupling efficiencies of peptide synthesis. Insufficient coupling yields of peptides will result in lowering overall yields of the desired products. In order to overcome such problems as (1) limitation of amino acid components in peptides and (2) low overall yields of the products, fragment coupling on solid support has been attempted (Zubin et al. 1999; Zatsepin et al. 2002).

Recently, we reported a universal method to prepare oligonucleotide-peptide conjugates by solid phase fragment condensation (SPFC) (Scheme 1) (Kubo et al. 2001, 2003). The strategy of SPFC is that a DNA fragment having a free amino group prepared on CPG support is reacted with α,ω-diisocyanatoalkane or carbonyl diimidazole (CDI) and then with a peptide fragment bearing a single reactive amino group. The resulting oligonucleotide-peptide conjugate covalently linked to solid support is

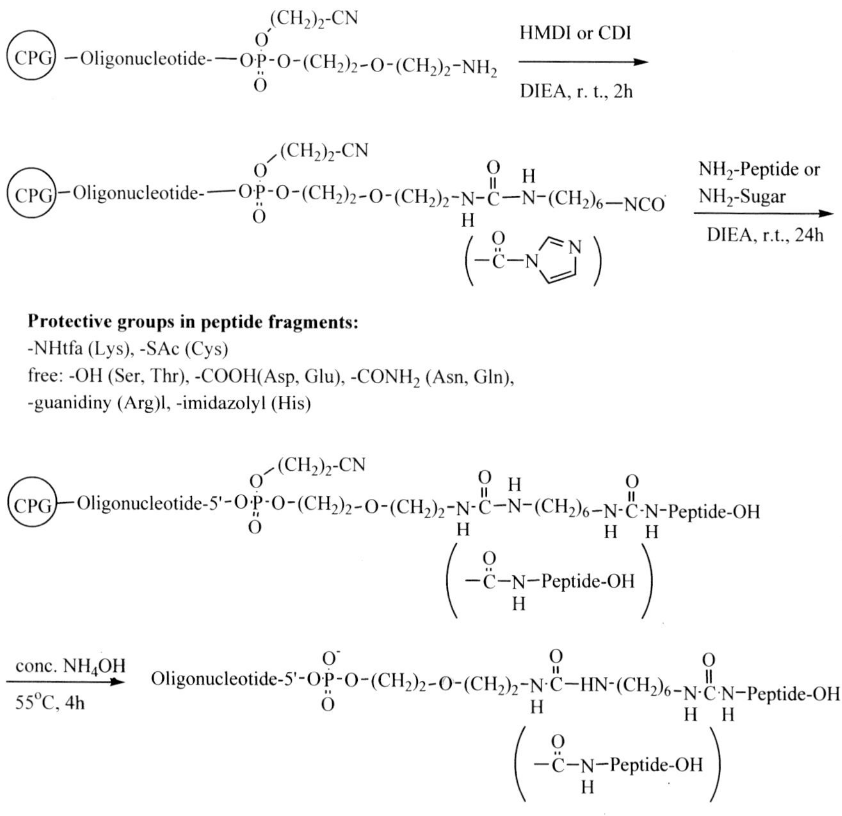

SCHEME 1. Synthesis of Oligonucleotide-Peptide Conjugates by SPFC

cleaved from CPG and deprotected by treatment with ammonia. The isolated yields were 10%–30% and the products were well characterized by MALDI-TOF MS and RPHPLC.

As shown in Table 1, peptide moieties in the conjugates could contain aspartic acid (D), glutamic acid (E), lysine (K), methionine (M), aspergine (N), glutamine (Q), arginine (R), serine (S), threonine (T), and tryptophan (W) as well as other non-protected hydrophobic amino acids. Except for lysine, all the side chain residues were deprotected prior to the coupling reaction. ε-Amino groups of lysine were protected with trifluoroacetyl groups, which were removed at the final treatment with ammonia. It is advantageous to use amino groups of peptide fragments for SPFC because these can be easily introduced at desired positions of peptides and selectively protected by both acid- and base-labile protective groups. SPFC will allow the selective linkage of oligonucleotides to peptides at the desired positions. Peptide moieties in the conjugates synthesized are derived from the nuclear localization signal (NLS) of SV40 T-antigen (Newmeyer et al. 1986), the NLS of HIV-1 Tat protein (Dingwell et al. 1986), a nuclear export signal (NES) sequence of HIV-1 Rev protein (Dingwell et al. 1986), NES of MAPKK protein (GoldFarb et al. 1986) as well as designed amphiphilic peptides (Fujii et al. 2001; Kubo et al. 2001, Yokoyama et al. 2001, Yano et al. 2003). Peptides in **C4** and **C5** (also in **C15** and **C16**) were proven to form an antiparallel β-sheet and an amphipathic α-helical structure in the presence of DNA, respectively, and both could bind to and stabilize dsDNA (Fujii et al. 2000; Kubo et al. 2001). Galactosamine was also successfully conjugated to oligonucleotides without any protection (**C6, C17**).

3 Nuclear and Cytoplasmic Localization of Oligonucleotide-Peptide Conjugates

Recently, the mechanisms of intracellular transport of proteins have been revealed in detail, and some of the signal sequences responsible for nuclear-cytoplasmic transport of proteins have been identified (Kim et al. 1994; Goerlich et al. 1999).

Heitz and colleagues showed that oligonucleotide conjugated with a hydrophobic peptide derived from HIV-1 gp41 protein was effectively taken up into human fibroblast H9C2 cells (Chaloin et al. 1998). Behr and co-workers showed that plasmid DNA conjugated with a NLS peptide derived from SV40 large-T-antigen was effectively taken up into cells and localized in the nucleus (Zanta et al. 1999).

Our group also studied cellular uptake and intracellular localization of DNA-signal peptide conjugates in detail by confocal laser fluorescence microscopy and flow-cytometry (Figs. 1, 2) (Kubo et al. 2003, 2005). Jurkat human leukemia cells were incubated with fluorescently labeled oligonucleotide-peptide conjugates **C1F-C6F** (Scheme 2) in 10% serum for 48 h. Microscopy clearly showed that cellular uptake of all of the oligonucleotide-NLS conjugates **C1F-C3F** was enhanced without any transfection reagents, whereas natural oligonucleotide **N1F** was hardly taken up into cells (Fig. 1a). It can be assumed that the basic characteristic features of NLSs enhanced ability of the conjugates to penetrate the cellular membrane. As expected, conjugates bearing NLS sequences of SV40 large-T-antigen (**C1F**) and HIV-1 Tat protein (**C2F**) were taken up and localized into the nucleus (Fig. 1b,c). The conjugate of phosphorothioate oligonucleotide and SV40 large-T-antigen NLS (**C3F**) was taken up much better and localized throughout the large nuclei of Jurkat cells (Fig. 1d).

TABLE 1. MALDI TOF-MS Data of Oligonucleotide-Peptide Conjugates
N1: 5′-TTTTTCTCTCTCTCT-3′
N2: 5′-CAGTTAGGGTTAG-3′
S1: 5′-CsAsGsTsTsAsGsGsGsTsTsAsG-3′

#	Nucleotide	Peptide/Sugar (Origin)	Linker	Y (%)[a]	MALDI TOF-MS Experimental/calculated
C1	**N1**	-βA[b]PKKKRKV-OH (SV40 T antigen NLS)	HMDI	6.5	5637.77/5636.10
C2	**N1**	-βA[b]NSAAFEDLRVLS-OH (Influenza V nucleoprotein NLS)	CDI	10.7	6003.97/6001.78
C3	**N1**	-βA[b]LPPLERLTL-OH (HIV-1 Rev NES)	CDI	10.7	5799.53/5798.03
C4	**N1**	-βA[b]LRALLRALLRAL-OH (designed)	CDI	10.6	6064.46/6061.29
C5	**N1**	-βA[b]RLRLRLRLRL-OH (designed)	CDI	15.7	6051.81/6050.38
C6	**N1**	galactosamine	CDI	27.0	59322.67/5930.19
C7	**N2**	-βA[b]NSAAFEDLRVLS-OH (Influenza V nucleoprotein NLS)	CDI	7.3	5602.72/5600.56
C8	**N2**	-βA[b]PKKKRKV-OH (SV40 T antigen NLS)	HMDI	21.5	5505.97/5505.34
C9	**N2**	-βA[b]RKKRRQRRRPP-OH (HIV-1 Tat NLS)	HMDI	10.7	6303.02/6301.98
C10	**N2**	-βA[b]LPPLERLTL-OH (HIV-1 Rev NES)	CDI	12.7	5614.30/5613.21

C11	**N2**	Ac-GPKKKRKVGK[c]- (SV40 T antigen NLS)	CDI	18.2	5384.26/5390.51
C12	**N2**	Ac-GRKKRRQRRRPPGGK[c]-CDI (HIV-1 Tat NLS)	10.1	6226.18/6226.40	
C13	**N2**	-βA[b]NSAAFEDLRVLS-OH (Influenza V nucleoprotein NLS)	CDI	7.3	5602.72/5613.30
C14	**N2**	Ac-LPPLERLTLGK[c]- (HIV-1 Rev NES)	CDI	20.7	5461.60/5460.59
C15	**N2**	-βA[b]LRALLRALLRAL-OH (designed)	CDI	15.4	5815.13/5814.89
C16	**N2**	-βA[b]RLRLRLRLRL-OH (designed)	CDI	19.3	5659.38/5657.86
C17	**N2**	galactosamine	CDI	27.9	4400.03/4401.69
C18	**S1**	Ac-GPKKKRKVGK[c]- (SV40 T antigen NLS)	CDI	30.5	5642.63/5599.25

[a] Isolated yield determined by absorbance at 260 nm after HPLC purification.
[b] β-amino group is linked to oligonucleotide.
[c] ε-amino group of Lys is linked to oligonucleotide.

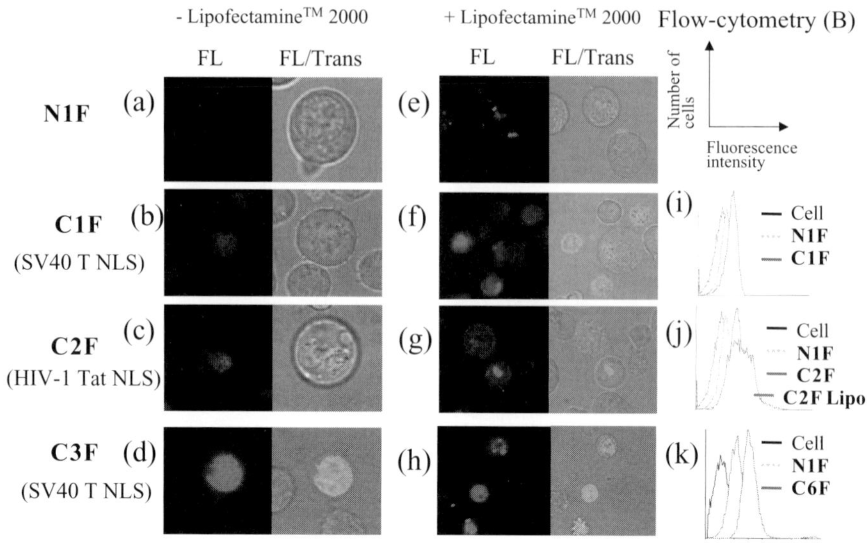

FIG. 1. Nuclear localization of oligonucleotide-peptide conjugates

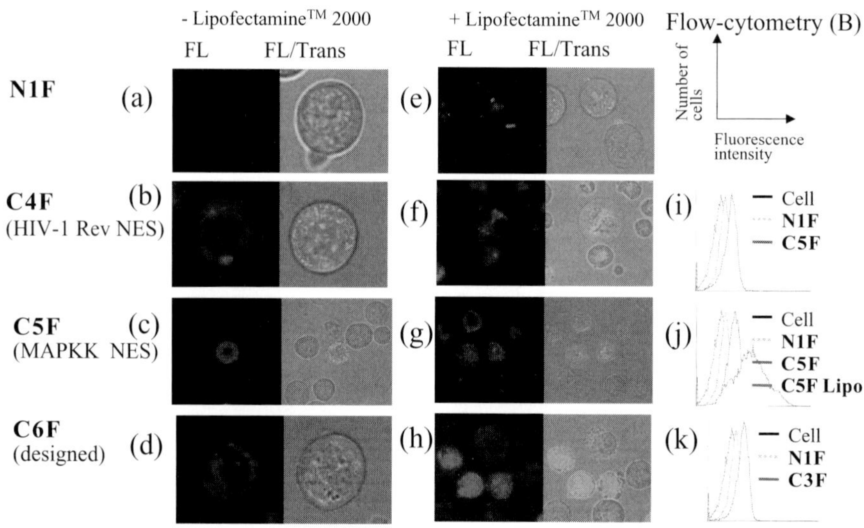

FIG. 2. Cytoplasmic localization of oligonucleotide-peptide conjugates

N1F: FITC-5'-CAGTTAGGGTTAG-3'

C1F: **N1**-SV40 T ant NLS

3'-GATTGGGATTGAC-5'-$OPO_3CH_2CH_2OCH_2CH_2NHCONH(CH_2)_4CHCONH$-**PKKKRKV**-OH

NHCOCH$_2$CH$_2$NH-FITC

C2F: **N1**-HIV-1 Tat NLS

3'-GATTGGGATTGAC-5'-$OPO_3CH_2CH_2OCH_2CH_2NHCONH(CH_2)_4CH(COOH)NH$-**QPPRRRQRRK KRG**-$COCH_2CH_2NH$-FITC

C3F: **S1**-SV40 T ant NLS

3'-s(GATTGGGATTGAC)-5'-$OPO_3CH_2CH_2OCH_2CH_2NHCONH(CH_2)_4CH(COOH)NH$-**GVKRKKKPG** -$COCH_2CH_2NH$-FITC

C4F: **N1**-HIV-1 Rev NES

3'-GATTGGGATTGAC-5'-$OPO_3CH_2CH_2OCH_2CH_2NHCONH(CH_2)_4CHCONH$-**LPPLERLTL**-OH

NHCOCH$_2$CH$_2$NH-FITC

C5F: **N1**-MAPKK NES

3'-GATTGGGATTGAC-5'-$OPO_3CH_2CH_2OCH_2CH_2NHCONH(CH_2)_4CH(COOH)NH$-**GEDLELEELKK QA**-$COCH_2CH_2NH$-FITC

C6F: **N1**-designed peptide

3'-GATTGGGATTGAC-5'-$OPO_3CH_2CH_2OCH_2CH_2NHCONH(CH_2)_4CHCONH$-**LRALLRALLRAL**-OH

NHCOCH$_2$CH$_2$NH-FITC

SCHEME 2. Structures of FITC Labeled Oligonucleotide-Peptide Conjugates

When the conjugates were mixed with Lipofectamine 2000 and incubated with cells in 10% serum for 48h, different features of cellular uptake were observed (Figure 1e–h). The natural olignucleotide **N1F** was taken up into cells but to a lesser extent (Fig. 1e). Small fluorescent particles could be observed in the cells, indicating that a large portion of the oligonucleotides remained in the endosomes after entry into cells. By contrast, oligonucleotide-NLS conjugates (**C1F-C3F**) were taken up much more efficiently than **N1F** and escaped from endosomes very rapidly. Flow-cytometric analysis also showed that use of the transfection reagent slightly enhanced cellular uptake of the conjugate **C2F** (Fig. 1j). It should be pointed out that use of the transfection reagent seemed to disturb control of the localization of the conjugates in the nuclei, perhaps by disturbing the interaction of importin-α with NLS peptides that are rich in basic amino acids.

Cellular uptake and intracellular localization of oligonucleotide-NES peptide conjugates **C4F-C6F** were analyzed by microscopy and flow cytometry (Fig. 2). It can be clearly seen that cellular uptake of the oligonucleotide-NES conjugates was greatly enhanced without any transfection reagents compared with uptake of the natural oligonucleotide **N1F** (Fig. 2a). It is likely that the hydrophobic characteristics of NES peptides facilitated penetration of the conjugates through the cellular membrane. It should be noted that conjugates bearing either a NES sequence of HIV-1 Rev protein (**C4F**) or MAPKK (**C5F**) localized throughout the cytoplasm (Fig. 2b, c). Curiously, conjugate **C6F**, bearing a designed peptide with cationic amphipathic features, localized in the cytoplasm, just like the oligopnucleotide-NES conjugates **C4F** and **C5F** (Fig.

2d). It can be assumed that the sequence of the designed peptide, with leucine residues every three amino acids, was recognized as a NES signal by an exportin. The use of the transfection reagent Lipofectamine 2000 enhanced cellular uptakes of the conjugate (Fig. 2j) but seemed to disturb the precise control of intracellular localization of the conjugates (Fig. 2f–h).

Thus oligonucleotide-peptide conjugates were shown to be taken up effectively into cells in the absence of any transfection reagents. Controlled nuclear localization was achieved by oligonucleotide-NLS conjugates and cytoplasmic localization was achieved by oligonucleotide-NES conjugates.

4 Mechanisms of Cellular Uptake of Oligonucleotide-Peptide Conjugates

To investigate the mechanisms of the entry of the oligonucleotide-peptide conjugates, the effect of an endocytosis inhibitor cytochalasin B on cellular uptake were studied (Simeoni et al. 2003). The results are shown in Fig. 3. Cellular uptake of **N1F** using Lipofectamine 2000 was inhibited by the addition of cytochalasin B (Fig. 3a–c), which indicated that natural oligonucleotide was taken up into cells by an endocytotic mechanism. By contrast, the entries of both the oligonucleotide-NLS conjugate **C2F** and the oligonucleotide-NES conjugate **C5F** were not affected by the addition of cytochalasin B in the presence or absence of the transfection reagent (Fig. 3d–f and g–i, respectively). These results strongly indicate that entry of the oligonucleotide conjugates is not dependent on an endocytotic mechanism. Further investigations to identify the detailed mechanism involved in this process are now in progress in our laboratory.

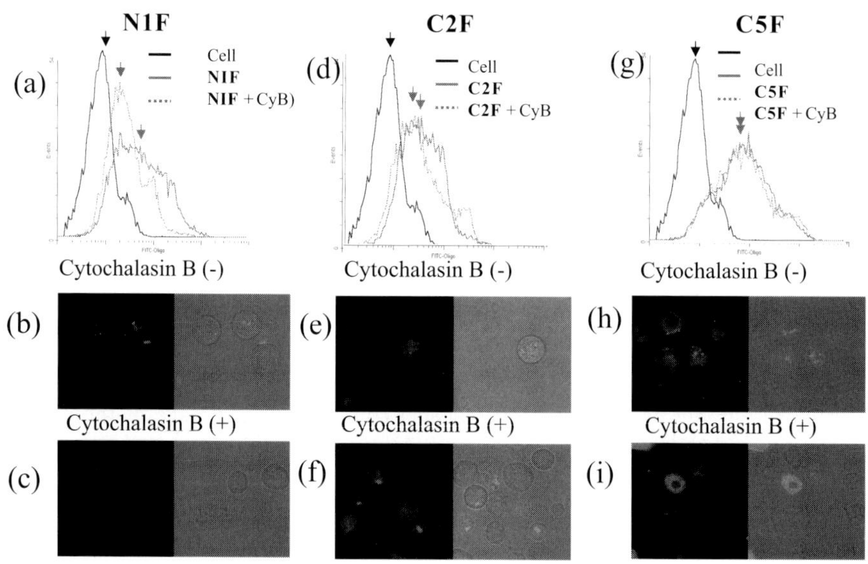

FIG. 3. Inhibitory effect of endocytotic entry by cytochalasin

We have already found that an antisense oligonucleotide-NLS conjugate inhibited telomerase in human leukemia cells by 98% while an antisense oligonucleotide-NES conjugate did not inhibit it at all. We also have found that an antisense oligonucleotide-NES conjugate suppressed the expression of the gene *bcr/abl* on Philadelphia chromosome by 86%, while an oligonucleotide-NLS conjugate suppressed expression by only 20% (Bakalova et al. 2004). These results prove that intracellular localization of antisense oligonucleotides is extremely important for their activities (Kuwabara et al. 1999; Warashima et al. 2001). Furthermore, we confirmed that oligonucleotide-peptide conjugates were much more resistant against cellular nucleases, that they could bind to target RNA with higher affinities, that the hybrids between them and complimentary RNAs could activate RNase H, and that they showed little cytotoxicity. The detailed data will be published elsewhere in the near future.

These findings strongly suggest that oligonucleotide-peptide conjugates are promising candidates for effective genetic medicines of the next generation. We believe that new, intelligent oligonucleotides can be created by linking oligonucleotides to natural and unnatural molecules, forming compounds that are never found in nature.

References

Antopolsky M, Azhayeva E, Tengvall U, Auriola S, Jaaskelainen I, Ronkko S, Honkakoski P, Urtti A, Lonnberg H, Azhayev A (1999) Peptide-Oligonucleotide Phosphorothioate Conjugates with Membrane Translocation and Nuclear Localization Properties. Bioconjugate Chem 10:598–606

Antopolsky M, Azhayev A (2000) Stepwise Solid-Phase Synthesis of Peptide-Oligonucleotide Phosphorothioate Conjugates Employing Fmoc Peptide Chemistry. Tetrahedron Lett 41:9113–9117

Antopolsky M, Azhayeva E, Tengvall U, Azhayev A (2002) Toward a General Method for the Stepwise Solid-Phase Synthesis of Peptide-Oligonucleotide Conjugates. Tetrahedron Lett 43:527–530

Bakalova R, Ohba H, Zhelev Z, Kubo T, Fujii M, Ishikawa M, Shinohara Y, Baba Y (2004) Atypical Protein-kinase Cζ, but Neither Conventional Ca^{2+}-Dependent Protein-Kinase C Isoenzymes nor Ca^{2+}-Calmodulin, Participates in Regulation of Telomerase Activity in Burkitt's Lymphoma Cells. Cancer Chemother Pharmacol 54:161–172

Chaloin L, Vidal P, Lory P, Mery J, Lautredou N, Divita G, Heitz F (1998) Design of Carrier Peptide-Oligonucleotide Conjugate with Rapid Membrane Translocation and Nuclear Localization Properties. Biochem Biophys Res Commun 243:601–608

Crooke ST (2004) Progress in Antisense Technology. Annu Rev Med 55:61–95

Dingwell C, Laskey RA (1986) Protein Import into the Cell Nucleus. Annu Rev Cell Biol 2: 367–390

Eisele F, Owen DJ, Waldmann H (1999) Peptide Conjugates as Tools for the Study of Biological Signal Transduction. Bioorg Med Chem 7:193–234

Fischer PM, Krausz E, Lane DP (2001) Cellular Delivery of Impermeable Effector Molecules in the Form of Conjugates with Peptides Capable of Mediating Membrane Translocation. Bioconjugate Chem 12:825–841

Fujii M, Yokoyama K, Kubo T, Ueki R, Abe S, Goto K, Niidome T, Aoyagi H, Iwakuma K, Ando S, Ono S (2001) Design, Synthesis and Characterization of DNA Binding Peptides. Peptide Science 2000:109–112

Goerlich D, Kutay U (1999) Transport between the Cell Nucleus and the Cytoplasm. Annu Rev Cell Dev Biol 15:607–660

Goldfarb DS, Gariepy J, Schoolnik G, Komberg RD (1986) Synthetic Peptides as Nuclear Localization Signals. Nature 322:641–642

Haralambidis J, Duncan L, Tregear GW (1987) The Solid Phase Synthesis of Oligonucleotide Containing a 3′-Peptide Moiety. Tetrahedron Lett 28:5199–5202

Kim NW, Piatyszek MA, Prowse KR, Harley CB, West MD, Ho PLC, Coviello GM, Wright WE, Weinrich SL, Shay JW (1994) Specific Association of Human Telomerase Activity with Immortal Cells and Cancer. Science 266:2011–2015

Kubo T, Fujii M (2001) Specific Binding and Stabilization of DNA and Phosphorothioate DNA by Amphiphilic α-Helical Peptides. Nucleosides Nucleotides & Nucleic Acids 20: 1313–1316

Kubo T, Dubey K, Fujii M (2001) A Novel Approach for the Solid Phase Synthesis of DNA-Peptide Conjugates. Nucleosides Nucleotides & Nucleic Acids 20:1321–1324

Kubo T, Morikawa M, Ohba H, Fujii M (2003) Synthesis of DNA-Peptide Conjugates by Solid-Phase Fragment Condensation. Org Lett 5:2623–2626

Kubo T, Bakalova R, Ohba H, Fujii M (2003) Antisense Efects of DNA-Peptide Conjugates. Nucleic Acids Research Suppl 3:179–180

Kubo T, Takamori K, Kanno K, Bakalova R, Ohba H, Matsukisono M, Akebiyama Y, Fujii M (2005) Efficient Cleavage of RNA, Enhanced Cellular Uptake, and Controlled Intracellular localization of Conjugate DNAzymes. Bioorg Med Chem Lett 15:167–170

Kurreck J (2003) Antisense Technologies. Eur J Biochem 270:1628–1644

Kuwabara T, Warashima M, Nakayama A, Ohkawa J, Taira K (1999) $tRNA^{Val}$-Heterodimeric Maxizymes with High Potential as Gene Inactivating Agents: Simultaneous Cleavage at Two Sites in HIV-1 tat mRNA in Cultured Cells. Proc Natl Acad Sci USA 96:1886–1891

Manoharan M (2002) Oligonucleotide Conjugates as Potential Antisense Drugs with Improved Uptake, Biodistribution, Targeted Delivery, and Mechanism of Action. Antisense & Nucelic Acid Drug Development. 12:103–128

Newmeyer DD, Lucocq JM, Buerglin TR, DeRobertis EM (1986) Assembly in vitro of Nuclei Active in Nuclear Protein Transport: ATP is Required for Nucleoplasmin Accumulation. EMBO J 5:5001–510

Simeoni F, Morris MC, Heitz F and Divita G (2003) Insight into the Mechanism of the Peptide-based Gene Delivery System MPG: Implications for Delivery of siRNA into Mammalian Cells. Nucleic Acids Research 31:2717–2724

Soukchareun S, Tregear GW, Haralambidis J (1995) Preparation and Characterization of Antisense Oligonucleotide-Peptide Hybrids Containing Viral Fusion Peptides. Bioconjugate Chem 6:43–53

Stetsenko DA, Gait MJ (2000) Efficient Conjugation of Peptides to Oligonucleotides by "Native Ligation". J Org Chem 65:4900–4908

Stetsenko DA, Gait MJ (2002) Total Stepwise Solid-Phase Synthesis of Oligonucleotide-($3' \rightarrow N$)-Peptide Conjugates. Org Lett 4:3259–3262

Tung CH (2000) Preparation and Applications of Peptide-Oligonucleotide conjugates. Bioconjugate Chem 11:605–618

Warashima M, Kuwabara T, Kato Y, Sano M, Taira K (2001) RNA-Protein Hybrid Ribozyme That Efficiently Cleave Any mRNA Independently of the Structure of the Target RNA. Proc Natl Acad Sci USA 98:5572–5577

Yano M, Kubo T, Yokoyama K, Ueki R, Sasaki K, Anno Y, Ohba H, Fujii M (2003) Control of Intracellular Delivery of Oligonucleotides by Conjugation with Signal Peptides. Nucleosides Nucleotides & Nucleic Acids 22:1367–1369

Yokoyama K, Kubo T, Fujii M (2002) Synthesis and Biological Activity of DNA-NLS Conjugate. Peptide Science 2001:313–316

Zanta MA, Belguise-Valladier P, Behr JP (1999) Gene Delivery: A Single Nuclear Localization Signal Peptide is Sufficient to Carry DNA to the Cell Nucleus. Proc Natl Acad Sci USA 96:91–96

Zatsepin TS, Stetsenko DA, Arzummanov AA, Momanova EA, Gait MJ, Oretskaya TS (2002) Synthesis of Peptide-Oligonucleotide Conjugates with Single and Multiple Peptides

Attached to 2′-Aldehyde through Thiazolidine, Oxime and Hydrazine Linkages. Bioconjugate Chem 13:822–830

Zubin EM, Romanova EA, Volkov EM, Tashlitsky VN, Korshunova GA, Shabarova ZA, Oretskaya TS (1999) Oligonucleotide-Peptide Conjugates as Potential Antisense Agents. FEBS Lett 456:59–62

Zubin EM, Romanova EA, Oretskaya TS (2002) Modern Methods for the Synthesis of Peptide-Oligonucleotide Conjugates. Russ Chem Rev 71:239–264

Functional Nucleotide Sequences Capable of Promoting Non-viral Genetic Transfer

OSAM MAZDA[1] and TSUNAO KISHIDA[1,2]

1 Introduction

The major disadvantages of non-viral gene delivery systems are the low efficiency of transfection and transient expression of the transgene. These shortages can be overcome, at least partly, by employing plasmid vectors with particular functional sequences, such as Epstein-Barr virus (EBV)-derived genetic elements (EBNA1 gene and oriP) and the *sleeping beauty* (*SB*) transposable element. Transfection experiments with EBV-based plasmid vectors strongly suggest that the performance of non-viral vectors (gene delivery methods/materials) per se is not as low as widely believed, in terms of the rate of plasmid DNA trafficking from the extracellular space into the cytoplasm. The inability of of plasmid DNA to be transported into the nucleus is the critical hurdle in conventional non-viral transfection, but it may be cleared by using an EBV-based plasmid vector albeit not only by modifying non-viral vectors. This concept may have significant implications for development of novel non-viral gene delivery systems that are applicable to a variety of purposes, including functional genomics and molecular therapeutics.

2 Plasmid Vectors as an Important Component of Non-viral Gene Delivery Systems

Viral vectors consist of recombinant viral particles that are capable of infecting target cells, while non-viral gene delivery systems do not make use of any genetically mod-

[1] Department of Microbiology, Kyoto Prefectural University of Medicine, Kamigyo-ku, Kyoto 602-8566, Japan
[2] Louis Pasteur Center for Medical Research, Kyoto 606-8225, Japan

ified infectious viruses. Being free from virus-associated undesired effects, e.g. potential generation of replication-competent retroviruses and a high immunogenicity due to adenoviral proteins, non-viral gene delivery systems may offer promising measures not only in genetically modifying cultured cells and animals but also in treating patients (Niidome and Huang 2002; Mazda 2002). Another advantage of non-viral systems is their large-scale, affordable manufacture. The major problem that needs to be solved is the low efficiency of non-viral systems in transferring and expressing exogenous genes in target cells, compared with the transduction efficiency obtained using their viral counterparts. This is the reason why a technical breakthrough is required in order to develop non-viral gene delivery systems that are feasible for genetic engineering, functional genomics, and molecular therapeutics.

Generally, non-viral gene delivery systems consist of two components, both of which should be taken into account in developing superior non-viral systems:(1) nucleic acids carrying genetic information, most typically plasmid DNA (pDNA), and (2) gene delivery methods/materials (non-viral vector). Therefore, both the pDNA and the non-viral vectors (gene delivery methods/materials) must be improved to obtain ideal non-viral vector systems.

A variety of non-viral vectors have been devised, based on a range of chemical compounds and physical methods. Chemical compounds (carrier molecules) include synthetic and natural macromolecules, such as cationic lipids (Felgner and Ringold 1989; Rocha et al. 2002; Kumar et al. 2003) and cationic polymers (Tang et al. 1996; Kukowska-Latallo et al. 1996; Lemkine and Demeneix 2001) that interact with nucleic acid to form complexes (lipoplex, polyplex, etc.) to be endocytosed into cells. Physical methods, such as electroporation (Herweijer and Wolff 2003), particle bombardment (gene gun) (Cui and Mumper 2003), and other procedures (Plank et al. 2003; Hosseinkhani et al. 2003), promote transfer of nucleic acid into cells in an endosome-pathway-independent manner. "Naked pDNA" methods are another means to transfect exogenous genes into tissue/organs (Wolff et al. 1990).

However, the other component of the non-viral gene delivery systems, pDNA, has not been very extensively studied, in an effort to improve non-viral gene delivery systems. Indeed, nucleotide sequence drastically affects not only intensity but also longevity of transgene expression. More importantly, the use of particular nucleotide sequences drastically improves the efficacy of gene delivery, as will be discussed later. If ideal sequence elements are devised, potentially every non-viral vector may be significantly improved in terms of the transfection efficiency by combining them with plasmid vectors containing such elements.

3 Plasmid Vector Sequence Crucially Affects the Efficacy of Non-viral Gene Transfer

This section discusses the powerful impact of plasmid vector components on the efficiency of non-viral gene delivery by providing evidence from a very simple, but profound and evocative, experiment (Fig. 1).

pS.CD8α and pSES.CD8α are plasmid vectors harboring exactly the same expression cassette for murine CD8α cDNA as a reporter gene (Satoh et al. 1997). The pSES.CD8α possesses EBV nuclear antigen 1 (EBNA1) gene and oriP sequence, I

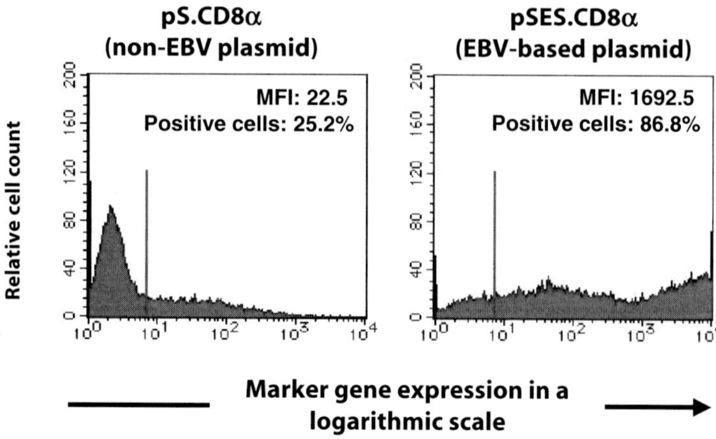

Fig. 1. Plasmid vector composition critically affects the efficacy of non-viral gene transfer. Transfection with an EBV-based plasmid vector resulted in an extremely high proportion of marker-gene-positive cells as well as a tremendously high intensity of marker gene expression on a single-cell basis (*right*). This indicates that a non-viral vector, in this case electroporation, operates at a considerably high efficiency (both plasmids were transfected by the same procedure), while the low efficiency of a conventional plasmid vector (*left*) can be attributed to the inability of the plasmid sequence to appropriately function inside the cells. See text for details

discussed later, while pS.CD8α is a conventional plasmid vector without the EBV sequences. These plasmid vectors were transfected by electroporation into KE cells, which are an EBNA1-expressing transformant derived from the K562 human leukemic cell line (Mazda et al. 1997). Two days after cultivation without any drug selection, cells were stained with a fluorochrome-conjugated anti-mouse CD8α antibody followed by flow-cytometric analysis.

Figure 1 shows the histograms. The KE cells that had been transfected with pS.CD8α expressed murine CD8α at a low level on the surface. This is not surprising because electroporation-mediated transfection into a cell line does not achieve very high efficiency without drug selection. In striking contrast, transfection with pSES.CD8α resulted in tremendously strong expression of the marker gene on the cell surfaces.

Comparing the two histograms in more detail, the results clearly reveal that the pSES.CD8α transfection succeeded in both (1) a higher transfection efficiency, in terms of the increase in percentage of CD8α-positive cells (87% (pSES.CD8α) vs. 25% (pS.CD8α)), and (2) a much stronger expression in each single CD8α-positive cell, as demonstrated by a dramatic elevation in the fluorescent intensities (1690 (pSES.CD8α) vs. 22.5 (pS.CD8α)).

This simple experiment provides insights into non-viral transfection. The two cell populations shown in Fig. 1 were transfected and cultured in exactly the same way, while the plasmids shared a common expression cassette for the marker gene. The only difference was the presence or absence of the EBV genomic elements (EBNA1 gene and oriP). It is quite unlikely that the electric pulse increased the permeability of plasma membrane in such a manner that only pSES.CD8α was effectively penetrated into cells while pS.CD8α selectively remained impermeable. Instead, the experimental result that nearly 90% of the pSES.CD8α-transfected cells expressed the transgene at a high level indicates that the electroporation procedure worked at an extremely high efficiency in this experimental setting, and that pS.CD8α also should have been transmitted from extracellular space into the cytoplasm at a comparable efficiency. The failure of pS.CD8α to result in high CD8α expression can thus ascribed to the different behaviors of the two plasmids after being introduced into the cells, namely intracellular distribution.

Therefore, the experimental data strongly suggest that the major obstacle to non-viral gene transfection, i.e., the failure of cytoplasmic pDNA to be transported into the nucleus (see below), can be overcome by employing an EBV-based plasmid vector, but not by improving the electroporation procedure, which already has reached maximum efficacy. Similar results can be obtained by electro-transfecting various types of cells with plasmid vectors with or without EBNA1 gene/oriP (Mazda et al. 1997; Hirai et al. 1997; Satoh et al. 1998; Tomiyasu et al. 1998).

Interestingly, non-viral vectors other than electroporation also give similar outcomes. Indeed, it has frequently been shown that differences of several-fold to several-thousand-fold between the transfection efficiencies of EBV-based and conventional (non-EBV-based) plasmid vectors are obtained when they were transfected into various cells in culture by means of cationic lipids (Satoh et al. 1997; Harada et al. 2000; Asada et al. 2002) or cationic polymers (Tomiyasu et al. 1998; Harada et al. 2000; Maruyama-Tabata et al. 2000; Tanaka et al. 2000; Ohashi et al. 2001; Iwai et al. 2002; Nakanishi et al. 2003). Significant difference in transfection rates were also obtained by transfection in vivo into various tissues or tumors by means of cationic polymers (Maruyama-Tabata et al. 2000; Iwai et al. 2002; Nakanishi et al. 2003), electroporation (Kishida et al. 2001; Ohashi et al. 2002), and naked DNA methods (Tomiyasu et al. 1998; Cui et al. 2001).

The experimental evidence indicates that the efficacies of non-viral vectors are not as poor as widely believed. Successful transfection with EBV-based plasmid vectors strongly suggest that non-viral delivery systems enables transfer of pDNA into cytoplasm at a considerable rate, although the efficiencies vary according to the types of delivery systems, target cells, etc.

This notion raises a fundamental question as to whether or not "transfection efficiency" can accurately be assessed using the widely accepted experimental systems, which are typically as follows: (1) appropriate target cells are transfected with pDNA carrying a marker gene by means of a delivery system to be examined, (2) cultivation of the cells for an appropriate period, (3) the cells, cell extracts, or culture supernatants are collected and the expression level of the marker gene is measured. The procedures are modified depending on the purpose of the study, e.g. to examine transfection efficiency in vivo, appropriate organs/tissues are substituted for cells in step (1), and instead of (3), the organs/tissues are collected to evaluate marker gene product. Apart from such variations, the efficiencies of non-viral gene transfer systems are assessed by performing experiments that are based on the same concept as described above.

However, interpreting the data obtained from such experiments requires consideration of the following: Several independent steps are critically involved in marker

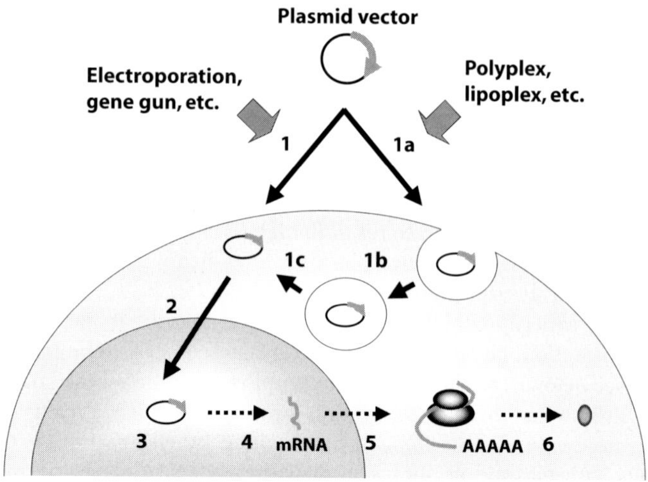

FIG. 2. Multiple steps are involved in successful non-viral gene transfection.
The performance of a non-viral vector cannot be simply evaluated by the magnitude of marker gene expression, which depends on efficiencies of multiple steps. Indeed, intra-nuclear delivery of pDNA (*step 2*) is the most critical step that determines the rate of non-viral transfection. Use of the EBV-based plasmid vector may elevate the transfection rate by accelerating this step. See text for details

gene expression after transfection (Fig. 2). These include: (1) transfer of pDNA from the extracellular space to the cytoplasm through the plasma membrane, (2) transfer of pDNA from the cytoplasm to the nucleus through the nuclear pore complex (NPC), (3) maintenance of pDNA in the nucleus, (4) transcription, (5) maturation of mRNA and its transfer from the nucleus to the cytoplasm, and (6) translation and post-translational modification of the protein. When polyplexes or lipoplexes are used as the non-viral vectors, additional steps are usually required instead of (1), i.e., (1a) transfer of pDNA from the extracellular space to the endosome/lysosome compartment by endocytosis, (1b) maintenance of pDNA in the compartment, (1c) release of pDNA from the endosome/lysosome compartment to the cytoplasm.

Although the efficiency of step (1) may reflect the capability of the non-viral vector of interest, the efficiencies of the other steps depend on different factors. For example, the rate of step (4) has been attributed mainly to the transcriptional activity of the promoter/enhancer element contained in the plasmid. Therefore, the expression level of a marker gene is proportional to the product of the efficiencies of all these steps, but not simply to the performance of the non-viral vector tested.

Actually, step (2) is the most critical process, in which efficiency crucially determines marker gene expression in many non-viral gene delivery systems (Lechardeur and Lukacs 2002; Hebert 2003)(see below). Without ameliorating this step, non-viral gene delivery cannot be improved, because transfer of pDNA from the extracellular space to the cytoplasm has succeeded at a considerably high rate as demonstrated in Fig. 1 and by other, similar experimental evidence for a variety of non-viral vectors, as discussed above. Basically, non-viral vectors are not capable of promoting step (2), which is the reason why plasmid vector sequences are extremely important.

From another point of view, in order to devise or modify non-viral vectors, use of EBV-based plasmid vectors instead of conventional plasmid vectors may greatly help in assessing the capability of novel non-viral vectors, because the efficiency of step (2) should be sufficiently high so that the efficiency of step (1) is measured at high sensitivity and accuracy.

4 Epstein-Barr Virus-Based Plasmid Vectors

The rationale for EBV-based plasmid vectors, as well as concrete examples of their use in non-viral gene delivery systems, has been reviewed elsewhere (Mazda 2000, 2002). Here, the EBV system is briefly explained in order to support the concepts presented above. The EBV-based plasmid vector contains the EBV nuclear antigen 1 (EBNA1) gene and oriP sequence, which were originally described as elements to support replication of the EBV genome (double-stranded circular DNA of approximately 172 kb) in latently infected human cells (Adams 1987; Haase and Calos 1991; Yates and Guan 1991). EBNA1 is a nuclear phosphoprotein that binds specific motifs in the oriP sequence, conducting DNA replication in concert with cellular cofactors. Therefore, EBV-based plasmid vectors are maintained in transfected cells as an extrachromosomal circular DNA (episome). Some groups have developed autonomously replicative artificial chromosomes, so that transgene expression persists for a sufficiently long period of time (Sun et al. 1994; Westphal et al. 1998; Kelleher et al. 1998; Black and Vos 2002; Stoll et al. 2001; Stoll and Calos 2002; Sclimenti et al. 2003). The episomal maintenance is particularly important when target cells proliferate after transfection.

EBNA1 also facilitates trafficking of oriP-bearing plasmid DNA from the cytosol to the nucleus (Ambinder et al. 1991; Fischer et al. 1997) and its maintenance in the nucleus (Jankelevich et al. 1992). Another function of EBNA1 is the activation of transcription through binding to oriP (Gahn and Sugden 1995; Puglielli et al. 1996; Reisman and Sugden 1986; Sugden and Warren 1989; Wysokenski and Yates 1989). We hypothesized that these characteristics explain why plasmid vectors with EBNA1 gene and oriP enable high-level gene transfer and expression.

We then estimated the contribution of each activity of the multifunctional viral elements to the high efficiency of transfection. We determined that transfer of pDNA from the cytoplasm into the nucleus is actually the most critical step, and that the efficacies are quite different between conventional and EBV-based plasmid vectors (Kishida et al., in preparation). This is consistent with previous reports indicating that most pDNA molecules introduced into a cell are entrapped and degraded in the cytoplasmic and endo-lysosomal compartments without successful transport into the nucleus, and this barrier is the critical obstacle of conventional non-viral gene delivery strategies (reviewed in Lechardeur and Lukacs 2002). We also found that replication of pDNA does not contribute to the high rate of transfection of EBV-plasmids, although this function plays key roles in prolonged transgene expression in cultured human cells.

Taking advantage of the high transfection efficiency, preclinical gene therapy studies were carried out with EBV-based plasmid vectors in a variety of animal model systems. Briefly, subcutaneous tumor transplants such as melanoma (Asada et al. 2002; Kishida et al. 2001), hepatocellular carcinoma (Iwai et al. 2002), prostate cancer (Nakanishi et al. 2003), and Ewing's sarcoma (Maruyama-Tabata et al. 2000) were treated with cytokine-gene-transfected tumor vaccine (Asada et al. 2002) or intratumoral delivery of cytokine genes (Kishida et al. 2001), a suicide gene (Maruyama-Tabata et al. 2000; Iwai et al. 2002), or an apoptosis-inducing gene (Nakanishi et al. 2003). Metastatic malignancies were also attempted, including hepatic metastasis of melanoma (Asada et al. 2002; Kishida et al. 2003a), lung metastasis of lymphoma (Kishida et al. 2003b), and hepatic metastasis and peritoneal carcinomatosis of reticulum cell sarcoma (Itokawa et al. 2004). Therapeutic/prophylactic experiments were also performed against nonmalignant disorders including cardiomyopathy (genetic transfer of the β2-adrenergic receptor into failing cardiomyocytes) (Tomiyasu et al. 2000), diabetes mellitus (intrahepatic delivery of the insulin gene under the control of a glucose-responsive promoter) (Yasutomi et al. 2003), and acute herpes simplex virus type 1 (HSV1) infection (DNA vaccination using a viral antigen gene with/without cytokine genes as adjuvants) (Cui, et al. 2003, 2005). Significant therapeutic outcomes were obtained from all these trials due to the high efficacy of the non-viral delivery systems employing the EBV-based plasmid vectors.

5 The Sleeping Beauty Transposable Element

In the folllowing, DNA transposons, and especially the *sleeping beauty* (*SB*) element, are discussed as another example of using a plasmid vector with particular nucleotide sequences as a component of non-viral gene delivery systems. Although EBV-based plasmid vectors are maintained as episomes in target cells, resulting in prolonged expression of the transgene, employment of a transposable element enables chromosomal integration of the transgene, providing an alternative strategy to prolonging the expression period obtained after non-viral gene delivery.

Unlike retrotransposons, which replicate and jump in a copy-and-paste manner, DNA transposons are excised from their original genomic location and subsequently reinserted elsewhere in the genome (a cut-and-paste process). DNA transposons have been utilized for insertional mutagenesis as well as germline transgenesis in invertebrate organisms including *Drosophila* and *C. elegans*, but similar systems were not available in vertebrates until recently. While some DNA transposons in invertebrates, such as the Tc3 element in nematode (Raz et al. 1998) and *Minos* in fly (Zagoraiou et al. 2001), were shown to function in cells of vertebrate species, a more efficient DNA transposon for vertebrates was developed by "awakening" a salmonid fish transposon belonging to the Tc1/mariner superfamily, which are molecular fossils inactivated millions of years ago by the accumulation of mutations (Ivics et al. 1997). Ivics et al. constructed a synthetic transposon based on phylogenetic data so that inactivating mutations were eliminated. The reconstructed element, the *SB* transposon, precisely worked in fish as well as in mouse and human cells (Ivics et al. 1997).

The structure of *SB* and the mechanisms of transposition have been described elsewhere in detail (Izsvak and Ivics 2004; Ivics and Izsvak 2004). Briefly, natural *SB* consists of a single open reading frame encoding an enzymatic polypeptide (transposase), which is flanked by two terminal inverted repeats (IRs) at both ends. The transposase binds the recognition sequences in the IRs in a highly sequence-specific fashion, leading to mobilization of the transposon. *SB* was engineered to transpose a gene of interest, by inserting the transgene segment between the IRs while supplying the transposase either in *trans* (Izsvak et al. 2000; Dupuy et al. 2002) or in *cis* (Mikkelsen et al. 2003). In the original *SB*, the gene to be mobilized should be less than several kilo base pairs in size, since the efficiency of transposition declines with increasing the length (Izsvak et al. 2000).

More recently, the *SB* transposon was further improved through a mutagenesis approach so that the modified version is capable of transposing DNA segments more than 10 kilo base pairs in size (Zayed et al. 2004). The transposable activity was also augmented, particularly when combined with the DNA-bending, high-mobility-group protein, HMGB1, which is a host-encoded cofactor of *SB* transposition (Zayed et al. 2003).

Plasmid vectors harboring the *SB* transposon can be combined with variety of non-viral vectors, including cationic lipid (Liu, L. et al. 2004b), cationic polymer (Belur et al. 2003) and other carrier molecules (Harris et al. 2002; Zayed et al. 2003, 2004), while microinjection (Fischer et al. 2001; Dupuy et al. 2002; Horie et al. 2003), polybrane shock (Ortiz-Urda et al. 2003) and naked DNA (Yant et al. 2000; Montini et al. 2002; Belur et al. 2003) procedures have also been used. *SB* transposes not only in vertebrate cells in culture, but also in somatic (Yant et al. 2000; Montini et al. 2002; Belur et al. 2003; Liu et al. 2004a, Liu, 2004b) and germline (Fischer, S. E. et al. 2001; Dupuy et al. 2002; Horie et al. 2003) tissues of mice in vivo. The major advantage of using *SB* is the prolonged expression of the transgene in transfected cells and tissues due to chromosomal integration of the recombinant transposon, while the integration sites in host chromosomes are almost random (Vigdal et al. 2002; Roberg-Perez et al. 2003).

Some studies have suggested the feasibility of the *SB* system for gene therapy, based on therapeutic experiments in which *SB*-bearing plasmid vectors were administered to model animals (reviewed in Izsvak and Ivics 2004; Ivics and Izsvak 2004). Yant et al. treated hemophilic mice with a rapid intravenous administration of a *SB* construct carrying the α1-antitrypsin gene together with a transposase construct, partially ameliorating the bleeding phenotype for more than 5 months (Yant et al. 2000). Montini et al. transfected fumarylacetoacetate hydrolase (FAH) gene knockout mice with a mixture of a FAH gene-containing transposon construct and a transposase expression plasmid, as a therapeutic model against hereditary tyrosinemia type I, resulting in long-lasting expression of the therapeutic gene in the liver (Montini et al. 2002). Ortiz-Urda et al. performed ex vivo transfection experiments in which a *SB* vector encoding both laminin 5 and a selectable antibiotic marker was delivered into keratinocytes obtained from patients with junctional epidermolysis bullosa (JEB), a lethal hereditary skin disorder (Ortiz-Urda et al. 2003). After drug selection, the genetically modified cells were transplanted into immune-deficient mice, so that the skin tissue regenerated. Long-term expression of laminin 5 at a therapeutic level and phenotypic correction of the skin (hemidesmosome formation and blistering) were obtained, strongly suggesting that *SB*-mediated genetic engineering of self-renewing cells results in stable maintenance of transgene expression in the progeny cells (Ortiz-Urda et al. 2003).

6 Conclusions

Currently, significant advances in the development of non-viral vectors are being made by means of recent technological innovations, including nanotechnology (Vijayanathan et al. 2002). Employment of functional genetic elements, such as EBNA1/oriP and the *SB* transposon, may further improve their efficacy, compensating for the shortcomings of non-viral systems. EBV-based plasmid vectors are transported into and maintained in the nucleus as well as replicated as episomes, overcoming the problems of the low transfection rate and the transience of transgene expression. The *SB* transposon enables chromosomal integration of a transgene in transfected cells, potentially enabling its permanent expression, although random integration into the host genome may cause aberrant expression and/or silencing of cellular genes that locate close to the integration site. Both the EBV-based plasmid vector and *SB* have been shown to work not only in vitro but also in vivo. Plasmid vector sequences should be focused on as an important component of non-viral gene delivery systems, while their combination with other devises, such as polypeptides containing a nuclear localization signal (reviewed in Hebert 2003), may also be useful in developing promising strategies.

References

Adams A (1987) Replication of latent Epstein-Barr virus genomes in Raji cells. J Virol 61:1743–1746

Ambinder RF, Mullen MA, Chang YN, Hayward GS, Hayward SD (1991) Functional domains of Epstein-Barr virus nuclear antigen EBNA-1. J Virol 65:1466–1478

Asada H, Kishida T, Hirai H, Satoh E, Ohashi S, Takeuchi M, Kubo T, Kita M, Iwakura Y, Imanishi J, Mazda O (2002) Significant antitumor effects obtained by autologous tumor cell vaccine engineered to secrete interleukin-12 (IL-12) and IL-18 by means of the EBV/lipoplex. Mol Ther 5:609–616

Belur LR, Frandsen JL, Dupuy AJ, Ingbar DH, Largaespada DA, Hackett PB, Scott McIvor R (2003) Gene insertion and long-term expression in lung mediated by the Sleeping Beauty transposon system. Mol Ther 8:501–507

Black J, Vos JM (2002) Establishment of an oriP/EBNA1-based episomal vector transcribing human genomic beta-globin in cultured murine fibroblasts. Gene Ther 9:1447–1454

Cui FD, Kishida T, Ohashi S, Asada H, Yasutomi K, Satoh E, Kubo T, Fushiki S, Imanishi J, Mazda O (2001) Highly efficient gene transfer into murine liver achieved by intravenous administration of naked Epstein-Barr virus (EBV)-based plasmid vectors. Gene Ther 8:1508–1513

Cui FD, Asada H, Kishida T, Itokawa Y, Nakaya T, Ueda Y, Yamagishi H, Gojo S, Kita M, Imanishi J, Mazda O (2003) Intravascular naked DNA vaccine encoding glycoprotein B induces protective humoral and cellular immunity against herpes simplex virus type 1 infection in mice. Gene Ther 10:2059–2066

Cui FD, Asada H, Jin ML, Kishida T, Shin-Ya M, Nakaya T, Kita M, Ishii M, Iwai M, Okanoue T, Imanishi J, Mazda O (2005) Cytokine genetic adjuvant facilitates prophylactic intravascular DNA vaccine against acute and latent herpes simplex virus infection in mice. Gene Ther 12:160–168

Cui Z, Mumper RJ (2003) Microparticles and nanoparticles as delivery systems for DNA vaccines. Crit Rev Ther Drug Carrier Syst 20:103–137

Dupuy AJ, Clark K, Carlson CM, Fritz S, Davidson AE, Markley KM, Finley K, Fletcher CF, Ekker SC, Hackett PB, Horn S, Largaespada DA (2002) Mammalian germ-line transgenesis by transposition. Proc Natl Acad Sci U S A 99:4495–4499

Felgner PL, Ringold GM (1989) Cationic liposome-mediated transfection. Nature 337: 387–388

Fischer N, Kremmer E, Lautscham G, Mueller-Lantzsch N, Grasser FA (1997) Epstein-Barr virus nuclear antigen 1 forms a complex with the nuclear transporter karyopherin alpha2. J Biol Chem 272:3999–4005

Fischer SE, Wienholds E, Plasterk RH (2001) Regulated transposition of a fish transposon in the mouse germ line. Proc Natl Acad Sci U S A 98:6759–6764

Gahn TA, Sugden B (1995) An EBNA-1-dependent enhancer acts from a distance of 10 kilobase pairs to increase expression of the Epstein-Barr virus LMP gene. J Virol 69: 2633–2636

Haase SB, Calos MP (1991) Replication control of autonomously replicating human sequences. Nucleic Acids Res 19:5053–5058

Harada Y, Iwai M, Tanaka S, Okanoue T, Kashima K, Maruyama-Tabata H, Hirai H, Satoh E, Imanishi J, Mazda O (2000) Highly efficient suicide gene expression in hepatocellular carcinoma cells by Epstein-Barr virus-based plasmid vectors combined with polyamidoamine dendrimer. Cancer Gene Ther 7:27–36

Harris JW, Strong DD, Amoui M, Baylink DJ, Lau KH (2002) Construction of a Tc1-like transposon Sleeping Beauty-based gene transfer plasmid vector for generation of stable transgenic mammalian cell clones. Anal Biochem 310:15–26

Hebert E (2003) Improvement of exogenous DNA nuclear importation by nuclear localization signal-bearing vectors: a promising way for non-viral gene therapy? Biol Cell 95: 59–68

Herweijer H, Wolff JA (2003) Progress and prospects: naked DNA gene transfer and therapy. Gene Ther 10:453–458

Hirai H, Satoh E, Osawa M, Inaba T, Shimazaki C, Kinoshita S, Nakagawa M, Mazda O, Imanishi J (1997) Use of EBV-based Vector/HVJ-liposome complex vector for targeted gene therapy of EBV-associated neoplasms. Biochem Biophys Res Commun 241:112–118

Horie K, Yusa K, Yae K, Odajima J, Fischer SE, Keng VW, Hayakawa T, Mizuno S, Kondoh G, Ijiri T, Matsuda Y, Plasterk RH, Takeda J (2003) Characterization of Sleeping Beauty transposition and its application to genetic screening in mice. Mol Cell Biol 23: 9189–9207

Hosseinkhani H, Aoyama T, Ogawa O, Tabata Y (2003) Ultrasound enhances the transfection of plasmid DNA by non-viral vectors. Curr Pharm Biotechnol 4:109–122

Itokawa Y, Mazda O, Ueda Y, Kishida T, Asada H, Cui FD, Fuji N, Fujiwara H, Shin-Ya M, Yasutomi K, Imanishi J, Yamagishi H (2004) Interleukin-12 genetic administration suppressed metastatic liver tumor unsusceptible to CTL. Biochem Biophys Res Commun 314:1072–1079

Ivics Z, Izsvak Z (2004) Transposable elements for transgenesis and insertional mutagenesis in vertebrates: a contemporary review of experimental strategies. Methods Mol Biol 260:255–276

Ivics Z, Hackett PB, Plasterk RH, Izsvak Z (1997) Molecular reconstruction of Sleeping Beauty, a Tc1-like transposon from fish, and its transposition in human cells. Cell 91:501–510

Iwai M, Harada Y, Tanaka S, Muramatsu A, Mori T, Kashima K, Imanishi J, Mazda O (2002) Polyethylenimine-mediated suicide gene transfer induces a therapeutic effect for hepatocellular carcinoma *in vivo* by using an Epstein/Barr virus-based plasmid vector. Biochem Biophys Res Commun 291:48–54

Izsvak Z, Ivics Z (2004) Sleeping beauty transposition: biology and applications for molecular therapy. Mol Ther 9:147–156

Izsvak Z, Ivics Z, Plasterk RH (2000) Sleeping Beauty, a wide host-range transposon vector for genetic transformation in vertebrates. J Mol Biol 302:93–102

Jankelevich S, Kolman JL, Bodnar JW, Miller G (1992) A nuclear matrix attachment region organizes the Epstein-Barr viral plasmid in Raji cells into a single DNA domain. EMBO J 11:1165–1176

Kelleher ZT, Fu H, Livanos E, Wendelburg B, Gulino S, Vos JM (1998) Epstein-Barr-based episomal chromosomes shuttle 100 kb of self- replicating circular human DNA in mouse cells. Nat Biotechnol 16:762–768.

Kishida T, Asada H, Satoh E, Tanaka S, Shinya M, Hirai H, Iwai M, Tahara H, Imanishi J, Mazda O (2001) *In vivo* electroporation-mediated transfer of interleukin-12 and interleukin-18 genes induces significant antitumor effects against melanoma in mice. Gene Ther 8:1234–1240

Kishida T, Asada H, Itokawa Y, Cui FD, Shin-Ya M, Gojo S, Yasutomi K, Ueda Y, Yamagishi H, Imanishi J, Mazda O (2003a) Interleukin (IL)-21 and IL-15 genetic transfer synergistically augments therapeutic antitumor immunity and promotes regression of metastatic lymphoma. Mol Ther 8:552–558

Kishida T, Asada H, Itokawa Y, Yasutomi K, Shin-Ya M, Gojo S, Cui FD, Ueda Y, Yamagishi H, Imanishi J, Mazda O (2003b) Electrochemo-gene therapy of cancer: intratumoral delivery of interleukin-12 gene and bleomycin synergistically induced therapeutic immunity and suppressed subcutaneous and metastatic melanomas in mice. Mol Ther 8:738–745

Kukowska-Latallo JF, Bielinska AU, Johnson J, Spindler R, Tomalia DA, Baker JR, Jr. (1996) Efficient transfer of genetic material into mammalian cells using Starburst polyamidoamine dendrimers. Proc Natl Acad Sci U S A 93:4897–4902

Kumar VV, Singh RS, Chaudhuri A (2003) Cationic transfection lipids in gene therapy: successes, set-backs, challenges and promises. Curr Med Chem 10:1297–1306

Lechardeur D, Lukacs GL (2002) Intracellular barriers to non-viral gene transfer. Curr Gene Ther 2:183–194

Lemkine GF, Demeneix BA (2001) Polyethylenimines for in vivo gene delivery. Curr Opin Mol Ther 3:178–182

Liu G, Aronovich EL, Cui Z, Whitley CB, Hackett PB (2004a) Excision of Sleeping Beauty transposons: parameters and applications to gene therapy. J Gene Med 6:574–583

Liu L, Sanz S, Heggestad AD, Antharam V, Notterpek L, Fletcher BS (2004b) Endothelial targeting of the Sleeping Beauty transposon within lung. Mol Ther 10:97–105

Maruyama-Tabata H, Harada Y, Matsumura T, Satoh E, Cui F, Iwai M, Kita M, Hibi S, Imanishi J, Sawada T, Mazda O (2000) Effective suicide gene therapy *in vivo* by EBV-based plasmid vector coupled with polyamidoamine dendrimer. Gene Ther 7:53–60

Mazda O (2000) Application of Epstein-Barr virus (EBV) and its genetic elements to gene therapy. In: Cid-Arregui A, Garcia-Carranca A (Eds) Viral Vectors: Basic Science and Gene Therapy. Eaton Publishing, Natick, MA. pp 325–337

Mazda O (2002) Improvement of non-viral gene therapy by Epstein-Barr virus (EBV)-based plasmid vectors. Curr Gene Ther 2:379–392

Mazda O, Satoh E, Yasutomi K, Imanishi J (1997) Extremely efficient gene transfection into lympho-hematopoietic cell lines by Epstein-Barr virus-based vectors. J Immunol Methods 204:143–151

Mikkelsen JG, Yant SR, Meuse L, Huang Z, Xu H, Kay MA (2003) Helper-Independent Sleeping Beauty transposon-transposase vectors for efficient non-viral gene delivery and persistent gene expression in vivo. Mol Ther 8:654–665

Montini E, Held PK, Noll M, Morcinek N, Al-Dhalimy M, Finegold M, Yant SR, Kay MA, Grompe M (2002) In vivo correction of murine tyrosinemia type I by DNA-mediated transposition. Mol Ther 6:759–769

Nakanishi H, Mazda O, Satoh E, Asada H, Morioka H, Kishida T, Nakao M, Mizutani Y, Kawauchi A, Kita M, Imanishi J, Miki T (2003) Non-viral genetic transfer of Fas ligand induced significant growth suppression and apoptotic tumor cell death in prostate cancer in vivo. Gene Ther 10:434–442

Niidome T, Huang L (2002) Gene therapy progress and prospects: non-viral vectors. Gene Ther 9:1647–1652

Ohashi S, Kubo T, Ikeda T, Arai Y, Takahashi K, Hirasawa Y, Takigawa M, Satoh E, Imanishi J, Mazda O (2001) Cationic polymer-mediated genetic transduction into cultured human chondrosarcoma-derived HCS-2/8 cells. J Orthop Sci 6:75–81

Ohashi S, Kubo T, Kishida T, Ikeda T, Takahashi K, Arai Y, Terauchi R, Asada H, Imanishi J, Mazda O (2002) Successful genetic transduction *in vivo* into synovium by means of electroporation. Biochem Biophys Res Commun 293:1530–1535

Ortiz-Urda S, Lin Q, Yant SR, Keene D, Kay MA, Khavari PA (2003) Sustainable correction of junctional epidermolysis bullosa via transposon-mediated non-viral gene transfer. Gene Ther 10:1099–1104

Plank C, Anton M, Rudolph C, Rosenecker J, Krotz F (2003) Enhancing and targeting nucleic acid delivery by magnetic force. Expert Opin Biol Ther 3:745–758

Puglielli MT, Woisetschlaeger M, Speck SH (1996) oriP is essential for EBNA gene promoter activity in Epstein-Barr virus- immortalized lymphoblastoid cell lines. J Virol 70: 5758–5768

Raz E, van Luenen HG, Schaerringer B, Plasterk RH, Driever W (1998) Transposition of the nematode Caenorhabditis elegans Tc3 element in the zebrafish Danio rerio. Curr Biol 8:82–88

Reisman D, Sugden B (1986) trans activation of an Epstein-Barr viral transcriptional enhancer by the Epstein-Barr viral nuclear antigen 1. Mol Cell Biol 6:3838–3846

Roberg-Perez K, Carlson CM, Largaespada DA (2003) MTID: a database of Sleeping Beauty transposon insertions in mice. Nucleic Acids Res 31:78–81

Rocha A, Ruiz S, Coll JM (2002) Improvement of DNA transfection with cationic liposomes. J Physiol Biochem 58:45–56

Satoh E, Hirai H, Inaba T, Shimazaki C, Nakagawa M, Imanishi J, Mazda O (1998) Successful transfer of ADA gene *in vitro* into human peripheral blood $CD34^+$ cells by transfecting EBV-based episomal vectors. FEBS Lett 441:39–42

Satoh E, Osawa M, Tomiyasu K, Hirai H, Shimazaki C, Oda Y, Nakagawa M, Kondo M, Kinoshita S, Mazda O, Imanishi J (1997) Efficient gene transduction by Epstein-Barr-virus-based vectors coupled with cationic liposome and HVJ-liposome. Biochem Biophys Res Commun 238:795–799

Sclimenti CR, Neviaser AS, Baba EJ, Meuse L, Kay MA, Calos MP (2003) Epstein-Barr virus vectors provide prolonged robust factor IX expression in mice. Biotechnol Prog 19: 144–151

Stoll SM, Calos MP (2002) Extrachromosomal plasmid vectors for gene therapy. Curr Opin Mol Ther 4:299–305

Stoll SM, Sclimenti CR, Baba EJ, Meuse L, Kay MA, Calos MP (2001) Epstein-Barr Virus/Human Vector Provides High-Level, Long-Term Expression of alpha(1)-Antitrypsin in Mice. Mol Ther 4:122–129.

Sugden B, Warren N (1989) A promoter of Epstein-Barr virus that can function during latent infection can be transactivated by EBNA-1, a viral protein required for viral DNA replication during latent infection. J Virol 63:2644–2649

Sun TQ, Fernstermacher DA, Vos JM (1994) Human artificial episomal chromosomes for cloning large DNA fragments in human cells [published erratum appears in Nat Genet 1994 Dec;8(4):410]. Nat Genet 8:33–41

Tanaka S, Iwai M, Harada Y, Morikawa T, Muramatsu A, Mori T, Okanoue T, Kashima K, Maruyama-Tabata H, Hirai H, Satoh E, Imanishi J, Mazda O (2000) Targeted killing of carcinoembryonic antigen (CEA)-producing cholangiocarcinoma cells by polyamidoamine dendrimer-mediated transfer of an Epstein-Barr virus (EBV)-based plasmid vector carrying the CEA promoter. Cancer Gene Ther 7:1241–1250

Tang MX, Redemann CT, Szoka FC, Jr. (1996) *In vitro* gene delivery by degraded polyamidoamine dendrimers. Bioconjug Chem 7:703–714

Tomiyasu K, Satoh E, Oda Y, Nishizaki K, Kondo M, Imanishi J, Mazda O (1998) Gene transfer *in vitro* and *in vivo* with Epstein-Barr virus-based episomal vector results in markedly high transient expression in rodent cells. Biochem Biophys Res Commun 253:733–738

Tomiyasu K, Oda Y, Nomura M, Satoh E, Fushiki S, Imanishi J, Kondo M, Mazda O (2000) Direct intra-cardiomuscular transfer of beta2-adrenergic receptor gene augments cardiac output in cardiomyopathic hamsters. Gene Ther 7:2087–2093

Vigdal TJ, Kaufman CD, Izsvak Z, Voytas DF, Ivics Z (2002) Common physical properties of DNA affecting target site selection of sleeping beauty and other Tc1/mariner transposable elements. J Mol Biol 323:441–452

Vijayanathan V, Thomas T, Thomas TJ (2002) DNA nanoparticles and development of DNA delivery vehicles for gene therapy. Biochemistry 41:14085–14094

Westphal EM, Sierakowska H, Livanos E, Kole R, Vos JM (1998) A system for shuttling 200-kb BAC/PAC clones into human cells: stable extrachromosomal persistence and long-term ectopic gene activation. Hum Gene Ther 9:1863–1873

Wolff JA, Malone RW, Williams P, Chong W, Acsadi G, Jani A, Felgner PL (1990) Direct gene transfer into mouse muscle in vivo. Science 247:1465–1468.

Wysokenski DA, Yates JL (1989) Multiple EBNA1-binding sites are required to form an EBNA1-dependent enhancer and to activate a minimal replicative origin within oriP of Epstein-Barr virus. J Virol 63:2657–2666

Yant SR, Meuse L, Chiu W, Ivics Z, Izsvak Z, Kay MA (2000) Somatic integration and long-term transgene expression in normal and haemophilic mice using a DNA transposon system. Nat Genet 25:35–41

Yasutomi K, Itokawa Y, Asada H, Kishida T, Cui FD, Ohashi S, Gojo S, Ueda Y, Kubo T, Yamagishi H, Imanishi J, Takeuchi T, Mazda O (2003) Intravascular insulin gene delivery as potential therapeutic intervention in diabetes mellitus. Biochem Biophys Res Commun 310:897–903

Yates JL, Guan N (1991) Epstein-Barr virus-derived plasmids replicate only once per cell cycle and are not amplified after entry into cells. J Virol 65:483–488

Zagoraiou L, Drabek D, Alexaki S, Guy JA, Klinakis AG, Langeveld A, Skavdis G, Mamalaki C, Grosveld F, Savakis C (2001) In vivo transposition of Minos, a Drosophila mobile element, in mammalian tissues. Proc Natl Acad Sci U S A 98:11474–11478

Zayed H, Izsvak Z, Walisko O, Ivics Z (2004) Development of hyperactive sleeping beauty transposon vectors by mutational analysis. Mol Ther 9:292–304

Zayed H, Izsvak Z, Khare D, Heinemann U, Ivics Z (2003) The DNA-bending protein HMGB1 is a cellular cofactor of Sleeping Beauty transposition. Nucleic Acids Res 31: 2313–2322

Shielding of Cationic Charge of the DNA Complex to Avoid Nonspecific Interactions for In Vivo Gene Delivery

Yoshiyuki Koyama

1 Introduction

Although a number of attempts to transfect cells in vitro using non-viral vectors have to some extent succeeded, very limited success has been reported for in vivo gene transfection. The barriers against in vivo application of artificial vectors can be attributed to (1) the difference in the physiological characteristics between cultured cell lines and cells in the intact organs, and (2) the difficulty in targeting the DNAs to the target-cell surfaces.

Transformed cell lines have a different physiology than their in vivo counterparts, and are generally known to be more easily transfected than intact cells in tissue or primary cell lines. A difference in the intracellular trafficking of a plasmid complex in cultured cells vs. primary cells was observed (Vitiello et al. 1998). For example, a promoter that is active in cultured cell line is not always effective in cells of the original tissue (Hafenrichter et al. 1994). Thus, consideration of physiological factors is very important for successful in vivo gene therapy, as was detailed in a book edited by J.A. Wolff (1994).

Gene transportation to the surface of target cells forms a significant barrier that must be overcome for efficient gene transfer in vivo (Fig. 1). The development of delivery system for low-molecular-weight drugs has greatly aided progress in plasmid DNA delivery. However, the efficient delivery of such giant polyanions to target tissues or cells involves a number of additional obstacles, such as size-limiting constraints, and nonspecific interactions with blood components, cells, or extracellular matrices.

In this chapter, a strategy to overcome the barriers to gene delivery for successful in vivo gene transfer is described.

Otsuma Women's University, 12 Sanbancho, Chiyoda-ku, Tokyo 102-8357, Japan

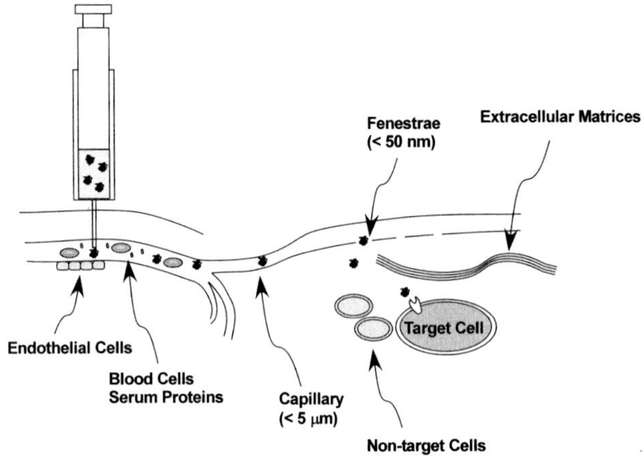

FIG. 1. Barriers to the transportation of DNA complexes

2 Interaction of DNA Complexes with Blood Components or Extracellular Matrices

2.1 The Positively Charged Surface of the Complex Invites Nonspecific Interactions

Although a variety of cationic lipids and polymers have facilitated effective gene transfer in cell culture, efficient gene delivery in vivo is still the major challenge in gene therapy. One of the major obstacles to the use of cationic lipids and polymers in in vivo gene therapy is the uncontrolled and undefined interaction of the DNA complexes with blood cells, serum proteins, and extracellular matrices, which occurs before they reach the target cell surfaces.

Plasmid DNA is a rigid giant polyanion, but it can be condensed to a small complex in the presence of excess cations. This lack of equilibrium in the charge balance always results in the poly-ion complexes being positively charged, at least on their surfaces.

The positive surface charge of the plasmid/cationic lipid or polymer complexes facilitates their electrostatic binding to the cell surface, leading to efficient internalization and gene expression by the cultured cells in vitro. Condensation with cations also protects plasmids from enzymatic degradation and inactivation. However, these positively charged complex surfaces also account for undesirable interaction with blood components and extracellular matrices (Fig. 2). For example, when the cationic plasmid complexes are injected into the blood stream, they coagulate blood cells, whose membranes are negatively charged (Maruyama et al. 2004; Kircheis et al. 2001). In addition, there are abundant negatively charged proteins, such as albumin, in the serum and in body fluid that are able to bind to the cationic plasmid complexes and cause the formation of large aggregates (Dash et al. 1999), thus preventing efficient delivery of the DNA complex. Positively charged DNA complex was also shown to activate the serum complement system (Plank et al. 1996), which may not directly hinder

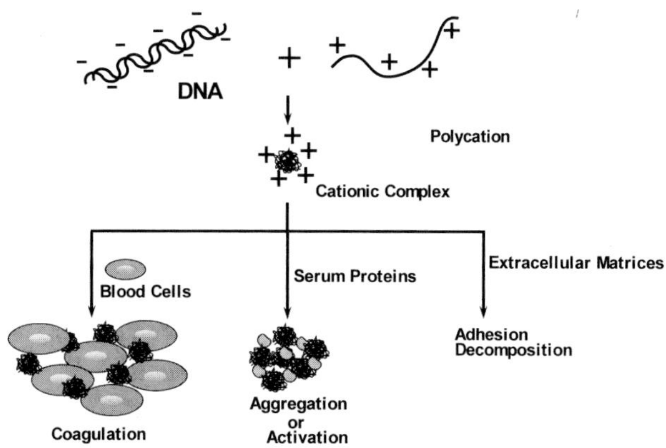

FIG. 2. DNA complexes have a positively charged surface that invites nonspecific interactions with blood components and extracellular matrices

gene expression, but should be considered with respect to applications in human gene therapy.

Immune cells recognize the plasmid DNA complex as foreign material. Consequently, the intravenous injection of cationic liposome/DNA complex has been reported to lead to a high level of cytokine production (Li et al. 1999). Unmethylated CpG dinucleotides in the plasmid also induce immune reactions (Krieg et al. 1995), which represent a significant obstacle for in vivo gene therapy.

If, nonetheless, the plasmid complex reaches the target tissue and extravasates out of the blood vessels, it then encounters the extracellular matrices, comprising sulfated or carboxyl-containing polysaccharides such as heparan sulfate, chondroitin sulfates, and hyaluronic acid. These anionic polysaccharides bind to the cationic plasmid complex, and some of them have been shown to decompose the complex by competing with the plasmid molecule (Ruponen et al. 1999). Not only polysaccharides in the matrices but also scaffold proteins, collagen and fibronectin attach to the positive surface of the complex. After immersion in a collagen-sponge, DNA/polycation complexes became immobilized in the matrix and could not be released (Scherer et al. 2002).

2.2 Protection of the Plasmid Complex from Undesired Interactions

2.2.1 Targeting Delivery Requires Suppression of Nonspecific Interactions

Tissue- or cell-specific ligands have been employed in targeting systems for low-molecular-weight drugs. Ligand-guided delivery of plasmid has also been similarly attempted. Galactose-bearing polycations, such as poly(L-lysine) (PLL) (Plank et al. 1992), and polyethyleneimine (PEI) (Zanta et al. 1997), were prepared, and their complexes with plasmid showed high in vitro transfection efficiency in hepatoma cells via asialoglycoprotein-receptor-mediated endocytosis. However, in vivo gene expression of such sugar-bearing vectors in the liver after injection into mouse tail vein has often failed, although portal-vein injection achieved a relatively high level of transgene expression (Morimoto et al. 2003). The low-level expression after systemic administration seems to be caused by entrapment of the plasmid complex before it reached the liver, which shows that suppression of nonspecific interactions is a prerequisite for efficient in vivo target-specific plasmid delivery.

2.2.2 Shielding of the Positive Surface Charge

As described above, most undesired nonspecific interactions of the DNA/polycation complexes are due to their positive surface charge. In order to achieve stability in the blood stream and target specificity, the positive charge should be shielded by hydrophilic polymers, such as polyethylene glycol (PEG).

Surface modification by PEG was adopted to prepare "stealth" liposomes, which were not recognized by the reticuloendothelial system after systemic administration, and thus persisted in the blood circulation (Woodle 1992; Maruyama 2000). "Stealth" liposomes have also been applied to in vivo gene transfection. A lipid-PEG conjugate was grafted onto a cationic liposome, and a PEG-shielded complex with plasmid was formed. After injection into mouse tail vein, such complexes showed diminished interaction with blood components and the reticuloendothelial system, leading to effective transgene expression in target tumor cells (Maurer et al. 1999; Yu et al. 2004).

The strategy of shielding the positive charge by PEG-grafting has also been applied to polycation/DNA complexes (Fig. 3). Ogris et al. (1999) prepared PEGylated DNA complexes by covalently grafting PEG onto pre-assembled PEI/DNA complexes. Improved plasma circulation and high-level gene expression in tumors after systemic administration into tumor-bearing mice were obtained. In addition, PEGylation reduced the in vivo toxicity of the PEI/DNA complex. Moreover, hydrophilic polymers, such as PEG, stabilize the dispersion of the condensed DNA particles, preventing aggregation and precipitation. This size-stabilizing effect will be discussed later.

Post-PEGylation of DNA complexes is, however, an additional time-consuming process. Furthermore, PEGylation of the complex surfaces is a semi-heterogeneous solid-surface reaction, and the degree of PEGylation is difficult to be finely controlled.

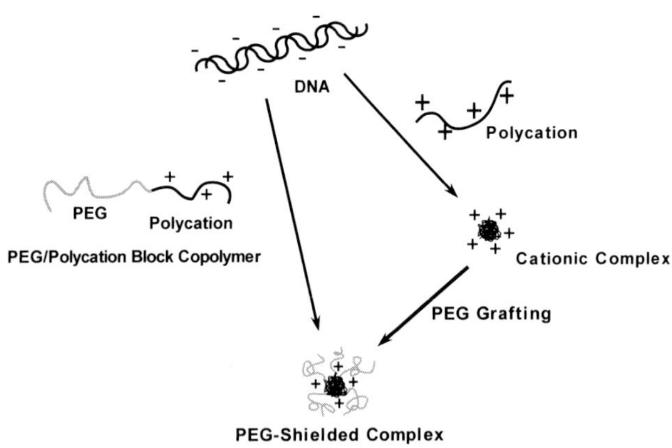

FIG. 3. Preparation of PEG-shielded DNA complexes

An alternative strategy is to synthesize a conjugate made up of PEG and a condensing agent prior to mixing with DNA.

Copolymers of polylysine-PEG (Wolfert et al. 1996; Katayose et al. 1997), PEI-PEG (Erbacher et al. 1999; Vinogradov et al. 1999), and polyspermine-PEG (Kabanov et al. 1995) were prepared and examined for their DNA-condensing and transfection abilities. In some cases, the hydrophilic part of the copolymer appeared to hinder DNA condensation, but the addition of extra unmodified polycation enabled efficient DNA condensation, and compacted complexes were obtained (Kircheis et al. 2001).

Polycation-PEG copolymers and DNA form "polymer micelle" type complex, with an inner core of water-insoluble polyion complex and an outer shell of the polyhydrophile (see the chapter by K. Kataoka, this volume). In experiments with mice, PEG-surrounded plasmid complex micelles promoted high-level gene-expression in tumors without visible toxicity (Ogris et al. 1999). Recently, a clinical trial with PLL-PEG/plasmid polymer micelles was carried out in cystic fibrosis patients, and improved chloride transport was observed in eight of 12 patients, with no adverse effects (Davis 2004).

2.2.3 Protective Polyanion Coating of Polycation/DNA Complexes: A Ternary Polyion Complex Vector System

The addition of polyanions, such as heparin or polyacrylic acid, to a polycation/DNA complex usually causes decomposition of the complex by competitive dissociation of the DNA molecule by the polyanion. However, when polyacrylic acid was added to a PEI/DNA complex at a polyanion/polycation charge ratio of less than one, a ternary complex, the surface of which was recharged to negative, was obtained. PEI/plasmid/polyacrylic acid ternary complexes showed relatively high-level gene-expression in lung after injection into mouse tail vein (Trubetskoy et al. 2003).

Certain polyanions were found not to decompose the polycation/DNA complex even at a high polyanion/polycation ratio, but they attached to the cationic surface of the complex to form a polycation/DNA/polyanion ternary polyion complex without significant loss of transfection efficiency. Finsinger et al (2000) reported the synthesis of a new class of polyanion, based on PEG, that could be assembled with positively charged PEI/DNA complexes. These anionic PEG derivatives effectively reduced the zeta potential of the DNA complexes and protected them against interaction with serum proteins. Complement activation was also evidently diminished by the protective polyanion coating.

Another type of anionic PEG derivatives, having carboxylic acid-side chains (PEG-C), were developed in my laboratory (Koyama et al. 2003). PEG-C coated the polycation/DNA complexes without dissociation of the DNA molecule (Fig. 4). The coating effectively protected the DNA complexes from albumin-induced aggregation (Fig. 5) and prevented blood-cell coagulation (Maruyama et al. 2004). Making use of a succinic acid derivative, a PEG derivative with twice the amount of carboxylic acid as PEG-C was obtained. This succinic-acid-pendant PEG afforded more stable ternary complex with polycation and DNA, and showed higher protective properties even at concentrations of albumin as high as those in serum (40 μg/ml) (Sakae et al. 2004).

Recently, we found that hyaluronic acid could also form a ternary complex with polycations and DNA. The resulting complex had a negatively charged surface, and

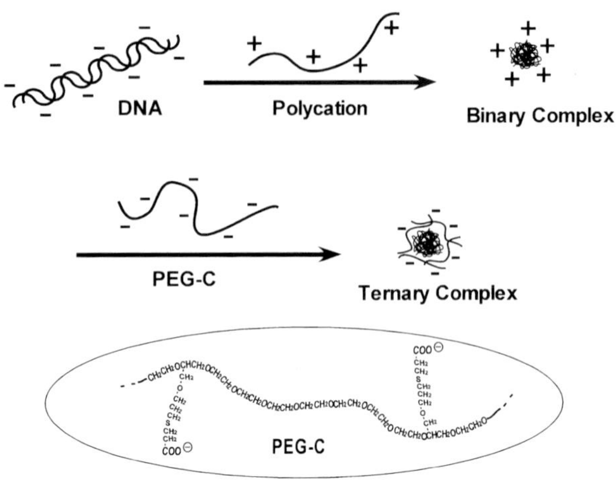

FIG. 4. Formation of polycation/DNA/PEG-C ternary polyion complexes

FIG. 5. PEG-C coating protects DNA complexes from albumin (BSA)-induced aggregation

showed high stability and gene expression in serum-containing medium (Ito et al. 2004). The very low toxicity of hyaluronic acid suggests the possibility of clinical use of the polysaccharide in gene therapy.

The ability of polyanions to suppress nonspecific interactions of the DNA complexes with proteins also enabled their sustained release from a swollen collagen sponge (Scherer et al. 2002) or from a fibrinogen matrix (Schillinger et al. 2004). These biodegradable DNA-releasing devices were very effective not only for sustained gene expression in cultured cells (Scherer et al. 2002) but also as implantable transfection systems for gene therapy of an osteochondral defect in the rabbit knee joint (Schillinger et al. 2004).

2.2.4 Ligand Introduction to Shielded DNA Complexes: Targetable Vectors

Shielding of the positive surface charge of the DNA complexes may prevent their nonspecific interaction with blood cells or serum proteins. However, it may also reduce the electrostatic interaction of the cationic DNA complex with the negatively charged target cell surface. Therefore, the introduction of a ligand for binding of the complex to the surface of the target cells would be required for efficient cellular uptake of the DNA complex. Moreover, the specificity of the ligand may allow cell-selective transfection.

Kursa et al. (2003) prepared transferrin-PEG-PEI conjugates (Tf-PEG-PEI), in which a ligand, transferrin, was connected to PEI through PEG-bridging. A mixture of Tf-PEG-PEI, PEG-PEI, and PEI (1:6:3 in mol) was complexed with a plasmid encoding tumor necrosis factor (TNF)-α, and injected into the tail vein of tumor-bearing mice. Administration of the formulation five to seven times led to strong tumor necrosis without systemic toxicity. Transferrin is a negatively charged protein and was suggested to act not only as a ligand but also as a shielding hydrophile (Kircheis et al. 2001). Freshly prepared transferrin-bearing non-PEGylated complexes (DNA/Tf-PEI + PEI) were indeed stable small particles and showed high transfection efficiency. However, they aggregated upon freeze-thawing, and their transfection ability in vivo was diminished, while the PEGylated formation retained its size and transfection efficiency after freeze-thawing (Kursa et al. 2003; Ogris et al. 2003).

2.2.5 Ligand-Bearing Ternary Complexes

Ternary complexes with anionic PEG-coating often led to even higher transgene expression than obtained with non-coated polycation/plasmid binary complexes (Koyama et al. 2003). The enhancing effect of the PEG-coating can be attributed to: (1) the high dispersion stability of the ternary complex, (2) the proton-sponge effect of the carboxylate ions (Koyama et al. 2003), and (3) the transcription-enhancing effect of polyanions (Yamashita et al. 2004), which will be discussed below.

In some cases, however, the anionic coating resulted in diminished gene expression, most probably due to the hindered deposition of the shielded complex onto the target cell surface. In order to avoid this problem, we synthesized PEG derivatives having both carboxylic acid and galactose side chains. These adhered to the pre-complexed polycation/DNA particles, and the resulting galactose-bearing ternary complexes showed strong gene expression in HepG2 cells (Koyama et al. 2002; Maruyama et al. 2004). The effect of sugar- and cell-specificity on transfection efficiency (Fig. 6) and competitive inhibition by asialofetuin confirmed the receptor-mediated mechanism of the ligand-bearing ternary complexes. The β-Galactose side chain is recognized not only by hepatocytes but also by several other types of malignant cells; thus, a high transfection efficiency by the galactose-bearing ternary complexes on various cultured tumor cell lines was obtained (unpublished). Breast cancer cells that had been inoculated into the foot pad of mice efficiently expressed the reporter gene after peritumoral injection of PEI/plasmid/galactose-PEG-C complexes.

Introduction of a ligand into the polycation element of a ternary complex was shown to compensate for the loss of cell binding by the polyanion coating. Tf-PEI/plasmid/anionic PEG ternary complexes showed much higher transgene expression than those without transferrin residues or Tf-PEI/plasmid binary complexes without anionic coating (Finsinger et al. 2000).

Hyaluronic acid (HA) functions as protective coating, as described above. Furthermore, a number of malignant cells are known to express the receptor to this polysaccharide. HA complexed with polycation and plasmid DNA was, thus, expected to facilitate binding of the complexes to receptor-expressing cells. The transfection efficiencies of these PEI/plasmid/HA complexes were examined with CHO cells, which express CD44, a receptor for HA, on their surface membrane, and with CD44-negative COS-7 cells (Ito et al. 2004). Enhanced gene-expression was observed only with CHO cells, while HA reduced expression in the COS-7 cells. Diminished transgene expression by pre-addition of excess HA also supports a receptor-mediated transfection mechanism of HA-containing complexes.

3 Size Control of DNA Complexes

Naked DNA plasmids are random coils with radii in the micrometer range in water. While direct injection of naked DNA can induce transgene expression in skeletal muscle (Wolff et al. 1990) and in tumor (Vile and Hart 1993), efficient gene transfection has usually been achieved by complexing plasmids with cationic lipids or polycations. These cations effectively condense DNA molecules into small compacted particles. By using the appropriate cations, the outer diameter of the plasmids can be reduced to below 100 nm. At very low concentrations ([DNA] <1 μM in base units) in pure water, small unimolecular complexes, with respect to the plasmid molecule, with a size of 30–50 nm are obtained. However, aggregation of the complex occurs in more concentrated solutions or at physiological salt concentrations, resulting in the formation of polydisperse large particles, typically from 150 nm to several micrometers. This severely limits their diffusion and permeation through the capillaries (~5 μm) or the fenestrae (30–50 nm), thereby affecting their body distribution. Endocytosis is also a size-limiting process. Size control and dispersion stabilization of the DNA complexes under physiological conditions are, thus, essential to achieve efficient and target-specific gene delivery in vivo.

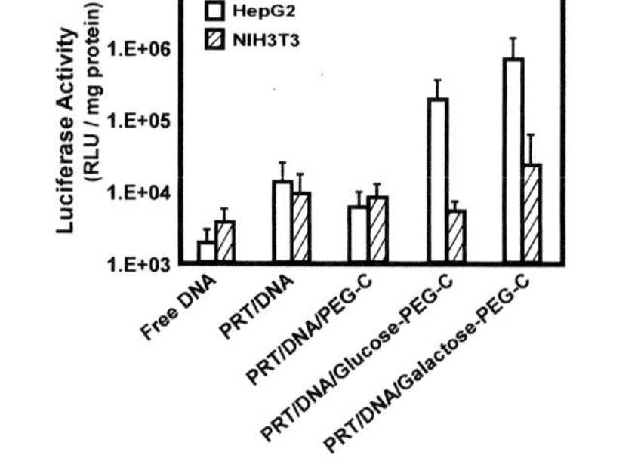

FIG. 6. Cell-specific transfection by polycation/plasmid/sugar-PEG-C ternary complexes. Terplexes with galactose-bearing PEG-C mediate transgene expression in HepG2 cells, but not in NIH3T3 cells

3.1 Dimerizable Cationic Detergent: Preparation of "Frozen" Lipid/DNA Complexes

Cationic lipids have been widely used as non-viral gene vectors, and their complexes with plasmids displayed high transfection efficiency in a variety of cultured cell lines in vitro. The cationic lipids/DNA complexes, however, tend to stick together into a large aggregate. Although cationic detergents can be used to condense individual plasmid molecule into discrete particles, consisting of a single plasmid (Melnikov et al. 1995), the cationic detergent/plasmid complex does not mediate gene transfection.

Blessing et al. (1998) synthesized a thiol-group-containing cationic detergent in order to prepare single molecular plasmid particles. The cationic detergent inside the particle was then dimerized into a lipid by oxidizing the thiol group (Fig. 7). This procedure afforded "frozen" lipid/plasmid particles, consisting of unimolecular plasmid. The particles were very homogeneous spheres, with a diameter of 23 nm, and exhibited good diffusion behavior, but gene expression was not as high as expected, probably due to problems with intracellular trafficking to nucleus (Zuber et al. 2003).

3.2 Cross-Linking of Polycation/Plasmid Complexes

DNA complexes with PLL, polyallylamine, or histone H1 was cross-linked by simple addition of dimethyl-3,3′-dithiobispropionimidate (DTBP), which reacts with two amino groups in the polycation and connects them through a disulfide bond. The resulting "caged" DNA complexes were very stable in high salt solutions and did not aggregate after over 7 days in the concentrated NaCl solution (Trubetskoy et al. 1999). However, they did lose transfection ability in vitro, in spite of the cross-linking through a degradable S-S linkage. Oupicky et al. (2001) reported that controlled light cross-linking of PLL/DNA complexes by DTBP did not significantly change their transfection efficiency. They also prepared the PEGylated formulation of the DTBP-cross-linked complexes, which exhibited a ten-fold increased plasma circulation following intravenous administration to mice (Oupicky et al. 2001).

Polyion complex of polycations with antisense oligonucleotide (ODN) is generally unstable because of the relatively few charged groups in the ODN molecule. Stabilized complex micelle containing ODN was prepared by cross-linking of the core as follows (Kakizawa et al. 1999). PEG-PLL block copolymer having thiol groups was prepared by partial modification of the amino groups by N-succinimidyl-3-(2-pyridyldithio)-propionate (SPDP), followed by reduction with dithiothreitol. It was mixed with ODN to form the polyion complex micelles, and the core was then cross-linked through S-S bonds. The obtained cross-linked micelles had sufficient colloidal stability, and could

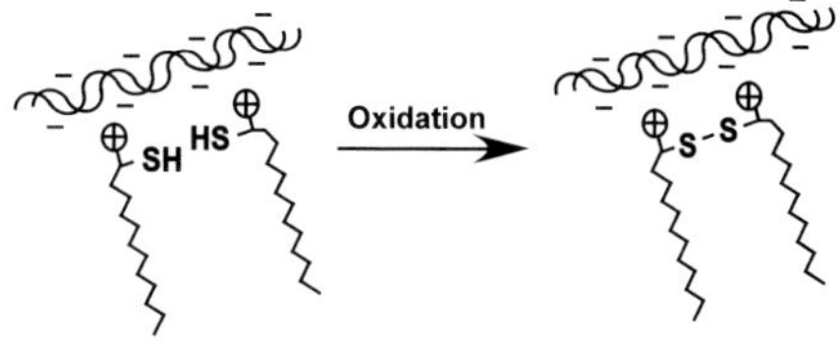

FIG. 7. Thiol-group-containing cationic detergent oxidized into lipids after complexing with DNA molecules results in the formation of "frozen" lipoplexes

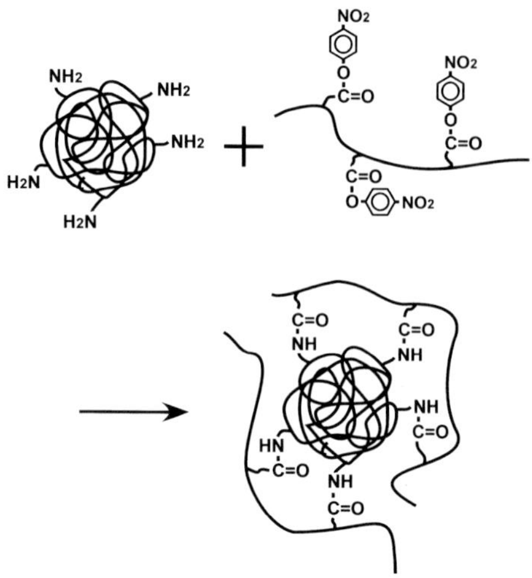

FIG. 8. Surface cross-linking of DNA complexes by multivalent polymers having active ester side chains

also dissociate to release ODN in the presence of gluthathion at a concentration comparable to the intracellular environment (Kakizawa et al. 2001).

PLL/DNA polyion complexes were also cross-linked by the addition of a multivalent reactive hydrophilic polymer with active ester side chains (Dash et al. 2000; Fisher et al. 2000) (Fig. 8). The surface-cross-linked complexes had an average diameter of 90 nm and reduced zeta potential, and were resistant to serum proteins. Introduction of a ligand provided a means of tissue-specific targeting and transfection.

3.3 PEG Shielding Stabilizes Complex Dispersion

Electrostatically complexed polycation/DNA particles are likely to aggregate at relatively high concentration ([DNA] >50 μg/ml); but, such highly concentrated solutions of DNA complexes are usually required for in vivo gene therapy. Aggregation is accelerated in high salt solution by Coulomb interactions between the polycation and DNA, presumably leading to interparticle cross-bridging by the polycation (Trubetskoy et al. 1999).

A size stabilizing effect has often been reported for PEG-shielded polyion complexes. Ogris et al. (1999) showed that the PEGylated PEI/DNA complexes prepared at a low concentration ([DNA] 20 μg/ml) in deionized water retained their size, below 100 nm after addition of PBS to the complex suspension, while non-PEG ylated PEI/DNA complexes rapidly aggregated soon after PBS addition. PEGylated small particles could be concentrated on microconcentrators to 400 μg DNA/ml, and 90% of the DNA complexes were recovered without a significant increase in their diameters after the concentration.

Ternary complexes prepared by anionic PEG addition to polycation/DNA complexes also had a higher size stability than naked polycation/DNA binary complexes (Finsinger et al. 2000; Maruyama et al. 2004). Protamine/plasmid complexes prepared in pure water abruptly increased in size upon the addition of PBS, but anionic PEG-coated ternary complexes maintained a size below 200 nm in the high salt solution.

Pre-addition of the anionic PEG derivative (PEG-C) to the plasmid prior to complexing with polycation also effectively prevented aggregation in a highly concentrated solution. When a plasmid solution at 1650 µg/ml was mixed with PEI in water at N/P ratio = 10, a large white precipitate appeared soon after mixing. By contrast, plasmid mixed with PEG-C prior to complexing with PEI did not form an aggregate with the polycation, and the solution remained nearly clear after ternary complex formation. Fluorescence microscopy revealed that pre-addition of PEG-C before polycation addition did not interfere with complex formation of the plasmid, and the polycation preferentially bound to the DNA molecule regardless of the presence of PEG-C. Furthermore, no significant difference was seen in the in vitro transfection efficiency between pre-addition and post-addition of PEG-C.

4 Transcription-Activating Effect of Polyanions and Polyampholytes: Ionized PEGs as Artificial HMG Proteins

Anionic PEG derivatives (PEG-C) form a ternary complex with polycation and DNA as described above. In some cases, PEG-C-containing terplexes showed even higher transgene expression (Fig. 9), though their electrostatic binding to cells had to be suppressed (Koyama et al. 2003). Fine dispersion of the PEG-C-containing terplexes seemed not to be very important in in vitro transfection, although particle size is an important factor for in vivo transfection. Therefore other mechanisms of PEG-C enhancement of gene expression efficiency need to be explored.

Non-viral vectors, such as cationic lipids or polymers, should form stable complexes with plasmid for efficient transport and to avoid enzymatic degradation. But, plasmid DNA must be dissociated, at latest before transcribed in the cell nucleus. A complex that is too stable and that cannot dissociate is, thus, less effective at the transcription stage. PEG-C is thought to relax polycation/plasmid complexes by competitive binding to the polycation; but, unlike other common polyanions, PEG-C is not able to completely dissociate the plasmid molecule. Relaxation of the complex would facilitate access of transcription factors to the plasmid. Loosened plasmid complex is more efficiently transcribed than unrelaxed tight complexes.

The relative transcription efficiency of polycation/plasmid/PEG-C ternary complexes was examined using *E. Coli* RNA polymerase. The amount of consumed nucleotide triphosphate was evaluated using γ-fluorescence-labeled UTP (UTP γ-AmNS), in which the fluorophore is quenched by the nucleotide. The recovered fluorescence after transcription thus represents the quantity of pyrophosphate derivative produced, which should be proportional to the amount of synthesized RNA (Dunkak et al. 1996).

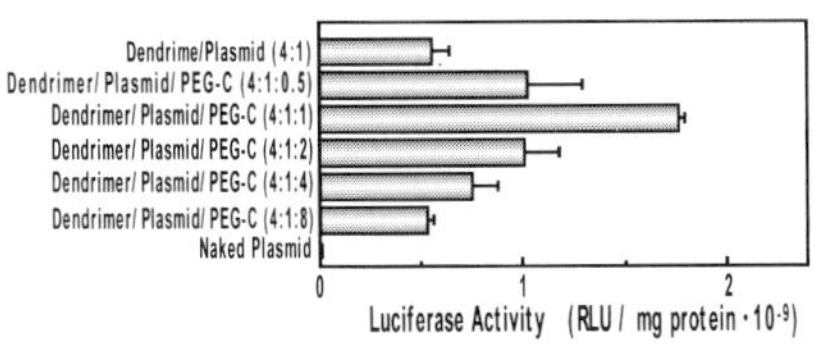

FIG. 9. PEG-C enhances gene expression mediated by poly-L-lysine dendrimers in Cos-7 cells. The mixing ratio is given in parentheses as N:P:COOH (in moles)

Cell-free transcription experiments revealed that PEG-C evidently enhanced the transcription efficiency of the polycation/plasmid complexes depending on the dose ratio of the polyanion. Such transcription-activating functional compounds also exist in nature. High-mobility-group (HMG) proteins are non-histone nuclear DNA-binding proteins that relax chromatin by competing with histone H1, and accelerate transcription (Varga-Weisz et al. 1994). An enhancing effect of HMG proteins on cationic-lipids-mediated transfection was also observed (Namiki et al. 1998; Morishita et al. 2000). HMG proteins not only have DNA-binding cationic regions, but also an anionic amino-acid-rich region near the C-terminus that is thought to be necessary to enhance transcription (Aizawa et al. 1994). By mimicking the amphoteric structure of HMG proteins, PEG derivatives having both cationic and anionic pendants (PEG-A/C) were synthesized.

PEG-A/C having six amino- and 12 carboxyl-pendants was prepared, mixed with plasmid and PEI, and stable ternary complexes were obtained. These exhibited even higher transcription efficiency than those with PEG-C, and the high potential of the water-soluble polyampholyte to enhance the transcription efficiency of the compacted DNA molecule was confirmed (Yamashita et al. 2004). In in vitro transfection experiments using cultured cells, PEI/plasmid/PEG-A/C terplexes actually resulted in higher levels of gene expression than obtained with PEI/plasmid binary complexes or those with PEG-C (Fig. 10).

Ionized PEG derivatives, such as PEG-C and PEG-A/C are thus thought to function as artificial HMG proteins, besides exhibiting shielding and dispersion-stabilizing effects. Further appropriate functionalized PEG-containing systems are now being developed in order to achieve effective in vivo gene therapeutic formulation.

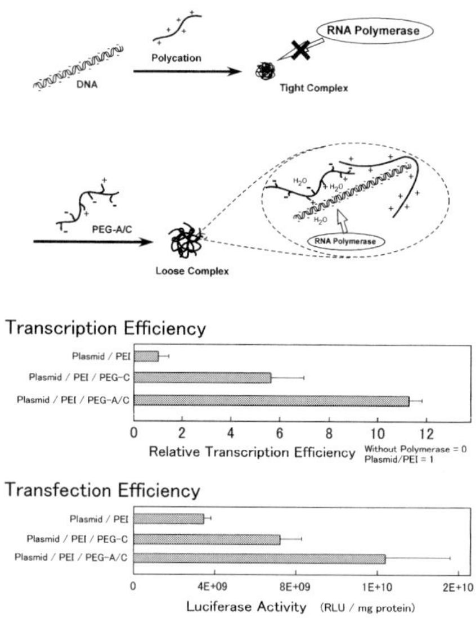

FIG. 10. Amphoteric PEG derivatives enhance transcription and transfection efficiencies; and thus function as artificial HMG proteins. Transcription efficiency was evaluated at a mixing ratio of PEI : plasmid : PEG-C (or PEG-A/C) = 8 : 1 : 15 N : P : COOH (in moles). Transfection experiments were performed at the same N : P : COOH ratio on CHO cells

References

Aizawa S, Nishino H, Saito K, Kimura K, Shirakawa H, Yoshida M (1994) Stimulation of transcription in cultured cells by high mobility group protein 1: essential role of the acidic carboxyl-terminal region. Biochemistry 33:14690–14695

Blessing T, Remy JS, Behr JP (1998) Monomolecular collapse of plasmid DNA into stable virus-like particles. Proc Natl Acad Sci U S A 95:1427–1431

Dash PR, Read ML, Barrett LB, Wolfert MA, Seymour LW (1999) Factors affecting blood clearance and in vivo distribution of polyelectrolyte complexes for gene delivery. Gene Ther 6:643–650

Dash PR, Read ML, Fisher KD, Howard KA, Wolfert M, Oupicky D, Subr V, Strohalm J, Ulbrich K, Seymour LW (2000) Decreased binding to proteins and cells of polymeric gene delivery vectors surface modified with a multivalent hydrophilic polymer and retargeting through attachment of transferrin. J Biol Chem 275:3793–3802

Davis PB (2004) Stabilized, Compacted DNA Nanoparticles Transfect Airway Epithelium in Vivo. Executive Summary of the American Society of Gene Therapy's 7th Annual Meeting SS303

Dunkak KS, Otto MR, Beechem JM (1996) Real-time fluorescence assay system for gene transcription: simultaneous observation of protein/DNA binding, localized DNA melting, and mRNA production. Anal Biochem 243:234–244

Erbacher P, Bettinger T, Belguise-Valladier P, Zou S, Coll JL, Behr JP, Remy JS (1999) Transfection and physical properties of various saccharide, poly(ethylene glycol), and antibody-derivatized polyethylenimines (PEI). J Gene Med 1:210–222

Finsinger D, Remy JS, Erbacher P, Koch C, Plank C (2000) Protective Copolymers for Non-viral Gene Vectors: Synthesis, Vector Characterization and Application in Gene Delivery. Gene Ther 7:1183–1192

Fisher KD, Ulbrich K, Subr V, Ward CM, Mautner V, Blakey D, Seymour LW (2000) A versatile system for receptor-mediated gene delivery permits increased entry of DNA into target cells, enhanced delivery to the nucleus and elevated rates of transgene expression. Gene Ther 7:1337–1343

Hafenrichter DG, Wu X, Rettinger SD, Kennedy SC, Flye MW, Ponder KP (1994) Quantitative evaluation of liver-specific promoters from retroviral vectors after in vivo transduction of hepatocytes. Blood 84:3394–3404

Ito T, Tanaka NI, Sato T, Yang Z, Niidome T, Kawano T, Koyama Y (2004) Hyaluronic Acid as a Self-Assembled Coating of Plasmid/Polycation Complexes for Cell-Specific Gene Delivery. Molecular Ther 9:S321–322

Kabanov AV, Vinogradov SV, Suzdaltseva YG, Alakhov VYu (1995) Water-soluble block polycations as carriers for oligonucleotide delivery. Bioconjug Chem 6:639–643

Kakizawa Y, Harada A, Kataoka K (1999) Environment-sensitive stabilization of core-shell structured polyion complex micelle by reversible cross-linking of the core through disulfide bond. J Am Chem Soc 121:11247–11248

Kakizawa Y, Harada A, Kataoka K (2001) Glutathione-sensitive stabilization of block copolymer micelles composed of antisense DNA and thiolated poly(ethylene glycol)-block-poly(L-lysine): a potential carrier for systemic delivery of antisense DNA. Biomacromolecules 2:491–497

Katayose S, Kataoka K (1997) Water-soluble polyion complex associates of DNA and poly(ethylene glycol)-poly(L-lysine) block copolymer. Bioconjug Chem 8:702–707

Kircheis R, Wightman L, Schreiber A, Robitza B, Rossler V, Kursa M, Wagner E (2001) Polyethylenimine/DNA complexes shielded by transferrin target gene expression to tumors after systemic application.Gene Ther 8:28–40

Koyama Y, Yamada E, Ito T, Mizutani Y, Yamaoka T (2002) Sugar-Containing Polyanions as a Self-Assembled Coating of Plasmid/Polycation Complexes for Receptor-Mediated Gene Delivery. Macromol Biosci 2:251–256

Koyama Y, Ito T, Matsumoto H, Tanioka A, Okuda T, Yamaura N, Aoyagi H, Niidome T (2003) Novel poly(ethylene glycol) derivatives with carboxylic acid pendant groups: synthesis and their protection and enhancing effect on non-viral gene transfection systems. J Biomater Sci Polym Ed 14:515–531

Krieg AM, Yi AK, Matson S, Waldschmidt TJ, Bishop GA, Teasdale R, Koretzky GA, Klinman DM (1995) CpG motifs in bacterial DNA trigger direct B-cell activation. Nature 374:546–549.

Kursa M, Walker GF, Roessler V, Ogris M, Roedl W, Kircheis R, Wagner E (2003) Novel shielded transferrin-polyethylene glycol-polyethylenimine/DNA complexes for systemic tumor-targeted gene transfer. Bioconjug Chem 14:222–231

Li S, Wu SP, Whitmore M, Loeffert EJ, Wang L, Watkins SC, Pitt BR, Huang L (1999) Effect of immune response on gene transfer to the lung via systemic administration of cationic lipidic vectors. Am J Physiol. 276:L796–804

Maruyama K (2000) In vivo targeting by liposomes. Biol Pharm Bull 23:791–799

Maruyama K, Iwasaki F, Takizawa T, Yanagie H, Niidome T, Yamada E, Ito T, Koyama Y (2004) Novel receptor-mediated gene delivery system comprising plasmid/protamine/sugar-containing polyanion ternary complex.Biomaterials 25:3267–3273

Maurer N, Mori A, Palmer L, Monck MA, Mok KW, Mui B, Akhong QF, Cullis PR (1999) Lipid-based systems for the intracellular delivery of genetic drugs. Mol Membr Biol 16:129–140

Melnikov SM, Sergeyev SG, Yoshikawa K (1995) Discrete coil-globule transition of large DNA induced by cationic surfactant. J. Am. Chem. Soc 117:2401–2408

Morimoto K, Nishikawa M, Kawakami S, Nakano T, Hattori Y, Fumoto S, Yamashita F, Hashida M (2003) Molecular weight-dependent gene transfection activity of unmodified and galactosylated polyethyleneimine on hepatoma cells and mouse liver. Mol Ther 7:254–261

Morishita R, Gibbons GH, Tomita N, Zhang L, Kaneda Y, Ogihara T, Dzau VJ (2000) Antisense oligodeoxynucleotide inhibition of vascular angiotensin-converting enzyme expression attenuates neointimal formation: evidence for tissue angiotensin-converting enzyme function. Arterioscler Thromb Vasc Biol 20:915–922

Namiki Y, Takahashi T, Ohno T (1998) Gene transduction for disseminated intraperitoneal tumor using cationic liposomes containing non-histone chromatin proteins: cationic liposomal gene therapy of carcinomatosa. Gene Ther 5:240–246

Ogris M, Brunner S, Schuller S, Kircheis R, Wagner E (1999) PEGylated DNA/transferrin-PEI complexes: reduced interaction with blood components, extended circulation in blood and potential for systemic gene delivery. Gene Ther 6:595–605

Ogris M, Walker G, Blessing T, Kircheis R, Wolschek M, Wagner E (2003) Tumor-targeted gene therapy: strategies for the preparation of ligand-polyethylene glycol-polyethylenimine/DNA complexes. J Control Release 91:173–181

Oupicky D, Carlisle RC, Seymour LW (2001) Triggered intracellular activation of disulfide crosslinked polyelectrolyte gene delivery complexes with extended systemic circulation in vivo. Gene Ther 8:713–724

Plank C, Zatloukal K, Cotten M, Mechtler K, Wagner E (1992) Gene transfer into hepatocytes using asialoglycoprotein receptor mediated endocytosis of DNA complexed with an artificial tetra-antennary galactose ligand. Bioconjug Chem 3: 533–539

Plank C, Mechtler K, Szoka FC Jr, Wagner E (1996) Activation of the complement system by synthetic DNA complexes: a potential barrier for intravenous gene delivery. Hum Gene Ther 7:1437–1446

Ruponen M, Yla-Herttuala S, Urtti A (1999) Interactions of polymeric and liposomal gene delivery systems with extracellular glycosaminoglycans: physicochemical and transfection studies. Biochim Biophys Acta 8:331–341

Sakae M, Ito T, Tanaka NI, Niidome T, Yanagie H, Koyama Y (2004) Novel Receptor-Mediated Gene Transfection Systems Comprising Plasmid/Polycation/RGD-PEG-COOH Ternary Complexes. Molecular Ther 9:S319

Scherer F, Schillinger U, Putz U, Stemberger A, Plank C (2002) Nonviral vector loaded collagen sponges for sustained gene delivery in vitro and in vivo. J Gene Med 4:634–643

Schillinger U, Wexel G, Hacker C, Kullmer M, Stemberger A, Plank C (2004) Copolymer-Protected Vector Mediated BMP-2 Gene Delivery to Articular Chondrocytes in a 3-Dimensional Injectable Fibrin Matrix for Tissue Engineering. Molecular Ther 9:S321

Trubetskoy VS, Loomis A, Slattum PM, Hagstrom JE, Budker VG, Wolff JA (1999) Caged DNA does not aggregate in high ionic strength solutions.Bioconjug Chem 10:624–628

Trubetskoy VS, Wong SC, Subbotin V, Budker VG, Loomis A, Hagstrom JE, Wolff JA (2003) Recharging cationic DNA complexes with highly charged polyanions for in vitro and in vivo gene delivery. Gene Ther 10:261–271

Varga-Weisz P, van Holde K, Zlatanova J (1994) Competition between linker histones and HMG1 for binding to four-way junction DNA: implications for transcription. Biochem Biophys Res Commun 203:1904–1911

Vile RG, Hart IR (1993) In vitro and in vivo targeting of gene expression to melanoma cells. Cancer Res 53:962–967

Vinogradov S, Batrakova E, Li S, Kabanov A (1999) Polyion complex micelles with protein-modified corona for receptor-mediated delivery of oligonucleotides into cells. Bioconjug Chem 10:851–860

Vitiello L, Bockhold K, Joshi PB, Worton RG (1998) Transfection of cultured myoblasts in high serum concentration with DODAC:DOPE liposomes. Gene Ther 5:1306–1313

Wolfert MA, Schacht EH, Toncheva V, Ulbrich K, Nazarova O, Seymour LW (1996) Characterization of vectors for gene therapy formed by self-assembly of DNA with synthetic block co-polymers. Hum Gene Ther 7:2123–2133

Wolff JA, Malone RW, Williams P, Chong W, Acsadi G, Jani A, Felgner PL (1990) Direct gene transfer into mouse muscle in vivo. Science. 1990 247:1465–1468

Wolff JA (Ed) (1994) Gene Therapeutics: Methods and Applications of Direct Gene Transfer Birkhauser, Boston

Woodle MC, Lasic DD (1992) Sterically stabilized liposomes. Biochim Biophys Acta 1113: 171–199

Yamashita M, Ito T, Koyama Y (2004) Analysis of DNA/Polycation/Polyanion ternary polyion complexes and their application to gene therapy. Polymer Prepr Japan 53:5156

Yu W, Pirollo KF, Rait A, Yu B, Xiang LM, Huang WQ, Zhou Q, Ertem G, Chang EH (2004) A sterically stabilized immunolipoplex for systemic administration of a therapeutic gene. Gene Ther in press

Zanta MA, Boussif O, Adib A, Behr JP (1997) In vitro gene delivery to hepatocytes with galactosylated polyethylenimine. Bioconjug Chem 8:839–844

Zuber G, Zammut-Italiano L, Dauty E, Behr JP (2003) Targeted gene delivery to cancer cells: directed assembly of nanometric DNA particles coated with folic acid. Angew Chem Int Ed Engl 42:2666–2669

In Vivo Gene Transfer by Ligand-Modified Gene Carriers

Shigeru Kawakami and Mitsuru Hashida

1 Introduction

Efficient in vivo gene transfer relies on the development of vectors that offer the efficient delivery of genes to target cells, high transfection efficiencies, and the persistence of transgene expression. Although the gene-transfer efficacy of the current non-viral vector systems is lower than that of viral vectors, for safety reasons, the former approach is useful for many applications. Furthermore, non-viral vectors have advantages over viral ones with respect to controlling the biodistribution of a gene, because the physicochemical properties that determine the tissue disposition of the gene carrier, such as the particle size, electric charge, and specific ligand, can be easily controlled in these vectors. Among the various types of non-viral vectors, cationic liposomes and cationic polymer-mediated gene transfection seem to be the most promising approaches because of the relatively high transfection efficiencies of such vectors and the ability to target gene delivery by their chemical modification (Mahato et al. 1997; Huang and Li 1997; Sagara and Kim 2002). Cationic carriers condense pDNA to form particles (100–200 nm) based on electrostatic interactions and protect it from degradation. In most cases, after intravenous injection of cationic carriers/pDNA complexes, the highest levels of gene expression occur in the lung because the lung capillaries are the first "traps" to be encountered. Therefore, the development of carrier systems that can escape from undesired tissue uptake and exhibit target-cell-specific gene expression are urgently required.

For cell-specific delivery, the receptor-mediated endocytosis (RME) systems of various cell types are potentially useful, and a number of gene delivery systems have been developed to introduce foreign DNA into specific cells by RME. In developing a strategy for efficient and safe in vivo gene deliver, however, a variety of aspects, such as the construction of sophisticated carrier systems, and the pharmacokinetics and physicochemical properties of pDNA, need to be considered (Takakura and Hashida

Department of Drug Delivery Research, Graduate School of Pharmaceutical Sciences, Kyoto University, Sakyo-ku, Kyoto 606-8501, Japan

1996). In this chapter, the pharmacokinetics of intravenously administered pDNA, with or without non-viral gene carriers, are discussed. This is followed by a review of the current receptor-mediated gene delivery systems developed for use in vivo.

2 Biodistribution Characteristics of Naked pDNA

It has been reported that intravenously administered naked pDNA does not lead to gene expression. Nonetheless, a strategy for establishing non-viral gene carrier systems, however, must consider the in vivo disposition characteristics of naked pDNA. Kawabata et al. (1995) demonstrated that intravenously injected ^{32}P-labeled-pDNA is rapidly eliminated from the plasma by a mechanism involving extensive uptake by the liver . Pharmacokinetic analysis demonstrated that hepatic clearance is almost identical to the plasma flow rate in the liver, suggesting highly effective elimination by this organ. In addition, pDNA is taken up preferentially by liver non-parenchymal cells via a scavenger-receptor-mediated process that is specific for polyanions. In vitro binding and uptake studies using cultured mouse peritoneal macrophages demonstrated that the binding of pDNA was significantly inhibited by polyinosinic acid (poly [I]) and dextran sulfate, which are substrates of the scavenger receptors, but not by polycytidylic acid (poly [C]), dextran, or EDTA, which are not substrates of this receptor. These data suggest that pDNA is taken up by macrophages via a mechanism mediated by a receptor like the macrophage scavenger receptor. Involvement of scavenger receptors (SRs) in the hepatic uptake of pDNA has also been supported by single-pass rat perfusion experiments using [^{32}P]-pDNA (Yoshida et al. 1996).

The class A scavenger receptor (SRA), the best characterized SR, recognizes a wide variety of anionic macromolecules based on their three-dimensional structure and seems likely to be responsible for pDNA uptake. Recently, Takakura et al. reported an in vitro study of [^{32}P]-pDNA binding and uptake using cultured CHO cells expressing SRA (CHO (SRA) cells) and peritoneal macrophages from SRA-knockout mice (Takakura et al. 1999). [^{32}P]-pDNA binding and uptake by CHO (SRA) cells were minimal and almost identical to that by wild-type CHO cells. Macrophages from the knockout mice showed pronounced pDNA binding and uptake, as did the control macrophages. In both types of macrophages, [^{32}P]-pDNA binding was significantly inhibited by pDNA, poly [I], and dextran sulfate, but not by poly [C] or Acetylated low density lipoprotein (Ac-LDL). These results provide direct evidence that SRA is not responsible for any significant binding and subsequent uptake of pDNA by mouse peritoneal macrophages. Instead, pDNA binding and uptake by mouse peritoneal macrophages are mediated by a specific mechanism involving defined polyanions. These findings form an important basis for further studies to elucidate the mechanisms of pDNA uptake by macrophages. It is clear, however, that the strong anionic charge of pDNA should be neutralized in order for it to escape being recognized by scavenger receptor-like mechanisms. For this reason, cationic liposomes and cationic polymers have been developed.

3 Biodistribution Characteristics of pDNA Complexed with Cationic Liposomes

Cationic liposome and polymers condense pDNA to form particles based on electrostatic interactions and protect it from degradation. For subsequent gene expression, the complex must contain an excess of cationic charges After intravenous administration of the pDNA complexes, the immediate effect of erythrocytes on is to induce aggregation of the complexes (Sakurai et al. 2001). Large aggregates (>400 nm) are readily entrapped in the lung capillaries. Mahato et al. (1995) demonstrated the disposition characteristics of [^{32}P]-pDNA-cationic liposome complex after intravenous administration in mice. Rapid clearance of [^{32}P]-pDNA from the circulation was observed, with extensive accumulation in the lung and liver. In addition, [^{32}P]-pDNA complexed with cationic liposomes was predominantly taken up by liver nonparenchymal cells and uptake was inhibited by the pre-administration of dextran sulfate, suggesting the involvement of a phagocytic process. The attachment of a ligand that can be recognized by a specific mechanism would endow a vector with the ability to target a specific population of cells. However, while intravenous administration of pDNA complexed with either cationic liposome or polymers led to gene expression in various tissues, the level of gene expression in the lung was extremely high (Zhu et al. 1993; Song et al. 1997; Goula et al. 1998; Uyechi et al. 2001). Therefore, in order to improve the delivery of pDNA to target cells, such as hepatocytes, macrophages, and tumors, several ligands have been used.

4 Biological Barriers to Gene Delivery

Many in vitro studies on gene delivery have not been predictive of in vivo functionality; therefore, most formulations are effective in vitro but fail to function in vivo. Cell culture models do not replicate many biological factors important for in vivo gene delivery. For example, under in vitro conditions of cellular proliferation, the charge of the complex, its size , and its interaction with erythrocytes are substantially different than under in vivo conditions. The following conditions must be taken into consideration in designing a gene delivery system for use in vivo: (1) The carrier should form a stable complex that can deliver intact DNA in the blood circulation; (2) the cationic nature of the complexes should be considered in order to allow escape from nonspecific biodistribution via electrostatic interactions; (3) the size of the complexes should be considered regarding access to target cells; (4) the carrier should contain appropriate ligands in order to ensure a high affinity for cellular receptors (Fig. 1).

4.1 Stability of pDNA Complexes in the Blood

When naked pDNA into administered intravenously to mice, it is rapidly degraded by DNAase and/or due to uptake by Kupffer cells; and thus gene expression is prevented. Both cationic liposomes and/or polymer improve the stability of pDNA by condensing it to particles of defined size by electrostatic interactions. Furthermore, condensed pDNA is protected from both degradation in the blood stream and nonspecific interaction with cell surfaces.

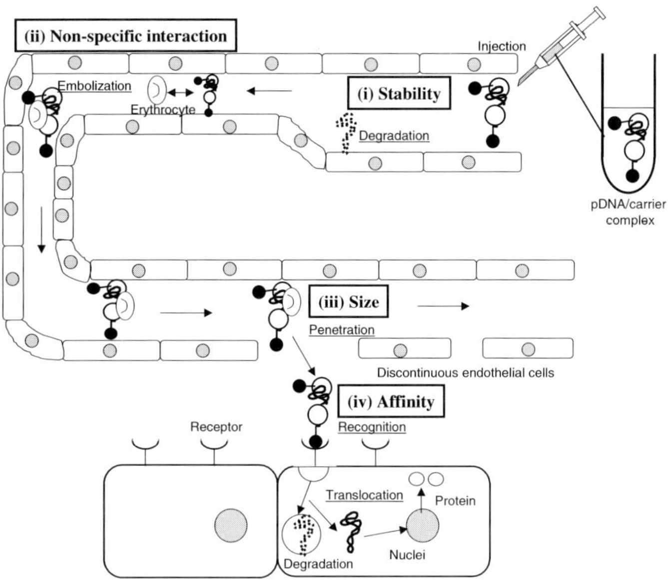

FIG. 1. Barriers to in vivo gene transfer by ligand-modified gene carriers

4.2 The Effect of Cationic Charge on In Vivo Gene Delivery

Yang and Huang (1997) reported that neutralization of the excess positive charge in the pDNA/cationic liposome complex by negatively charged serum proteins likely reduces the transfection efficiency in vitro. They demonstrated that this problem can be overcome by increasing the cationic charge ratio of pDNA/cationic liposome complexes, and that the optimal charge ratio (−:+) was 1.0:4.0 for efficient transfection, even in the presence of 20% serum. After intravenous administration, gene expression in the lung of pDNA/cationic liposome complexes carrying a higher cationic charge is enhanced due to electrostatic interactions. However, the cationic charge of pDNA/cationic liposome complexes can also result in non-specific biodistribution via electrostatic interactions.

We have evaluated the effect of cationic charge on asialoglycoprotein-receptor-mediated gene transfection systems using pDNA/galactosylated cationic liposome complexes administered intraportally (Kawakami et al. 2000a; Fumoto et al. 2003). When pDNA/galactosylated cationic liposome complexes were prepared at a charge ratio (−:+) of 1.0:2.3 and/or 1.0:3.1, selective gene expression in the liver was obtained, whereas at a charge ratio of 1.0:7.0, gene expression in the lung exceeded that in the liver, suggesting the highly non-specific interactions. Therefore, a cationic charge ratio (−:+) of pDNA to galactosylated cationic liposomes of 1.0:2.3 and/or 1.0:3.1 seems to be optimal for receptor-mediated in gene transfection.

We have also evaluated the effect of cationic charge on mannose receptor-mediated gene transfection systems using pDNA/mannosylated cationic liposome complexes administered intravenously (Kawakami et al. 2004). The transfection efficiencies in liver and spleen, which express mannose receptors on their cell surfaces, after intravenous administration of complexes with a charge ratio (−:+) of 1.0:2.3 and/or 1.0:3.1 were higher than those in the lung. When complexes were formed at a charge ratio (−:+) of 1.0:4.7, the transfection efficiency in the lung was higher. These results confirm that complexes at a charge ratio (−:+) of 1.0:2.3 and/or 1.0:3.1 are optimized for receptor-mediated gene delivery systems using ligand-modified gene carriers.

4.3 The Effect of Size on In Vivo Gene Delivery

The size of the complexes is an important factor for controlling targeted gene delivery systems because the structure of the capillary wall varies greatly in different organs and tissues. Because of its large molecular weight, pDNA dose not effectively penetrate endothelial and epithelial barriers and can hardly extravasate from the vascular to the interstitial space. Following complex formation with cationic carriers, not only is pDNA condensed but its cellular uptake is also enhanced via electrostatic interactions.

pDNA/carrier complexes are often prepared in a non-ionic solution due to their well-known tendency to aggregate out of solution as the salt concentration is increased (Ogris et al. 1998). Aggregation during lipoplex formation in ionic solution might be due to neutralization of the surface positive charge of the lipoplex intermediates by the associated counter-ions. pDNA complexed with cationic liposomes at a charge ratio (−:+) of 1.0:2.3 is well condensed (100–200 nm) when the complexes are prepared in non-ionic solvent, such as dextrose and sucrose, compared with ionic solvent (Kawakami et al. 2000a).

5 In Vivo Receptor-Mediated Gene Delivery

The endothelium is a monolayer of metabolically active cells that mediate the bidirectional exchange of fluid between plasma and interstitial fluid. Thus, the endothelium has a profound influence on the extravasation of macromolecules. Discontinuous capillaries, also known as sinusoidal capillaries, are common in the liver, spleen, and bone marrow.

Tumor tissues are characterized by increased interstitial pressure, which may retard the extravasation of macromolecules. In addition, a lack of functional lymphatic drainage results in the passive accumulation of macromolecules. Capillary vessels in a human tumor inoculated into SCID mice are permeable even to liposomes up to 400 nm in diameter (Yuan et al. 1995). Hence, ligand-modified targeted gene delivery systems can be readily applied in liver, spleen, and tumors because of the size factor. In vivo receptor-mediated gene delivery systems are listed in Table 1.

5.1 Asialoglycoprotein Receptor-Mediated Gene Transfection

Hepatocytes exclusively express large numbers of high-affinity cell-surface receptors that bind and subsequently internalize asialoglycoproteins. In order to achieve liver-parenchymal-cell-specific gene transfection, a galactose moiety is introduced onto either cationic polymers or cationic liposomes.

TABLE 1. In vivo receptor-mediated gene delivery

Receptor	System	Results
Asialoglycoprotein		
Wu 1988	Asialoorosomucoid-polylysine	Expression in liver after i.v. injection
Perales 1994	Galactose-polylysine	Expression in liver after i.v. injection
Hara 1995	Asialofetuin-liposome	Expression in liver after intraportal injection
Kawakami 2000	Galactose-liposome	Expression in hepatocytes after intraportal injection
Nishikawa 2000	Galactose-polyornithine-HA2	Expression in hepatocytes after i.v. injection
Morimoto 2003	Galactose-PEI	Expression in hepatocytes after intraportal injection
Mannose		
Kawakami 2000	Mannose-liposome	Expression in non-parenchymal cells after i.v. injection
Kawakami 2004	Mannose-liposome	Expression in non-parenchymal cells after i.v. injection; effect of cationic charge
Hattori 2004	Mannose-liposome	Enhancement of immune responses by DNA vaccination
Transferrin		
Ogris 1999	Transferrin-PEG-PEI (800 kDa)	Expression in cancer cells (s.c.) after i.v. injection
Kircheis 2001	Transferrin-PEI (22 kDa)	Expression in cancer cells (s.c.) after i.v. injection
Kursa 2003	Transferrin-PEG-PEI(22 kDa)	Expression in cancer cells (s.c.) after i.v. injection
Folate		
Hofman 2002	Folate-liposome	Expression in cancer cells (s.c.) after i.v. injection
Reddy 2002	Folate-liposome	Efficient expression in intraperitoneal cancer cells after intraperitoneal injection

In the late 1980s, Wu et al. (1988) demonstrated successful in vivo gene transfer to liver using poly-L-lysine linked with asialo-orosomucoid (Chowdhury et al. 1993). Successful in vivo gene expression after intravenous injection has been also reported for glycosylated poly-L-lysine (Perales et al 1994). While these fusogenic peptides could be promising materials for enhancing in vivo gene expression, their transfection efficacy is low and must be improved for their successful use in gene therapy. Nishikawa et al. (2000) demonstrated that galactosylated poly-L-ornithine conjugated with a fusogenic peptide was very effective in improving the level of gene transfection after intravenous administration in mice.

In general, the transfection efficacy of cationic liposomes is higher than that of cationic polymers. Therefore, by using galactosylated cationic liposomes, effective hepatocyte targeting might be achieved. Liposomes can be galactosylated by coating them with either glycoproteins or galactose-conjugated synthetic lipids. As for targeted gene delivery by liposomes, Hara et al. reported that asialofetuin-labeled liposomes encapsulating pDNA were taken up by cultured hepatocytes via asialoglycoprotein receptor-mediated endocytosis and that the highest levels of hepatic gene expression were obtained after intraportal injection with a preload of EDTA (Hara et al. 1995). However, the introduction of asialoglycoproteins to liposomes is complicated and there are a number of problems associated with the carriers, such as reproducibility and immunogenicity. Therefore, low-molecular-weight glycolipids appear to be more promising due to their low immunogenicity and high reproducibility. Remy et al. (1995) reported the feasibility of using galactose-presenting lipopolyamine vectors for targeted gene transfer into hepatoma cells under in vitro conditions. Inclusion of galactose residues in the electrically neutral complex increased transgene expression to nearly the value obtained with a large excess of cationic liposomes alone. The authors suggested that the galactose-presenting DNA particles avoid interacting with serum proteins because of their electrical neutrality.

Successful in vivo gene delivery systems require a thorough understanding of the pharmacokinetics and physicochemical properties of the complexes as well a theoretical design for galactosylated lipids as ligands for binding to target cells. Based on these considerations, we synthesized cholesten-5-yloxy-*N*-(4-((1-imino-2-D-thiogalactosylethyl)amino)alkyl)for mamide (Gal-C4-Chol), which possesses both the cationic charge necessary for pDNA binding and galactose residues as targetable ligands for binding to liver parenchymal cells. In vivo gene transfer was tested by optimizing the pharmacokinetics and physicochemical properties of the complexes (Kawakami et al. 2000a). The galactosylated lipid Gal-C4-Chol was specially designed as a modified cationic lipid because galactose residues can be stably fixed on liposomal membranes under in vivo conditions. The radioactivity in the liver from the Gal-C4-Chol liposome/[^{32}P] pDNA complexes was about 75% of the dose as early as 1 min after intraportal administration. Furthermore, hepatic gene expression of pDNA complexed with Gal-C4-Chol liposomes was more than a ten-fold greater than that of pDNA complexed with conventional cationic liposomes. When intrahepatic cellular levels of gene expression were examined, the expression by liver parenchymal cell (PC) of pDNA complexed with Gal-C4-Chol liposomes was significantly higher than that of liver non-parenchymal cells (NPC). By contrast, gene expression of PC and NPC of conventional cationic liposomes was almost the same. In addition, when an excess of galactosylated bovine serum albumin (Gal-BSA) was intravenously injected 5 min prior to injection of pDNA complexed with Gal-C4-Chol liposomes; gene expression in the liver was significantly reduced, suggesting that uptake occurred by asialoglycoprotein receptor-mediated endocytosis. Although the size of the Gal-C4-Chol liposome/pDNA complexes is about 120 nm, the highest level of gene expression was observed in the lung. We previously reported that intravenously administered pDNA/cationic liposome complexes interact with erythrocytes (Sakurai et al. 2001), which suggests that the Gal-C4-Chol liposome/pDNA complexes are aggregated by non-specific interaction with erythrocytes. Recently, Eliyahu et al. (2002) also reported that the medium (i.e. plasma and serum) and/or modification of cationic liposomes with 1% polyethyleneglycol lipids reduced the aggregation of pDNA/cationic liposome complexes in the presence of erythrocytes. As a consequence, polyethyleneglycol coating should enhance the cell-specificity of pDNA complexed with Gal-C4-Chol liposomes, even after intravenous administration. However, further studies on the interaction with blood components and on the synthesis of polyethyleneglycol-grafted glycosylated lipids for cell-selective gene delivery are needed.

5.2 Mannose Receptor-Mediated Gene Transfection

Macrophages are important targets for the gene therapy of diseases such as Gaucher's disease and human immunodeficiency virus (HIV) infection, but gene transfection of these cells is not easy. Non-viral vectors offer advantages for in vivo gene delivery because they are simpler and safer than viral systems. While the addition of DEAE-dextran is one method used for gene delivery to macrophages in vitro, this method is generally not suitable for in vivo use due to problems associated with cellular toxicity, low efficiency, and non-specific biodistribution. Erbacher et al. (1996) investigated the suitability of various glycosylated poly(L-lysine) derivatives for introducing pDNA into human-monocyte-derived macrophages and found that mannosylated poly(L-lysine) exhibited high transfection activity. The authors also reported that transfection activity was markedly enhanced in the presence of chloroquine due to the prevention of endosomal and/or lysosomal degradation of pDNA after mannose-receptor-mediated endocytosis. However, for in vivo application, it is difficult to use chloroquine, which limits the use of this approach.

One of the most promising non-viral gene delivery systems developed so far involves cationic liposomes because of their high in vivo transfection efficiency. Recently, we synthesized a novel mannosylated cholesterol derivative, Man-C4-Chol, for mannose receptor-mediated gene transfection of macrophages (Kawakami et al. 2000b, 2001), which express large numbers of mannose receptors on their surfaces. In primary cultured mouse peritoneal macrophages, pDNA complexed with Man-C4-Chol liposomes had a higher transfection activity than pDNA complexed with conventional cationic liposomes. The presence of 20 mM mannose significantly inhibited the transfection efficiency of the pDNA/Man-C4-Chol liposome complexes, suggesting that they are recognized and taken up by mannose receptors on macrophages. Gene transfection in macrophages was further enhanced by incorporating PEI into these complexes (pDNA/Man-C4-Chol-PEI-complexes), since it was observed that PEI has a pH-buffering effect in endosomes as well as DNA-condensing activity (Sato et al. 2001). In mouse peritoneal macrophages, the uptake and transfection activities of pDNA/Man-C4-Chol-PEI-complexes were, respectively, two- and six-fold higher than those of pDNA/ Man-C4-Chol liposome complexes. The presence of 1 mg mannan/ml significantly inhibited both the uptake and transfection efficiency of the complexes, suggesting a mechanism of mannose receptor-mediated endocytosis.

As for in vivo gene transfection, the highest level of gene expression was observed in the liver after intravenous injection of pDNA/Man-C4-Chol liposome complexes in mice. In the liver, expression was higher in NPC and was significantly reduced by pretreatment with mannosylated bovine serum albumin. These results suggest that pDNA complexed with mannosylated liposomes exhibits high transfection activity in liver NPC due to recognition by mannose receptors. In contrast to results obtained with Gal-C4-Chol liposome/pDNA complexes, cell-selective gene transfection could be achieved by the intravenous administration of Man-C4-Chol liposome/pDNA complexes. This phenomenon could be explained by the fact that the discontinuous capillaries in liver and spleen allow macrophages to come in contact with the complexes without passing through the sinusoids (100–200 nm). Hence, mannosylated gene carriers are effective for NPC-selective gene transfection.

Mannosylated liposomes also have potential applications in DNA vaccine therapy, because antigen-encoded pDNA must be efficiently transfected into dendritic cells, which express a large number of mannose receptors. Recently, Hattori et al. (2004) demonstrated the targeted delivery of DNA vaccine by Man-C4-Chol liposomes.

5.3 Transferrin Receptor-Mediated Gene Transfection

Transferrin, an iron-binding glycoprotein, is a well-studied ligand for tumor targeting. Iron-loaded transferrin is recognized by and binds to transferrin receptors on cell surfaces. In rapidly dividing cells, expression the receptor is elevated due to an increased cellular need for iron, while on the surfaces of malignant cells expression is often unregulated. Thus, transferrin has been used as a tumor-targeting ligand for several drug delivery systems.

Recently, a transferrin-linked polyethylenimine for tumor-selective gene transfection was developed (Ogris et al. 1999; Kircheis et al. 2001). In order to block undesired, non-specific interactions with blood components or non-target cells, the surface charge of the complexes was masked by either covalently attached hydrophilic polyethylene glycol or a higher density of attached transferrin. After intravenous injection, gene expression in the tumors was approximately 100-fold higher than in other tissues. More recently, the intravenous injection of PEG-PEI-transferrin containing pDNA encoding for tumor necrosis factor (TNF-α) was shown to inhibit tumor growth in murine tumor models (Kursa et al. 2003).

5.4 Folate Receptor-Mediated Gene Transfection

The folate receptor is overexpressed in a large fraction of human tumors, but is only minimally distributed in normal tissues. Therefore, this receptor has been used as a tumor-targeting ligand for several drug delivery systems. Recently, Hofland et al. (2002) synthesized folate-PEG-lipid derivatives for preparing folate-modified cationic liposomes. After intravenous injection of the folate-liposome complexes, gene expression in the tumors was not changed while that in the lungs was reduced compared with conventional complexes. After intraperitoneal injection into murine disseminated peritoneal tumors, however, the folate-liposome complex formulations produced an approximately ten-fold increase in tumor-associated gene expression, as compared with conventional complexes (Reddy et al. 2002).

References

Chowdhury NR, Wu CH, Wu GY, Yerneni PC, Bommineni VR, Chowdhury JR (1993) Fate of DNA targeted to the liver by asialoglycoprotein receptor-mediated endocytosis in vivo. Prolonged persistence in cytoplasmic vesicles after partial hepatectomy. J Biol Chem 268:11265–11271

Eliyahu H, Servel N, Domb AJ, Barenholz Y (2002) Lipoplex-induced hemagglutination: potential involvement in intravenous gene delivery. Gene Ther 9:850–858

Erbacher P, Bousser MT, Raimond J, Monsigny M, Midoux P, Roche AC (1996) Gene transfer by DNA/glycosylated polylysine complexes into human blood monocyte-derived macrophages. Hum Gene Ther 7:721–729

Fumoto S, Nakadori F, Kawakami S, Nishikawa M, Yamashita F, Hashida M (2003) Analysis of hepatic disposition of galactosylated cationic liposome/plasmid DNA complexes in perfused rat liver. Pharm Res 20:1452–1459

Goula D, Benoist C, Mantero S, Merlo G, Levi G, Demeneix BA (1998) Polyethylenimine-based intravenous delivery of transgenes to mouse lung. Gene Ther 5:1291–1295

Hara T, Aramaki Y, Takada S, Koike K, Tsuchiya S (1995) Receptor-mediated transfer of pSV2CAT DNA to mouse liver cells using asialofetuin-labeled liposomes. Gene Ther 2:784–788

Hattori Y, Kawakami S, Suzuki S, Yamashita F, Hashida M (2004) Enhancement of immune responses by DNA vaccination through targeted gene delivery using mannosylated cationic liposome formulations following intravenous administration in mice. Biochem Biophys Res Commun 317:992–999

Hofland HE, Masson C, Iginla S, Osetinsky I, Reddy JA, Leamon CP, Scherman D, Bessodes M, Wils P (2002) Folate-targeted gene transfer in vivo. Mol Ther 5:739–744

Huang L, Li S (1997) Liposomal gene delivery: a complex package. Nat Biotechnol 15:620–621

Kawabata K, Takakura Y, Hashida M (1995) The fate of plasmid DNA after intravenous injection in mice: involvement of scavenger receptors in its hepatic uptake. Pharm Res 12:825–830

Kawakami S, Fumoto S, Nishikawa M, Yamashita F, Hashida M (2000a) In vivo gene delivery to the liver using novel galactosylated cationic liposomes. Pharm Res 17:306–313

Kawakami S, Sato A, Nishikawa M, Yamashita F, Hashida M (2000b) Mannose receptor-mediated gene transfer into macrophages using novel mannosylated cationic liposomes. Gene Ther 7:292–299

Kawakami S, Sato A, Yamada M, Yamashita F, Hashida M (2001) The effect of lipid composition on receptor-mediated in vivo gene transfection using mannosylated cationic liposomes in mice. S.T.P. Pharma Sci 11:117–120

Kawakami S, Hattori Y, Lu Y, Higuchi Y, Yamashita F, Hashida M (2004) Effect of cationic charge on receptor-mediated transfection using mannosylated cationic liposome/plasmid DNA complexes following the intravenous administration in mice. Pharmazie 59:405–408

Kircheis R, Wightman L, Schreiber A, Robitza B, Rossler V, Kursa M, Wagner E (2001) Polyethylenimine/DNA complexes shielded by transferrin target gene expression to tumors after systemic application. Gene Ther 8:28–40

Kursa M, Walker GF, Roessler V, Ogris M, Roedl W, Kircheis R, Wagner E (2003) Novel shielded transferrin-polyethylene glycol-polyethylenimine/DNA complexes for systemic tumor-targeted gene transfer. Bioconjug Chem 14:222–231

Mahato RI, Kawabata K, Nomura T, Takakura Y, Hashida M (1995) Physicochemical and pharmacokinetic characteristics of plasmid DNA/cationic liposome complexes. J Pharm Sci 84:1267–1271

Mahato RI, Takakura Y, Hashida M (1997) Nonviral vectors for in vivo gene delivery: physicochemical and pharmacokinetic considerations. Crit Rev Ther Drug Carrier Syst 14:133–172

Nishikawa M, Yamauchi M, Morimoto K, Ishida E, Takakura Y, Hashida M (2000) Hepatocyte-targeted in vivo gene expression by intravenous injection of plasmid DNA complexed with synthetic multi-functional gene delivery system. Gene Ther 7:548–555

Ogris M, Steinlein P, Kursa M, Mechtler K, Kircheis R, Wagner E. (1998) The size of DNA/transferrin-PEI complexes is an important factor for gene expression in cultured cells. Gene Ther 5:1425–1433

Ogris M, Brunner S, Schuller S, Kircheis R, Wagner E (1999) PEGylated DNA/transferrin-PEI complexes: reduced interaction with blood components, extended circulation in blood and potential for systemic gene delivery. Gene Ther 6:595–605

Perales JC, Ferkol T, Beegen H, Ratnoff OD, Hanson RW (1994) Gene transfer in vivo: sustained expression and regulation of genes introduced into the liver by receptor-targeted uptake. Proc Natl Acad Sci USA 91:4086–4090

Reddy JA, Abburi C, Hofland H, Howard SJ, Vlahov I, Wils P, Leamon CP (2002) Folate-targeted, cationic liposome-mediated gene transfer into disseminated peritoneal tumors. Gene Ther 9:1542–1550

Remy JS, Kichler A, Mordvinov V, Schuber F, Behr JP (1995) Targeted gene transfer into hepatoma cells with lipopolyamine-condensed DNA particles presenting galactose ligands: a stage toward artificial viruses. Proc Natl Acad Sci USA 92:1744–1748

Sagara K, Kim SW (2002) A new synthesis of galactose-poly(ethylene glycol)-polyethylenimine for gene delivery to hepatocytes. J Control Release 79:271–281

Sakurai F, Nishioka T, Saito H, Baba T, Okuda A, Matsumoto O, Taga T, Yamashita F, Takakura Y, Hashida M (2001) Interaction between DNA-cationic liposome complexes and erythrocytes is an important factor in systemic gene transfer via the intravenous route in mice: the role of the neutral helper lipid. Gene Ther 8:677–686

Sato A, Kawakami S, Yamada M, Yamashita F, Hashida M (2001) Enhanced gene transfection in macrophages using mannosylated cationic liposome-polyethylenimine-plasmid DNA complexes. J Drug Target 9:201–207

Song YK, Liu F, Chu S, Liu D (1997) Characterization of cationic liposome-mediated gene transfer in vivo by intravenous administration. Hum Gene Ther 8:1585–1594

Takakura Y, Hashida M (1996) Macromolecular carrier systems for targeted drug delivery: pharmacokinetic considerations on biodistribution. Pharm Res 13:820–831

Takakura Y, Takagi T, Hashiguchi M, Nishikawa M, Yamashita F, Doi T, Imanishi T, Suzuki H, Kodama T, Hashida M (1999) Characterization of plasmid DNA binding and uptake by peritoneal macrophages from class A scavenger receptor knockout mice. Pharm Res 16:503–508

Uyechi LS, Gagne L, Thurston G, Szoka FC Jr (2001) Mechanism of lipoplex gene delivery in mouse lung: binding and internalization of fluorescent lipid and DNA components. Gene Ther 8:828–836

Wu GY, Wu CH (1988) Receptor-mediated gene delivery and expression in vivo. J Biol Chem 263:14621–14624

Yang JP, Huang L (1997) Overcoming the inhibitory effect of serum on lipofection by increasing the charge ratio of cationic liposome to DNA. Gene Ther 4:950–960

Yoshida M, Mahato RI, Kawabata K, Takakura Y, Hashida M (1996) Disposition characteristics of plasmid DNA in the single-pass rat liver perfusion system. Pharm Res 13: 599–603

Yuan F, Dellian M, Fukumura D, Leunig M, Berk DA, Torchilin VP, Jain RK (1995) Vascular permeability in a human tumor xenograft: molecular size dependence and cutoff size. Cancer Res 55:3752–3756

Zhu N, Liggitt D, Liu Y, Debs R (1993) Systemic gene expression after intravenous DNA delivery into adult mice. Science 261:209–211

Optimizing Polyplexes into Synthetic Viruses for Tumor-Targeted Gene Therapy

Ernst Wagner, Julia Kloeckner, and Manfred Ogris

1 Introduction

Specificity for the target tissue is a key issue in the development of gene medicines such as novel anticancer drugs. Many current gene vector formulations suffer from weak to non-existing targeting capacity and give rise to significant toxicities. In order to provide vectors with the ability to distinguish between target and non-target tissue, they have to be modified with cell-binding ligands that recognize target-specific cellular receptors. Already more than 15 years ago, efforts were made to realize this concept of targeted delivery; Wu and Wu (1988) reported successful systemic delivery of asialoglycoprotein-polylysine/DNA complexes to the hepatocyte-specific asialoglycoprotein receptor. Meanwhile this concept has been applied for many vector types (Wickham 2003; Schatzlein 2003; Wagner et al. 2004).

Non-viral vectors are based on formulations with chemically defined cationic DNA carriers and have low immunogenicity, as they can be generated protein-free or by using non-immunogenic proteins and peptides. A weakness of non-viral vectors is their low efficiency in intracellular delivery, which currently is partly compensated for by the administration of large dosages of vector. However, the use of this approach is also limited due to the significant toxicities of current vector formulations.

This chapter reviews the characteristics of targeted DNA/polymer complexes ("polyplexes") developed in our laboratory and their preclinical therapeutic application in mouse tumor models. For example, DNA was complexed with polyethylenimine conjugates into polyplexes of virus-like dimensions. The conjugates contain cell-binding ligands for receptor-mediated endocytosis (transferrin or epidermal growth factor) and polyethylene glycol (for masking the surface charge of polyplexes). Upon systemic administration of these formulations, tumor-specific gene expression and therapeutic efficacy were demonstrated in several murine tumor models.

Pharmaceutical Biology—Biotechnology, Department of Pharmacy, Ludwig-Maximilians-Universitaet Muenchen, Butenandtstrasse 5-13, 81377 Munich, Germany

In addition, ongoing strategies for optimizing targeted polyplexes into therapeutic synthetic virus-like gene vectors are reviewed. Key concepts for optimization include: (1) the use of biodegradable polymers and better purification protocols for polyplexes, and (2) the incorporation of bioresponsive, smart elements that, in a dynamic manner, present virus-like delivery functions during the appropriate phase of the gene delivery process.

2 Shielded Ligand-PEI Polyplexes for Tumor-Targeted Gene Transfer

2.1 Receptor Targeting

Numerous cell-targeting ligands, including peptides (Ziady et al. 1999), growth factors (Wolschek et al. 2002), antibodies (Merdan et al. 2003), or other proteins (Wu and Wu 1988;Wagner et al. 1990), have been incorporated into polyplexes after chemical conjugation to cationic polymers (Schatzlein 2003). Polycations commonly employed include polylysines, polyethylenimines, dendrimers, and other cationic polymers (see De Smedt et al. 2000; Han et al. 2000). Polyethyleinimine (PEI) is a polymer with exceptional high transfection efficiency (Zou et al. 2000). For this reason our group has investigated receptor-targeted PEI polyplexes using transferrin (Tf) or epidermal growth factor (EGF) as ligands for receptors over-expressed in a variety of cancers.

Despite successful in vitro target specificity, as shown in many cases using ligand-containing polyplexes, only a few studies have reported successful in vivo targeting (Wu and Wu 1988; Perales et al. 1994; Kircheis et al. 2001). The presence of targeting ligands is required but not sufficient for in vivo targeting, and interaction with other (non-target) cells, plasma components, or the extracellular matrix must also be taken into account.

In many cases, the systemic administration of cationic polyplexes resulted in acute and severe toxicity. Non-specific interactions of cationic polyplexes with blood are the major reasons for these effects (Plank et al. 1996; Chollet et al. 2002; Verbaan et al. 2001). Such non-specific interactions of polyplexes followed by aggregation with blood components and subsequent trapping of polyplex aggregates in the lung capillaries are considered to be responsible for the observed toxicity and a preferential transfection of lung tissue. Therefore, to make ligand-mediated targeting effective, the positive surface charges of polyplexes have to be masked.

2.2 Shielding of Polyplex Surface Charge

Attachment of hydrophilic polymers like polyethylene glycol (PEG) to the surface of liposomes was previously reported to shield the latter from undesired binding activity in the blood, resulting in prolonged circulation times. We and others applied the same concept to shield targeted DNA polyplexes, and several strategies have been developed for the attachment of PEG to polyplexes (Ogris et al. 2003). The hydrophilic polymer was covalently coupled to the DNA-binding polycation either before polyplex formation (pre-PEG-ylation) (Kursa et al. 2003; Wolschek et al. 2002; Verbaan et al. 2004) or after the polyplex formation (post-PEG-ylation) (Ogris et al. 1999;

Verbaan et al. 2004). Shielding by PEG increased solubility and provided stability to freeze-thawing (Ogris et al. 2003), reduced the toxicity, and extended the circulation time of polyplexes in blood.

Instead of PEG-ylation, alternative approaches have been followed to mask polyplex surface charges. Another hydrophilic polymer, hydroxypropyl methacrylate, has been attached to the DNA polyplex surface, either in a stable or bioreducible manner (Carlisle et al. 2004; Dash et al. 2000). A further approach utilized transferrin (a serum protein and therefore well adapted to the requirements for blood circulation) for both surface shielding and targeting (Kircheis et al. 2001).

2.3 Systemic Tumor Targeting

Targeting tumors via the systemic route presents an interesting opportunity to attack tumor metastases. In order to reach this goal, interactions with blood components and healthy tissues have to be avoided, and targeting ligands have to be incorporated to improve the efficiency, as outlined above.

Intravenous injection of Tf-containing polyplexes shielded by PEG or an higher density of Tf resulted in gene transfer into distant subcutaneous tumors of mice (Ogris et al. 1999; Kursa et al. 2003; Kircheis et al. 2001). Marker-gene expression levels in Neuro2A neuroblastoma tumor tissues were 100-fold higher than in other organs (Fig. 1). This specificity of expression was also verified by in vivo imaging of luciferase reporter gene expression (Hildebrandt et al. 2003). Hepatocellular carcinoma overexpress the EGF receptor. Consistently, EGF-PEG-coated polyplexes were successfully applied systemically for targeting human hepatocellular carcinoma xenografts in SCID mice (Wolschek et al. 2002). Several factors contribute to the targeting effect; the unique hyperpermeability of the tumor vasculature and inadequate lymphatic drainage results in enhanced uptake and retention of particles in the tumor, the so-called EPR-effect (Maeda 2001); active targeting is mediated by ligand binding to the receptors over-expressed in the target tumors. In addition, replicating tumor cells are more readily transfected compared to non-dividing normal cells.

The therapeutic efficacy of tumor-targeted polyplexes was demonstrated by repeated systemic administration of Tf-coated polyplexes coding for tumor necrosis factor alpha (TNF-alpha) into tumor-bearing mice (Kursa et al. 2003; Kircheis et al. 2002). The treatment induced tumor necrosis and inhibition of tumor growth in four murine tumor models of different tissue origin. Since gene expression of TNF-alpha was localized in the tumor, no systemic TNF-related toxicities were observed.

3 Development of Novel Polymers as DNA Carriers

The in vivo efficiencies of current vector formulations are not satisfactory for several reasons: (1) polyplexes are toxic at doses only slightly higher than the effective dosage; (2) upon systemic administration, only a tiny fraction of the vector dose reaches the target site; and (3) the in vivo gene transfer activities are low. Important aspects for further optimization include the assembly and purification of polyplexes into formulation with increased storage stability. Polyplex aggregates and potentially toxic non-bound cationic polymers have to be removed (Boeckle et al. 2004). New polycationic

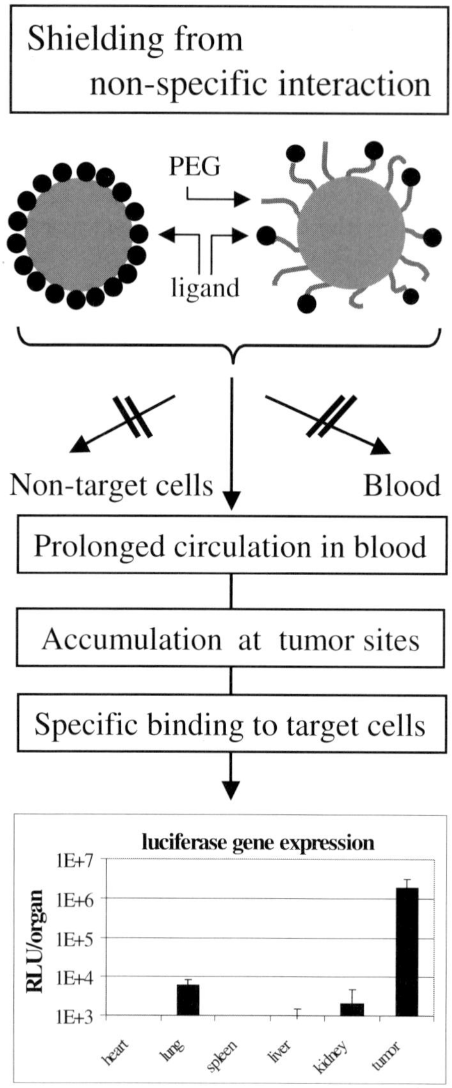

FIG. 1. Surface shielded polyplexes for systemic tumor targeting. Polyplexes with cell-binding ligands (e.g. Tf, EGF) can specifically interact with target tumor cells expressing the according receptor. Non-specific interactions with blood components and non-target cells have to be blocked by shielding the surface charge of the polyplex, e.g. by PEG-ylation. A prolonged circulation of polyplexes in the blood leads to their accumulation at tumor sites followed by uptake and gene expression in tumor cells (Kircheis et al. 2001)

carriers have to be applied with backbones that mediate higher transfection efficiencies than obtained with existing carriers, and that are more biocompatible and biodegradable.

3.1 Novel Polymers with Improved Characteristics for Tumor Targeting

Toxicity is an inherent property of polycationic carriers that is associated with their avidity to bind many biological materials in an unspecific manner. Neutralization of the positive polymer charges by polyplex formation and subsequent purification of the polyplexes reduces their toxicity, but does not eliminate the problem. Toxicity increases with the degree of polymerization of the carrier molecule. In fact, the toxicity of many low-molecular-weight polymers is much less than that of their high-molecular-weight counterparts (Kramer et al. 2004); however, they do not provide the optimum stability of DNA polyplexes required for in vivo administration.

Recently, novel low-molecular-weight dendritic polymers were developed that show very encouraging properties for systemic targeted gene transfer (Kawano et al. 2004). Apparently, the dendrimer structure of polymers contributes to stabilizing the polyplexes at a low polymer molecular weight, which prevents undesired unspecific effects such as blood aggregation and gene expression in the lung. Passive tumor targeting was observed with dendrimers based on dendritic poly(L-lysine) of the sixth generation (Okuda et al. 2004). Despite the absence of shielding agents (such as PEG), polyplexes circulated in the blood stream for 3 h upon intravenous injection; at 1 h after administration, DNA particles were observed in the tumor (Kawano et al. 2004). In another approach, low-molecular-mass (1.7 kDa) polypropylenimine (PPI) dendrimers (Zinselmeyer et al. 2002) were applied for tumor targeting. In a therapeutic strategy related to that of Kircheis et al. (2002), intravenous injection of TNF-alpha plasmid PPI polyplex formulations delivered gene expression to solid tumors and resulted in complete regression of A431 tumors (Uchegbu et al. 2004).

3.2 Biodegradable Polymers

In addition to acute toxicity, the long-term fate of the cationic polymeric carrier remaining in the host is also a major concern; therefore, biodegradable polymers would be advantageous. Partial or complete degradation of the polycation results in less positive charges per molecule, which is also expected to reduce unspecific interactions and toxicity. Two main concepts are being explored in the development of novel biodegradable cationic polymers: either the synthesis of polymers that have biodegradable bonds within their repetitive monomeric units, or the generation of biodegradable polymer conjugates in which low-molecular-weight polymers with low toxicity are cross-linked into larger polycationic carriers by conjugation with biodegradable linkers.

Lim et al. (2002) described a new biodegradable polymer of a branched network of amino esters (n-PAE) that has transfection efficiencies similar to that of the commonly used branched 25-kDa PEI, but with minimal cytotoxicity. The network structure provides multiple ternary and primary amines for DNA binding and endosomal buffering. The branched structure apparently also tunes polyester degradation to intermediate stability, whereas linear amino-modified polyesters are hydrolyzed too quickly.

New biodegradable PEI derivatives were synthesized by cross-linking 0.8-kDa low-molecular-mass PEI with three different bioreversible linkers: two different disulfide linkers (Gosselin et al. 2001), and diacrylate-ester linkers (Forrest et al. 2003). Gene transfer properties were similar to that of the commonly used 25-kDa PEI, but the toxicity was far lower. Apart from the difference in the biodegradable linkage (disulfide versus ester), the conjugates differ in the type of amino group formed at the linkage sites at either primary or secondary PEI amines. Comparing the three different forms of cross-linked low-molecular-mass PEI in our laboratory, we observed the highest efficiency using the hexanediol diacrylate–cross-linked PEI (Kloeckner et al., our unpublished results).

4 Bioresponsive Elements for Optimizing Polyplexes into Synthetic Viruses

Current polyplexes are still very inefficient as compared to viral vectors. Thus, viruses might present ideal natural examples from which to learn how polyplexes can be optimized into "synthetic viruses" (Wagner et al. 2004). One unique property of viruses is their dynamic manner in responding to the biological micro-environment. Similarly, polyplexes should also be able to alter their structure during the gene delivery process, in order to make them most effective for the subsequent gene delivery steps.

For example, a PEG shield in polyplexes may enhance systemic delivery to the target cells, but reduces gene expression activity within the target cells; apparently an irreversible PEG shield hampers intracellular processing. In a virus-like polyplex, PEG shielding should be presented in a bioresponsive manner. Ideally, after entry into the target cell and delivery into the endosomal vesicle, the polyplex should release the PEG shield so that the cationic surface of the polyplex is exposed for efficient destabilization of the endosomal membrane. For this purpose we introduced bioresponsive PEG-polycation conjugates with pH-labile linkages, which make use of the acidic milieu of the endosomes (Walker et al. 2004). DNA particles shielded with these bioreversible PEG conjugates lose their PEG shield at endosomal pH and display up to 100-fold higher gene transfer activity than polyplexes with the analogous stable PEG shields (Fig. 2).

The acidic endosomal pH may also result in activation of membrane disrupting functions required within the endosome (but not at the cell surface or outside the cell). The acidification may trigger acidic membrane-active peptides, such as derived from influenza virus HA2 by pH-specific conformational changes (Mechtler and Wagner 1997). Alternatively, the low pH may induce the removal of a masking group,

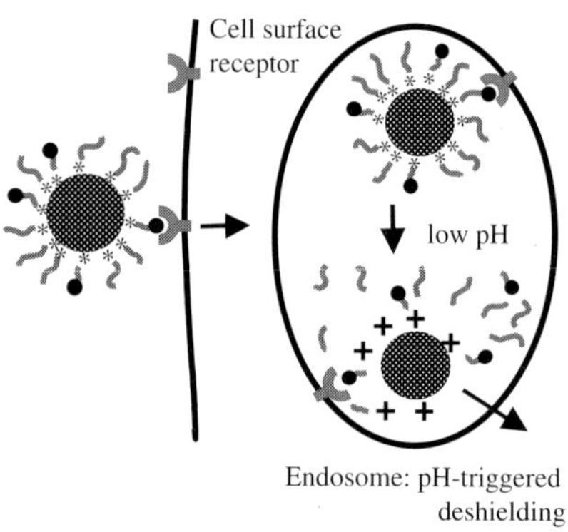

FIG. 2. Bio-reversible shielding of polyplexes: towards synthetic viruses. Following receptor-mediated uptake of ligand-coated, PEG-shielded polyplexes into endosomal vesicles of target cells, the bioresponsive PEG-polycation conjugate is cleavable within the acidic pH of the endosome, facilitating subsequent escape of polyplexes. This results in strongly enhanced gene expression as compared to polyplexes containing the same amount of stable PEG-polycation conjugates (Walker et al. 2004)

such as described in the example by Rozema et al. (2003), in which melittin is unmasked by cleavage of the protective groups from lysine residues, which recovers the lytic activity of the melittin peptide.

The intracellular reducing environment may contribute to polyplex activation and disassembly by cleaving disulfide-bridged cationic carriers. Triggered by the reducing environment, a lytic form of listeriolysin O was released from an inactive precursor form (Saito et al. 2003). Several reports have described bioresponsive vectors that combine extracellular stability of DNA, by polycationic disulfide-bond-containing "cages," with rapid intracellular release of the DNA upon cleavage of these cages (Trubetskoy et al. 1999; Miyata et al. 2004; Carlisle et al. 2004). These examples demonstrate that incorporation of delivery functions that are presented in a bioresponsive fashion can strongly improve polyplex efficiency.

5 Conclusions

Receptor-targeted polyplex systems have been established that display encouraging in vivo activity. In particular, systemic targeting of distant tumors was demonstrated, and therapeutic effects have been obtained in several mouse tumor models. In the future, polyplexes have to be developed into virus-like complexes, consisting of biodegradable materials and designed to contain bioresponsive elements that present virus-like delivery functions in a controlled manner.

Acknowledgment. The contributions by Carolin Fella, Greg Walker, Julia Fahrmeir, Sabine Boeckle, Silke van der Piepen, and Wolfgang Roedl are greatly appreciated. Many thanks also to Olga Brück for skilful assistance in preparing the manuscript.

References

Boeckle S, von Gersdorff K, van der Piepen S, Culmsee C, Wagner E, Ogris M (2004) Purification of polyethylenimine polyplexes highlights the role of free polycations in gene transfer. J Gene Med 6:1102–1111

Carlisle RC, Etrych T, Briggs SS, Preece JA, Ulbrich K, Seymour LW (2004) Polymer-coated polyethylenimine/DNA complexes designed for triggered activation by intracellular reduction. J Gene Med 6:337–344

Chollet P, Favrot MC, Hurbin A, Coll JL (2002) Side-effects of a systemic injection of linear polyethylenimine-DNA complexes. J Gene Med 4:84–91

Dash PR et al. (2000) Decreased binding to proteins and cells of polymeric gene delivery vectors surface modified with a multivalent hydrophilic polymer and retargeting through attachment of transferrin. J Biol Chem. 275:3793–3802

De Smedt SC, Demeester J, Hennink WE (2000) Cationic polymer based gene delivery systems. Pharm Res 17:113–26

Forrest ML, Koerber JT, Pack DW (2003) A degradable polyethylenimine derivative with low toxicity for highly efficient gene delivery. Bioconjug Chem 14:934–940

Gosselin MA, Guo W, Lee RJ (2001) Efficient gene transfer using reversibly cross-linked low molecular weight polyethylenimine. Bioconjug Chem 12:989–994

Han S, Mahato RI, Sung YK, Kim SW (2000) Development of biomaterials for gene therapy. Mol Ther 2:302–317

Hildebrandt IJ, Iyer M, Wagner E, Gambhir SS (2003) Optical imaging of transferrin targeted PEI/DNA complexes in living subjects. Gene Ther 10:758–764
Kawano T, Okuda T, Aoyagi H, Niidome T (2004) Long circulation of intravenously administered plasmid DNA delivered with dendritic poly(L-lysine) in the blood flow. J Control Release 99:329–337
Kircheis R et al. (2002) Tumor-targeted gene delivery of tumor necrosis factor-alpha induces tumor necrosis and tumor regression without systemic toxicity. Cancer Gene Ther 9:673–680
Kircheis R et al. (2001) Polyethylenimine/DNA complexes shielded by transferrin target gene expression to tumors after systemic application. Gene Ther 8:28–40
Kramer M et al. (2004) Dendritic polyamines: simple access to new materials with defined treelike structures for application in nonviral gene delivery. Chembiochem 5:1081–1087
Kursa M et al. (2003) Novel Shielded Transferrin-Polyethylene Glycol-Polyethylenimine/DNA Complexes for Systemic Tumor-Targeted Gene Transfer. Bioconjug Chem 14:222–231
Lim YB, Kim SM, Suh H, Park JS (2002) Biodegradable, endosome disruptive, and cationic network-type polymer as a highly efficient and nontoxic gene delivery carrier. Bioconjug Chem 13:952–957
Maeda H (2001) The enhanced permeability and retention (EPR) effect in tumor vasculature: the key role of tumor-selective macromolecular drug targeting. Adv Enzyme Regul 41:189–207
Mechtler K, Wagner E (1997) Gene transfer mediated by influenza virus peptides: the role of peptide sequence. New J Chem 21:105–11
Merdan T et al. (2003) Pegylated polyethylenimine-fab' antibody fragment conjugates for targeted gene delivery to human ovarian carcinoma cells. Bioconjug Chem 14:989–996
Miyata K et al. (2004) Block catiomer polyplexes with regulated densities of charge and disulfide cross-linking directed to enhance gene expression. J Am Chem Soc 126:2355–2361
Ogris M, Brunner S, Schuller S, Kircheis R, Wagner E (1999) PEGylated DNA/transferrin-PEI complexes: reduced interaction with blood components, extended circulation in blood and potential for systemic gene delivery. Gene Ther 6:595–605
Ogris M, Walker G, Blessing T, Kircheis R, Wolschek M, Wagner E (2003) Tumor-targeted gene therapy: strategies for the preparation of ligand-polyethylene glycol-polyethylenimine/DNA complexes. J Control Release 91:173–181
Okuda T, Sugiyama A, Niidome T, Aoyagi H (2004) Characters of dendritic poly(L-lysine) analogues with the terminal lysines replaced with arginines and histidines as gene carriers in vitro. Biomaterials 25:537–544
Perales JC, Ferkol T, Beegen H, Ratnoff OD, Hanson RW (1994) Gene transfer in vivo: sustained expression and regulation of genes introduced into the liver by receptor-targeted uptake. Proc Natl Acad Sci U S A 91:4086–4090
Plank C, Mechtler K, Szoka FJ, Wagner E (1996) Activation of the complement system by synthetic DNA complexes: A potential barrier for intravenous gene delivery. Hum Gene Ther 7:1437–1446
Rozema DB, Ekena K, Lewis DL, Loomis AG, Wolff JA (2003) Endosomolysis by Masking of a Membrane-Active Agent (EMMA) for Cytoplasmic Release of Macromolecules. Bioconjug Chem 14:51–57
Saito G, Amidon GL, Lee KD (2003) Enhanced cytosolic delivery of plasmid DNA by a sulfhydryl-activatable listeriolysin O/protamine conjugate utilizing cellular reducing potential. Gene Ther 10:72–83
Schatzlein AG (2003) Targeting of Synthetic Gene Delivery Systems. J Biomed Biotechnol 2003:149–158
Trubetskoy VS, Loomis A, Slattum PM, Hagstrom JE, Budker VG, Wolff JA (1999) Caged DNA does not aggregate in high ionic strength solutions. Bioconjug Chem 10:624–628

Uchegbu IF, Dufes C, Elouzi A, Schatzlein AG (2004) Gene therapy with polypropylenimine dendrimers. Proceed 12th Inter Pharm Technol Symp 2004:11–12

Verbaan FJ, Oussoren C, Snel CJ, Crommelin DJ, Hennink WE, Storm G (2004) Steric stabilization of poly(2-(dimethylamino)ethyl methacrylate)-based polyplexes mediates prolonged circulation and tumor targeting in mice. J Gene Med 6:64–75

Verbaan FJ et al. (2001) The fate of poly(2-dimethyl amino ethyl)methacrylate-based polyplexes after intravenous administration. Int J Pharm 214:99–101

Wagner E, Kircheis R, Walker GF (2004) Targeted nucleic acid delivery into tumors: new avenues for cancer therapy. Biomed Pharmacother 58:152–161

Wagner E, Zenke M, Cotten M, Beug H, Birnstiel ML (1990) Transferrin-polycation conjugates as carriers for DNA uptake into cells. Proc Natl Acad Sci U S A 87:3410–3414

Walker GF et al. (2005) Toward synthetic viruses: Endosomal pH triggered deshielding of targeted polyplexes greatly enhances gene transfer in vitro and in vivo. Mol Ther 11:418–425.

Wickham TJ (2003) Ligand-directed targeting of genes to the site of disease. Nat Med 9:135–139

Wolschek MF et al. (2002) Specific systemic nonviral gene delivery to human hepatocellular carcinoma xenografts in SCID mice. Hepatology 36:1106–1114

Wu GY, Wu CH (1988) Receptor-mediated gene delivery and expression in vivo. J Biol Chem 262:14621–14624

Ziady AG, Ferkol T, Dawson DV, Perlmutter DH, Davis PB (1999) Chain length of the polylysine in receptor-targeted gene transfer complexes affects duration of reporter gene expression both in vitro and in vivo. J Biol Chem 274:4908–4916

Zinselmeyer BH, Mackay SP, Schatzlein AG, Uchegbu IF (2002) The lower-generation polypropylenimine dendrimers are effective gene-transfer agents. Pharm Res 19:960–967

Zou SM, Erbacher P, Remy JS, Behr JP (2000) Systemic linear polyethylenimine (L-PEI)-mediated gene delivery in the mouse. J Gene Med 2:128–134

Gene Transfer and Target Diseases

Mariko Harada-Shiba

1 Gene Transfer and Target Diseases

The first human gene-transfer protocol was developed in 1989, in an attempt to track lymphocytes in the immunologic treatment of melanoma and renal cell cancer. Although many attempts have been made in clinical trials of gene therapy, the FDA has not approved the marketing of any gene therapy agent.

For the successful therapeutic application of gene therapy, the delivery of several kinds of genes and nucleotides has been tested in animal models of genetic diseases as well as in patients to supply missing proteins to maintain cellular function, or to deliver proteins that induce proliferation or apoptosis of the cell, etc. For example, gene therapy has been applied to treating cardiovascular diseases, including coronary artery disease (CAD), peripheral artery disease (PAD), restenosis after vascular interventions and graft failure, hyperlipidemia, thrombosis, and cancer (e.g., lung, kidney, prostate, brain). In this chapter, the pathophysiology of several diseases and the application of gene delivery in their treatment will be described.

2 Cardiovascular Diseases

2.1 Coronary Artery Disease and Peripheral Artery Disease

Atherosclerosis is the most prevalent process that affects adult coronary and peripheral arteries. Atherosclerotic lesions narrow arteries, leading to a reduction of the arterial blood supply to the myocardium and skeletal muscle. Stimulation of collateral vessel formation by the use of gene therapy will help to increase perfusion of the ischemic tissues. Genes encoding growth factors, such as vascular endothelial growth factor (VEGF), fibroblast growth factor (FGF) and hepatocyte growth factor (HGF),

Department of Bioscience, National Cardiovascular Center Research Institute, 5-7-1 Fujishirodai, Suita, Osaka 565-8565, Japan have been successfully tested in animal models and clinical trials for therapeutic angiogenesis.

VEGFs are mitogenic and survival factors for endothelial cells, and promote angiogenesis and lymphangiogenesis. The VEGF family consists of six members, VEGF-A, -B, -C, -D, -E and placental growth factor (PLGF), which differ in their molecular masses and biological properties (Dvorak et al. 1995; Leung et al. 1989; Maglione et al. 1993; Joukov et al. 1996; Achen et al. 1998; Ogawa et al. 1998; Olofsson et al. 1996). The three receptors, VEGFR-1, VEGFR-2 and VEGFR-3, have tyrosine kinase activity. VEGF-A is the most well known member of the VEGF family, as it plays a crucial role in angiogenesis and vasculogenesis (Ferrara 2000, 2001). In addition, VEGF-A has several splice variants, of which two ($VEGF_{121}$ and $VEGF_{165}$) have been reported to be angiogenic in both animal models and clinical trials (Losordo et al. 1998; Vale et al. 2000). $VEGF_{121}$ is readily diffusible because it lacks a heparan-sulfate binding site, whereas $VEGF_{165}$ binds to the matrix after being secreted. VEGF-B, -C, -D, -E and PLGF also show angiogenic activity in animal models (Yoon et al. 2003; Rissanen et al. 2003; Kiba et al. 2003).

The FGF family has twenty-three members that share 30%–70% identical amino acid sequences. These growth factors act directly on vascular cells and induce endothelial cell growth and angiogenesis. Among the FGFs, FGF-1, -2, -4 and -5 have been tested for their angiogenic activity in animal models and are also the subjects of clinical trials in the therapy of cardiovascular disease and Atherosclerosis obliterans (ASO) (Ueno et al. 1997b; Javerzat et al. 2002; Grines et al. 2003). FGFs are multifunctional proteins that act through various alternatively spliced isoforms with four tyrosine kinase receptors, FGFR-1, -2, -3 and -4 (Galzie et al. 1997; Ornitz et al. 1996).

HGF is another multifunctional growth factor that stimulates the proliferation and migration of endothelial cells. In rabbit, rat and mouse ischemia models, HGF was reported to stimulate angiogenesis (Morishita et al. 1999, 2002; Morishita 2004; Hayashi et al. 1999). The efficacy and safety of intramuscular injection of naked human HGF plasmid was recently demonstrated in clinical trials in 22 patients with PAD or Berger's disease (Morishita 2004).

Angiopoietins (Angs) are also growth factors for vascular development. The Ang family has four members, Ang-1, -2, -3, -4, all of which bind to Tie-2, a tyrosine kinase receptor (Ward and Dumont 2002). Ang-1/Tie-2 and VEGF/VEGFR2 are crucial for the mobilization and recruitment of hematopoietic stem cells and the recruitment of circulating endothelial progenitor cells (Hattori et al. 2001). Ang-1 decreases the inflammatory response and promotes vascular maturity; thus, a combination of VEGF and Ang-1 may be a good strategy for therapeutic angiogenesis (Siddiqui et al. 2003; Yamauchi et al. 2003).

Hypoxia-inducible transcription factor (HIF)-1α can activate several genes involved in angiogenesis, such as VEGF, VEGFR-2, IGF-2 and erythropoietin (Levy et al. 1995). Adenovirus-mediated HIF-1α gene therapy is currently in clinical testing for the treatment of myocardial ischemia.

2.2 Restenosis After Vascular Interventions and Vein Graft Failure

The occlusion of arteries after balloon angioplasty, stenting, or the failure of bypass vein graft is a major factor that determines the prognosis of peripheral and coronary artery disease. Smooth muscle cell proliferation, remodeling, matrix deposition, thrombosis, and platelet and leukocyte adhesion may all play a role in the development of arterial restenosis in these settings (Topol and Serruys 1998). In order to decrease vascular cell proliferation, various gene therapy strategies have been employed. Antiproliferative strategies designed for the treatment of experimental cardiovascular disease can be grouped into two main categories: (1) antisense approaches, ribozymes, and transcription-factor decoy strategies to inactivate positive cell cycle regulators; (2) overexpression of negative regulators of cell growth.

Transfection to arterial smooth muscle cells with thymidine kinase combined with gancyclovir, antisense oligonucleotides and ribozymes against cell cycle regulators, c-*myb*, c-*myc*, *cdc*-2, *cdk*-2, *ras*, *bcl*-x, and decoy constructs against transcription factors, such as E2F and NFκB, have all been shown to inhibit neointimal proliferation (Morishita et al. 1993, 1997; Indolfi et al. 1995; Pollman et al. 1998; Burgess et al. 1995; Suzuki et al. 1997). Inhibition of the cell cycle by transfection of genes encoding the non-phosphorylated forms of the retinoblastoma gene products p21, p27, p53, or of the growth arrest homeobox gene (*gax*) has been reported in animal models (Tanner et al. 1998; Yonemitsu et al. 1998; Chang et al. 1995a,b; Smith et al. 1997).

The transfection of genes encoding growth factors, including VEGF and HGF, results in decreased neointima formation in experimental animals (Hiltunen et al. 2000; Laitinen et al. 1997). The rapid regeneration of endothelial cells by growth factors restored the secretion of nitric oxide, C-type natriuretic peptide (CNP) and prostacyclin I_2, which have anti-proliferative effects on smooth muscle cells. The local expression of CNP suppressed neointimal formation in injured arteries of rats and vascular remodeling in porcine coronary arteries (Ueno et al. 1997a; Morishige et al. 2000).

2.3 Hyperlipidemia

Some hyperlipidemias are congenital, caused by a monogenic disorder. The strategies of gene therapy in hyperlipidemia are divided into three groups. (1) to supply the defective gene to correct the dyslipidemia, (2) to overexpress proteins involved in lipid metabolism, and (3) miscellaneous approaches.

Several studies have applied the first therapy strategy to the treatment of monogenic hyperlipidemias. Familial hypercholesterolemia (FH) is caused by defect in the LDL receptor, which results in severe hypercholesterolemia beginning at birth, cutaneous and tendon xanthomas, and atherosclerosis in childhood (Goldstein JL 2001). Trials of LDL or VLDL receptor gene delivery to the liver were carried out in WHHL rabbits and FH patients (Grossman et al. 1994, 1995; Kozarsky et al. 1994; Pakkanen et al. 1999; Lebherz et al. 2004; Kankkonen et al. 2004). The Apo A1 gene encoding a protein necessary for HDL synthesis and involved in reverse cholesterol transport, was delivered in patients deficient in this gene (Benoit et al. 1999). Apo E gene transfer was shown to be successful in treating *apoE* knockout mice (Cioffi et al. 1999; Okamoto et al. 2002; Gough and Raines 2003; Harris et al. 2002). Lipoprotein lipase (LPL), which catalyzes triglyceride formation in chylomicrons and VLDL, were delivered in LPL-deficient animals (Excoffon et al. 1997; Liu et al. 2000). Delivery of the gene encoding lecithin cholesterol acyl transferase (LCAT) corrected dyslipidemia in patients deficient in the gene and in hypoalphalipoproteinemia patients (Brousseau et al. 1998; Seguret-Mace et al. 1996).

The overexpression of proteins involved in the regulation of lipid metabolism has been tested in several systems. Hepatic expression of the catalytic subunit of apolipoprotein B mRNA editing enzyme (Apobec-1) reduced serum LDL-cholesterol levels in normal and WHHL rabbits (Greeve et al. 1996; Kozarsky et al. 1996). Expression of secreted "decoy" human-macrophage scavenger receptors (MSR) inhibited foam-cell formation in murine macrophages (Jalkanen et al. 2003a,b). Overexpression of LPL in transgenic WHHL rabbits improved hyperlipidemia and obesity (Koike et al. 2004). Finally, the long-term expression of human apo A-1 increased HDL size and inhibited atherosclerosis progression in LDLR knockout mice (Belalcazar et al. 2003).

2.4 Thrombosis

Gene transfer has a number of cardiac and systemic applications disease conditions, such as acute coronary syndromes, restenosis following percutaneous coronary intervention (PCI) and venous grafts, and thrombotic states. The strategies for gene therapy of thrombosis are: (1) to inhibit the coagulation pathway or platelet aggregation, (2) to activate fibrinolysis, and (3) to modulate endothelial function.

In order to target the coagulation cascade, transfer of the hirudin gene led to reduced intimal hyperplasia in a rat carotid-artery injury model (Rade et al. 1996). Gene delivery of thrombomodulin, a cell-surface glycoprotein of endothelial cells that binds thrombin, was reported to reduce thrombus formation in a rabbit model of stasis-induced arterial thrombosis (Waugh et al. 1999). In a rabbit carotid shear-stress-induced model of thrombosis, overexpression of the tissue-factor pathway inhibitor (TFPI) gene reduced thrombus formation (Nishida et al. 1999). Overexpression of cyclooxygenase-1, an inhibitor of platelet aggregation, increased production of the antiplatelet prostaglandin prostacyclin and reduced thrombus production in a porcine model of balloon-injury-induced carotid thrombosis (Zoldhelyi et al. 2001). Transfer of the nitric oxide (NO) synthase gene was reported to reduce arterial thrombosis in a rat carotid injury model and in a porcine coronary artery balloon-injury model (von der Leyen et al. 1995).

The targeted activation of fibrinolysis was achieved by overexpressing either recombinant tissue-type plasminogen activator (rTPA) or surface-anchored urokinase in endothelial cells and then seeding these cells into grafts in order to increase fibrinolytic activity. Reduced local platelet and fibrin deposition were observed while systemic markers of coagulation and fibrinolysis remained unchanged (Dichek et al. 1996).

Overexpression of NO synthase modulates endothelial function, inhibiting platelet adhesion after arterial injury (Yan et al. 1996).

2.5 Primary Pulmonary Hypertension

Primary pulmonary hypertension is a rare but life-threatening disease that causes right ventricular failure and death. The average survival from the time of diagnosis is 2.8 years (Nagaya 2004). In order to reduce pulmonary vascular resistance, the transfer of genes encoding eNOS, calcitonin gene-related peptide (CGRP), and prostacyclin synthase (PGIS) has been shown to be effective in model animals (Champion et al. 1999, 2000; Christman et al. 1992; Tuder et al. 1999; Nagaya et al. 2000).

2.6 Cerebral Vascular Disease

Several possible targets for gene therapy in treating cerebral vascular disease have been proposed: (1) prevention of vasospasm after subarachnoid hemorrhage (SAH), (2) protection against brain damage after ischemic stroke, (3) stimulation of collateral vessel formation in areas at risk of ischemia, (4) prevention of restenosis after angioplasty of the carotid and vertebrobasilar arteries, (5) inhibition of thrombosis.

Vasospasm is a serious problem after SAH and, currently, there is no effective method of prevention. Vasospasm is a great potential target of gene therapy because it occurs several days after the occurrence of SAH, so that there is sufficient time to deliver a gene (Toyoda et al. 2003). Moreover, the risk of vasospasm is highest during the 2 to 3 weeks after SAH, a short enough period of time to allow transient gene expression. The mechanisms of vasospasm after SAH may include impaired endothelium-dependent vasorelaxation, production of endothelium-derived contracting factors (endothelin, etc.), and impaired activity of potassium channels in cerebral vessels.

The in vivo transfer of the gene encoding endothelial NOS improved the NO mediated relaxation of the basilar arteries in vitro after experimental SAH (Onoue et al. 1998). Vasospasm in transgenic mice that overexpressed CuZn-SOD or EC-SOD was less severe after experimental SAH (Kamii et al. 1999); McGirt et al. 2002). Vascular contraction was inhibited after SAH by intracisternal administration of preproendothelin-1 antisense oligoDNA, which reduced production of endothelin peptide (Onoda et al. 1996).

CGRP has potent activity in opening potassium channels, hyperpolarizing arterial muscle, and dilating arteries. After SAH, CGRP was shown to depleted from nerves supplying cerebral arteries. The genetic transfer of prepro-CGRP prevented vasospasm in rabbits after experimental SAH (Nozaki et al. 1989; Edvinsson et al. 1991).

The therapeutic targets and genes used for gene therapy of vascular diseases are summarized in Table 1.

TABLE 1. Therapeutic target and genes used for gene therapy in vascular disease

Therapeutic target	Treatment genes
Therapeutic angiogenesis	VEGF-A, -B, -V, -D, -E, FGF-1, -2, -4, -5, angiopoetin-1, HGF, MCP-1, PDGF, eNOS, iNOS, adrenomedullin
Restennosis, vein-graft failure	VEGF-A, C, eNOS, iNOS, COX, Thymidine kinase, CNP Fas ligand, p16, p21, p27, p53, NFkB and E2F decoys, cdk-2, cdc-2, c-myb, c-myc, ras, bcl-x, PCNA antisense oligonucleotides Ribozimes, Bloking PDGF or TGF-β expression or their receptors
Atherosclerosis, hyperlipidaemia	LDL receptor, VLDL receptor, apoA-1, Lipoprotein lipase, Hepatic lipase, LCAT, apoB, Lipid transfer proteins, Lp(a) inhibition, Soluble scavenger-receptor decoy, Soluble VCAM or ICAM, SOD, PAF-AH
Thrombosis	Hirudin, tPA, thrombomodulin, COX, TFPI
Pulmonary hypertension	Prepro-calcitonin gene related peptide, ANP, eNOS, prostacyclin syntase, VEGF-A, adrenomeullin
Vasospasm after SAH	Endothelial NOS, ECSOD, CuZnSOD, Antisense preproendothelin-1, Prepro-CGRP

3 Cancer

Cancer is a genetic disease in which individual cells have mutations in genes related to growth control and apoptosis. In addition, cancer cells also have support systems to promote invasion and metastasis. Tumor growth is a result of the interaction of cancer cells with their microenvironment, including the extracellular matrix, immune system cells and cells involved in angiogenesis; therefore, each of these steps can serve as a target in controlling tumor growth. The strategies for cancer gene therapy can be directed at the tumor itself or at the host. Strategies directed at the tumor involve killing the tumor cells or slowing down their growth by, for example, introducing tumor suppressor genes, suppressing protooncogenes, and inducing prodrug/suicide, or apoptosis. Strategies direct at the host involve the inhibition of angiogenesis, the protection of normal tissues and increasing immunity.

3.1 Tumor Suppressor Genes

In order to target growth regulation in cancer cells, tumor suppressors, that inactivate the growth of some tumors, can be introduced (Bookstein et al. 1993). Since mutations in *p53* are widespread in human cancer, this gene may be the first target for the genetic therapy of cancer using tumor suppressors. The *p53* gene encodes a transcription factor involved in the regulation of the cell cycle and apoptosis. Gene transfer of *p53* to *p53*-defective cells resulted in cessation of cell growth or the induction of apoptosis (Yen et al. 2000; Horio et al. 2000). The transfer of other tumor suppressor genes, such as the retinoblastoma (*Rb*), *p16*, *pTEN* and *mda-7* genes, also effected suppression of tumor growth, (Demers et al. 1998; Jarrard et al. 1997; Lu et al. 1997).

3.2 Protooncogenes

Protooncogenes are activated by overexpression due to gene amplification, point mutations, modification of regulatory elements leading to increased transcription, and rearrangements. The products of these genes include growth factors or their receptors (EGF, EGFR), signal transduction proteins (ras, PI3 kinase), transcription factors (myc, fos, jun) (Isaacs et al. 1995; Konishi et al. 1997; Roylance et al. 1997), and suppressors of apoptosis (bcl-2). Gene therapy strategies to correct protooncogene activation include the transfer of dominant negative gene products, antisense oligodeoxynucleotides, ribozymes, and small interference RNAs. For example, disruption of overexpressed c-*myc* using anti-sense c-*myc* resulted in a 94.5% reduction of tumor size in prostate cancer (Steiner et al. 1998).

3.3 Prodrug/Suicide

The mechanism of prodrug/suicide gene therapy is based on the difference in mitotic activity between normal and cancer cells. The cells are transduced with the herpes simplex virus thymidine kinase (HSV-TK) gene, and are subsequently killed by treatment with the drug acyclovir. The transduced enzyme, HSV-TK, phosphorylates the prodrug into the antiviral compund gancyclovir triphosphate, an inhibitor of DNA synthesis, which leads to cell death (Ayala et al. 2000; Koeneman et al. 2000; Shalev et al. 2000; Martiniello-Wilks et al. 1998; Hall et al. 1999).

3.4 Apoptosis Induction

The triggering of programmed cell death in tumor cells without affecting normal cells is an attractive therapeutic approach in cancer gene therapy. Ligands that induce apoptosis include tumor necrosis factor (TNF)-α, FasL (Hedlund et al. 1998; Hedlund et al. 1999), and TRAIL (Griffith and Broghammer 2001; Voelkel-Johnson et al. 2002). As TRAIL has the lowest activity in normal tissue, it is an especially promising new therapeutic approach to cancer therapy.

3.5 Inhibition of Angiogenesis

Regulation of the angiogenic switch depends on the local balance between activators and inhibitors. Angiogenesis is triggered by the release of angiogenic stimulators by tumor cells, either as a result of genetic alterations or through activation of the physiologic response to hypoxia, which activates various inducible factors, e.g., transcription factors that trigger the transcription of genes encoding angiogenic stimulators. Many angiogenic factors have already been identified, including VEGF (Ferrara 1999), vascular permeability factor (VPF), FGF, EGF, platelet-derived endothelial growth factor (PD-EGF), PDGF, insulin-like growth factors (IGFs) (Trojan et al. 1994), interleukin-8 (IL-8), transforming growth factor α and β (TGF-α and -β) (Maggard et al. 2001), heparin growth factor, granulocyte colony stimulating factor (GMCSF), E-selectin (Tang et al. 2004), and TNF-α (Claesson-Welsh 2003; Brieger et al. 2003), and used for anti-angiogenic gene therapy of cancers. The goal of this type of therapy is to switch the balance between angiogenic factors and angiogenic inhibitors in the tumor microenvironment to the anti-angiogenic phenotype.

3.6 Increased Immunity

Gene therapy targeted to the immune system has a good likelihood of success as a cancer therapy. In this approach, an antitumor immune response in the host is created, either by immunotherapy, vaccination with cytokine genes, including IL-2 (Belldegrun et al. 2001; Kawakita et al. 1997; Moody et al. 1994; Toloza et al. 1996), IL-4, IL-7, IL-12 (Sanford et al. 2001; Hull et al. 2000; Nasu et al. 1999), GM-CSF (Simons et al. 1999), M-CSF, and interferons.

4 Conclusions

The therapeutic target and genes used for gene therapy in cancer are summarized in Table 2.

Table 2. Therapeutic target and genes used for gene therapy in cancer

Aim of Therapeutic Strategy	Target Genes
Induction of tumor suppressor genes	p53, Rb, p16, pTEN, mda-7
Suppression of Protooncogene	EGF, EGFR, ras, PI3 kinase, c-myc, c-fos, jun, Bcl-2
Prodrug/Suiside	HSV-tk (ganciclovir)
Apoptosis Induction	TNFa, Fas ligand
Induction of apoptosis	Bcl-2, c-myc, c-raf, MDM-2, IGF-II, STAT-3 p21, p53, Fas ligand, c-cam, Caspase7 TRAIL, Bax
Inhibition of angiogenesis	Suppression of VEGF, VPF, FGF, EGF, PD-EGF, IGFs IL-9, TGF-a and b, GM-CSF, E-selection, TNF-a
Immunotherapy	IL-2, IL-4, IL-7, IL-12, GM-CSF, Interferons

References

Achen MG, Jeltsch M, Kukk E, Makinen T, Vitali A, Wilks AF, Alitalo K, Stacker SA (1998) Vascular endothelial growth factor D (VEGF-D) is a ligand for the tyrosine kinases VEGF receptor 2 (Flk1) and VEGF receptor 3 (Flt4). Proc Natl Acad Sci USA 95: 548–553

Ayala G, Wheeler TM, Shalev M, Thompson TC, Miles B, Aguilar-Cordova E, Chakraborty S, Kadmon D (2000) Cytopathic effect of in situ gene therapy in prostate cancer. Hum Pathol 31:866–870

Belalcazar LM, Merched A, Carr B, Oka K, Chen KH, Pastore L, Beaudet A, Chan L (2003) Long-term stable expression of human apolipoprotein A-I mediated by helper-dependent adenovirus gene transfer inhibits atherosclerosis progression and remodels atherosclerotic plaques in a mouse model of familial hypercholesterolemia. Circulation 107:2726–2732

Belldegrun A, Tso CL, Zisman A, Naitoh J, Said J, Pantuck AJ, Hinkel A, deKernion J, Figlin R (2001) Interleukin 2 gene therapy for prostate cancer: phase I clinical trial and basic biology. Hum Gene Ther 12:883–892

Benoit P, Emmanuel F, Caillaud JM, Bassinet L, Castro G, Gallix P, Fruchart JC, Branellec D, Denefle P, Duverger N (1999) Somatic gene transfer of human ApoA-I inhibits atherosclerosis progression in mouse models. Circulation 99:105–110

Bookstein R, MacGrogan D, Hilsenbeck SG, Sharkey F, Allred DC (1993) p53 is mutated in a subset of advanced-stage prostate cancers. Cancer Res 53:3369–3373

Brieger J, Bedavanija A, Lehr HA, Maurer J, Mann WJ (2003) Expression of angiogenic growth factors in acoustic neurinoma. Acta Otolaryngol 123:1040–1045

Brousseau ME, Wang J, Demosky SJ, Jr., Vaisman BL, Talley GD, Santamarina-Fojo S, Brewer HB, Jr., Hoeg JM (1998) Correction of hypoalphalipoproteinemia in LDL receptor-deficient rabbits by lecithin:cholesterol acyltransferase. J Lipid Res 39:1558–1567

Burgess TL, Fisher EF, Ross SL, Bready JV, Qian YX, Bayewitch LA, Cohen AM, Herrera CJ, Hu SS, Kramer TB (1995) The antiproliferative activity of c-myb and c-myc antisense oligonucleotides in smooth muscle cells is caused by a nonantisense mechanism. Proc Natl Acad Sci U S A 92:4051–4055

Champion HC, Bivalacqua TJ, D'Souza FM, Ortiz LA, Jeter JR, Toyoda K, Heistad DD, Hyman AL, Kadowitz PJ (1999) Gene transfer of endothelial nitric oxide synthase to the lung of the mouse in vivo. Effect on agonist-induced and flow-mediated vascular responses. Circ Res 84:1422–1432

Champion HC, Bivalacqua TJ, Toyoda K, Heistad DD, Hyman AL, Kadowitz PJ (2000) In vivo gene transfer of prepro-calcitonin gene-related peptide to the lung attenuates chronic hypoxia-induced pulmonary hypertension in the mouse. Circulation 101: 923–930

Chang MW, Barr E, Lu MM, Barton K, Leiden JM (1995a) Adenovirus-mediated overexpression of the cyclin/cyclin-dependent kinase inhibitor, p21 inhibits vascular smooth muscle cell proliferation and neointima formation in the rat carotid artery model of balloon angioplasty. J Clin Invest 96:2260–2268

Chang MW, Barr E, Seltzer J, Jiang YQ, Nabel GJ, Nabel EG, Parmacek MS, Leiden JM (1995b) Cytostatic gene therapy for vascular proliferative disorders with a constitutively active form of the retinoblastoma gene product. Science 267:518–522

Christman BW, McPherson CD, Newman JH, King GA, Bernard GR, Groves BM, Loyd JE (1992) An imbalance between the excretion of thromboxane and prostacyclin metabolites in pulmonary hypertension. N Engl J Med 327:70–75

Cioffi L, Sturtz FG, Wittmer S, Barut B, Smith-Gbur J, Moore V, Zupancic T, Gilligan B, Auerbach R, Gomez F, Chauvin F, Antczak M, Platika D, Snodgrass HR (1999) A novel endothelial cell-based gene therapy platform for the in vivo delivery of apolipoprotein E. Gene Ther 6:1153–1159

Claesson-Welsh L (2003) Signal transduction by vascular endothelial growth factor receptors. Biochem Soc Trans 31:20–24

Demers GW, Harris MP, Wen SF, Engler H, Nielsen LL, Maneval DC (1998) A recombinant adenoviral vector expressing full-length human retinoblastoma susceptibility gene inhibits human tumor cell growth. Cancer Gene Ther 5:207–214

Dichek DA, Anderson J, Kelly AB, Hanson SR, Harker LA (1996) Enhanced in vivo antithrombotic effects of endothelial cells expressing recombinant plasminogen activators transduced with retroviral vectors. Circulation 93:301–309

Dvorak HF, Brown LF, Detmar M, Dvorak AM (1995) Vascular permeability factor/vascular endothelial growth factor, microvascular hyperpermeability, and angiogenesis. Am J Pathol 146:1029–1039

Edvinsson L, Ekman R, Jansen I, McCulloch J, Mortensen A, Uddman R (1991) Reduced levels of calcitonin gene-related peptide-like immunoreactivity in human brain vessels after subarachnoid haemorrhage. Neurosci Lett 121:151–154

Excoffon KJ, Liu G, Miao L, Wilson JE, McManus BM, Semenkovich CF, Coleman T, Benoit P, Duverger N, Branellec D, Denefle P, Hayden MR, Lewis ME (1997) Correction of hypertriglyceridemia and impaired fat tolerance in lipoprotein lipase-deficient mice by adenovirus-mediated expression of human lipoprotein lipase. Arterioscler Thromb Vasc Biol 17:2532–2539

Ferrara N (1999) Molecular and biological properties of vascular endothelial growth factor. J Mol Med 77:527–543

Ferrara N (2000) VEGF: an update on biological and therapeutic aspects. Curr Opin Biotechnol 11:617–624

Ferrara N (2001) Role of vascular endothelial growth factor in regulation of physiological angiogenesis. Am J Physiol Cell Physiol 280:C1358–C1366

Galzie Z, Kinsella AR, Smith JA (1997) Fibroblast growth factors and their receptors. Biochem Cell Biol 75:669–685

Goldstein JL HH, Brown MS (2001) Familial hypercholesterolemia. Vol chap. 120. McGraw-Hill, New York

Gough PJ, Raines EW (2003) Gene therapy of apolipoprotein E-deficient mice using a novel macrophage-specific retroviral vector. Blood 101:485–491

Greeve J, Jona VK, Chowdhury NR, Horwitz MS, Chowdhury JR (1996) Hepatic gene transfer of the catalytic subunit of the apolipoprotein B mRNA editing enzyme results in a reduction of plasma LDL levels in normal and watanabe heritable hyperlipidemic rabbits. J Lipid Res 37:2001–2017

Griffith TS, Broghammer EL (2001) Suppression of tumor growth following intralesional therapy with TRAIL recombinant adenovirus. Mol Ther 4:257–266

Grines C, Rubanyi GM, Kleiman NS, Marrott P, Watkins MW (2003) Angiogenic gene therapy with adenovirus 5 fibroblast growth factor-4 (Ad5FGF-4): a new option for the treatment of coronary artery disease. Am J Cardiol 92:24N–31N

Grossman M, Rader DJ, Muller DW, Kolansky DM, Kozarsky K, Clark BJ, 3rd, Stein EA, Lupien PJ, Brewer HB, Jr., Raper SE (1995) A pilot study of ex vivo gene therapy for homozygous familial hypercholesterolaemia. Nat Med 1:1148–1154

Grossman M, Raper SE, Kozarsky K, Stein EA, Engelhardt JF, Muller D, Lupien PJ, Wilson JM (1994) Successful ex vivo gene therapy directed to liver in a patient with familial hypercholesterolaemia. Nat Genet 6:335–341

Hall SJ, Mutchnik SE, Yang G, Timme TL, Nasu Y, Bangma CH, Woo SL, Shaker M, Thompson TC (1999) Cooperative therapeutic effects of androgen ablation and adenovirus-mediated herpes simplex virus thymidine kinase gene and ganciclovir therapy in experimental prostate cancer. Cancer Gene Ther 6:54–63

Harris JD, Schepelmann S, Athanasopoulos T, Graham IR, Stannard AK, Mohri Z, Hill V, Hassall DG, Owen JS, Dickson G (2002) Inhibition of atherosclerosis in apolipoprotein-E-deficient mice following muscle transduction with adeno-associated virus vectors encoding human apolipoprotein-E. Gene Ther 9:21–29

Hattori K, Dias S, Heissig B, Hackett NR, Lyden D, Tateno M, Hicklin DJ, Zhu Z, Witte L, Crystal RG, Moore MA, Rafii S (2001) Vascular endothelial growth factor and angiopoietin-1 stimulate postnatal hematopoiesis by recruitment of vasculogenic and hematopoietic stem cells. J Exp Med 193:1005–1014

Hayashi S, Morishita R, Nakamura S, Yamamoto K, Moriguchi A, Nagano T, Taiji M, Noguchi H, Matsumoto K, Nakamura T, Higaki J, Ogihara T (1999) Potential role of hepatocyte growth factor, a novel angiogenic growth factor, in peripheral arterial disease: down-regulation of HGF in response to hypoxia in vascular cells. Circulation 100:II301–II308

Hedlund TE, Duke RC, Schleicher MS, Miller GJ (1998) Fas-mediated apoptosis in seven human prostate cancer cell lines: correlation with tumor stage. Prostate 36:92–101

Hedlund TE, Meech SJ, Srikanth S, Kraft AS, Miller GJ, Schaack JB, Duke RC (1999) Adenovirus-mediated expression of Fas ligand induces apoptosis of human prostate cancer cells. Cell Death Differ 6:175–182

Hiltunen MO, Laitinen M, Turunen MP, Jeltsch M, Hartikainen J, Rissanen TT, Laukkanen J, Niemi M, Kossila M, Hakkinen TP, Kivela A, Enholm B, Mansukoski H, Turunen AM, Alitalo K, Yla-Herttuala S (2000) Intravascular adenovirus-mediated VEGF-C gene transfer reduces neointima formation in balloon-denuded rabbit aorta. Circulation 102: 2262–2268

Horio Y, Hasegawa Y, Sekido Y, Takahashi M, Roth JA, Shimokata K (2000) Synergistic effects of adenovirus expressing wild-type p53 on chemosensitivity of non-small cell lung cancer cells. Cancer Gene Ther 7:537–544

Hull GW, McCurdy MA, Nasu Y, Bangma CH, Yang G, Shimura S, Lee HM, Wang J, Albani J, Ebara S, Sato T, Timme TL, Thompson TC (2000) Prostate cancer gene therapy: comparison of adenovirus-mediated expression of interleukin 12 with interleukin 12 plus B7-1 for in situ gene therapy and gene-modified, cell-based vaccines. Clin Cancer Res 6:4101–4109

Indolfi C, Avvedimento EV, Rapacciuolo A, Di Lorenzo E, Esposito G, Stabile E, Feliciello A, Mele E, Giuliano P, Condorelli G (1995) Inhibition of cellular ras prevents smooth muscle cell proliferation after vascular injury in vivo. Nat Med 1:541–545

Isaacs WB, Bova GS, Morton RA, Bussemakers MJ, Brooks JD, Ewing CM (1995) Molecular biology of prostate cancer progression. Cancer Surv 23:19–32

Jalkanen J, Leppanen P, Narvanen O, Greaves DR, Yla-Herttuala S (2003a) Adenovirus-mediated gene transfer of a secreted decoy human macrophage scavenger receptor (SR-AI) in LDL receptor knock-out mice. Atherosclerosis 169:95–103

Jalkanen J, Leppanen P, Pajusola K, Narvanen O, Mahonen A, Vahakangas E, Greaves DR, Bueler H, Yla-Herttuala S (2003b) Adeno-associated virus-mediated gene transfer of a secreted decoy human macrophage scavenger receptor reduces atherosclerotic lesion formation in LDL receptor knockout mice. Mol Ther 8:903–910

Jarrard DF, Bova GS, Ewing CM, Pin SS, Nguyen SH, Baylin SB, Cairns P, Sidransky D, Herman JG, Isaacs WB (1997) Deletional, mutational, and methylation analyses of CDKN2 (p16/MTS1) in primary and metastatic prostate cancer. Genes Chromosomes Cancer 19:90–96

Javerzat S, Auguste P, Bikfalvi A (2002) The role of fibroblast growth factors in vascular development. Trends Mol Med 8:483–489

Joukov V, Pajusola K, Kaipainen A, Chilov D, Lahtinen I, Kukk E, Saksela O, Kalkkinen N, Alitalo K (1996) A novel vascular endothelial growth factor, VEGF-C, is a ligand for the Flt4 (VEGFR-3) and KDR (VEGFR-2) receptor tyrosine kinases. Embo J 15:290–298

Kamii H, Kato I, Kinouchi H, Chan PH, Epstein CJ, Akabane A, Okamoto H, Yoshimoto T (1999) Amelioration of vasospasm after subarachnoid hemorrhage in transgenic mice overexpressing CuZn-superoxide dismutase. Stroke 30:867–871; discussion 872

Kankkonen HM, Vahakangas E, Marr RA, Pakkanen T, Laurema A, Leppanen P, Jalkanen J, Verma IM, Yla-Herttuala S (2004) Long-term lowering of plasma cholesterol levels in LDL-receptor-deficient WHHL rabbits by gene therapy. Mol Ther 9:548–556

Kawakita M, Rao GS, Ritchey JK, Ornstein DK, Hudson MA, Tartaglia J, Paoletti E, Humphrey PA, Harmon TJ, Ratliff TL (1997) Effect of canarypox virus (ALVAC)-mediated cytokine expression on murine prostate tumor growth. J Natl Cancer Inst 89:428–436

Kiba A, Sagara H, Hara T, Shibuya M (2003) VEGFR-2-specific ligand VEGF-E induces non-edematous hyper-vascularization in mice. Biochem Biophys Res Commun 301:371–377

Koeneman KS, Kao C, Ko SC, Yang L, Wada Y, Kallmes DF, Gillenwater JY, Zhau HE, Chung LW, Gardner TA (2000) Osteocalcin-directed gene therapy for prostate-cancer bone metastasis. World J Urol 18:102–110

Koike T, Liang J, Wang X, Ichikawa T, Shiomi M, Liu G, Sun H, Kitajima S, Morimoto M, Watanabe T, Yamada N, Fan J (2004) Overexpression of lipoprotein lipase in transgenic Watanabe heritable hyperlipidemic rabbits improves hyperlipidemia and obesity. J Biol Chem 279:7521–7529

Konishi N, Cho M, Yamamoto K, Hiasa Y (1997) Genetic changes in prostate cancer. Pathol Int 47:735–747

Kozarsky KF, Bonen DK, Giannoni F, Funahashi T, Wilson JM, Davidson NO (1996) Hepatic expression of the catalytic subunit of the apolipoprotein B mRNA editing enzyme (apobec-1) ameliorates hypercholesterolemia in LDL receptor-deficient rabbits. Hum Gene Ther 7:943–957

Kozarsky KF, McKinley DR, Austin LL, Raper SE, Stratford-Perricaudet LD, Wilson JM (1994) In vivo correction of low density lipoprotein receptor deficiency in the Watanabe heritable hyperlipidemic rabbit with recombinant adenoviruses. J Biol Chem 269:13695–13702

Laitinen M, Zachary I, Breier G, Pakkanen T, Hakkinen T, Luoma J, Abedi H, Risau W, Soma M, Laakso M, Martin JF, Yla-Herttuala S (1997) VEGF gene transfer reduces intimal thickening via increased production of nitric oxide in carotid arteries. Hum Gene Ther 8:1737–1744

Lebherz C, Gao G, Louboutin JP, Millar J, Rader D, Wilson JM (2004) Gene therapy with novel adeno-associated virus vectors substantially diminishes atherosclerosis in a murine model of familial hypercholesterolemia. J Gene Med 6:663–672

Leung DW, Cachianes G, Kuang WJ, Goeddel DV, Ferrara N (1989) Vascular endothelial growth factor is a secreted angiogenic mitogen. Science 246:1306–1309

Levy AP, Levy NS, Wegner S, Goldberg MA (1995) Transcriptional regulation of the rat vascular endothelial growth factor gene by hypoxia. J Biol Chem 270:13333–13340

Liu G, Ashbourne Excoffon KJ, Wilson JE, McManus BM, Rogers QR, Miao L, Kastelein JJ, Lewis ME, Hayden MR (2000) Phenotypic correction of feline lipoprotein lipase deficiency by adenoviral gene transfer. Hum Gene Ther 11:21–32

Losordo DW, Vale PR, Symes JF, Dunnington CH, Esakof DD, Maysky M, Ashare AB, Lathi K, Isner JM (1998) Gene therapy for myocardial angiogenesis: initial clinical results with direct myocardial injection of phVEGF165 as sole therapy for myocardial ischemia. Circulation 98:2800–2804

Lu S, Tsai SY, Tsai MJ (1997) Regulation of androgen-dependent prostatic cancer cell growth: androgen regulation of CDK2, CDK4, and CKI p16 genes. Cancer Res 57: 4511–4516

Maggard M, Meng L, Ke B, Allen R, Devgan L, Imagawa DK (2001) Antisense TGF-beta2 immunotherapy for hepatocellular carcinoma: treatment in a rat tumor model. Ann Surg Oncol 8:32–37

Maglione D, Guerriero V, Viglietto G, Ferraro MG, Aprelikova O, Alitalo K, Del Vecchio S, Lei KJ, Chou JY, Persico MG (1993) Two alternative mRNAs coding for the angiogenic factor, placenta growth factor (PlGF), are transcribed from a single gene of chromosome 14. Oncogene 8:925–931

Martiniello-Wilks R, Garcia-Aragon J, Daja MM, Russell P, Both GW, Molloy PL, Lockett LJ, Russell PJ (1998) In vivo gene therapy for prostate cancer: preclinical evaluation of two different enzyme-directed prodrug therapy systems delivered by identical adenovirus vectors. Hum Gene Ther 9:1617–1626

McGirt MJ, Parra A, Sheng H, Higuchi Y, Oury TD, Laskowitz DT, Pearlstein RD, Warner DS (2002) Attenuation of cerebral vasospasm after subarachnoid hemorrhage in mice overexpressing extracellular superoxide dismutase. Stroke 33:2317–2323

Moody DB, Robinson JC, Ewing CM, Lazenby AJ, Isaacs WB (1994) Interleukin-2 transfected prostate cancer cells generate a local antitumor effect in vivo. Prostate 24:244–251

Morishige K, Shimokawa H, Yamawaki T, Miyata K, Eto Y, Kandabashi T, Yogo K, Higo T, Egashira K, Ueno H, Takeshita A (2000) Local adenovirus-mediated transfer of C-type natriuretic peptide suppresses vascular remodeling in porcine coronary arteries in vivo. J Am Coll Cardiol 35:1040–1047

Morishita R (2004) Perspective in progress of cardiovascular gene therapy. J Pharmacol Sci 95:1–8

Morishita R, Gibbons GH, Ellison KE, Nakajima M, Zhang L, Kaneda Y, Ogihara T, Dzau VJ (1993) Single intraluminal delivery of antisense cdc2 kinase and proliferating-cell nuclear antigen oligonucleotides results in chronic inhibition of neointimal hyperplasia. Proc Natl Acad Sci U S A 90:8474–8478

Morishita R, Nakamura S, Hayashi S, Taniyama Y, Moriguchi A, Nagano T, Taiji M, Noguchi H, Takeshita S, Matsumoto K, Nakamura T, Higaki J, Ogihara T (1999) Therapeutic angiogenesis induced by human recombinant hepatocyte growth factor in rabbit hind limb ischemia model as cytokine supplement therapy. Hypertension 33:1379–1384

Morishita R, Sakaki M, Yamamoto K, Iguchi S, Aoki M, Yamasaki K, Matsumoto K, Nakamura T, Lawn R, Ogihara T, Kaneda Y (2002) Impairment of collateral formation in lipoprotein(a) transgenic mice: therapeutic angiogenesis induced by human hepatocyte growth factor gene. Circulation 105:1491–1496

Morishita R, Sugimoto T, Aoki M, Kida I, Tomita N, Moriguchi A, Maeda K, Sawa Y, Kaneda Y, Higaki J, Ogihara T (1997) In vivo transfection of cis element "decoy" against nuclear factor-kappaB binding site prevents myocardial infarction. Nat Med 3:894–899

Nagaya N (2004) Drug therapy of primary pulmonary hypertension. Am J Cardiovasc Drugs 4:75–85

Nagaya N, Yokoyama C, Kyotani S, Shimonishi M, Morishita R, Uematsu M, Nishikimi T, Nakanishi N, Ogihara T, Yamagishi M, Miyatake K, Kaneda Y, Tanabe T (2000) Gene transfer of human prostacyclin synthase ameliorates monocrotaline-induced pulmonary hypertension in rats. Circulation 102:2005–2010

Nasu Y, Bangma CH, Hull GW, Lee HM, Hu J, Wang J, McCurdy MA, Shimura S, Yang G, Timme TL, Thompson TC (1999) Adenovirus-mediated interleukin-12 gene therapy for prostate cancer: suppression of orthotopic tumor growth and pre-established lung metastases in an orthotopic model. Gene Ther 6:338–349

Nishida T, Ueno H, Atsuchi N, Kawano R, Asada Y, Nakahara Y, Kamikubo Y, Takeshita A, Yasui H (1999) Adenovirus-mediated local expression of human tissue factor pathway inhibitor eliminates shear stress-induced recurrent thrombosis in the injured carotid artery of the rabbit. Circ Res 84:1446–1452

Nozaki K, Kikuchi H, Mizuno N (1989) Changes of calcitonin gene-related peptide-like immunoreactivity in cerebrovascular nerve fibers in the dog after experimentally produced subarachnoid hemorrhage. Neurosci Lett 102:27–32

Ogawa S, Oku A, Sawano A, Yamaguchi S, Yazaki Y, Shibuya M (1998) A novel type of vascular endothelial growth factor, VEGF-E (NZ-7 VEGF), preferentially utilizes KDR/Flk-1 receptor and carries a potent mitotic activity without heparin-binding domain. J Biol Chem 273:31273–31282

Okamoto Y, Kihara S, Ouchi N, Nishida M, Arita Y, Kumada M, Ohashi K, Sakai N, Shimomura I, Kobayashi H, Terasaka N, Inaba T, Funahashi T, Matsuzawa Y (2002) Adiponectin reduces atherosclerosis in apolipoprotein E-deficient mice. Circulation 106:2767–2770

Olofsson B, Pajusola K, Kaipainen A, von Euler G, Joukov V, Saksela O, Orpana A, Pettersson RF, Alitalo K, Eriksson U (1996) Vascular endothelial growth factor B, a novel growth factor for endothelial cells. Proc Natl Acad Sci U S A 93:2576–2581

Onoda K, Ono S, Ogihara K, Shiota T, Asari S, Ohmoto T, Ninomiya Y (1996) Inhibition of vascular contraction by intracisternal administration of preproendothelin-1 mRNA antisense oligoDNA in a rat experimental vasospasm model. J Neurosurg 85:846–852

Onoue H, Tsutsui M, Smith L, Stelter A, O'Brien T, Katusic ZS (1998) Expression and function of recombinant endothelial nitric oxide synthase gene in canine basilar artery after experimental subarachnoid hemorrhage. Stroke 29:1959–1965; discussion 1965–1966

Ornitz DM, Xu J, Colvin JS, McEwen DG, MacArthur CA, Coulier F, Gao G, Goldfarb M (1996) Receptor specificity of the fibroblast growth factor family. J Biol Chem 271: 15292–15297

Pakkanen TM, Laitinen M, Hippelainen M, Kallionpaa H, Lehtolainen P, Leppanen P, Luoma JS, Tarvainen R, Alhava E, Yla-Herttuala S (1999) Enhanced plasma cholesterol lowering effect of retrovirus-mediated LDL receptor gene transfer to WHHL rabbit liver after improved surgical technique and stimulation of hepatocyte proliferation by combined partial liver resection and thymidine kinase–ganciclovir treatment. Gene Ther 6:34–41

Pollman MJ, Hall JL, Mann MJ, Zhang L, Gibbons GH (1998) Inhibition of neointimal cell bcl-x expression induces apoptosis and regression of vascular disease. Nat Med 4:222–227

Rade JJ, Schulick AH, Virmani R, Dichek DA (1996) Local adenoviral-mediated expression of recombinant hirudin reduces neointima formation after arterial injury. Nat Med 2:293–298

Rissanen TT, Markkanen JE, Gruchala M, Heikura T, Puranen A, Kettunen MI, Kholova I, Kauppinen RA, Achen MG, Stacker SA, Alitalo K, Yla-Herttuala S (2003) VEGF-D is the strongest angiogenic and lymphangiogenic effector among VEGFs delivered into skeletal muscle via adenoviruses. Circ Res 92:1098–1106

Roylance R, Spurr N, Sheer D (1997) The genetic analysis of prostate carcinoma. Semin Cancer Biol 8:37–44

Sanford MA, Yan Y, Canfield SE, Hassan W, Selleck WA, Atkinson G, Chen SH, Hall SJ (2001) Independent contributions of GR-1+ leukocytes and Fas/FasL interactions to induce apoptosis following interleukin-12 gene therapy in a metastatic model of prostate cancer. Hum Gene Ther 12:1485–1498

Seguret-Mace S, Latta-Mahieu M, Castro G, Luc G, Fruchart JC, Rubin E, Denefle P, Duverger N (1996) Potential gene therapy for lecithin-cholesterol acyltransferase (LCAT)-deficient and hypoalphalipoproteinemic patients with adenovirus-mediated transfer of human LCAT gene. Circulation 94:2177–2184

Shalev M, Kadmon D, Teh BS, Butler EB, Aguilar-Cordova E, Thompson TC, Herman JR, Adler HL, Scardino PT, Miles BJ (2000) Suicide gene therapy toxicity after multiple and repeat injections in patients with localized prostate cancer. J Urol 163:1747–1750

Siddiqui AJ, Blomberg P, Wardell E, Hellgren I, Eskandarpour M, Islam KB, Sylven C (2003) Combination of angiopoietin-1 and vascular endothelial growth factor gene therapy enhances arteriogenesis in the ischemic myocardium. Biochem Biophys Res Commun 310:1002–1009

Simons JW, Mikhak B, Chang JF, DeMarzo AM, Carducci MA, Lim M, Weber CE, Baccala AA, Goemann MA, Clift SM, Ando DG, Levitsky HI, Cohen LK, Sanda MG, Mulligan RC, Partin AW, Carter HB, Piantadosi S, Marshall FF, Nelson WG (1999) Induction of immunity to prostate cancer antigens: results of a clinical trial of vaccination with irradiated autologous prostate tumor cells engineered to secrete granulocyte-macrophage colony-stimulating factor using ex vivo gene transfer. Cancer Res 59:5160–5168

Smith RC, Branellec D, Gorski DH, Guo K, Perlman H, Dedieu JF, Pastore C, Mahfoudi A, Denefle P, Isner JM, Walsh K (1997) p21CIP1-mediated inhibition of cell proliferation by overexpression of the gax homeodomain gene. Genes Dev 11:1674–1689

Steiner MS, Anthony CT, Lu Y, Holt JT (1998) Antisense c-myc retroviral vector suppresses established human prostate cancer. Hum Gene Ther 9:747–755

Suzuki J, Isobe M, Morishita R, Aoki M, Horie S, Okubo Y, Kaneda Y, Sawa Y, Matsuda H, Ogihara T, Sekiguchi M (1997) Prevention of graft coronary arteriosclerosis by antisense cdk2 kinase oligonucleotide. Nat Med 3:900–903

Tang NH, Chen YL, Wang XQ, Li XJ, Yin FZ, Wang XZ (2004) Cooperative inhibitory effects of antisense oligonucleotide of cell adhesion molecules and cimetidine on cancer cell adhesion. World J Gastroenterol 10:62–66

Tanner FC, Yang ZY, Duckers E, Gordon D, Nabel GJ, Nabel EG (1998) Expression of cyclin-dependent kinase inhibitors in vascular disease. Circ Res 82:396–403

Toloza EM, Hunt K, Swisher S, McBride W, Lau R, Pang S, Rhoades K, Drake T, Belldegrun A, Glaspy J, Economou JS (1996) In vivo cancer gene therapy with a recombinant interleukin-2 adenovirus vector. Cancer Gene Ther 3:11–17

Topol EJ, Serruys PW (1998) Frontiers in interventional cardiology. Circulation 98:1802–1820

Toyoda K, Chu Y, Heistad DD (2003) Gene therapy for cerebral vascular disease: update 2003. Br J Pharmacol 139:1–9

Trojan J, Johnson TR, Rudin SD, Blossey BK, Kelley KM, Shevelev A, Abdul-Karim FW, Anthony DD, Tykocinski ML, Ilan J (1994) Gene therapy of murine teratocarcinoma: separate functions for insulin-like growth factors I and II in immunogenicity and differentiation. Proc Natl Acad Sci U S A 91:6088–6092

Tuder RM, Cool CD, Geraci MW, Wang J, Abman SH, Wright L, Badesch D, Voelkel NF (1999) Prostacyclin synthase expression is decreased in lungs from patients with severe pulmonary hypertension. Am J Respir Crit Care Med 159:1925–1932

Ueno H, Haruno A, Morisaki N, Furuya M, Kangawa K, Takeshita A, Saito Y (1997a) Local expression of C-type natriuretic peptide markedly suppresses neointimal formation in rat injured arteries through an autocrine/paracrine loop. Circulation 96:2272–2279

Ueno H, Li JJ, Masuda S, Qi Z, Yamamoto H, Takeshita A (1997b) Adenovirus-mediated expression of the secreted form of basic fibroblast growth factor (FGF-2) induces cellular proliferation and angiogenesis in vivo. Arterioscler Thromb Vasc Biol 17:2453–2460

Vale PR, Losordo DW, Milliken CE, Maysky M, Esakof DD, Symes JF, Isner JM (2000) Left ventricular electromechanical mapping to assess efficacy of phVEGF(165) gene transfer for therapeutic angiogenesis in chronic myocardial ischemia. Circulation 102:965–974

Voelkel-Johnson C, King DL, Norris JS (2002) Resistance of prostate cancer cells to soluble TNF-related apoptosis-inducing ligand (TRAIL/Apo2L) can be overcome by doxorubicin or adenoviral delivery of full-length TRAIL. Cancer Gene Ther 9:164–172 von der Leyen HE, Gibbons GH, Morishita R, Lewis NP, Zhang L, Nakajima M, Kaneda Y, Cooke JP, Dzau VJ (1995) Gene therapy inhibiting neointimal vascular lesion: in vivo transfer of endothelial cell nitric oxide synthase gene. Proc Natl Acad Sci USA 92: 1137–1141

Ward NL, Dumont DJ (2002) The angiopoietins and Tie2/Tek: adding to the complexity of cardiovascular development. Semin Cell Dev Biol 13:19–27

Waugh JM, Yuksel E, Li J, Kuo MD, Kattash M, Saxena R, Geske R, Thung SN, Shenaq SM, Woo SL (1999) Local overexpression of thrombomodulin for in vivo prevention of arterial thrombosis in a rabbit model. Circ Res 84:84–92

Yamauchi A, Ito Y, Morikawa M, Kobune M, Huang J, Sasaki K, Takahashi K, Nakamura K, Dehari H, Niitsu Y, Abe T, Hamada H (2003) Pre-administration of angiopoietin-1 followed by VEGF induces functional and mature vascular formation in a rabbit ischemic model. J Gene Med 5:994–1004

Yan ZQ, Yokota T, Zhang W, Hansson GK (1996) Expression of inducible nitric oxide synthase inhibits platelet adhesion and restores blood flow in the injured artery. Circ Res 79:38–44

Yen N, Ioannides CG, Xu K, Swisher SG, Lawrence DD, Kemp BL, El-Naggar AK, et al. (2000) Cellular and humoral immune responses to adenovirus and p53 protein antigens in patients following intratumoral injection of an adenovirus vector expressing wild-type. P53 (Ad-p53). Cancer Gene Ther 7:530–536

Yonemitsu Y, Kaneda Y, Tanaka S, Nakashima Y, Komori K, Sugimachi K, Sueishi K (1998) Transfer of wild-type p53 gene effectively inhibits vascular smooth muscle cell proliferation in vitro and in vivo. Circ Res 82:147–156

Yoon YS, Murayama T, Gravereaux E, Tkebuchava T, Silver M, Curry C, Wecker A, Kirchmair R, Hu CS, Kearney M, Ashare A, Jackson DG, Kubo H, Isner JM, Losordo DW (2003) VEGF-C gene therapy augments postnatal lymphangiogenesis and ameliorates secondary lymphedema. J Clin Invest 111:717–725

Zoldhelyi P, Chen ZQ, Shelat HS, McNatt JM, Willerson JT (2001) Local gene transfer of tissue factor pathway inhibitor regulates intimal hyperplasia in atherosclerotic arteries. Proc Natl Acad Sci USA 98:4078–4083

Clinical Trials Using Non-viral Gene Delivery Systems

Hironobu Yanagie

1 Introduction

Many clinical trials of gene and antisense therapies are currently underway. The percentage of gene therapy clinical trials for cancer diseases, monogenetic diseases, vascular diseases, infectious diseases, other diseases, and gene marking in healthy volunteers is 66% ($n = 656$), 9.4% ($n = 93$), 8.1% ($n = 80$), 6.6% ($n = 65$), 2.9% ($n = 29$), and 5.3% ($n = 52$), 1.2% ($n = 12$), respectively (Gene Therapy Clinical Trials Worldwide 2004). Approximately 38% of all cancer gene therapy trials use various forms of immune-modulatory agents, e.g., cytokines (24%, $n = 241$) and antigens (14%, $n = 136$), administered either systemically or locally. Twelve percent of the trials use tumor suppressor, and 7.5% suicide gene therapy. Further approaches, such as the insertion of multidrug resistance genes in stem cells, represent ~5.7% ($n = 56$). The vast majority of gene therapy clinical trials are in phase I and only a very few have progressed to phase II. The phases of gene therapy clinical trials are as follows: phase I, 63%; phase I/II, 21%; phase II, 13%; phase II/III, 1.1%; and phase III, 1.7%. The majority of these studies have demonstrated that gene therapy is generally feasible using either viral or non-viral strategies. Vectors used in gene therapy clinical trials include retrovirus, adenovirus, naked/plasmid DNA, lipofection, herpes simplex virus, adeno-associated virus, accounting for 27% ($n = 263$), 26% ($n = 258$), 15% ($n = 150$), 8.6% ($n = 85$), 3% ($n = 30$), and 2.5% ($n = 25$), respectively. Cationic polymers have only been used in animal models and have not advanced into clinical trials due to various problems. While cancer gene therapy has not fulfilled the high expectations, remarkable efforts have been made to optimize gene delivery systems in vitro and in vivo by variously modifying the features of the vectors. However, additional research to elucidate the factors influencing opsonization, reticuloendothelial system (RES) uptake, and other forms of elimination in vivo is necessary.

Department of Intellectual Property, Incubation Project: Cancer Metastasis Inhibition, Research Center for Advanced Science and Technology, The University of Tokyo, 4-6-1 Komaba, Meguro-ku, Tokyo 153-8904, Japan

Delivery at the subcellular level remains inefficient as well. While endocytotic uptake of polyplexes occurs rather efficiently, release from the endosome is still one of the primary reasons for the low efficiency of non-viral gene and oligonucleotide therapies. For polyplexes, there is some evidence that the majority of the nucleic acid remains entrapped in such vesicles. Although some success has been obtained with fusogenic peptides or other endosomolytic agents, such as chloroquine, the applicability of these structures in vivo is limited. Subcellular trafficking at this stage probably bears great potential for further advances in non-viral vectors. However, it should be noted that the release mechanism, as far as is known to date, leads to the release of endosomal or lysosomal contents into the cytosol.

Recent advanced clinical trials of non-viral gene therapy are listed in Tables 1 and 2 (Gene Therapy Clinical Trials Worldwide, updated July 1 2004).

2 Clinical Trials Using Cationic Lipid Gene Transfer

Since the 1980s, it has been frequently reported that liposomes bearing positive charges on their surface afford a more efficient delivery of their entrapped components into the cells, compared to other types of liposomes. In 1987, Felgner et al. demonstrated that cationic liposomes consisting of *N*-[1,(2,3-doileyloxy)propyl]-*N,N,N*-tri-methylammonium chloride (DOTMA) and dioleoyl phosphatidylethanolamine (DOPE) provide a highly efficient and convenient means to deliver nucleic acids and proteins into various cell types, and offer several advantages over in vivo gene transfection (Felgner et al. 1987). Cationic liposomes have a high potential to effectively deliver large molecules, such as DNA, RNA, and proteins, and are expected to be one of the most promising vectors for human gene therapy (Felgner 1989). However, they also have significant toxicity when administered in large quantities to cells. Since toxicity varies with the cell type, duration of exposure, and density of the cell culture, cationic liposomes should be used within optimal concentrations.

In the following, several examples of gene therapy using non-viral vector systems are discussed.

2.1.1 Allovectin-7

Allovectin-7 is a gene transfer product consisting of the gene encoding the allogenic MHC class I protein human leukocyte antigen (HLA)-B7 heavy chain and β2-microglobulin genes inserted into a simplified eukaryotic expression vector (pBR322) and complexed with a cationic lipid mixture 1,2-dimyristyloxypropyl-3-dimethyl-hydroxyethyl ammonium bromide/dioleoyl-phosphatidyl-ethanolamine (DMRIE/DOPE). It is being developed as an immunotherapy approach for a variety of malignancies, with special focus on melanoma, and head and neck cancers (Nabel 1993a; Clark 2000; Stopeck 1997). In 1993, Stopeck et al.evaluated the responses and survival of patients with metastatic melanoma treated by direct intratumoral injection with Allovectin-7 (Stopeck 1993).

Expression of the HLA-B7 and β2-microglobulin genes in Allovectin-7 is driven by the Rous sarcoma virus (RSV) long terminal repeat promoter. The two genes are separated by a cap-independent translational enhancer, an internal ribosomal entry site that permits coexpression of both genes from a single promoter. The cationic lipid mixture DMRIE/DOPE aids in the uptake of DNA by the tumor. HLA-B7 antigen is infrequently expressed in the United States population (about 20% of the Caucasian population) and thus allows for a potential allogenic immune response in most patients. Loss of or mutations in β2-microglobulin have also been reported as possible mechanisms for deficient MHC class I expression in tumor cells (D'Urso 1991). Thus, the β2-microglobulin gene was also included in Allovectin-7 to allow expression of the complete MHC class I complex on the tumor cell surface. Allovectin-7 therefore provides several potential immune-stimulating functions: expression of a foreign and highly immunogenic cell surface protein (i.e., HLA-B7 in patients negative for this antigen), an increased antigen-presentation signal (antigen presented in the context of HLA-B7), and replacement of β2-microglobulin for increased surface expression of the patient's own MHC-class I molecules. The lack of effective therapies for patients with advanced melanoma and the toxicities associated with standard chemotherapeutic regimens and nonspecific immunotherapies make targeted therapy using DNA-based gene transfer an attractive alternative approach. Immune-based gene therapies make use of the host immune response to recognize and destroy cancerous cells specifically. In this scenario, non-transfected tumor cells are subject to immune attack in response to locally high concentrations of immunostimulating cytokines or by virtue of shared tumor antigens recognized by the immune effector cells (Bocchia 2000).

Several functional mechanisms have been proposed for Allovectin-7(Stopeck 2001): (1) Allo-immune responses against injected tumor cells expressing the allo-antigen HLA-B7 may explain the responses in HLA-B7-negative patients. Alternatively, reintroducing HLA-B7 into patients whose tumors have lost the ability to present antigens because of deficient MHC class I expression may lead to responses in HLA-B7-positive patients. The higher rate of regression of injected tumors observed in HLA-B7-negative patients [one complete response (CR), three partial responses (PRs), and four minor responses (MRs)] than in HLA-B7-positive patients (a single MR) suggests that the generation of an allo-immune response is the predominant mechanism accounting for the observed clinical responses. (2) A second possible mechanism involves deficient MHC class I expression in tumor cells because of deletion of or mutation of the β2-microgloblin gene. Reintroducing β2-microgloblin may increase the surface expression of the patient's own MHC class I molecules and restore antigen presentation. (3) Unmethylated bacterial CpG motifs in the backbone of the Allovectin-7 plasmid may be responsible for inducing an immune response with antitumor activity. CpG motifs in bacterial DNA have been reported to trigger direct B-cell activation by inducing B-cells to proliferate and secrete immunoglobulin (Krieg 1995). Bacterial CpG motifs have been shown to promote IgG, interleukin (IL)-12, IL-18, and interferon (IFN)-α, IFN-β, and IFN-γ production, all of which foster a Th1 response and enhance cell-mediated immunity in gene-vaccinated animals (Roman 1997).

Fifty-two patients with metastatic melanoma were enrolled in a phase II study of Allovectin-7. A dose of 10 μg was chosen because responses had been seen at this dose level in prior phase I studies, although a clear dose response was not observed. Therapy consisted of six intratumoral injections of 10 μg of Allovectin-7 over a 9-week period with a follow-up evaluation at week 11. Treatment was well tolerated. Treatment-related adverse events were mild to moderate, with the most frequent side

TABLE 1

Traial ID	Indication	Phaze	Gene	Vector	Gene delivery	Administration route
US-234	Melanoma	III	HLA-B7/Beta 2-Microglobulin	Lipofection	In vivo	Intratumoral
CA-004	Melanoma	II	HLA-B7/Beta 2-Microglobulin	Lipofection	In vivo	Intratumoral
CA-005	Melanoma, renal cell cancer, non-Hodgkin's lymphoma	II	HLA-B7/Beta 2-Microglobulin	Lipofection	In vivo	Intratumoral
US-438	Squamous Cell Carcinoma of the Head and Neck	II	E1A	Lipofection		Intratumoral
US-432	Squamous Cell Carcinoma of the Head and Neck	II	HLA-B7/Beta 2-Microglobulin	Lipofection		Intratumoral
US-431	Melanoma	II	HLA-B7/Beta 2-Microglobulin	Lipofection		Intratumoral
US-416	Ovarian cancer	II	E1A	Lipofection		Intraperitoneal
US-415	Ovarian cancer	II	E1A	Lipofection		Intraperitoneal
US-391	Renal Cell Carcinoma	II	Interleukin-2 (IL-2)	Lipofection	In vivo	Intradermal
US-323	Squamous Cell Carcinoma of the Head and Neck	II	Interleukin-2 (IL-2)	Lipofection	In vivo	Intratumoral
US-312	Prostate Cancer	II	Interleukin-2 (IL-2)	Lipofection	In vivo	Intratumoral
US-270	Squamous Cell Carcinoma of the Head and Neck	II	HLA-B7/Beta 2-Microglobulin	Lipofection	In vivo	Intratumoral
US-259	Renal cell cancer	II	Interleukin-2 (IL-2)	Lipofection	In vivo	Intratumoral
US-246	Squamous Cell Carcinoma of the Head and Neck	II	HER-2	Lipofection	In vivo	Intratumoral
US-233	Melanoma	II	HLA-B7/Beta 2-Microglobulin	Lipofection	In vivo	Intratumoral
US-191	Squamous Cell Carcinoma of the Head and Neck	II	HLA-B7/Beta 2-Microglobulin	Lipofection	In vivo	Intratumoral
US-168	Melanoma	II	HLA-B7/Beta 2-Microglobulin	Lipofection	In vivo	Intratumoral
US-142	Head and Neck Cancer	II	HLA-B7/Beta 2-Microglobulin	Lipofection	In vivo	Intratumoral
US-121	Renal cell cancer	II	HLA-B7/Beta 2-Microglobulin	Lipofection	In vivo	Intratumoral
US-072	Melanoma, renal cell cancer, colorectal cancer, breast cancer, lymphoma	II	HLA-B7/Beta 2-Microglobulin	Lipofection	In vivo	Intratumoral

US-064	Colorectal carcinoma, renal cell carcinoma, melanoma, breast cancer, lymphoma	II	HLA-B7/Beta 2-Microglobulin	Lipofection	In vivo	Intratumoral
US-519	Pancreas cancer	II	Granulocyte-macrophage colony stimulating factor (GM-CSF)	Naked/Plasmid DNA	In vitro	Intradermal
US-475	Adenocarcinoma of the pancreas	II	Granulocyte-macrophage colony stimulating factor (GM-CSF)	Naked/Plasmid DNA	In vitro	Intradermal
US-393	Non-Small Cell Lung Cancer	II	Transforming Growth Factor-α (TGF-α)	Naked/Plasmid DNA	In vitro	Subcutaneous
US-378	Squamous cell carcinoma of the head and neck	II	Interferon-gamma (IFN-g)/ Interleukin-2 (IL-2)/ Polyvinylpyrrolidone	Naked/Plasmid DNA	In vivo	Intratumoral
US-351	Squamous cell carcinoma of the head and neck	II	Interferon-gamma (IFN-g)/ Interleukin-2 (IL-2)/ Polyvinylpyrrolidone	Naked/Plasmid DNA	In vivo	Intratumoral
US-254	Melanoma	II	Melanoma antigen gp100	Naked/Plasmid DNA	In vivo	Intradermal and intramuscular
UK-071	Ano-Genital neoplasia III	II	Human papilloma virus (HPV) E6 and E7	Naked/Plasmid DNA		
DE-033	Pancreatic carcinoma	II	Cytochrome p450	Naked/Plasmid DNA	In vitro	
DE-032	Mesothelioma	II	Interleukin-2 (IL-2)	Naked/Plasmid DNA	In vitro	
EG-001	Post-hepatitis liver cancer	II	p53	Naked/Plasmid DNA	In vivo	Intratumoral
US-568	Intermittent Claudication Secondary to Peripheral Arterial Disease	II	Poloxamer 188/Del-1	Lipofection	In vivo	Intramuscular
US-645	Peripheral Artery Disease	II	Fibroblast growth factor (FGF)	Naked/Plasmid DNA	In vivo	Intramuscular
US-567	Ischemic Myocardium	II	Vascular endothelial growth factor (VEGF)	Naked/Plasmid DNA	In vivo	Intramyocardial
US-502	Severe Peripheral Artery Occlusive Disease	II	Fibroblast growth factor (FGF)	Naked/Plasmid DNA		Intramuscular

Table 1. *Continued*

Traial ID	Indication	Phaze	Gene	Vector	Gene delivery	Administration route
US-384	Coronary Artery Disease	II	Vascular endothelial growth factor (VEGF)	Naked/Plasmid DNA	In vivo	Intramyocardial
US-383	Coronary Artery Disease	II	Vascular endothelial growth factor (VEGF)	Naked/Plasmid DNA	In vivo	Intramyocardial
US-354	Coronary Artery Disease	II	Vascular endothelial growth factor (VEGF)	Naked/Plasmid DNA	In vivo	Intramyocardial
US-316	Coronary Artery Disease	II	Vascular endothelial growth factor (VEGF)	Naked/Plasmid DNA	In vivo	Intramyocardial
US-294	Coronary Artery Disease	II	Vascular endothelial growth factor (VEGF)	Naked/Plasmid DNA	In vivo	Intramyocardial
UK-069	Peripheral artery occlusive disease	II	Fibroblast growth factor (FGF)	Naked/Plasmid DNA		
CH-028	Severe Peripheral Artery Occlusive Disease(PAOD)	II	Fibroblast growth factor (FGF)	Naked/Plasmid DNA		Intramuscular
XX-002	Refractory angina pectoris	II	Vascular endothelial growth factor (VEGF)	Naked/Plasmid DNA	In vivo	Intramyocardial
DE-050	Peripheral artery occlusive disease	II	Fibroblast growth factor (FGF)	Naked/Plasmid DNA	In vivo	
US-568	Intermittent Claudication Secondary to Peripheral Arterial Disease	II	Poloxamer 188/Del-1	Lipofection	In vivo	Intramuscular
DE-011	glioblastoma	I/II	Herpes simplex virus thymidine kinase (HSV-TK)	Lipofection	In vivo	Intratumoral
DE-002	Melanoma	I/II	Interleukin-2 (IL-2)	Lipofection	In vivo	Intradermal
DE-003	Renal cell cancer	I/II	Interleukin-2 (IL-2)	Lipofection	In vivo	Intradermal
DE-004	Melanoma	I/II	Interleukin-2 (IL-2)	Lipofection	In vivo	Intradermal
SG-002	Ovarian and cervical cancer with cutaneous metastases	I/II	HLA-A2 or -B13 or H-2K	Lipofection	In vivo	Intratumoral
US-582	Ovarian cancer	I/II	DC-Chol-DOPE /E1A	Lipofection	In vivo	Intraperitoneal
US-466	Non-Small Cell Lung Cancer	I/II	Manganese Superoxide Dismutase (MnSOD)	Lipofection	In vivo	Intraesophageal

US-357	Cancer Immunotherapy	I/II	Interleukin-2 (IL-2)	Lipofection	In vivo	Intratumoral
US-352	Prostate cancer	I/II	Interleukin-2 (IL-2)	Lipofection	In vivo	Intratumoral
US-210	Melanoma	I/II	HLA-B7/Beta 2-Microglobulin	Lipofection	In vivo	Intratumoral
US-169	Melanoma, renal cell carcinoma, sarcoma	I/II	Interleukin-2 (IL-2)	Lipofection	In vivo	Intratumoral
US-108	Renal cell cancer, melanoma	I/II	HLA-B7/Beta 2-Microglobulin	Lipofection	In vivo	Subcutaneous
US-071	Metastatic renal cell carcinoma	I/II	HLA-B7/Beta 2-Microglobulin	Lipofection	In vivo	Intratumoral
UK-015	Cystic fibrosis	I/II	Cystic fibrosis transmembrane conductance regulator (CFTR)	Lipofection	In vivo	Intranasal
UK-009	Cystic fibrosis	I/II	Cystic fibrosis transmembrane conductance regulator (CFTR)	Lipofection	In vivo	Intranasal
UK-008	Cystic fibrosis	I/II	Cystic fibrosis transmembrane conductance regulator (CFTR)	Lipofection	In vivo	Intranasal
UK-003	Cystic fibrosis	I/II	Cystic fibrosis transmembrane conductance regulator (CFTR)	Lipofection	In vivo	Intranasal
UK-084	Tetanus	I/II	HIV-1 Env	Naked/Plasmid DNA		
CH-004	HIV infection	I/II	HIV-1 Env/Rev	Naked/Plasmid DNA		
US-595	Cervical cancer	I/II	Human papilloma virus (HPV) 16 E7	Naked/Plasmid DNA	In vivo	Intramuscular
US-545	Prostate cancer	I/II	Immunoglobulin Heavy (H) Chain Telomerase Reverse Transcriptase (hTERT)	Naked/Plasmid DNA	In vivo	Intravenous
US-541	Sage IV Breast Carcinoma	I/II	NY-ESO-1	Naked/Plasmid DNA	In vivo	Intralymphnodal
US-490	Melanoma	I/II	Melanoma antigen Melan-A Melanoma antigen MART-1	Naked/Plasmid DNA	In vivo	Intralymphnodal
US-479	Acute Myelogenous Leukemia	I/II	Granulocyte-macrophage colony stimulating factor (GM-CSF)	Naked/Plasmid DNA	In vivo	Intradermal
US-472	Non-Small Cell Lung Cancer	I/II	Granulocyte-macrophage colony stimulating factor (GM-CSF)	Naked/Plasmid DNA		Intradermal
US-461	Melanoma	I/II	Interferon-alpha (IFN-a) Interleukin-12 (IL-12)	Naked/Plasmid DNA		Intramuscular

Table 1. *Continued*

Traial ID	Indication	Phaze	Gene	Vector	Gene delivery	Administration route
US-435	Leukemia, Multiple Myeloma	I/II	Granulocyte-macrophage colony stimulating factor (GM-CSF)	Naked/Plasmid DNA		Intradermal
US-408	B-Cell Chronic Lymphocytic Leukemia	I/II	Tumor idiotype	Naked/Plasmid DNA	In vivo	Intramuscular
US-392	Non-Hodgkin's B-Cell Lymphoma, Mantle Cell Lymphoma	I/II	Tumor idiotype Granulocyte-Macrophage Colony Stimulating Factor (GM-CSF)	Naked/Plasmid DNA	In vivo	Intradermal and intramuscular
US-332	Squamous Cell Carcinoma of the Head and Neck	I/II	Interleukin-12 (IL-12) Polyvinylpyrrolidone	Naked/Plasmid DNA	In vivo	Intratumoral
US-266	Squamous Cell Carcinoma of the Head and Neck	I/II	Interferon-alpha (IFN-a)	Naked/Plasmid DNA	In vivo	Intratumoral
US-200	Non-Hodgkin's B-Cell Lymphoma, Mantle Cell Lymphoma	I/II	Tumor idiotype	Naked/Plasmid DNA	In vivo	Intramuscular
UK-099	Carcinoma	I/II	CAP-1 peptide from CEA	Naked/Plasmid DNA		
UK-086	Melanoma	I/II	pSG2 Mel3	Naked/Plasmid DNA		
UK-073	Breast cancer	I/II	MUC-1	Naked/Plasmid DNA		
UK-065	Metastatic Melanoma	I/II	Melanoma antigen MART-1 Melanoma antigen gp100	Naked/Plasmid DNA		
UK-053	Chronic lymphocytic leukaemia	I/II		Naked/Plasmid DNA		
UK-052	Multiple myeloma	I/II		Naked/Plasmid DNA		
UK-020	B-cell lymphoma	I/II		Naked/Plasmid DNA		
UK-016	Breast cancer	I/II	Cytosine deaminase	Naked/Plasmid DNA	In vivo	Intratumoral
UK-007	Non-Hodgkin B-cell lymphoma	I/II	Specific anti-Idiotype	Naked/Plasmid DNA	In vivo	Intramuscular
PL-003	Glioblastoma, meningioma	I/II	Insulin-Like Growth Factor-1 (IGF-1)	Naked/Plasmid DNA	In vitro	Subcutaneous
PL-002	Hepatoma, colon cancer	I/II	Insulin-Like Growth Factor-1 (IGF-1)	Naked/Plasmid DNA	In vivo	Subcutaneous
DE-036	Breast cancer	I/II		Naked/Plasmid DNA	In vivo	
DE-030	Solid tumors	I/II		Naked/Plasmid DNA		
DE-025	Pancreatic cancer	I/II	Cytochrome p450	Naked/Plasmid DNA	In vitro	

DE-024	Breast cancer, Pancreatic cancer, gallbladder carcinoma	I/II		Naked/Plasmid DNA	In vitro	
DE-009	Inoperable pancreatic adenocarcinoma	I/II	Cytochrome p450	Naked/Plasmid DNA	In vitro	Intravenous
CA-003	Melanoma	I	HLA-B7/Beta 2-Microglobulin	Lipofection	In vivo	Intratumoral
US-591	Solid tumors	I	DOTAP-DOPE p53	Lipofection	In vivo	Intravenous
US-544	Melanoma	I	Herpes simplex virus thymidine kinase (HSV-TK)	Lipofection	In vivo	Intravenous
US-525	Ovarian cancer	I	DC-Chol-DOPE/E1A	Lipofection	In vivo	Intraperitoneal
US-540	Ovarian cancer	I	DC-Chol-DOPE/E1A	Lipofection	In vivo	Intraperitoneal
US-513	Non-Small Cell Lung Cancer	I	Cholesterol/Fus 1	Lipofection	In vivo	Intravenous
US-361	Non-Small Cell Lung Cancer	I	HLA-A1 or A2/B7.1 (CD80)	Lipofection	In vivo	Subcutaneous
US-409	Lung cancer	I	Interleukin-2 (IL-2)	Lipofection	In vivo	Intravenous
US-350	Ovarian cancer	I	E1A/DC-Chol-DOPE	Lipofection	In vivo	Intraperitoneal
US-286	Lung, Head and Neck	I	HLA-B7/Beta 2-Microglobulin	Lipofection	In vivo	Intratumoral
US-285	Head and Neck Squamous Cell Carcinoma	I	Epidermal growth factor receptor (EGFR) antisense	Lipofection	In vivo	Intratumoral
US-260	Melanoma	I	Interleukin-2 (IL-2)	Lipofection	In vivo	Intratumoral
US-244	Melanoma	I	Staph. Enterotoxin B Interleukin-2 (IL-2)	Lipofection	In vivo	Intratumoral
US-212	Melanoma	I	HLA-B7/Beta 2-Microglobulin Interleukin-2 (IL-2)	Lipofection	In vivo	Intratumoral
US-190	Squamous cell carcinoma of the head and neck	I	Interleukin-2 (IL-2)	Lipofection	In vivo	Intratumoral
US-184	Prostate cancer	I	Interleukin-2 (IL-2)	Lipofection	In vivo	Intratumoral
US-162	Solid tumors	I	E1A	Lipofection	In vivo	Intratumoral
US-156	Breast cancer	I	B7.1(CD80)	Lipofection	In vitro	Subcutaneous
US-161	Small cell lung cancer	I	B7.1(CD80)	Lipofection	In vitro	Subcutaneous
US-150	Melanoma	I	HLA-B7	Lipofection	In vivo	Intratumoral

TABLE 1. *Continued*

Traial ID	Indication	Phaze	Gene	Vector	Gene delivery	Administration route
US-137	Metastatic breast cancer, ovarian cancer overexpressing her-2/neu	I	E1A	Lipofection	In vivo	Intraperitoneal and intrapleural
US-132	Locally advanced or local metastatic prostate cancer	I	Interleukin-2 (IL-2)	Lipofection	In vivo	Intradermal
US-120	Melanoma	I	HLA-B7	Lipofection	In vivo	Intratumoral
US-095	Lymphomas, solid malignant tumors	I	Interleukin-2 (IL-2)	Lipofection	In vivo	Intratumoral
US-110	Refractory metastatic ovarian cancer	I	Interleukin-2 (IL-2)	Lipofection	In vivo	Intradermal
US-086	Breast cancer (refractory, recurrent or metastatic)	I	Interleukin-2 (IL-2)	Lipofection	In vitro	Subcutaneous
US-063	Metastatic melanoma	I	B7.1(CD80)	Lipofection	In vitro	Subcutaneous
US-053	Small cell lung cancer (limited stage)	I	Interleukin-2 (IL-2)/Neomycin resistance (NeoR)	Lipofection	In vitro	Subcutaneous
US-052	Glioblastoma	I	Antisense IGF-1	Lipofection	In vitro	Subcutaneous
US-045	Tumors	I	HLA-B7/Beta 2-Microglobulin	Lipofection	In vivo	Intratumoral
US-013	Melanoma, adenocarcinomas	I	HLA-B7/Beta 2-Microglobulin	Lipofection	In vivo	Intratumoral
SG-001	NSCLC cutaneous metastases or axillary lymph nodes, melanoma, breast cutaneous metastases or axillary lymph nodes	I	HLA-A2 or -B13 or H-2K	Lipofection	In vivo	Intraperitoneal
UK-038	Ovarian cancer	I	E1A HER-2/neu	Lipofection		
US-630	Prostate cancer	I	Human Telomerase Reverse Transcriptase (hTERT)	RNA transfer	In vitro	Intradermal
US-607	Renal Cell Carcinoma or Melanoma	I	Tumor RNA	RNA transfer	In vitro	Intravenous
US-596	Melanoma	I	Melanoma antigen MART-1 Tyrosinase/gp100/MAGE-3	RNA transfer	In vitro	Intralymphnodal
US-548	Renal Cell Carcinoma	I	Tumor RNA	RNA transfer	In vitro	Intravenous

US-510	Prostate cancer	I	Prostate specific antigen (PSA)	RNA transfer	In vitro	Intradermal
US-498	Hormone Refractory Prostate Carcinoma	I	hTERT	RNA transfer	In vitro	Intradermal
US-410	Prostate cancer	I	Tumor RNA	RNA transfer	In vitro	Intravenous
US-367	Renal Cell Carcinoma	I	Tumor RNA	RNA transfer	In vitro	Intravenous
US-320	CEA-Expressing Malignancies	I	Carcinoembryonic antigen (CEA)	RNA transfer	In vitro	Intravenous
US-306	Prostate Cancer	I	Prostate specific antigen (PSA)	RNA transfer	In vitro	Intravenous
US-179	Breast, ovary	I	Carcinoembryonic antigen (CEA)	RNA transfer	In vitro	Intravenous
US-289	Cystic fibrosis	I	Alpha-1 antitrypsin (ATA)	Lipofection	In vivo	Intranasal
US-186	Cystic fibrosis	I	Cystic fibrosis transmembrane conductance regulator (CFTR)	Lipofection	In vivo	Intranasal
US-170	Cystic fibrosis	I	Cystic fibrosis transmembrane conductance regulator (CFTR)	Lipofection	In vivo	Intranasal
US-127	Cystic fibrosis	I	Cystic fibrosis transmembrane conductance regulator (CFTR)	Lipofection	In vivo	Intranasal
US-070	Alpha-1-antitrypsin deficiency		Alpha-1 antitrypsin (ATA)	Lipofection	In vivo	Intranasal
US-066	Cystic fibrosis	I	Cystic fibrosis transmembrane conductance regulator (CFTR)	Lipofection	In vivo	Intranasal
UK-014	Cystic fibrosis	I	Cystic fibrosis transmembrane conductance regulator (CFTR)	Lipofection	In vivo	Intrabronchial
NZ-001	Canavan disease	I	Aspartoacylase (ASPA)	Lipofection	In vivo	Intracranial
US-247	Hemophilia A	I	Factor VIII	Naked/Plasmid DNA	In vivo	Intraperitoneal
FR-017	Duchenne/Becker Muscular distrophy	I	Dystrophin	Naked/Plasmid DNA		
UK-068	HIV infection	I	HIV-1 Gag Antisense Pol 1	Naked/Plasmid DNA		Intramuscular
JP-003	Glioblastoma, astrocytoma	I	Interferon-beta (IFN-β)	Lipofection	Intratumoral	Intratumoral
JP-010	Buerger's disease	I	HGF	Naked/Plasmid DNA		

TABLE 2. Antisense oligodeoxynucleotides against cancer in higher-stage clinical trials

Name	Application	Disease	Target	Stage
ISIS 3521	Parenteral	Non-small-cell lung cancer and others	PKC-α	III
G 3139	Subcutaneous	Melanoma	bcl-2	III
		B-cell lymphoma	bcl-2	II
ISIS 2503	Parenteral	Pancreatic cancer	H-ras	II
ISIS 5132	Parenteral	Ovarian cancer	raf	II
GTI 2040	Parenteral	Advanced or metastatic renal cell carcinoma	R2 protein or ribonucleotide reductase	II
GEM 231	Parenteral	Cancers	PKA	II

effects being ecchymosis, pruritus (and/or discomfort at the injection site), and pneumothraces. Regression of the injected lesion was observed in 185 patients, including one CR, three PRs, and five MRs. An overall response rate of 4% (two PRs) was documented, and nine patients (18%) maintained stable disease (SD) for at least 11 weeks. Six patients remained alive 25.1 to 39.4 months after their first injection, including two patients with local (injected tumor) responses and one patient with an overall disease PR.

Three phase I studies have been conducted assessing the safety and efficacy of various dosing schedules of Alovectin-7 in patients with metastatic melanoma, colon cancer, and renal cell carcinoma. A total of seven (28%) patients had a decrease of >25% in at least one tumor lesion. Five patients had shrinkage of their injected tumor (local response), whereas two had regression of at least one non-injected lesion (Hersh 2001). Allovectin-7 was administered by intratumoral injection into a single accessible tumor lesion. The lesion selected for injection was required to be >1 cm in longest diameter but <5 cm × 5 cm in size and accessible to direct needle injection. A CR was defined as the disappearance of all clinical evidence of the tumor. A PR required a >50% decrease in the size of all of the measurable lesions with no new lesions appearing, and SD was defined as <25% decrease or increase in the sizes of the tumor lesions with no new lesions appearing. MRs were defined as tumor regression >25% but <50%. All of the toxicities were graded according to the WHO Common Toxicity Criteria Grading Scale.

A local (injected) tumor response rate of 18% (9 of 51) was observed, with one CR, three PRs, and five MRs. Thirty-one of the 52 patients entered in the study completed all six injections of a cycle according to schedule. Maximum tumor response was achieved at a median of 11 weeks. One patient developed a PR in the injected tumor after retreatment. Stable injected lesions were observed at week 11 in 12 of 51 (24%) patients. An overall disease response rate of 4% (2 of 51) was observed. Two patients developed a PR in overall disease. A 34-year-old male presented with a single site of disease in mediastinal lymph nodes, a Karnofsky performance status (KPS) of 90%, and a normal lactate dehydrogenase (LDH) level. After two cycles of therapy, the 12.8 cm^2 lymph node regressed 85% to 1.5 cm^2. This patient has received no further melanoma therapy and remains in a stable PR 26 months after initiating Allovectin-7 therapy. The median time to progression for all of the evaluable nonresponders was 2.3 months. Patients with stable overall disease had a median time to progression of

5.3 months (range, 2.3–14.9 months). Six patients were alive at the last observation (range, 25.1 to 39.4 months from first injection), including two patients who developed local responses and one patient with a PR in overall disease. Nonresponding patients had a median time to death of 7.6 months.

Adverse events were further classified as study drug-related (24 events) or procedure-related (29 events). All of the study-drug-related toxicities were mild to moderate (no grade 3 or higher) and included pruritus and erythema at the injection site and general aches and pains. Procedure-related toxicities included mild to moderate pneumothraces, ecchymoses, and pain at injection sites. A composite of all of the evaluable patients (90 patients) with stage III or IV metastatic melanoma treated with Allovectin-7 showed a correlation between tumor response and site of disease (Stopeck 1997; Hersh 2001; Nabel 1993b). Thirty percent of all of the evaluable patients with cutaneous and or nodal disease developed a local response (CR, PR or PR). A local response rate of 30% and an overall disease response rate of 10% were reported in patients with disease limited to lung metastases. Only 16% of patients with visceral disease had local responses, with 3% of patients developing an overall disease response.

A phase III trial comparing Allovectin-7 plus dacarbazine versus dacabazine in untreated patients with metastatic melanoma has been completed, and preliminary results are soon to be reported. Phase I/II data also indicate promising activity of Allovectin-7 in patients with advanced refractory head and neck squamous cell carcinoma, with 10% of 60 patients achieving PR and 23% SD after one cycle of treatment. Trials of Allovectin-7 as an adjuvant treatment in earlier stages of disease evolution are planned (Galanis 2002).

Rubin et al. (1997) completed a phase I study testing the feasibility and toxicity of immunotherapy of hepatic metastases from colorectal carcinoma by direct gene transfer of HLA-B7. Eligible patients were HLA-B7-negative, immunocompetent by PHA lymphocyte stimulation, and had at least two measurable hepatic lesions on CT scan for measurement of response of the injected lesion, as well as evaluation of possible distant response. Under ultrasonographic guidance, the hepatic lesions were injected with Allovectin-7. Eligible patients were injected on two schedules. On the first schedule, patients received an injection on day 1 and the injected lesion was biopsied to determine transfection every 2 weeks for 8 weeks. Doses were escalated from 10 to 50 to 250 μg with three patients treated at each level. The second schedule included multiple injections of 10 μg. Injections were administered on days 1 and 15 (3 patients), or on days1, 15, and 29 (3 patients). A total of 15 patients have completed treatment. Plasmid DNA was detected in 14 of 15 patients (93%) by PCR. In five of 15 patients (33%), mRNA was also detected. HLA-B7 was detected in five of eight patients (63%) by immunohistochemistry and in seven of 14 patients (50%) by fluorescence-activated cell sorting (FACS) analysis. There has been no serious toxicity directly attributable to allovectin-7. The results suggest that liposomal gene transfer by direct injection is feasible and non-toxic. Further studies will be necessary in order to establish the therapeutic efficacy.

2.1.2 IFN-β Gene Therapy Using TMAG Cationic Lipid

Yagi et al. (1993) found that cationic liposomes consisting of *N*-(α-trimethylammonioacetyl)-didodecyl-D-glutamate chrolide (TMAG), dilauroyl phosphatidylcholine

(DLPC), and DOPE (1:2:2 or 1:2:3, molar ratio) trap DNA with high efficiency and thus showed high potential to mediate DNA transfer to human glioma cells.

Yoshida et al. (1992a, 1992b). carried out studies on IFN-β gene therapy using their original cationic multilammelar liposomes. In April 2000, clinical trial of IFN-β gene therapy were started for patients with malignant glioma. This was the first clinical trial in the world (Yoshida 2003, 2004; Mizuno 1998; Natsume 1999, 2000). Malignant gliomas are highly lethal neoplasms that represent about 20% of all intracranial tumors in Japan. The Brain Tumor Registry of Japan reported 1-, 2-, and 5-year survival rates of 50.5, 18.5, and 7.8%, respectively. These tumors are too invasive to cure by surgical resection alone. Although patients undergo adjuvant therapy, including post-operative radiation therapy, chemotherapy, and immunologic therapy, their prognosis has remained poor. Thus, gene therapy is one of the most promising alternatives for treating malignant glioma. The safety and effectiveness of treating patients with malignant glioma (glioblastoma multiforme or anaplastic astrocytoma) by IFN-β gene therapy using multilammelar liposomes was demonstrated in a pilot clinical trial .

Gene therapy was based on four antitumor mechanisms induced by IFN-β gene transfer (Yoshida 2003); (1) Apoptosis of tumor cells, (2) growth inhibition of the cells by IFN-β produced by cells transfected with the gene, (3) induction of immune responses, and (4) increased secretion of chemokines. The apoptosis mechanism is thought to proceed via activation of DNase-γ and heat shock protein (HSP), which markedly induce this process. IFN-β gene transfer induces not only IFN-β but also proinflammatory cytokines such as IL-1β, tumor necrosis factor (TNF)-α, and IL-6. This cytokine cocktail strongly inhibits tumor growth, more than IFN-β alone, and also activates the host immune response through the cytokine network. In the third mechanism, IFN-β gene transfer markedly activates T cells, especially cytotoxic T lymphocytes. It was reported that glioma cells transfected with the IFN-β gene showed unique morphological changes, characterized by bleb formation, cytoplasmic shrinkage, and DNA laddering, culminating in apoptotic cell death (Yoshida 2003).

The clinical trials conducted by Yoshida's group consisted of reoperation and injection of liposomes containing the human IFN-β gene on days 0, 14, 17, 21, 24 and 28 (first case) or 0, 14, 21, and 28 (other case), and treatment evaluation 3 months after the first day of liposome administration (Yoshida 2003). In patients undergoing tumor resection, the surgical margin of the cavity was infiltrated with 1 ml of liposomes containing the human IFN-β gene at a concentration of 30 μg DNA/ml, evenly distributed at multiple sites. A repeated procedure of two to six injections was stereotactically performed under local anesthesia. Upon completion of the 28-day treatment, patients entered follow-up and were evaluated 3 months after the first injection. Subsequent evaluations were carried out every 3 months until the third year of the study, and are now being done annually until the study is terminated or the patient dies.

Clinical-grade liposomes were prepared at the human gene therapy vector-producing facility established at Nagoya University Hospital. Vectors produced in this facility are thoroughly tested in preclinical studies and receive regulatory approval before clinical trials. The preparation consists of a frozen concentrate of liposomes containing the human IFN-β gene. The pDRSV-IFN-β plasmid includes an inserted human IFN-β gene driven by the RSV promoter. Before administration, the preparation is rapidly thawed and then diluted to the desired concentration (30 μg of DNA per ml) with sterile phosphate-buffered saline.

Patients in the study had recurrent malignant gliomas and had previously undergone surgery. The tumors were diagnosed histologically as glioblastoma multiforme or anaplastic astrocytoma in accordance with the criteria of the World Health Organization (WHO, Geneva, Switzerland). All patients had failed to respond to standard therapy, including surgery, radiotherapy, chemotherapy, and immunotherapy. All of the tumors were localized to the supratentorial compartment, without dissemination via the cerebrospinal fluid (CSF). None of the patients had received anticancer therapy for at least 4 weeks. The primary objective was to determine the safety of the cationic liposomes containing the human IFN-β gene in patients with malignant gliomas. The second objective was to determine antitumor efficacy, as assessed by tumor response on magnetic resonance imaging (MRI), functional status, survival, and time to progression.

Sample obtained from all patients before injection contained no detectable IFN-β. After the injection, the highest concentration of IFN-β was, 24 IU/ml, and was detected in patient 1 10 days after the first injection. IL-1β was detected in patients 1 and 5, and TNF-α was detected in patients 1, 2 and 5. Each protein was detected a few days after injection, reaching maximum concentrations 10 days later and then decreasing gradually. IFN-β mRNA was detected in tumor-rich tissues (patients 1, 2, 3, and 5), whereas mRNA for TNF-α was detected in macrophage-rich tissues (patients 1 and 2), but not in tumor-rich tissues by RT-PCR.

In patient 1, after treatment, tumor growth ceased with little change in size over 10 weeks. Patients 2 and 3 each had a PR. Patient 5 had SD during the 10 weeks after the first injection. After therapy, tumor tissues showed dramatic changes in all patients. In patient 1, multinuclear giant cells disappeared rapidly and many tumor cells showed shrinkage or pyknotic change, reflecting apoptosis or necrosis. Simultaneously, MIB-1-positive cells were notably decreased. These alterations were also observed in the other patients. Immunohistochemistry identified many CD8+ lymphocytes and macropharges infiltrating into the tumor and surrounding brain, whereas few CD4+ lymphocytes or B lymphocytes were present. Notable cell infiltration was detected 2 weeks after injection in all patients; the infiltrates then gradually increased, persisting for at least 1 month after the first injection.

Adverse effects of treatment were limited to headache (NCI-CTC, grade I), brain edema, and various perioperative problems, including mild anemia (patient 1), controllable intracerebral hemorrhage at a distance from the tumor site (patient 2), and leakage of cerebrospinal fluid (CSF; patients 3 and 5). No direct toxicity attributable to the liposomes was seen. Laboratory data were normal or nearly normal. Intravenous administration of IFN-β in clinical trials has shown only minimal benefit as a postoperative adjuvant treatment for malignant glioma; tumors regressed in only 10 to 30% of treated patients, and no prolongation of survival was achieved. Yoshida et al. found that the antiproliferative effect was increased by treating cells with a small dose of TNF-α before transfection. Treatment with TNF-α followed by intratumoral injection of liposomes with entrapped pSV21IFN-β (150 nmol lipid, 3 μg DNA) had a remarkable effect.

This study demonstrated the feasibility and safety of IFN-β gene therapy for malignant glioma, an aggressive tumor that resists treatment, and potentially for other IFN-β sensitive diseases, such as medulloblastoma, malignant melanoma, and hepatitis.

2.2 Clinical Trials Using Naked DNA

2.2.1 Gene Therapy for Cardiovascular Diseases

The clinical consequences of peripheral arterial disease include pain on walking (claudication), pain at rest, and loss of tissue integrity in the distal ischemic limbs. Although many drugs and interventional devices contribute to the treatment of this disease, critical limb ischemia is estimated to develop in 500 to 1000 individuals per million per year. Recent progress in molecular biology has led to the development of gene therapy as a new strategy to treat a variety of cardiovascular diseases.

Vascular Endothelial Growth Factor

Therapeutic angiogenesis using vascular endothelial growth factor (VEGF) gene transfer to treat human patients with critical limb ischemia and myocardial ischemia has been reported (Baumgartner 1998, 2000; Isner 1998; Rosengraft 1999a, 1999b).

The first clinical trial using the VEGF gene was started in 1994, by J.M. Isner at Tufts University (Boston, MA, USA). An initial trial tested the efficacy of a hydrogel catheter with naked VEGF165 plasmid. Since most patients lack an appropriate target vascular lesion for catheter delivery, naked plasmid encoding the VEGF165 gene was intramuscularly injected. This clinical trial demonstrated clinical efficacy for treatment of peripheral arterial disease. Subsequently, numerous angiogenic growth factors, such as VEGF121, VEGF2, and β-FGF (basic fibroblast growth factor), have also been tested in clinical trials (Makinen 2002; Comerota 2002). Local catheter-mediated VEGF165 gene therapy in ischemic lower-limb arteries after percutaneous transluminal angioplasty (PTA) was also successful. Follow-up digital subtraction angiography revealed increased vascularity in the VEGF-treated groups distal to the gene transfer site and the region of the clinically most severe ischemia. A significant reduction in pain and aggregate ulcer size was detected after FGF gene transfer, associated with an increased transcutaneous oxygen pressure and ABI (ankle pressure index) (Isner 1998). A human gene therapy trial to treat coronary artery disease using the VEGF165 gene was also initiated by Isner's group (Losordo 1998; Vale 2000), in which naked plasmid containing the VEGF gene was intramuscularly injected into ischemic myocardium during a mini-operation. Similar to human trials in peripheral arterial disease, transfection of VEGF gene resulted in a marked increase in blood flow and improved clinical symptoms without apparent toxicity. More, recently, data from 13 consecutive patients with chronic stable angina clearly suggest that phVEGF165 gene therapy can successfully rescue foci of hibernating myocardium. A phase I clinical trial of direct myocardial gene transfer of naked DNA-encoding VEGF165, as sole therapy for refractory angina, was carried out in 30 patients with class 3 or 4 angina. Twenty-nine of the 30 patients reported reduced angina and decreased sublingual nitroglycerin consumption (Lathi 2001).

Prevention of peripheral artery restenosis after angioplasty using a hydrogel catheter to deliver naked VEGF165 plasmid DNA is being assessed in a human trial (Isner 1996). Preliminary results documented the success of this approach. Another clinical trial, in which the VEGF165 gene is transfected by cationic liposome or adenovirus with a catheter into the coronary artery, is underway in Finland (Latinen 2000). A recent report demonstrated the clinical safety of this strategy.

Hepatocyte Growth Factor

Morishita et al. have focused on hepatocyte growth factor (HGF), which is a potent angiogenic agent in mouse, rat and rabbit ischemia models (Taniyama 2001). The angiogenic activity of HGF is more potent than that of VEGF or FGF both in vitro and in vivo. Transfection of the human HGF gene by naked plasmid DNA or the HVJ (hemagglutinating virus of Japan)-liposome method resulted in a significant increase in blood flow. Even in models of high-risk conditions for atherosclerosis, such as a diabetic and high lipoprotein (a) concentrations, overexpression of HGF stimulates collateral formation and thus improves ischemic symptoms. Based on these results, Morishita et al. (2004a, 2004b) designed a human clinical trial of HGF gene therapy of peripheral artery disease. In a subanalysis of results from a phase I/IIa trial, it was demonstrated that intramuscular injection of naked HGF plasmid is safe and feasible.

Morishita et al. carried out another human clinical trial using intramuscular injection of naked human HGF plasmid. First, a small dose of the plasmid DNA (test injection; 0.4 mg plasmid DNA) was injected intramuscularly in order to examine acute or subacute allergy. After none of the patients exhibited an allergic reaction or anaphylaxis, a therapeutic dose (2 mg) of naked HGF plasmid DNA was intramuscularly injected 2 weeks after the test injection. HGF gene transfer has since been performed in 22 patients with peripheral arterial disease or Buerger disease of Fontaine grade III or IV who had failed conventional therapy. A reduction of pain was reported by 12 of 13 patients. An increase in ABI to >0.1 was observed in 11 of 17 patients (efficacy rate 65%), while a reduction of ischemic ulcer size of over 25% was observed in 18 of 25 patients (efficacy rate 72%). Unlike previous reports using VEGF gene, in which the VEGF protein level in the systemic circulation transiently peaked 1 to 3 weeks after gene transfer, the serum level of human HGF protein did not change during gene therapy. No acute severe complications or allergic events were observed in any patients. Two-month follow-up studies showed no evidence of the development of neoplasm or hemangioma. There was also no evidence of edema in patients transfected with the human HGF gene, in marked contrast to the VEGF trial in which 60% of patients developed moderate or severe edema in a phase I/IIa trial.

Four weeks after the initial injection, a second injection (2000 μg) was similarly administrated, yielding a total dose of 4000 μg plasmid DNA per patient. None of the test, initial, or second therapeutic injections of human HGF plasmid DNA induced an allergic or anaphylactic reaction; systemic or local inflammatory reactions were also not observed. Development of tumors or progression of diabetic retinopathy has not been occurred in any patient transfected with HGF plasmid DNA during the trial. Two-month follow-up studies showed no evidence of the development of neoplasm or hemangioma. Digital subtraction angiography of another patient with Buerger disease revealed a marked increase in peripheral blood flow and the formation of new blood vessels. The ABI was significantly increased. Transcutaneous PO_2 is an indicator of effectiveness in terms of angiogenesis and increased blood supply in the targeted ischemic lesions. The change in $TcPO_2$ after O_2 stimulation was significantly increased at 8 weeks compared with baseline. Two of 11 ulcers completely disappeared. In addition, pain at rest significantly improved in a time-dependent manner.

2.3 Antisense Gene Therapy

Antisense therapy is clinically much more advanced than gene therapy. Remarkably, the FDA has already approved one drug, Fomivirsen (Vitravene), a 21-mer phosphorothioate oligonucleotide that is being used to treat infections of the eye caused by human cytomegalovirus (HCMV). Fomivirsen inhibits the expression of proteins necessary for virus replication through an antisense mechanism and needs to be injected intravitenally (eye). With regard to cancer therapy, several phase I studies are underway. An antisense oligonucleotide for the treatment of melanoma and B-cell lymphoma targeting bcl-2 and an antisense inhibitor of protein kinase C-α in solid tumors are being investigated in phase III trials. Phase II clinical trials are in progress against raf-kinase, H-ras, R2-protein and DNA methyltransferase. Numerous other antisense approaches against cancer-related targets are also in clinical trials, and interest in antisense strategies is expected to increase as they offer several advantages. For example, it is easier to synthesize large amounts of pure antisense oligonucleotides, which are much shorter than genes. Furthermore, the antisense mechanism is applicable to a wide range of conditions and antisense oligonucleotides are physiologically well tolerated.

Progress in molecular biology has provided the opportunity for introducing or replacing missing or malfunctioning genes. Using these tools, cancer cells may be selectively destroyed via suicide strategies, or the expression of undesired genes may be blocked by antisense or ribozyme strategies. Immunological approaches have yielded promising therapeutic effects as well. The best clinical results, however, have been achieved by either local administration or the use of antisense oligonucleotides, as indicated by the large number of clinical trials currently taking place. These new techniques, however, lack efficient and specifically targeted delivery systems for nucleic acids and raise toxicity concerns (Table 3).

The results from a phase II trial using antisense c-myc to treat restenosis have been reported. Treatment with 10 mg phosphorothioate-modified antisense oligodeoxynucleotides (ODN) directed against c-myc does not reduce neointimal volume obstruction or the angiographic restenosis rate. This trial utilized intra-coronary infusion of antisense c-myc ODN without any vectors, and several issues, such as low transfection efficiency, may limit the efficacy of this form of treatment.

2.4 Decoy

Synthetic double-stranded (ds) DNA with high affinity for a target transcription factor may be introduced into target cells as a "decoy" *cis* element that binds the transcription factor and thus alters gene transcription (Morishita 1998). Transcription of ds ODN corresponding to the *cis* sequence will result in the attenuation of the authentic *cis-trans* interaction, leading to the removal of *trans*-factors from the endogeneous *cis*-element, with subsequent modulation of gene expression. Vascular smooth muscle cell (VSMC) proliferation is dependent on the coordinated activation of a series of cell cycle regulatory genes, which results in mitosis. Transfection of VSMC with a sufficient quantity of decoy ODN containing the E2F *cis*-element should allow effective binding of E2F, which would thus be prevented from *trans*-activating genes encoding essential cell-cycle regulatory proteins. Consequently, VSMC proliferation and neointimal formation would be inhibited.

TABLE 3

Traial ID	Trial title
US-234	A Controlled, Randomized Phase III Trial Comparing the Response to Dacarbazine With and Without Allovectin-7 in Patients with Metastatic Melanoma
CA-004	Gene Therapy in patients with melanoma
CA-005	Gene Therapy in patients with cancer
US-438	A Multicenter, Phase II Study of Intratumoral Injections of E1A-Lipid Complex and Re-Irradiation for Treatment of Patients with Recurrent Head and Neck Squamous Cell Carcinoma. Sponsor: Targeted Genetics
US-432	A Phase II Study of Safety and Efficacy of Allovectin-7 Immunotherapy for the Treatment of Primary Resectable Squamous Cell Carcinoma of the Oral Cavity or Oropharynx. Sponsor: Vical Inc.
US-431	A Phase II Study of High-Dose Allovectin-7 in Patients with Advanced Metastatic Melanoma. Sponsor: Vical Inc.
US-416	A Phase II Study of Intraperitoneal E1A-Lipid Complex for Patients with Advanced Epithelial Ovarian Cancer with HER-2/neu Overexpression. Sponsor: Targeted Genetics Corporation
US-415	A Phase II Study of Intraperitoneal E1A-Lipid Complex for Patients without Advanced Epithelial Ovarian Cancer without HER-2/neu Overexpression. Sponsor: Targeted Genetics Corporation
US-391	Phase II Study of Leuvectin in Patients with Metastatic Renal Cell Carcinoma
US-323	A Multi-Center, Open-Label, Study of the Safety and Efficacy of Multiple Intratumoral Injections of hIl-2 with Unresectable or Recurrent/Refractory Squamous Cell Carcinoma of the Head and Neck.
US-312	Phase II Study Evaluating the Safety and Efficacy of Neoadjuvant Leuvectin Immunotherapy for Plasmid (1.8 mg) Formulated with DOTMA/Cholesterol [Ratio 1:0.5 (–/+)] Liposomes in Patients the Treatment of Prostate Cancer
US-270	Phase II Study of the Safety, Efficacy, and Effect on Quality of Life of Allovectin-7 Immunotherapy for the Treatment of Recurrent or Persistent Squamous
US-259	Phase II Study of Direct Gene Transfer of IL-2 Plasmid DNA/DMRIE/DOPE Lipid Complex (Leuvectin) as an Immunotherapeutic Regimen in Patients with Metastatic Renal Cell Carcinoma
US-246	A Multicenter Phase II Study of E1A Lipid Complex for the Intratumoral Treatment of Patients with Recurrent Head and Neck Squamous Cell Carcinoma
US-233	Phase II Study of Direct Gene Transfer of HLA-B7 Plasmid DNA/DMRIE/DOPE Lipid Complex (Allovectin-7) as an Immunotherapeutic Agent in Patients
US-191	Phase II Study of Immunotherapy by Direct Gene Transfer with Allovectin-7 for the Treatment of Recurrent or Metastatic Squamous Cell Carcinoma of the Head and Neck
US-168	Phase II Study of Immunotherapy of Metastatic Melanoma by Direct Gene Transfer
US-142	Allovectin-7 in the Treatment of Squamous Cell Carcinoma of the Head and Neck

TABLE 3. *Continued*

Traial ID	Trial title
US-121	Phase I Study of HLA-B7 Plasmid DNA/DMRIE/DOPE Lipid Complex as an Immunotherapeutic Agent in Renal Cell Carcinoma by Direct Gene Transfer with Concurrent Low Dose Bolus IL-2 Protein Therapy
US-115	Phase II Study of Immunotherapy of Metastatic Cancer by Direct Gene Transfer
US-072	Phase I Study of Immunotherapy of Malignant Melanoma by Direct Gene Transfer
US-064	Phase I Study of Immunotherapy of Advanced Colorectal Carcinoma by Direct Gene Transfer into Hepatic Metastases
US-519	A Phase II Trial of CG8020 and CG2505 in Patients with Nonresectable or Metastatic Pancreatic Cancer
US-475	A Safety and Efficacy Trial of Lethally Irradiated Allogeneic Pancreatic Tumor Cells Transfected with the GM-CSF Gene in Combination with Adjuvant Chemotherapy for the Treatment of Adenocarcinoma of the Pancreas
US-393	Phase II Study of a TGF-b2 Antisense Gene Modified Allogeneic Tumor Cell Vaccine in Patients with Non-Curable Non-Small Cell Lung Cancer
US-378	A Multicenter, Open-Label, Multiple Administration, Study of the Safety, Tolerability and Efficacy of IFN-a/IL-12 Combination Gene Therapy in Patients with Squamous Cell Carcinoma of the Head and Neck (SCCHN)
US-351	An Open-Label, Multiple Administration, Study of the Safety, Tolerability, and Efficacy of IFN-a Gene Medicine in Patients with Malignant Angioendothelioma
US-254	Immunization of Patients with Metastatic Melanoma Using DNA Encoding the GP100 Melanoma Antigen
UK-071	A phase II, multicentre, double-blind, placebo controlled, dose finding study of ZYC101 a in the treatment of high-grade squamous intra-epithelial lesions of the uterine cervix
DE-033	Phase II Study in patients with inoperable pancreatic cancer to evaluate response rate and clinical benefit of a cell therapy with encapsulated cells synthesising cytochrome P450 CYP2B1 enzyme wich concerts ifosfamide to ist active metabolites
DE-032	Phase II Randomized study of non-specific immunotherapy of malignant mesothelioma by repeated intratumoral injection of a vero cell producing human IL-2 comparing two dose levels
EG-001	Direct p53 DNA injection in primary hepatocellular carcinomas
US-568	A Phase II Multi-Center, Double-Blind, Placebo-Controlled, Trial of VLTS-589 in Subjects with Intermittent Claudication Secondary to Peripheral Arterial Disease
US-645	Double-Blind, Randomized, Placebo-Controlled, Parallel Group and Dose-Finding, Multicentric, Safety and Efficacy Study with Intramuscular Injections of NV1FGF in Subjects with Intermittent Claudication
US-567	A Multicenter, Randomized, Double-Blind, Dose Ranging Placebo Controlled Study Evaluating Defined Doses of Percutaneously Delivered pVGl.1 (VEGF2) (Placebo, 2, 200, or 2000 μg) in "No Option" Patients with Class III or IV Angina with an Option for Patients to Receive Active Treatment at Month 6 if they Experience a Treatment Failure
US-502	A Phase II, Randomized, Double-Blind, Placebo Controlled, Parallel Group, Efficacy and Safety Study of Different Doses and Schedules of Administration of NV1FGF in Patient s with Severe Peripheral Artery Occlusive Disease

US-384	A Double-Blind, Placebo-Controlled, Continuation Study of Intramyocardial pVGI.1 (VEGF2) Administered by Percutaneous Cardiac Catheterization in Patients with Class III or IV Angina
US-383	A Phase IIb Multicenter, Randomized, Controlled Study of Direct Intramyocardial Injection of pVGI.1 (VEGF2) Versus Maximum Medical Therapy in Patients with Class III or IV Angina
US-354	A Placebo-Controlled, Dose-Escalating Study of Intramyocardial Vascular Endothelial Growth Factor 2 (VEGF2) Gene Therapy Administered Using Percutaneous Cardiac Catheterization in Patients with Class III or IV Angina
US-316	Multicenter, Randomized, Single-Blind, Placebo-Controlled, Dose-Escalating Study of Intramyocardial Vascular Endothelial Growth Factor 2 (VEGF2) Gene Therapy Administered Using Percutaneous Cardiac Catheterization in Patients with Refractory and Stable Exertional Angina Who Are Not Candidates for Revascularization Procedures
US-294	A Multicenter, Open-Label, Dose-Escalating Study of Intramyocardial Vascular Endothelial Growth Factor 2 (VEGF-2) Gene Therapy in Refractory Patients with Stable Exertional Angina Who Are Not Candidates for Revascularization Procedures
UK-069	A phase II randomised double-blind, placebo-controlled, parallel group, efficay and safety Study of NV1FGF in patients with severe peripheral artery occlusive disease
CH-028	Gene therapy in patients with severe peripheral artery occlusive disease (PAOD)
XX-002	EUROINJECT ONE
DE-050	A phase II, randomized, double-blind, placebo-controlled parallel group, efficacy and safety study of NV1FGF in patients with severe peripheral artery occlusive disease.
US-568	A Phase II Multi-Center, Double-Blind, Placebo-Controlled, Trial of VLTS-589 in Subjects with Intermittent Claudication Secondary to Peripheral Arterial Diseas
DE-011	Imaging-guided convection-enhanced delivery and gene therapy of glioblastoma
DE-002	Gene Therapy in Patients with Melanoma
DE-003	Gene Therapy in Patients with Renal Cell Cancer
DE-004	Induction of tumor-specific cytotoxic T lymphocytes by immunization with autologous tumor cells and interleukin-2 gene transfected fibroblasts.
SG-002	Phase I study of immunotherapy of cutaneous metastases of human carcinoma using allogeneic and xenogeneic MHC DNA-liposome complexes
US-582	A Phase 1/2 Randomized Study of Intraperitoneal tgDCC-E1A and Intravenous Paclitaxel in Women with Platinum-Resistant Ovarian Cancer
US-466	Concurrent Chemotherapy (Paclitaxel and Carboplatin) and Thoracic Radiotherapy with Swallowed Manganese Superoxide Dismutase (MnSOD) Plasmid Liposome (PL) Protection in Patients with Locally Advanced Stage III Non-Small Cell Lung Cancer. A Phase I-II Study.

TABLE 3. *Continued*

Traial ID	Trial title
US-357	Protocol for Retreatment with Leuvectin Immunotherapy for Cancer
US-352	Phase I/II Study Evaluating the Safety and Efficacy of Leuvectin Immunotherapy for the Treatment of Locally Recurrent Prostate Cancer Following Radiation Therapy
US-210	Compassionate Use Protocol for Retreatment with Allovectin-7 Immunotherapy for Metastatic Cancer by Direct Gene Transfer
US-169	Phase I/II Trial of Interleukin-2 DNA/DMRIE/DOPE Lipid Complex as an Immunotherapeutic Agent in Cancer by Direct Gene Transfer
US-108	Adoptive Cellular Therapy of Cancer Combining Direct HA-B7/α-2 Microglobulin Gene Transfer with Autologous Tumor Vaccination for the Generation of Vaccine-Primed Anti-CD3 Activated Lymphocytes
US-071	Phase I Study of Immunotherapy for Metastatic Renal Cell Carcinoma by Direct Gene Transfer into Metastatic Lesions
UK-015	Gene Therapy Research for Cystic Fibrosis
UK-009	Towards gene therapy for cystic fibrosis
UK-008	Towards gene therapy for cystic fibrosis
UK-003	Gene Therapy for Cystic Fibrosis. Assessment of the Safety and Efficacy of Liposome-Mediated DNA Transfer to the Nasal Epithelium
UK-084	A pilot study of the safety and immunogenicity of a candidate HIV-1 clade A DNA vaccine, pTHr.HIVA, given by needle injection into the deltoid muscle in HIV-1 seropositive subjects receiving highly active antiretroviral therapy
CH-004	Gene Therapy in Patients with HIV Infection
US-595	A Phase I/II Clinical Trial of pNGVL4a-Sig/E7 (detox)/HSP70 for the Treatment of Patients with HPV16+ Cervical Intraepithelial Neoplasia 2/3 (CIN2/3)
US-545	A Phase I/II, Escalating Dose, Open Label Evaluation of Safety, Feasibility and Tolerability of Transgenic Lymphocyte Immunization Vaccine (TLI) in Subjects with Histologically Proven Prostate Adenocarcinoma
US-541	A Phase I/II Study of Intranodal Delivery of Synchrovax BPL Vaccine, an Epitope Synchronization Plasmid DNA Vaccine, in Sage IV Breast Carcinoma Patients
US-490	A Pilot Phase I/II Study of Intranodal Delivery of a Plasmid DNA (Synchrotope MA2M) in Stage IV Melanoma Patients
US-479	Vaccination in Peripheral Stem Cell Transplant Setting for Acute Myelogenous Leukemia: The Use of Autologous Tumor Cells with an Allogeneic GM-CSF Producing Bystander Cell Line.
US-472	Phase I/II Study of Vaccination with Irradiated Autologous Lung Tumor Cells Mixed with a GM-CSF Secreting Bystander Cell Line (Lung Bystander GVAXR) in Advanced Non-Small Cell Lung Cancer. Sponsor: Cell Genesys, Inc.

US-461	Phase I/II Multi-Center, Open-Label, Multiple Administration Trial of the Safety, Tolerability, and Efficacy of an IFN-alpha/IL-12 Plasmid-Based Therapeutic. Sponsor: Valentis, Inc.
US-435	Vaccination in Peripheral Stem Cell Transplant Setting for Multiple Myeloma: The Use of Autologous Tumor Cells with an Allogeneic GM-CSF Producing Bystander Cell Line. Sponsor: Cell Genesys, Inc.
US-408	A Phase I/II Study of Idiotypic Vaccination for Chronic Lymphocytic Leukemia using a Genetic Approach
US-392	Phase I/II Study of Vaccine Therapy for B-Cell Lymphoma Utilizing Plasmid DNA Coding for Tumor Idiotype
US-332	A Multi-Center, Open-Label, Multiple Administration, Rising Dose Study of the Safety, Tolerability, and Efficacy of IL-12 Gene Medicine in Patients with Unresectable or Recurrent/Refractory Squamous Cell Carcinoma of the Head and Neck (SCCHN)
US-266	A Multi-Center, Open-Label, Multiple Administration, Rising Dose Study of the Safety, Tolerability, and Efficacy of IFN-alpha Gene Medicine in Patients with Unresectable or Recurrent/Refractory Squamous Cell Carcinoma of the Head and Neck (SCCHN)
US-200	A Phase I/II Study of Vaccine Therapy for B-Cell Lymphoma Utilizing Plasmid DNA Coding for Tumor Idiotype
UK-099	A phase I/II study of DNA vaccination with a CEA/pDOM fusion gene in patients with carcinoma expressing CEA
UK-086	Phase I/II study to determine the optimum dose and dosing regimen then to assess the efficacy of a poly-epitope pharmaccine (therapeutic vaccine) involving pSG2, Mel 3 and MVA.Mel3, in patients with stage II or Stage IV metastatic melanoma
UK-073	The use of cDNA vaccine encoding the human MUC-1 gene in the treatment of patients with advanced breast cancer—a phase I/II study
UK-065	A phase I/II study of immunotherapy for patients with metatstatic melanoma using dendritic cells transfected with a plasmid encoding two melanoma antigens
UK-053	Phase I/II study of idiotypic vaccination for chronic lymphocytic leukaemia using a genetic approach (CLLIFT)
UK-052	Phase I/II study of idiotypic vaccination for multiple myeloma using a genetic approach (MMIFTT)
UK-020	A Phase I/II pilot study of idiotypic vaccination for follicular B-cell lymphoma using a genetic approach (i.m.)
UK-016	Genetic prodrug activation therapy for breast cancer
UK-007	A pilot study of idiotypic vaccination for follicular B-cell lymphoma using a genetic approach
PL-003	IGF-I (Insulin like growth factor 1) triple helix cellular therapy of brain tumors
PL-002	IGF-I (Insulin like growth factor 1) triple helix cellular therapy of digestive tube tumors
DE-036	Phase I-II Trial: Induction of a systemic specific immune response against specific tumor antigens after vaccination with a genetic modified HLA.A2+ breast cancer cell line
DE-030	Gene therapy in patients with solid tumours
DE-025	Microencapsulated CYP2B1-transfected cell-mediated treatment of advanced inoperable pancreatic carcinoma, a phase-II clinical trial

Table 3. *Continued*

Traial ID	Trial title
DE-024	Vaccination with autologe-mucinen-CDNA (MUC1)-transfected dendritic cells in patients with breastcancer or pancreas carcinoma
DE-009	Cell therapy using microencapsulated 293 cells transfected with a gene construct expressing CYP2B1, an ifosfamide converting enzyme, instilled intra-arterially in patients with advanced-stage pancreatic carcinoma: a phase I/II study
CA-003	Gene Therapy in patients with melanoma
US-591	An Open-Label Safety Study of Escalating Doses of SGT-53 for Systemic Injection in Patients with Advanced Solid Tumor Malignancies
US-544	A Phase I Study to Evaluate the Safety and Pharmacokinetics of Pro-1, a Liposome-Encapsulated Thymidine Kinase Gene Formulation, in Patients with Stage IV Metastatic Melanoma
US-525	A Phase I Dose Escalation Study of Intraperitoneal tgDCC-E1A and Intravenous Carboplatin for Treatment of Recurrent, Platinum-Sensitive Ovarian Cancer
US-540	A Phase I Dose Escalation Study of Intraperitoneal tgDCC-E1A and Intravenous Paclitaxel in Women with Platinum-Resistant Ovarian Cancer
US-513	Phase I Study of Intravenous DOTAP:Cholesterol-Fus 1 Liposome Complex (DOTAP:Chol-Fus 1) in Patients with Advanced Non-Small Cell Lung Cancer (NSCLC) Previously Treated with Chemotherapy
US-361	Elicitation of a Cellular Immune Response in Patients with Non-Small Cell Lung Cancer by Immunogenic Tumor Cell Vaccination—A Phase I Study
US-409	A Phase I, Multi-Center, Open-Label, Dose-Escalation Study of the Safety and Tolerability of Intravenously Administered VLTS-587 in Patients with Solid Tumors and the Presence of Metastases or Primary Cancer in the Lungs
US-350	A Phase I Dose Escalation Study of Intraperitoneal E1A-Lipid Complex (1:3) with Combination Chemotherapy in Women with Epithelial Ovarian Cancer
US-286	Phase I Study of HLA-B7/beta2M Plasmid DNA/DMRIE/DOPE Lipid Complex (Allovectin-7) by Direct Gene Transfer with Concurrent Low-Dose Subcutaneous IL-2 Protein Therapy as an Immunotherapeutic Regimen in Lung and Head and Neck Cancers
US-285	A Phase I Trial of Intratumoral Antisense EGFR DNA and DC-Chol Liposomes in Advanced Oral Squamous Cell Carcinoma
US-260	Phase I Study of HLA-B7/b2M Plasmid DNA/DMRIE/DOPE Lipid Complex (Allovectin-7) by Direct Gene Transfer with Concurrent Low-Dose Subcutaneous IL-2 Protein Therapy as an Immunotherapeutic Regimen in Malignant Melanoma

US-244	A Phase I Study Using Direct Combination DNA Injections for the Immunotherapy of Metastatic Melanoma
US-212	Phase I Study of Direct Gene Transfer of HLA-B7 Plasmid DNA/DMRIE/DOPE Lipid Complex (Allovectin-7) with IL-2 Plasmid DNA/DMRIE/DOPE Lipid Complex (Leuvectin) as an Immunotherapeutic Regimen in Patients with Metastatic Melanoma
US-190	A Double-Blind, Placebo-Controlled, Single Rising-Dose Study of the Safety and Tolerability of Formulated hIL-2 Plasmid in Patients with Squamous Cell Carcinoma of the Head and Neck (SCCHN)
US-184	A Phase I Study Evaluating the Safety and Efficacy of Interleukin-2 Gene Therapy Delivered by Lipid Mediated Gene Transfer (Leuvectin) in Prostate Cancer Patients.
US-162	Phase I Multicenter Study of Intratumoral E1A Gene Therapy for Patients with Unresectable or Metastatic Solid Tumors that Overexpress HER-2/neu
US-156	Phase I Trial Using a CD80-Modified Allogeneic Breast Cancer Line to Vaccinate HLA-A2-Positive Women with Breast Cancer
US-161	Treatment of Small Cell Lung Cancer Patients In Partial Remission Or At Relapse With B7-1 Gene-Modified Autologous Tumor Cells As A Vaccine With Systemic Interferon Gamma
US-150	Evaluation of Intratumoral Gene Therapy with HLA-B7/DMRIE/DOPE plus Subcutaneous Low Dose Il-2
US-137	Phase I Study of E1A Gene Therapy for Patients with Metastatic Breast or Ovarian Cancer that Overexpresses Her-2/neu
US-132	A Phase I Study of Autologous Human Interleukin-2 (IL-2) Gene Modified Tumor Cells in Patients with locally Advanced or Metastatic Prostate Cancer
US-120	Phase I Study of Tumor-Infiltrating Lymphocytes Derived from In Vivo HLA-B7 Gene Modified Tumors in the Adoptive Immunotherapy of Melanoma
US-095	Phase I Trial of Interleukin-2 Plasmid DNA/DMRIE/DOPE Lipid Complex as an Immunotherapeutic Agent in Solid Malignant Tumors or Lymphomas Systemic Interferon Gamma by Direct Gene Transfer
US-110	A Phase I Study of Autologous Human Interleukin-2 (IL-2) Gene Modified Tumor Cells in Patients with Refractory Metastatic Ovarian Cancer
US-086	A Pilot Study of Autologous Human Interleukin-2 Gene Modified Tumor Cells in Patients with Refractory or Recurrent Metastatic Breast Cancer
US-063	Phase I Trial of B7-Transfected Lethally Irradiated Allogeneic Melanoma Cell Lines to Induce Cell Mediated Immunity Against Tumor-Associated Antigens Presented by HLA-A2 or HLA-A1 in Patients with Stage IV Melanoma

TABLE 3. *Continued*

Traial ID	Trial title
US-053	Phase I Study of Transfected Cancer Cells Expressing the Interleukin-2 Gene Product in Limited Stage Small Cell Lung Cancer
US-052	Gene Therapy for Human Brain Tumors Using Episome-Based Antisense cDNA Transcription of Insulin-Like Growth Factor I
US-045	Immunotherapy for Cancer by Direct Gene Transfer into Tumors
US-013	Immunotherapy of Malignancy by In Vivo Gene Transfer into Tumors
SG-001	Phase I study of immunotherapy of cutaneous metastases of human carcinoma using allogeneic and xenogeneic MHC DNA-liposome complexes.
UK-038	A multiple ascending dose study evaluating the safety and the gene transduction into malignant cells after the administration of EIA-lipd complex by intra-peritoneal administration in patients with epithelial ovarian cancer who over express HER-2/neu
US-630	Active Immunotherapy with Mature, Human Telomerase Reverse Transcriptase RNA-Transfected, Autologous Dendritic Cells with or without the IL-2 Diphtheria Toxin Conjugate Denileukin Difitox (ONTAK) in Subjects with Metastatic Prostate Cancer
US-607	A Pilot Study of Mature, Total-Tumor-RNA-Transfected, Donor-Derived Dendritic Cell Therapy in Patients Who Have Undergone Nonmyeloablative Allogeneic Stem Cell Transplantation for Metastatic Renal Cell Carcinoma or Metastatic Melanoma
US-596	Phase I Open Labeled Non-Randomized Study of RNA-DC as Tumor Vaccine in Patients with Melanoma
US-548	Active Immunotherapy with Mature, Tumor RNA-Transfected, Autologous Dendritic Cells with or without the IL2-Diphtheria Toxin Conjugate Denileukin Difitox (Ontak) in Patients with Metastatic Renal Cell Carcinoma
US-510	Pilot Study evaluating the Migratory Patterns of Immature and In Vitro Matured Dendritic Cells Transfected with RNA Encoding PSA in Patients with Metastatic Prostate Cancer
US-498	Phase I Study of Active Immunotherapy of Metastatic, Hormone Refractory Prostate Carcinoma using Autologous Mature Dendritic Cells (DC) Transfected with RNA Encoding Human Telomerase Reverse Transcriptase (hTERT).
US-410	A Safety and Feasibility Study of Active Immunotherapy in Patients with Metastatic Prostate Carcinoma Using Autologous Dendritic Cells Pulsed with Antigen Encoded in Amplified Autologous Tumor RNA
US-367	Active Immunotherapy of Metastatic Renal Cell Carcinoma Using Autologous Dendritic Cells Transfected with Autologous Renal Tumor RNA

US-320	Pilot Study of CEA RNA-Loaded, FLT3 Ligand-Mobilized Peripheral Blood Antigen Presenting Cells for Patients with Metastatic Malignancies Expressing CEA
US-306	Safety and Feasibility Study of Active Immunotherapy in Patients with Hormone Refractory Prostate Cancer Using Autologous Dendritic Cells Pulsed with RNA Encoding Prostate Specific Antigen, PSA
US-179	A Phase I Study of Active Immunotherapy With Carcinoembryonic Antigen RNA-Pulsed Autologous Human Cultured Dendritic Cells In Patients with Metastatic Malignancies Expressing Carcinoembryonic Antigen
US-289	Expression of an Exogenously Delivered Human Alpha-1 Antitrypsin Gene in Nasal Epithelium of Patients with Cystic Fibrosis
US-186	A Double-Blind, Placebo Controlled, Dose Ranging Study to Evaluate the Safety and Biological Efficacy of the Lipid-DNA Complex GR213487B in the Nasal Epithelium of Adult Patients with Cystic Fibrosis
US-170	Safety and Efficiency of Gene Transfer of Aerosol Administration of a Single Dose of a Cationic Lipid/DNA Formulation fo the Lungs and Nose of Patients with Cystic Fibrosis
US-127	Cationic Lipid Mediated Gene Transfer of CFTR: Safety of a Single Administration to the Nasal Epithelia
US-070	Expression of an Exogenously Administered Human Alpha-1-Antitrypsin Gene in the Respiratory Tract of Humans
US-066	Gene Therapy for Cystic Fibrosis Using Cationic Liposome Mediated Gene Transfer: A Phase I Trial of Safety and Efficacy in the Nasal Airway
UK-014	Gene therapy for Cystic Fibrosis Delivery to nasal epithelium and lung by nebulisation of the pCFICFTR/#67
NZ-001	Gene Therapy in Patients with Canavan Disease
US-247	A Phase I Safety Study of Autologous Transfected Human Fibroblasts Producing Human Factor VIII in Patients with Severe Hemophilia A
FR-017	Gene therapy for Duchenne/Becker muscular dystrophy
UK-068	A phase I/II study of DNA vaccination against a CMV/FrC of tetanus toxin fusion gene in allograft donors and recipients
JP-003	Gene Therapy in patients with glioblastoma and astrocytoma
JP-010	Gene therapy in patients with Buerger's disease

Morishita et al. treated five patients with E2F decoy ODN. No side effects were observed for up to 6 months, although the clinical outcome has not yet been evaluated. In 1996, clinical application of "decoy" ODN against E2F by V.J. Dzau at Harvard University (Cambridge, MA; USA) was approved by the FDA to treat neointimal hyperplasia in vein bypass grafts, which have a 10-year failure rate of up to 50%. The Project in Ex-Vivo Vein Graft Engineering Via Transfection was the first clinical trial using genetic engineering techniques to inhibit cell-cycle activation in vein grafts. This prospective, randomized, controlled trial demonstrated the safety and biologic efficacy of intra-operative transfection of human bypass vein grafts with E2F decoy ODN in a high-risk human patient population with peripheral arterial occlusion. Successful inhibition of graft occlusion, accompanied by selective inhibition of PCNA and c-myc expression (Mann 1999), were demonstrated. PREVENT II, a randomized, double-blind, placebo-controlled trial, has been set up to investigate the effect of E2F decoy ODN in preventing autologous vein graft failure after coronary artery bypass surgery.

References

Baumgartner I, Pieczek A, Manor O (1998) Constitutive expression of phVEGF 165 after intramusclar gene transfer promotes collateral vessel development in patients with critical limb ischemia. Circulation 97:1114–1123

Baumgartner I, Rauh G, Pieczek A (2000) Lower-extremity edema associated with gene transfer of naked DNA encoding vascular endotherial growth factor. Ann Intern med 132:880–884

Bocchia M, Bronte V, Colombo MP, De Vincentiis A, Di Nicola M, Forni G, Lanata L, Lemoli RM, Massaia M, Rondelli D, Zanon P, Tura S (2000) Antitumor vaccination: where we stand. Haematologica 85:1172–1206

Clark PR, Stopeck AT, Ferrari M, Parker SE, Hersh EM (2000) Studies of direct intratumoral gene transfer using cationic lipid-complexed plasmid DNA. Cancer Gene Ther 7:853–860

Comerota AJ, Throm RC, Miller KA (2002) Naked plasmid DNA encoding fibroblast growth factor type I for the treatment of end-stage unreconstructible lower extremity ischemia: preliminary results of a phase I trial. J Vasc Surg 35:930–936

D' Urso CM, Wang ZG, Cao Y, Tatake R, Zeff RA, Ferrone S (1991) Lack of HLA class I antigen expression by cultured melanoma cell FO-1 due to a defect in B2m gene expression. J Clin Investig 87:284–292

Felgner PL, Gadek TR, Holm M, Roman R, Chan HW, Wenz M, Northrop JR, Ringold GM, Danielson M (1987) Lipofection: A highly efficient, lipid-mediated DNA-transfection procedure. Proc Natl Acad Sci USA 84:7413–7414

Felgner PL, Ringold GM (1989) Cationic liposome-mediated transfection. Nature 337: 387–388

Galanis E (2002) Technology evaluation: Allovectin-7, Vical, Curr Opin Mol Ther 4(1):80–87

Gene Therapy Clinical Trials Worldwide (www.wiley.co.uk/genmed/clinical), The Journal of Gene Medicine, 2004

Hersh EM, Stopeck AT, Unger EC, Warneke JA, Silver HK, Klasa R, Doroshow JH, Rubin J, Galanis E, Rhinehart JJ (2001) Phase II study of intratumoral gene transfer of Allovectin-7 (plasmid DNA encoding HLA-B7 complexed with cationic lipid) in HLA-B7 negative patients with metastatic melanoma. Cancer Gene Ther

Isner JM, Walsh K, Rosenfield K (1996) Clinical protocol: arterial gene therapy for restenosis. Human Gene Ther 7:989–1011

Isner JM, Baumgartner I, Rauh G (1998) Treatment of thromboangiitis obliterans (Buerger's disease) by intramusclar gene transfer of vascular endothelial growth factor: preliminary clinical results. J Vasc Surg 28:964–973
Krieg AM, Yi AK, Matson S, Waldschmidt TJ, Bishop GA, Teasdale R, Koretzky GA, Klinman DM (1995) CpG motifs in bacterial DNA trigger direct B-cell activation. Nature 374: 546–549
Lathi KG, Vale PR, Losordo DW (2001) Gene therapy with vascular endothelial growth factor for inoperable coronary artery disease: anesthetic management and results. Anesth Analg 92:19–25
Latinen M, Hartikainen J, Hiltunen MO (2000) Catheter-mediated vascular endothelial growth factor gene transfer to human coronary arteries after angioplasty. Human gene Ther 11:263–270
Losordo DW, Vale PR, Symes JF (1998) Gene therapy for myocardial angiogenesis: initial clinical results with direct myocardial injection of phVEGF 165 as sole therapy for myocardial ischemia. Circulation 98:2800–2804
Makinen K, Manninen H, Hedman M (2002) Increased vascularity detected by digital subtraction angiography after VEGF gene transfer to human lower limb artery: a randomized, placebo-controlled, double-blinded phase II study. Mol Ther 6:127–133
Mann MJ, Whitemore AD, Donaldson MC (1999) Ex-vivo gene therapy of human vascular bypass grafts with E2F decoy: the PREVENT single-centre, randomized, controlled trial. Lancer 354:1493–1498
Mizuno M, Yoshida J (1998) Effect of human interferon-β gene transfer upon human glioma transplanted into nude mouse brain involves induced natural killer cells. Cancer Immunol Immunother 47:227–232
Morishita R, Higaki J, Tomita N, Ogihara T (1998) Application of transcription factor "decoy" strategy as means of gene therapy and study of gene expression in cardiovascular disease. Circ Res 82:1023–1028
Morishita R (2004a) Perspective in progress of cardiovascular Gene Therapy. J Pharmacol Sci 95:1–8
Morishita R, Aoki M, Hashiya N, Makino H, Yamasaki K, Azuma J, Sawa Y, Matsuda H, Kaneda Y, Ogihara T (2004b) Safety evaluation of clinical gene therapy using hepatocyte growth factor to treat peripheral arterial disease. Hypertension 44:203–209
Nabel GJ, Nabel EG, Yang ZY, Fox BA, Plautz GE, Gao X, Huang L, Shu S, Gordon D, Chang AE (1993a) Direct gene transfer with DNA-liposome complexes in melanoma: expression, biologic activity, and lack of toxicity in humans. Proc Natl Acad Sci USA 90: 11307–11311
Nabel EG, Nabel GJ (1993b) Direct gene transfer: basic studies and human therapies. Thromb Haemostasis 70:202–203
Natsume A, Mizuno M, Ryuke Y, Yoshida J (1999) Antitumor effect and cellular immunity activation by nature interferon-β gene transfer against intracerebral glioma in mouse. Gene Ther 6:1626–1633
Natsume A, Tsujimura K, Mizuno M, Takahashi T, Yoshida J (2000) IFN-β gene therapy induces systemic antitumor immunity against malignant glioma. J Neurooncol 47: 117–124
Roman M, Martin-OrozcoE, Goodman JS, Nguyen MD, Sato Y, Ronaghy A, Kornbluth RS, Richman DD, Carson DA, Raz E (1997) Immunostimulatory DNA sequences function as T helper-1-promoting adjuvants. Nat Med 3:849–854
Rosengart TK, Lee LY, Patel Sr, Sanborn TA, Parikh M, Bergman GW, Hachamovitch R, Szulc M, Kligfield PD, Okin PM, Hahn RT, Devereux RB, Post MR, Hackett NR, Foster T, Grasso TM, Lesser ML, Isom OW, Crystal RG (1999a) Angiogenesis gene therapy: phase I assessment of direct intramyocardial administration of an adenovirus vector expressing VEGF 121 cDNA to individuals with clinically significant severe coronary artery disease. Circulation 100:468–474

Rosengart TK, Lee LY, Patel SR, Kligfield PD, Okin PM, Hackett NR, Isom OW, Crystal RG (1999b) Six-month assessment of a phase I trial of angiogenic gene therapy for the treatment of coronary artery disease using direct intramyocardial administration of an adenovirus vector expressing the VEGF 121 cDNA. Ann Surg 230:466–470

Rubin J, Galanis E, Pitot HC, Richardson RL, Burch PA, Charboneau JW, Reading CC, Lewis BD, Starhl S, Akporiaye ET, Harris DT (1997) Phase I study of immunotherapy of hepatic metastases of colorectal carcinoma by direct gene transfer of an allogenic histocompatibility antigen, HLA-B7. Gene Therapy 4(5):419–425

Stopeck AT, Nicholson AC, Mancini FP, Hajjar DP (1993) Cytokine regulation of low density lipoprotein receptor gene transcription in HepG2 cells. J Biol Chem 268:17489–17494

Stopeck AT, Hersh EM, Akporiaye ET, Harris DT, Grogan T, Unger E, Warneke J, Schluter SF, Stahl S (1997) Phase I study of direct gene transfer of an allogenic histocompatibility antigen, HLA-B7, in patients with metastatic melanoma. J Clin Oncol 15:341–349

Stopeck AT, Jones A, Hersh EM, Thompson JA, Finucane DM, Gutheil JC, Gonzalez R (2001) Phase II Study of direct intralesional gene transfer of Allovectin-7, an HLA-B7/ββ2-microglobulin DNA-liposome complex, in patients with metastatic melanoma. Clinical Cancer Research 7:2285–2291

Taniyama Y, Morishita R, Aoki M, Nakagami H, Yamamoto K, Yamazaki K, Matsumoto K, Nakamura T, Kaneda Y, Ogihara T (2001) Therapeutic angiogenesis induced by human hepatocyte growth factor gene in rat and rabbit hind limb ischemia models: preclinical study for treatment of peripheral arterial disease. Gene Ther 8:181–189

Vale PR, Losordo DW, Milliken CE (2000) Left ventricular electromechanical mapping to assess efficacy of phVEGF 165 gene transfer for therapeutic angiogenesis in chronic myocardial ischemia. Circulation 102:965–974

Yagi K, Noda H, Kurono M, Ohishi N (1993) Efficient gene transfer with less than cytotoxicity by means of cationic multilamellar liposomes. Biochem Biophys Res Commun 196:1042–1048

Yoshida J, Mizuno M, Yagi K (1992a) Cytotoxicity of human β-interferon produced human glioma cells transfected with its gene by means of liposomes. Biochemistry International 28:1055–1061

Yoshida J, Mizuno M, Yagi K (1992b) Antitumor effect of endogenous human β-interferon on malignant glioma and augmentation of the effect by tumor necrosis factor-a. J Clin Biochem Nutr 77:78–83

Yoshida J, Mizuno M (2003) Clinical gene therapy for brain tumors. Liposomal delivery of anticancer molecule to glioma. J Neuro-Oncology 65:261–267

Yoshida J, Mizuno M, Fujii M, Kajita Y, Nakahara N, hatano M, Saito R, Nobayashi M, Wakabayashi T (2004) Human gene therapy for malignant gliomas (glioblastoma multifirme and anaplastic astrocytoma) by in vivo transduction with human interferon β gene using cationic liposomes. Human Gene Therapy 15:77–86

3. Current Protocols of Gene Delivery

Evaluation of Size and Zeta Potential of DNA/Carrier Complexes

Yoshiyuki Hattori and Mitsuru Hashida

1 Introduction

To date, many kinds of non-viral gene vectors, such as cationic liposomes, cationic polymers, e.g., poly-L-lysine, polyethyleneimine (PEI), and cationic dendrimers, have been developed (Niidome and Huang 2002). These non-viral vectors have been a great advance clinically due to their low toxicity, low immunogenicity, and ease of preparation. And while some of them have already been applied to clinical use, the transfection activities are still low and have to be improved.

In order to achieve high and long-term gene expression with non-viral vectors, it is essential to understand the relationship between the physicochemical properties of non-viral vectors and their transfection activities. Many studies, including ours, have demonstrated that the physicochemical properties impact on several processes such as distribution, interaction with biocomponents, absorption to target cells and intracellular trafficking (Li et al. 1998; Sakurai et al. 2000). Since the formation of cationic carrier/plasmid DNA (pDNA) complexes depends on the electrostatic interaction between the cationic charge, derived from the cationic vector, and the anionic charge, from the pDNA, the mixing ratio of these two components exerts effects on the physicochemical properties of the resultant complex, such as particle size and zeta potential.

The zeta potential of the cationic carrier/pDNA complex is important for gene transfection because cellular uptake of a complex is considered to be a non-specific process that is based on the interaction of the excess positive charge of the complex and the negatively charged cell membrane.

Furthermore, the cationic charge of the complex also determines its interaction with non-target cells and biocomponents that are negatively charged under in vitro conditions. Therefore, the cationic charge has to be optimized in order to yield optimum transfection activity in target cells under both in vitro and in vivo conditions.

Department of Drug Delivery Research, Graduate School of Pharmaceutical Sciences, Kyoto University, Sakyo-ku, Kyoto 606–8501, Japan

The particle size of the cationic carrier/pDNA complex also plays an important role in gene delivery to the target cell under in vitro and in vivo conditions. Especially under in vivo conditions, the particle size has to be well controlled because it governs many factors, such as biodistribution, gene expression, and toxicity.

This chapter focuses on preliminary methods for evaluating the particle size and zeta potential of cationic carrier/pDNA complexes, and discusses recent studies of the physicochemical properties of cationic carrier/pDNA complexes with respect to the development of efficient non-viral systems.

2 Materials

2.1 Plasmid DNA

pCMV-Luc was constructed by subcloning the *Hin*dIII/*Xba*I firefly luciferase cDNA fragment from the pGL3-control vector (Promega, Madison, WI, USA) into the polylinker of the pcDNA3 vector (Invitrogen, Carlsbad, CA, USA). Purity was confirmed by 1% (w/v) agarose gel electrophoresis followed by ethidium bromide staining; the DNA concentration was measured by UV absorption at 260 nm.

3 Methods

3.1 Preparation of Cationic Liposomes

N-[1-(2,3-dioleyloxy)propyl]-*n,n,n*-trimethylammonium (DOTMA) (Tokyo Chemical Industry, Tokyo, Japan) was mixed with dioleoylphosphatidylethanolamine (DOPE) (Avanti Polar-Lipids, Alabaster, AL, USA) in chloroform (5.0 ml) at a molar ratio of 1 to 1 and evaporated to dryness in a round-bottomed flask. The resulting lipid film was vacuum desiccated to remove any residual organic solvent and resuspended in 1 ml sterile water in a round-bottomed flask. After hydration, the dispersion was sonicated for 5 min in a bath sonicator and then passed ten times through an Extruder (Lipex Biomembrane, Vancouver, Canada) fitted with a 100-nm pore membrane followed by passage through a 450-nm filter for sterilization. The lipid concentration was determined by phosphorus analysis (Bartlett et al. 1959).

3.2 Preparation of Cationic Liposome/pDNA Complexes

1. The liposome dispersion (1 mg/ml) was diluted with 150 mM NaCl solution to 0.1 mg/ml and allowed to stand for 30 min.
2. An aliquot of liposome solution was mixed with diluted plasmid DNA solution and allowed to stand for 15 min.

3.3 Preparation of PEI/pDNA Complexes

1. Plasmid DNA was mixed with varying amounts of branched PEI (M.W.: 10,000) (Polysciences, Warrington, PA, USA) at a final concentration of 2 μg pDNA/ml in 5% dextrose solution (Morimoto et al. 2003).
2. The complex solution was allowed to stand at room temperature for 30 min.

3.4 Zeta Potential and Particle Sizes at Various Mixing Ratios

3.4.1 Measurement of Particle Size Using a Dynamic Laser Scattering Particle Size Analyzer

In this section, we describe a fundamental procedure for measuring the particle size of the complex using a dynamic light scattering spectrophotometer (LS-900, Otsuka Electronics, Osaka, Japan). In order to minimize contamination by dusts, the wash solution of the cells and diluents of the carrier complex (5% dextrose or 150 mM NaCl solution) were filtrated five times through a 450-nm filter.

1. pDNA/cationic carrier complexes were prepared as described above.
2. The cells were washed three times with wash solution.
3. The sample was loaded into the cell of the LS-900, the parameters required for measurement were adjusted, and the sample was measured.
4. Particle size distribution was determined using the histogram analytical method.

3.4.2 Measurement of Zeta Potential Using a Laser Electrophoresis Zeta-Potential Analyzer

The zeta potential of cationic carrier/pDNA complex was evaluated using the laser-Doppler electrophoresis method with an LEZA-700 (Otsuka Electronics) equipped with He-Ne ion laser (633 nm). The analysis of the zeta potential analysis was carried out at 25.0°C.

1. Cationic carrier/pDNA complexes were prepared as described above.
2. The cells were washed with 150 mM NaCl.
3. The sample was loaded into to the cell and the zeta potential was analyzed.

3.5 Determination of Particle Sizes After Exposure of Plasmid DNA/Cationic Liposome to Rat Serum

The physicochemical properties of the administered cationic carrier/pDNA complex formulation may change in vivo because of the presence of salts, proteins, lipids, carbohydrates, opsonins, or enzymes in the vascular or interstitial space. These interactions may impair resultant gene expression by the target cells. Since both cationic liposome and cationic polymer/pDNA complexes are cationic, as described in the Introduction, the complexes could interact with various biological components.

In the case of intravenous administration, the interaction of cationic carrier/pDNA complexes with serum proteins and red blood cells has major consequences for both the cationic liposome and the polymer (Sakurai et al. 2001a,b; Ogris et al. 1998). Therefore, the interaction of cationic carrier/pDNA complexes with biocomponents and their relationship to the physicochemical properties of complex has to be investigated. In the following, a protocol to evaluate the physicochemical properties of cationic liposome/pDNA complex after exposure to serum is described.

1. Rat serum was isolated from fresh whole blood obtained from male Wistar rats. Blood was collected from the vena cava under anesthesia without heparin treatment and allowed to stand for 2 h at 37°C and then overnight at 4°C. Serum was collected after centrifugation.

2. Lipoplexes were prepared as described above.
3. The lipoplex dispersions were mixed with 66.7% (v/v) rat serum [final serum concentration 33.3% (v/v)], the mixtures were incubated for 5 min and then 60-fold diluted with 150 mM NaCl before measurement.
4. Particle sizes and zeta potential of the complex were measured with DLS-900 and LEZA-700 as described above.

4 Results

4.1 Comparison of the Zeta Potential and Particle Sizes of Cationic Liposome/pDNA and Cationic Polymer/pDNA Complexes

The zeta potential and particle sizes of cationic carrier/pDNA complex, i.e. DOTMA/DOPE liposome/pDNA complex and PEI/pDNA complex, were measured in order to compare the various mixing ratios of pDNA to cationic carrier.

Figure 1a shows the zeta potential and particle sizes of DOTMA/DOPE liposome/pDNA complexes. At a low (1:1 and 1:2.5) or high (1:7.5 and 1:10) mixing ratio, the particle size of the complex was relatively small (150–250 nm). In contrast, at intermediate mixing ratios of 1:3.75 (charge ratio −0.84) and 1:5 (charge ratio +1.12),

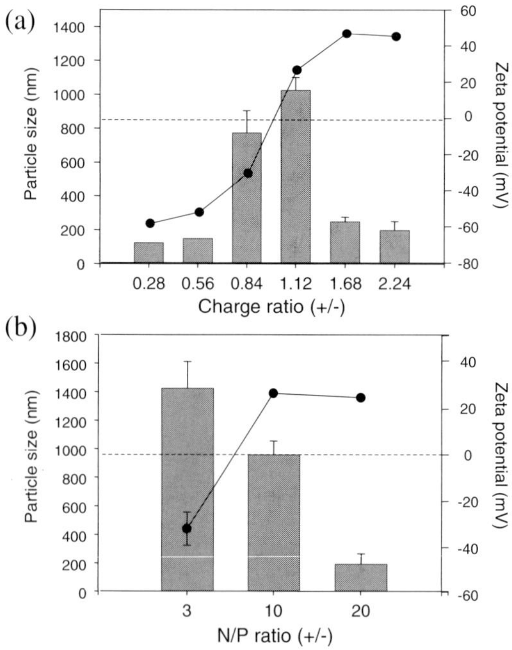

FIG. 1. Particle sizes and zeta potential of DOTMA/DOPE liposome (**a**) and branched PEI (**b**)/DNA complexes prepared at various mixing ratios. *Dotted line*, particle size; *hatched bar*, zeta potential. Each value represents the mean ± S.D.

larger complexes (800–1000 nm) were formed. Since, theoretically, the net charge becomes zero at a ratio of 1:4.47, this suggests that, at these mixing ratios, interaction between the complexes could easily occur due to their weak surface charge (small absolute zeta-potential values), resulting in the formation of larger complexes. At ratios of 1:7.5 and 1:10, the mean particle size decreased significantly and the zeta potential was high, almost equal to that of DOTMA/DOPE liposomes alone. These results suggest that plasmid DNA is covered or encapsulated by the liposomes.

Figure 1b shows the zeta potential and particle sizes of PEI/pDNA complexes. In the case of PEI, complexes were negatively charged at an N/P ratio of 3, and increasing the ratio to 10 or 20 resulted in the formation of a positively charged complex. The increase in the N/P ratio from 10 to 20 hardly changed the zeta potential of the complexes, suggesting the presence of free polymers in the mixtures. As for the particle sizes of PEI/DNA complexes, increasing the N/P ratio resulted in a decrease of particle size below 200 nm.

These results showed that, regardless of the cationic carrier, increasing the mixing ratio of cationic carrier to pDNA generally affects particle size and zeta potential in the same manner. Therefore, the mixing ratio may affect interactions with the cell membrane in the same manner, regardless of the cationic carrier.

The effects of the physicochemical properties of cationic carrier/pDNA complexes on transfection of cultured cells was evaluated by analyzing the effect of the mixing ratio of DOTMA/DOPE liposomes to pDNA on cellular association and intracellular trafficking. DOTMA/DOPE liposome/pDNA complexes were prepared at various mixing ratios in MBT-2, ECV304, NIH3T3, and RAW264 (Sakurai et al. 2000) in order to examine the cellular association of the complexes. The results showed that, in all types of cells, the cellular association of DOTMA/DOPE liposome/pDNA complexes increased in parallel with the amount of DOTMA/DOPE liposomes.

Oh et al. demonstrated that PEI/pDNA complexes prepared at an higher N/P ratio could be taken up by PaTu 8902 cells more efficiently than complexes prepared at a lower N/P ratio (Oh et al. 2002). These results indicate that cellular association is a zeta potential-dependent step in gene transfection, both in cationic liposomes and cationic polymers.

The particle size of cationic carrier/pDNA complexes is also important in in-vitro gene transfection. It has been reported that complexes consisting of larger cationic liposomes (Ross and Hui 1999) or cationic polymers/pDNA (Ogris et al. 1998) showed higher transfection activity to cultured cell. These reports suggested that larger complex associate efficiently with cells resulting in improved intracellular trafficking. Since the particle sizes of the complexes are strongly affected by ionic conditions, the particle size could be optimized not only by changing mixing ratio of cationic carrier to pDNA but also by choosing a different preparation medium (Kichler et al. 1998).

Transgene expression under in vivo condition is also affected by the particle size of the cationic carrier/pDNA complexes. Templeton et al. (1997) demonstrated that larger cationic liposomes led to higher gene expression after intravenous administration. However, since resolving the safety issue is essential for clinical applications, the influence of the particle size on toxicity to the body also has to be considered. It has been shown that larger PEI/pDNA complexes are potentially more lethal than small complexes (Ogris et al. 1999). Therefore, particle size has to be well considered, including understanding both its transfection activity and its toxicity.

4.2 Determination of Particle Sizes After Exposure of Plasmid DNA/Cationic Liposome to Rat Serum

Table 1 shows the physicochemical properties of DOTMA/DOPE liposome/pDNA complexes before and after exposure to rat serum. There were no differences in the particle sizes before and after exposure. By contrast, the zeta potential of the DOTMA/DOPE liposome/pDNA complexes changed to minus 10mV after exposure to rat serum. In addition, the effect of serum on the lung accumulation of cationic liposome/pDNA complexes using a lung perfusion system was evaluated. Pre-incubation of cationic liposome/pDNA complexes with serum reduced lung accumulation (Sakurai et al. 2001a, b), indicating that the interaction of cationic liposome/pDNA complex with serum could affect the biodistribution of the complexes.

Recent studies have demonstrated that the interaction of cationic carrier complexes with red blood cells (RBC) is a primary event that occurs after intravenous administration of cationic carrier/pDNA complexes. The complexes were shown to interact with RBC and the interaction was increased in a zeta-potential-dependent manner (Sakurai et al. 2001a, b). These interactions could explain why the highest level of gene expression occurred in the lungs after intravenous administration of cationic carrier/pDNA complexes. In the case of targeted delivery to organs other than the lung, this interaction is undesirable and has to be prevented.

It is known that cationic polymers like PEI also interact with plasma and RBC. Therefore, regardless of the cationic carrier, it is essential to control interaction of the cationic carrier/pDNA complexes with blood components in order to ensure efficient gene delivery to the target site Recently, Ogris et al. (1999) reported that PEG-ylation of transferrin-decorated PEI/pDNA complex strongly reduced plasma protein binding and erythrocyte aggregation and allowed successful transfer of pDNA to the tumor by intravenous administration without significant toxicity (Ogris et al. 1999). Such modifications for avoiding undesirable interactions caused by the physicochemical properties of the complexes offers a promising approach to efficient gene delivery.

Hence, the physicochemical characteristics of complexes, such as particle size and zeta potential, will no doubt play critical roles not only with respect to transfection activity in target cells but also regarding interactions with and toxicity to the body.

TABLE 1. The physicochemical properties of DOTMA/DOPE liposome/DNA complexes before and after exposure to rat serum. Complex dispersions were mixed with rat serum at a final serum concentration of 33.3%, and the mixtures were incubated for 5 min before measurement. Each value represents the mean ± S.D

Particle size (nm)		Zeta potential (mV)	
Before	After	Before	After
189.8 ± 22.7	234.4 ± 20.1	34.5 ± 1.2	−12.7 ± 0.6

References

Bartlett GR (1959) Phosphorus assay in column chromatography. J. Biol Chem 234:466–468

Kichler A, Zauner W, Ogris M, Wagner E. (1998) Influence of the DNA complexation medium on the transfection efficiency of lipospermine/DNA particles. Gene Ther 5:855–860

Li S, Rizzo MA, Bhattacharya S, Huang L (1998) Characterization of cationic lipid-protamine-DNA (LPD) complexes for intravenous gene delivery. Gene Ther 5:930–937

Morimoto K, Nishikawa M, Kawakami S, Nakano T, Hattori Y, Fumoto S, Yamashita F, Hashida M (2003) Molecular weight-dependent gene transfection activity of unmodified and galactosylated polyethyleneimine on hepatoma cells and mouse liver. Mol Ther 7:254–261

Niidome T, Huang L (2002) Gene therapy progress and prospects: nonviral vectors. Gene Ther 9:1647–1652

Oh YK, Suh D, Kim JM, Choi HG, Shin K, Ko JJ (2002) Polyethylenimine-mediated cellular uptake, nucleus trafficking and expression of cytokine plasmid DNA Gene Ther. 200 29: 1627–1632

Ogris M, Brunner S, Schuller S, Kircheis R, Wagner E. (1999) PEGylated DNA/transferrin-PEI complexes: reduced interaction with blood components, extended circulation in blood and potential for systemic gene delivery. Gene Ther 6:595–605

Ogris M, Steinlein P, Kursa M, Mechtler K, Kircheis R, Wagner E. (1998) The size of DNA/transferrin-PEI complexes is an important factor for gene expression in cultured cells. Gene Ther 5:1425–1433

Ross PC, Hui SW (1999) Lipoplex size is a major determinant of in vitro lipofection efficacy. Gene Ther 6:651–659

Sakurai F, Inoue R, Nishino Y, Okuda A, Matsumoto O, Taga T, Yamashita F, Takakura Y, Hashida M (2000) Effect of DNA/liposome mixing ratio on the physicochemical characteristics, cellular uptake and intracellular trafficking of plasmid DNA/cationic liposome complexes and subsequent gene expression. J Control Release 66:255–69

Sakurai F, Nishioka T, Saito H, Baba T, Okuda A, Matsumoto O, Taga T, Yamashita F, Takakura Y, Hashida M. (2001) Interaction between DNA-cationic liposome complexes and erythrocytes is an important factor in systemic gene transfer via the intravenous route in mice: the role of the neutral helper lipid. Gene Ther 8(9):677–686

Sakurai F, Nishioka T, Yamashita F, Takakura Y, Hashida M. (2001) Effects of erythrocytes and serum proteins on lung accumulation of lipoplexes containing cholesterol or DOPE as a helper lipid in the single-pass rat lung perfusion system. Eur J Pharm Biopharm 52: 165–72

Templeton NS, Lasic DD, Frederik PM, Strey HH, Roberts DD, Pavlakis GN (1997) Improved DNA: liposome complexes for increased systemic delivery and gene expression. Nat Biotechnol. 15:647–652

Observation of DNA/Carrier Complexes Under Fluorescence Microscopy

Tomoko Ito[1], Yoshiyuki Koyama[1], and Kenichi Yoshikawa[2]

1 Introduction

In water, plasmids take on a rather relaxed, elongated conformation, with a diameter of micrometers, which makes it difficult for cells to efficiently internalize the giant molecule (Zhou et al. 1994). Polycations or cationic lipids, which are used as non-viral vectors for gene therapy, condense plasmid molecule into a compact particle through electrostatic interactions as well as chemical effects such as hydrophobic interactions. The size and form of the resulting complex strongly affects the transfection efficiency. Thus, in non-viral gene therapy, it is important to examine the morphology of DNA complexes in order to achieve better results.

DNA molecules have been directly observed by electron microscopy. Recently, atomic force microscopy (AFM) has gained increasing popularity. While the resolution obtained with the fluorescence microscopy is lower than those of electron microscopy or of AFM (Yoshinaga et al. 2002), the latter two provide only static information on the structures of pretreated, and dried or fixed DNA molecules. By contrast, fluorescence microscopy can be carried out with the samples in aqueous solution, and affords information about the dynamic movement and conformation of DNA molecules in solution. Moreover, with fluorescence microscopy single DNA molecules can be directly observed (Yanagida et al. 1983), which provides important information that cannot be obtained with any other apparatus. For example, using fluorescence microscopy, DNA molecules were found to exhibit a first-order phase transition in their higher-ordered structures between the elongated coil and the compacted globule, by addition of a condensing agent (Yoshikawa et al. 1995). The hydrodynamic size of DNA complex particles in water can also be estimated by analyzing their Brownian motion (Mel'nikol et al. 1995). Thus, the observation of individual DNA molecules with fluorescence microscopy allows gene transfection to be monitored under natural conditions, including in cellular systems.

[1] Otsuma Women's University, 12 Sanbancho, Chiyoda-ku, Tokyo 102-8357, Japan
[2] Kyoto University, Kitasirakawa, Sakyo-ku, Kyoto 606-8502, Japan

2 Principle of Single Molecular Observation by Fluorescence Microscopy

In general, DNA is a very thin molecule, about 2 nm in diameter, but several DNA molecules, such as T4- or λ-DNA, have a length of tens of micrometers when fully stretched. When DNA molecules are stained by an appropriate fluorescence dye, the apparent diameter expands to about 0.6 μm due to blurring effects (Figs. 1, 2). Thus, in a ×1000 microscopy image, a single T4 DNA molecule would, thus appeared as a string 60 mm in length with a diameter of 0.6 mm (Minagawa et al. 1991). A compacted, very small DNA complex would be seen as a bright dot with a diameter of 0.6 mm (Fig. 3).

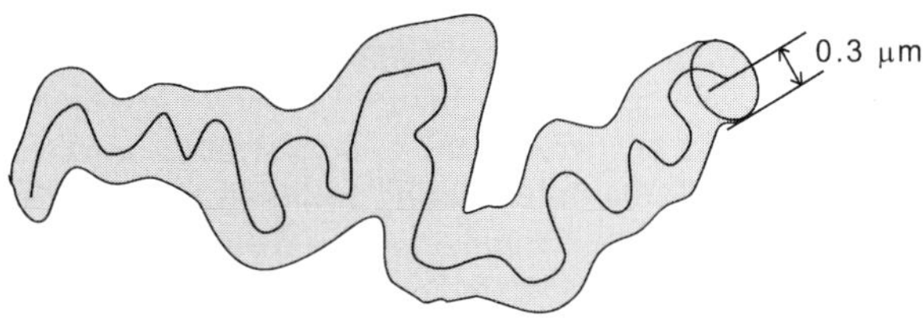

FIG. 1. Blurring effect

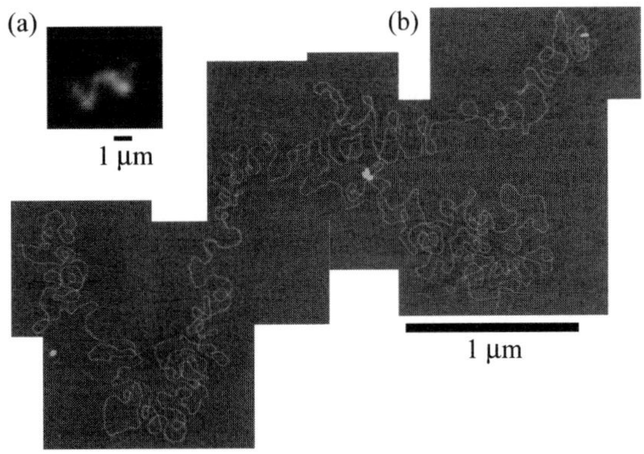

FIG. 2. Fluorescence (**a**) and atomic-force (**b**) micrographs of a single T4 DNA chain in the random coiled state absorbed onto a mica surface. These images show the very same DNA with different resolutions. (Images provided by T. Nakai, Kyoto University)

FIG. 3. Fluorescence microscopy of T4 DNA molecules. *Left* Elongated random coils, *right* compact globular complexes

3 Fluorescent Dyes for Observing DNA Molecules

The fluorescent dye 4′,6-diamidino-2-phenyl indole (DAPI) is often used for the observation of DNA molecules. DAPI binds to the minor groove of the DNA double helix but has little effect on the persistent length of the DNA molecular chain (Matsuzawa et al. 1994). In addition, if the concentration of DAPI is not too high, structural changes of the DNA are not obviously hindered. When the final concentration of DNA is less than 10 μM (in base), an equal concentration of DAPI is added to the DNA. If the concentration of DNA is higher, then the concentration of DAPI should be less than 10 μM in order to prevent the background from becoming overly bright.

DAPI binds reversibly and can permeate the cell membrane. For the observation of DNA molecules together with cells, an intercalating fluorescence dye, YOYO (Molecular Probes, Oregon USA), is recommended. This dye stably intercalates between the base pairs of DNA, and is not rapidly dissociated. It can, therefore, be used to examine intracellular trafficking of DNA complexes.

Since YOYO extends the persistent length of the DNA molecules, it should be added at a YOYO/DNA ratio of less than 1/5 in moles (on the basis of nucleotide), otherwise the condensing behavior of the DNA is strongly affected. It is also important to note that YOYO and other intercalating agents significantly retard the folding transition of DNA molecules into a compact state (Yoshinaga et al. 2001). Observation should be carried out with 3%–6 % (in volume) of 2-mercaptoethanol as an antioxidant.

4 Fluorescence Microscopy and the Recording System

An inverted fluorescence microscope equipped with a SIT TV camera, image processor, DVD recorder, and monitor is used for fluorescence microscopy observations. This system is schematically shown in Fig. 4. We use an IX70 inverted fluorescence microscope (Olympus, Japan) equipped with a ×100 oil-immersed objective lens, and the image is recorded on a DVD recorder at a sampling rate of 30 frames/s through the high-sensitivity Hamamatsu SIT TV camera (Hamamatsu Photonics, Japan). Using this system, we observed DNA molecules at concentrations ranging from 0.2 μM to as high as 120 μM, which is comparable to those used in in vivo experiments.

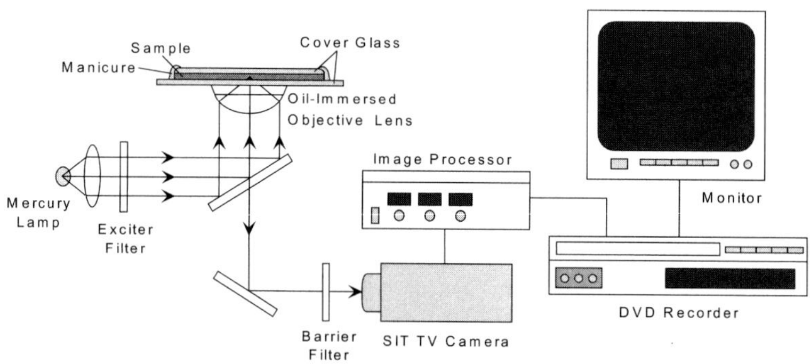

Fig. 4. The fluorescence microscopy system

5 Sample Preparation

A frame of 22 × 22 mm or less is drawn on a microscope cover-glass (30 × 40 mm, thickness: 0.12–0.17 mm) as a grid using a PapPen (Daido Sangyo, Japan). It is dried to serve as a spacer to separate the top and bottom glasses. The DNA sample solution (5–10 μl), containing a fluorescent dye and an antioxidant, is transferred drop-wise within the frame, covered with a smaller cover glass (22 × 22 mm), and then sealed by nail polish to prohibit drying out of the sample. Most DNA/polycation complexes are positively charged on their surfaces and thus readily attach to the negatively charged glass surfaces, while the spacer frame system allows the movement of the complexes in water to be observed over a period of minutes. A glass-bottomed culture dish (Glass Base Dish, Asahi Techno Glass, Japan) on which the cells have been grown can also be used to observe the interaction of DNA complexes with the cells.

6 Measurement of DNA Size in Solution

The light scattering method has been used to measure the size of the DNA complex particles; but with this method only the averaged size of the complexes is obtained. The size of an individual DNA molecule, however, can be obtained by fluorescence microscopy image analysis. Each DNA complex appears as bright dots in the microscopic image that can be measured. Nonetheless, the apparent sizes of the DNA complexes are larger than their real dimensions because of the blurring effects mentioned above. Instead, the hydrodynamic gyration radius can be evaluated by analyzing the randomly fluctuating Brownian motion of the complexes.

The images of the DNA complexes recorded on the DVD are transferred to a personal computer and analyzed by an animation analysis software (Move-tr32/2D: Library, Japan). The movement-track of the DNA by Brownian motion is followed every 1/30 of a second for 1 s. The mean square displacement (MSD) of the center of mass is plotted as a function of time (t) for each DNA complex (Fig. 5). The diffusion constant (D_G) is given as:

$$D_G = \mathrm{MSD}/4t$$

The hydrodynamic gyration radius, R_g, can be calculated according to the Einstein-Stokes' equation:

$$D_G = 0.2030 \frac{k_B T}{\sqrt{6}\eta_s R_g}$$

where k_B is the Boltzmann constant and η_s is the viscosity of the solvent.

In the case of extremely dilute solution, the viscosity of water is used as η_s.

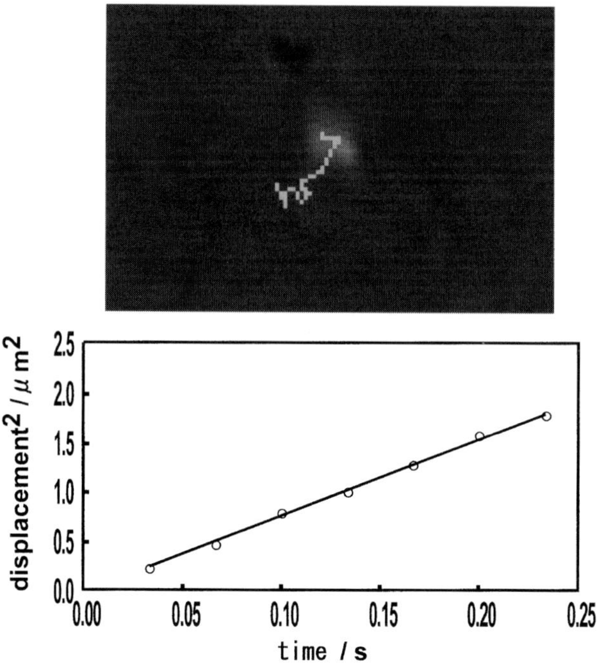

FIG. 5. Plot of the mean square displacement (MSD) of the DNA complexes vs. time

7 Observation of Each Component of the DNA Polyion Complex

When the DAPI-labeled DNA is complexed with a polycation attached to a second fluorescent dye, the DNA and the polycation can be observed separately by changing the excitation wavelength. The DNA/polycation ratio can then be roughly estimated by measuring the fluorescence intensity of each component.

We have studied a ternary polyion complex gene-expression system comprising DNA, polycation, and polyanion (see the chapter by Koyama, this volume). These three components were visualized with DAPI, FITC, and rhodamine, respectively. In each image generated with the corresponding excitation wavelength, the bright spots of the fluorophores were observed at the same position, visually confirming formation of the ternary polyion complex.

8 Observation of the DNA Interaction with Cells

Simultaneous observation of the DNA complexes and the cells by fluorescence microscopy is also possible. Attachment and internalization of DNA complex can be examined in detail. In a series of experiments by Ito et al. (2004), cells were cultivated in a glass-bottomed dish, and the fluorescently labeled DNA complexes were added. The cells were analyzed by phase microscopic image, and, under dark field, the DNA

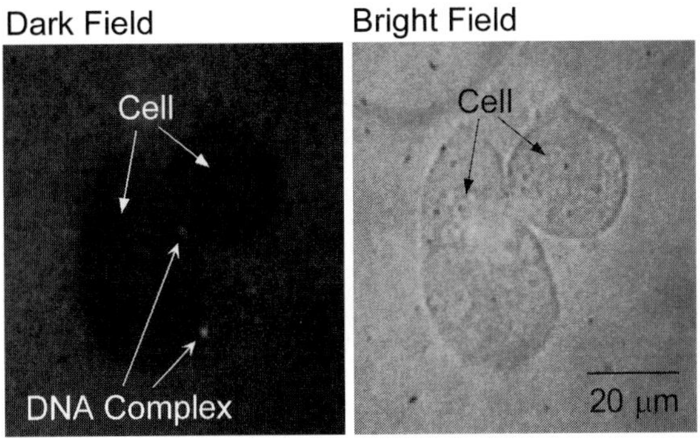

FIG. 6. DNA complexes coexistent with cells

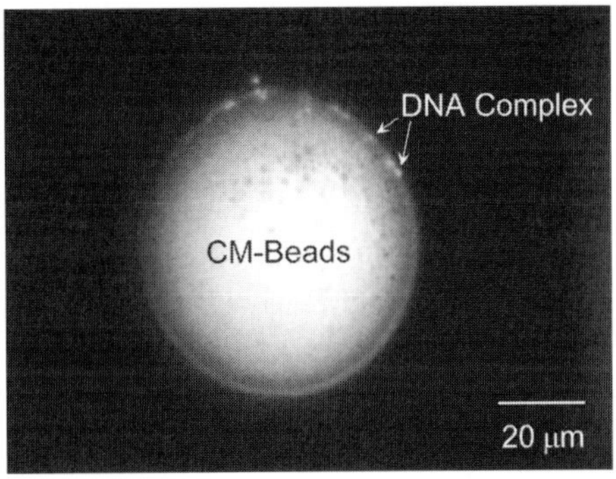

FIG. 7. Attachment of the positively charged DNA complexes onto carboxymethyl-Sephadex

complex is observed as bright dots. Irradiation with a small amount of visible light allows the shapes of the cells to be dimly observed as well. The binding behavior of the complexes onto the cell surface, and the trafficking of each component can be monitored by changing the excitation wavelength.

9 Attachment of the DNA Complexes onto Charged or Protein-Immobilized Beads

When the positively charged DNA complexes were incubated with anionic carboxymethyl-Sephadex (CM-Sephadex) microbeads, rapid disposition of the complex onto the beads surface was observed. Similarly, polyanion-coated complexes attached to cationic diethylaminoethyl Sephadex (DEAE-Sephadex) beads. Alteration of the surface potential of the complex could be easily and directly examined by microscopic observation of the adhesion behavior of the complexes to the surfaces of the charged beads (Koyama et al. 2002; Maruyama et al. 2004).

We have prepared sugar-bearing polyanions to coat positively charged DNA complexes. Recognition of the sugar residues on the DNA complex by lectins was examined using ricin-120- or concanavalin-A-agarose. Glucose-bearing complexes attached only to the concanavalin-A-agarose, while galactose-containing ones deposited onto ricin-120-agarose. Deposition was effectively inhibited by competition with excess free sugar, thereby confirming the sugar-specific adhesion of the complexes to the carbohydrate-binding proteins (Koyama et al. 2002).

References

Ito T, Tanaka NI, Sato T, Yang Z, Niidome T, Kawano T, Koyama Y (2004) Hyaluronic acid as a self-assembled coating of plasmid/polycation complexes for cell-specific gene delivery. Molecular Therapy 9:S321–S322

Koyama Y, Yamada E, Ito T, Mizutani Y, Yamaoka T (2002) Sugar-containing polyanions as a self-assembled coating of plasmid/polycation complexes for receptor-mediated gene delivery. Macromol. Biosci. 2:251–256

Maruyama K, Iwasaki F, Takizawa T, Yanagie H, Yamada E, Ito T, Koyama Y (2004) Novel receptor-mediated gene delivery system comprising plasmid/protamine/sugar-containing polyanion ternary complex. Biomaterials 25:3267–3273

Matsuzawa Y, Yoshikawa K (1994) Change of the higher order structure in a giant DNA induced by 4′,6-diamidino-2-phenylindole as a minor groove binder and ethidium bromide as an intercalator. Nucleosides & Nucleotides 13:1415–1423

Mel'nikol SM, Sergeyev VS, Yoshikawa K (1995) Discrete coil-globule transition of large DNA induced by cationic surfactant. J Am Chem Soc 117:2401–2408

Minagawa K, Matsuzawa Y, Yoshikawa K, Matsumoto M, Doi M (1991) Direct observation of the biphasic conformational change of DNA induced by cationic polymer. FEBS Letter 295:67–69

Yanagida M, Hiraoka Y, Katsura I (1983) Dynamic behaviors of DNA molecules in solution studied by fluorescence microscopy. Cold Spring Harbor Symp Quant Biol 47:177–187

Yoshikawa K, Matsuzawa Y (1995) Discrete phase transition of giant DNA dynamics of globule formation from a single molecular chain. Physica D84:220–227

Yoshinaga N, Akitaya T, Yoshikawa K (2001) Intercalating fluorescence dye YOYO-1 prevents the folding transition in giant duplex DNA. Biochemical and Biophysical Research Communications 286:264–267

Yoshinaga N, Yoshikawa K, Kidoaki S (2002) Multiscaling in a long semiflexible polymer chain in two dimension. J Chem Phys 116:9926–9929

Zhou X, Huang L (1994) DNA transfection mediated by cationic liposomes containing lipopolylysine: characterization and mechanism of action. Biochim. Biophys. Acta 1189: 195–203

Synthesis of Chemically Modified Chitosan and a Study of Its Gene Transfection Efficiency

Taku Satoh

1 Introduction

Chitosan (CT) has been widely explored with regard to its gene transfer efficiency, and a number of studies have revealed the superior characteristics of CT as a transfection reagent (see chapter 1–6). Moreover, numerous CT derivatives with enhanced transfection efficiency have been synthesized (Scheme 1). For example, *N*-trimethylated CT was prepared to increase the permeability of the intestinal epithelia (Thanou et al. 2002). Galactose residues were linked onto the CT backbone to recognize hepatocytes (Murata et al. 1996; Gao et al. 2003; Kim et al. 2004). Park et al. further modified galactosylated CT with hydrophilic polymers, such as dextran (Park et al. 2000), poly(ethylene glycol) (Park et al. 2001), and poly(vinyl pyrrolidone) (Park et al. 2003), to enhance the stability of the derivatives in aqueous media. Furthermore, urocanic-acid-modified CT was reported to rupture endosomes through the proton sponge mechanism (Kim et al. 2003). Hydrophobic modifications with deoxycholic acid (Kim et al. 2001) or alkyl groups (Liu et al. 2001) enabled these derivatives to control the particle size by self-assembly of substituents in aqueous media. In addition to galactose residues, the conjugation of transferrin and the KNOB domain of adenovirus (Mao et al. 2001) was attempted in order to provide cell-specific target ability through ligand-receptor interactions.

In addition, due to its high cationic charge density, polyethylenimine (PEI) has been one of the most successfully and widely studied cationic polymers reported to date for use in gene delivery (Bousiff et al. 1995). Therefore, in order to increase the cationic charge of biocompatible and biodegradable CT, we recently prepared 6-amino-6-deoxychitosan (6ACT), which contains two primary amino groups per repeating unit (Satoh et al. 2004). Evaluation of 6ACT acetate as a non-viral gene carrier revealed several basic characteristics: The cytotoxicity of 6ACT was as low as that of CT acetate. Complete complexation with plasmid DNA was observed when molar ratios of the amino groups of 6ACT to the phosphate groups of plasmid DNA (N/P) were adjusted at N/P > 2. The 6ACT/DNA complex was highly resistant against DNase I degradation.

Department of Applied and Bioapplied Chemistry, Graduate School of Engineering, Osaka City University, 3-3-138 Sugimoto, Sumiyoshi-ku, Osaka 558-8585, Japan

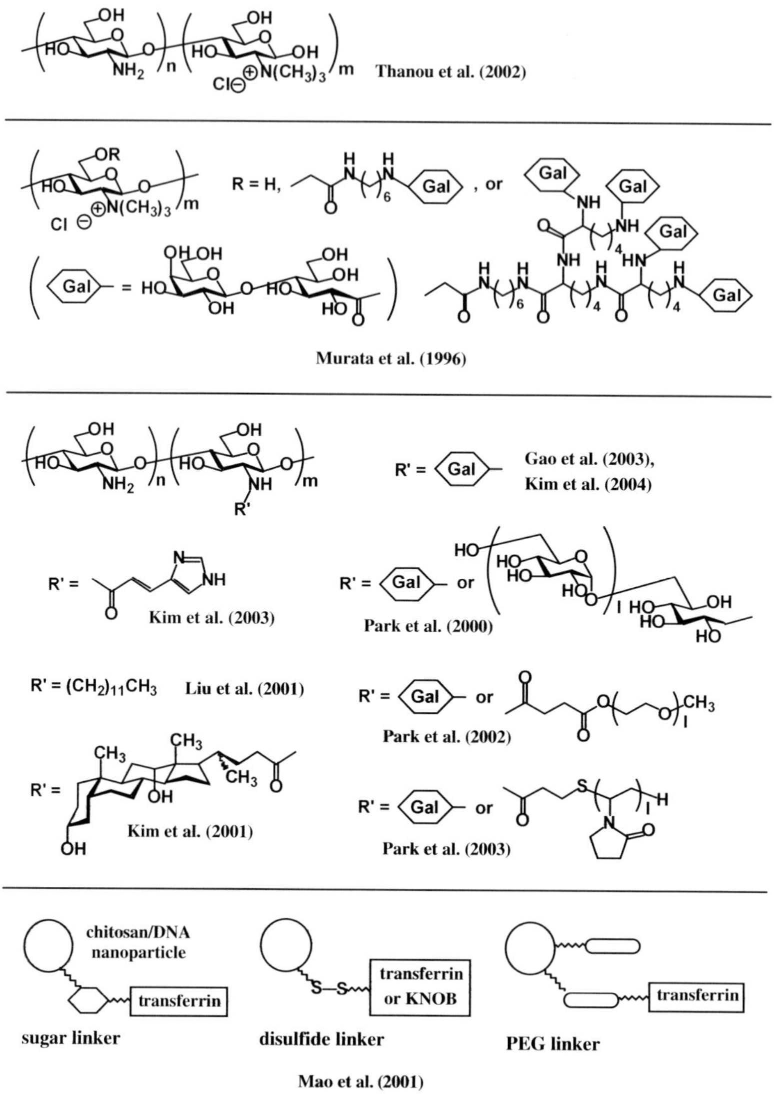

SCHEME 1. Chemical structure of representative non-viral gene carriers based on chitosan

The transfection efficiency of the 6ACT/DNA complexes into COS-1 cells was slightly affected by complexation pH, and under optimal conditions the efficiency was superior to that of CT acetate. These properties demonstrate the potential of 6ACT as a gene carrier with low cytotoxicity and high transfection efficiency. In this chapter, protocols for the preparation of 6ACT and the transfection of COS-1 cells with 6ACT/DNA complexes are provided.

2 Materials

2.1 Preparation of 6ACT

1. Partially *N*-acetylated CT Flonac C (Kyowa Tecnos, Chiba, Japan).
2. A Bruker ASX-300 spectrometer (Bruker Japan, Tsukuba, Japan) to record ^{1}H and ^{13}C NMR spectra at 300.13 and 75.48 MHz, respectively.
3. A HORIBA FT-210 spectrometer (Kyoto, Japan) to record IR spectra.
4. A Zetasizer Nano ZS (Malvern Instruments, Southborough, UK) to determine the average molecular weight of 6ACT acetate by static light scattering.

2.2 Preparation of Plasmid DNA

1. Plasmid DNA, pGL3-control (pGL3), which contains an SV40 promoter upstream of the luciferase gene (Promega, Madison, WI, USA).
2. After amplification in *Escherichia coli JM109*, pGL3 was purified using a QIAGEN Endofree Plasmid Mega Kit (Hilden, Germany).
3. Purified pGL3 was stored in 7 mM Tris buffer containing 1 mM EDTA (TE buffer). The purity was confirmed by agarose gel electrophoresis and the concentration was determined by UV absorption at 260 nm.

2.3 Cell Culture and Gene Transfer into Cells

1. COS-1 cells (an African Green Monkey kidney fibroblast cell line, ATCC accession no. CRL-1650; American Type Culture Collection, Rockville, MD, USA).
2. Dulbecco's modified Eagle medium (DMEM) (Nissui Pharmaceutical, Tokyo, Japan).
3. Fetal bovine serum (FBS) (JRH Biosciences, Lanexa, KS, USA).
4. The COS-1 cells were cultured in DMEM containing 10% FBS, 100 units penicillin/ml, and 100 µg streptomycin/ml in an incubator supplied with 5% CO_2 at 37°C.
5. Steady-Glo Luciferase Assay System (Promega).
6. Fluoroskan Ascent FL-S microplate luminometer (Thermo Labsystems).
7. Protein quantification Kit (Dojindo).
8. Multiskan Ascent BIF-S microplate reader (Thermo Labsystems).

3 Methods

3.1 Preparation of 6ACT

6-Amino-6-deoxychitosan was synthesized according to the reaction shown in Scheme 2 and is described below.

3.1.1 Completely *N*-Deacetylation of CT and Preparation of N-Phthaloylchitosan (**1**)

Completely *N*-deacetylated CT was obtained by treatment of partially *N*-deacetylated CT with 50% aqueous sodium hydroxide at 121°C for 120 min in the presence of a

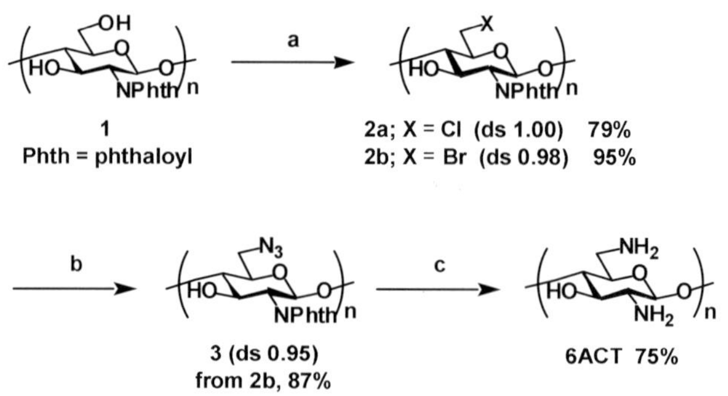

a) ***N*-halosuccinimide, triphenylphosphine, NMP, 80°C, 2 h.** ***b)*** **NaN_3, NMP, 80°C, 4 h.** ***c)*** **i) triphenylphosphine, NMP, rt, 12 h. ii) NH_2NH_2–H_2O, NMP–water (1:1), 100°C, 4 h.**

SCHEME 2. Synthetic scheme of 6-amino-6-deoxychitosan (6ACT)

small amount of sodium borohydride. The product was washed with water until the washings became pH-neutral. These sodium hydroxide treatment and purification steps were repeated to obtain a sufficient degree of deacetylation. CT used in this study had a degree of deacetylation of 0.99 (99% deacetylation) and a viscosity-average molecular mass of 110 kDa.

With regards to the *N*-phthaloylation of CT, the reader should consult the literature for references (Kurita et al. 2003, Satoh et al. 2003).

3.1.2 Preparation of 6-Deoxy-6-halo-*N*-Phthaloylchitosan (**2a–b**)

The reactive key intermediates, deoxy-halogenated derivatives **2a–b**, are typically prepared as follows:

1. Add *N*-cholosuccinimide (NCS) and triphenylphosphine (TPP) (each 3.4 mmol) to a mixture of *N*-phthaloylchitosan (**1**) (100 mg, 0.34 mmol sugar unit) and *N*-methyl-2-pyrrolidone (NMP, 10 ml) under cooling with ice–water.
2. Stir the mixture at 80°C for 2 h under nitrogen atmosphere and then pour into ethanol.
3. Collect the resultant precipitate by centrifugation and filtration, and wash with ethanol, acetone–water, and then acetone.
4. Dry the product under reduced pressure at 40°C (84 mg, 79% yield).

Chlorodeoxy derivative **2a** was obtained as a light-brown powder. The bromodeoxy derivative (**2b**) could also be prepared in a similar manner using *N*-bromosuccinimide (NBS) instead of NCS.

3.1.3 Preparation of 6-Azido-6-deoxy-N-phthaloylchitosan (***3***)

A precursor to the desired compound, 6-azido-6-deoxy-*N*-phthaloylchitosan (**3**), is prepared as follows:

1. Add sodium azide (926 mg, 14.2 mmol) to a solution of **2b** (degree of substitution with bromine, 0.98; 500 mg; 1.42 mmol of sugar unit) in NMP (50 ml).
2. Stir the mixture at 80°C for 4 h under a nitrogen atmosphere.
3. Filter the reaction mixture through cotton wool to crudely remove the salts and pour the filtrate into ethanol.
4. Collect the resultant precipitate by centrifugation and wash with ethanol–water, then acetone on a filter.
5. Dry the product under reduced pressure at 40°C (388 mg, 86% yield).

6-Azido-6-deoxy-*N*-phthaloylchitosan (**3**) was obtained as a light-brown powder. 6-Azidation of **2a** (degree of substitution with chlorine; 1.00) is obtained at a reaction temperature of 120°C under similar conditions.

3.1.4 Preparation of 6ACT

Azido derivative **3** can be converted into the final product, 6-amino-6-deoxychitosan (6ACT), by reduction of the 6-azido group and cleavage of the *N*-phthaloyl group simultaneously as follows:

1. Add TPP (496 mg, 1.89 mmol) to a solution of **3** (200 mg; 0.63 mmol sugar unit; degree of substitution with azido group, 0.95) in NMP (20 ml).
2. Stir the reaction solution at room temperature for 12 h under nitrogen atmosphere.
3. Add 4 M aqueous hydrazine monohydrate (20 ml) into the reaction solution.
4. Stir the reaction mixture at 100°C for 4 h.
5. Following evaporation of the water, pour the reaction mixture into ethanol.
6. Collect the resultant precipitate by centrifugation and wash with ethanol.
7. Dissolve the precipitate in neutral water and purify by ultrafiltration with a cut-off of 10 kDa.
8. Lyophilize the solution to obtain an ivory amorphous product (76 mg, 75% yield).

3.2 Protocol of Gene Transfer into COS-1 Cells

3.2.1 Molecular-Weight Fractionation of 6ACT Acetate and CT Acetate

Fractionation of 6ACT is carried out using the following procedure:

1. Dissolve 6ACT in 5% acetic acid.
2. Fractionate by ultrafiltration with cut-offs of 50 and 30 kDa.
3. Wash 6ACT acetate with water repeatedly to remove excess acetic acid, then lyophilize.

The average molecular mass of the 6ACT acetate used in this study was determined to be 36 kDa. CT acetate was prepared by dissolution of salt-free CT in 5% acetic acid and fractionated by the same procedure as used for 6ACT acetate.

3.2.2 Preparation of the Cationic Polymer/pGL3 Complex

Cationic polymer/pGL3 complexes are typically prepared for use in transfection assays on a 24-well plate as follows:

1. Dissolve a cationic polymer in 14 mM phosphate buffered saline (PBS) at its amine concentration of 7.7 mM.
2. Dilute the cationic polymer solution containing a predefined amount of amine with MilliQ water and adjust the solution to the predefined pH with 10 mM HCl or 0.2% $NaHCO_3$. The total volume of the solution is 87.5 μl for three wells in a transfection.
3. Dilute a solution of pGL3 in TE buffer (0.2 g/l, 17.5 μl) with MilliQ water and adjust the solution to the same pH for the cationic polymer solution with 10 mM HCl. The total volume of the solution is 87.5 μl for three wells in a transfection.
4. Add the cationic polymer solution into the pGL3 solution.
5. Mix the complex solution gently and allow to stand for 30 min at room temperature before use.

The recommended and optimized complexation with 6ACT for transfection is carried out with N/P = 2.5 and at pH 7.0.

3.2.3 Transfection

COS-1 cells were transfected using the cationic polymer/pGL3 complex system containing 6ACT acetate, CT acetate, and PEI as gene carriers. The procedure is as follows:

1. The day before transfection, seed 5×10^4 cells per well of a 24-well plate in 1 ml of growth medium.
2. Incubate the cells for 18–24 h.
3. Prepare cationic polymer/pGL3 complexes according to the protocol described in Sect. 3.2.2.
4. Replace the growth medium with 200 μl fresh medium/well. This medium had been 1.25-fold concentrated compared with normal growth medium, and preadjusted to pH 7.0 with 1 M HCl.
5. Add 50 μl complex solution/well and mix gently.
6. Incubate the cells with the complexes for 3 h.
7. Replace the growth medium containing the complexes with 1 ml normal fresh medium/well.
8. Incubate the cells further for 48 h.
9. Remove the growth medium and gently rinse the cells with 1 ml 140 mM PBS/well.
10. Assay luciferase activity in the cells using the Steady-Glo Luciferase Assay System.
11. After culture cells as same manner for luciferase assay, determine total protein content using Protein Quantification Kit.

3.3 Comments

3.3.1 Preparation of 6ACT

Halogen atoms were introduced into the C-6 position of **1** using NCS- or NBS-TPP reagent systems (Satoh and Sakairi 2003). Preliminary experiments revealed that the optimal conditions are halogenation reagents of 10 molar equivalents and a temperature of 80°C (Scheme 2, path a). The degree of substitution of the products was determined by elemental analyses (chloro derivative **2a**; 1.00 and bromo derivative **2b**; 0.98). ^{13}C NMR spectra of **2a–b** supported the notion that the reaction proceeded site-selectively at the C-6 positions (data not shown).

Azidation of **2b** with sodium azide readily proceeded in NMP at 80°C (Scheme 2, path b). The IR spectrum of the product showed a characteristic strong peak of the azido group at 2100 cm^{-1} (data not shown). The signal of the C-6 carbons shifted downfield from those of **2b** on the ^{13}C NMR spectrum (33.5 to 50.1 ppm) because of azido substitution. The degree of substitution calculated from the results of elemental analysis was 0.95 for the azido groups, with some remaining bromine (degree of substitution; 0.03). These results support the predicted structure of **3**. An increase in reaction temperature to 120°C was required for **2a** in order to obtain the same result.

The azido groups of **3** were reduced by reaction with TPP. The resultant phosphine imido group was then hydrolyzed with aqueous hydrazine monohydrate (Scheme 2, path c). The latter process simultaneously promoted cleavage of the *N*-phthaloyl group. The IR spectrum of the product showed the disappearance of the characteristic absorption bands for the azido (2100 cm^{-1}) and imido groups (1775, 1720, 1390, and 720 cm^{-1}). The NMR spectra of the product also supported the predicted structure of 6ACT (data not shown). Thus, a facile and practical synthetic route of a novel and non-natural poly(amino saccharide) from CT has been developed.

3.3.2 Evaluation of 6ACT Acetate as a Gene Carrier

The in vitro transfection efficiency of the 6ACT/pGL3 complexes was evaluated by luciferase assay using COS-1 cells and was compared with that of CT. The complexes were prepared at pH 6.5 and 7.0. The effects of the N/P ratio on transfection efficiency were examined in a range up to N/P = 20. Luciferase activity was assayed according to manufacturer's protocols and normalized against total protein content determined Protein Quantification Kit. The amount of luciferase activity expressed in the cells is summarized in Fig. 1. Maximum efficiency of the 6ACT/pGL3 system observed at

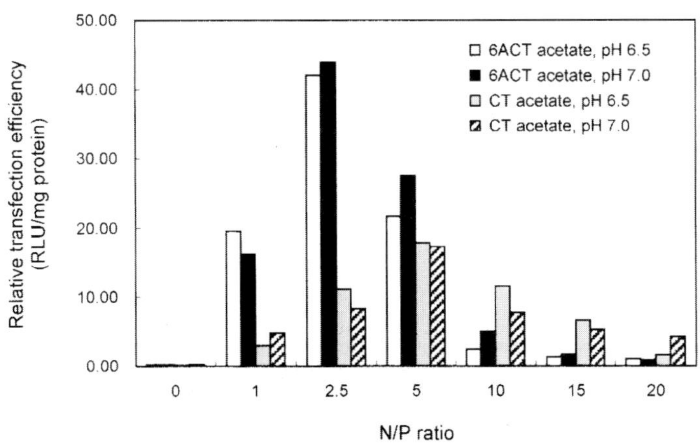

FIG. 1. Relative transfection efficiency of 6ACT and CT complexes with pGL3 for COS-1 cells. The 6ACT/pGL3 complexes prepared at pH 6.5 (*white bars*) and 7.0 (*black bars*), and CT/pGL3 complexes prepared at pH 6.5 (*gray bars*) and 7.0 (*striped bars*) were transfected in DMEM containing 10% fetal bovine serum at pH 7.0 for 3 h. Relative light unit (RLU) was normalized against total protein content

N/P = 2.5, whereas that of the CT/pGL3 was at N/P = 5. Under the optimal conditions, luciferase expression with 6ACT exceeded that obtained with CT. The slight difference of expression levels was found in the examination performed at pH 6.5 and 7.0. The results of this study suggest that 6ACT has potential for use as a gene carrier with high transfection efficiency under neutral conditions.

References

Boussif O, Lezoualc'h F, Zanta MA, Mergny MD, Scherman D, Demeneix B, Behr J-P (1995) A versatile vector for gene and oligonucleotide transfer into cells in culture and in vivo: polyethylenimine. Proc Natl Acad Sci USA 92:7297–7301

Gao S, Chen J, Cu X, Ding Z, Yang Y-H, Hua Z, Zhang J (2003) Galactosylated low molecular weight chitosan as DNA carrier for hepatocyte-targeting. Int J Pharm 255:57–68

Kim YH, Gihm SH, Park CR, Lee KY, Kim TW, Kwon IC, Chung H, Jeong SY (2001) Structural characteristics of size-controlled self-aggregates of deoxycholic acid-modified chitosan and their application as a DNA delivery carrier. Bioconjug Chem 12:932–938

Kim TH, Ihm JE, Choi YJ, Nah JW, Cho C (2003) Efficient gene delivery by urocanic acid-modified chitosan. J Control Release 93:389–402

Kim TH, Park IK, Nah JW, Choi YJ, Cho CS (2004) Galactosylated chitosan/DNA nanoparticles prepared using water-soluble chitosan as a gene carrier. Biomaterials 25:3783–3792

Kurita K, Ikeda H, Yoshid Y, Shimojoh M, Harata M (2002) Chemoselective protection of the amino groups of chitosan by controlled phthaloylation: facile preparation of a precursor useful for chemical modifications. Biomacromol 3:1–4

Liu WG, Yao KD, Liu OG (2001) Formation of a DNA/*N*-dodecylated chitosan complex and salt-induced gene delivery. J Appl Polym Sci 82:3391–3995

Mao H-Q, Roy K, Truong-Le VL, Janes KA, Lin KY, Wang Y, August JT, Leong KW (2001) Chitosan-DNA nanoparticles as gene carriers: synthesis, characterization and transfection efficiency. J Control Release 70:399–421

Murata J, Ohya Y, Ouchi T (1996) Possibility of application of quaternary chitosan having pendant galactose residues as gene delivery tool. Carbohydr Polym 29:69–74

Park YK, Park YH, Shin BA, Choi ES, Park YR, Akaike T, Cho CS (2000) Galactosylated chitosan-*graft*-dextran as hepatocyte-targeting DNA carrier. J Control Release 69:97–108

Park IK, Kim TH, Park YH, Shin BA, Choi ES, Chowdhury EH, Akaike T, Cho CS (2001) Galactosylated chitosan-*graft*-poly(ethylene glycol) as hepatocyte-targeting DNA carrier. J Control Release 76:349–362

Park IK, Ihm JE, Park YH, Choi YJ, Kim SI, Kim WJ, Akaike T, Cho CS (2003) Galactosylated chitosan (GC)-graft-poly(vinyl pyrrolidone) (PVP) as hepatocyte-targeting DNA carrier—Preparation and physicochemical characterization of GC-graft-PVP/DNA complex (1). J Control Release 86:349–359

Satoh T, Vladimirov L, Johmen M, Sakairi N (2003) Preparation and thermal dehydration of *N*-(carboxy)acyl chitosan derivative with high stereoregularity. Chem Lett 32:318–319

Satoh T, Nagasaki T, Sakairi N, Shinkai S (2004) 6-Amino-6-deoxychitosan. Preparation and Application as Plasmid Vector in COS-1 Cells. Chem Lett 33:340–341

Satoh T, Sakairi N (2003) Highly regioselective deoxyhalogenation at the C-6 positions of *N*-phthaloylchitosan with *N*-halosuccinimide–triphenylphosphine system. Chem Lett 32:1162–1163

Thanou M, Florea BI, Geldof M, Junginger HE, Borchard G (2002) Quaternized chitosan oligomers as novel gene delivery vectors in epithelial cell lines. Biomaterials 23:153–159

Newly Designed DNA Fragments for Gene Correction

HIROYUKI KAMIYA, HIROYUKI TSUCHIYA, and HIDEYOSHI HARASHIMA

1 Introduction

Gene correction therapy entails the conversion of a mutated gene to one with the normal (or desired) sequence (Richardson et al. 2002). The corrected genes should be properly expressed under the control of the original promoter, and respond appropriately to the intracellular and extracellular environments. When corrected, the "therapeutic effects" are expected to be life-long. Gain-of-function or predominant mutations, such as activated oncogenes, could be suitable subjects for gene correction, making it a highly attractive therapeutic strategy.

In the small-fragment homologous replacement (SFHR) approach to gene correction, a heat-denatured double-stranded (ds) PCR fragment that is hundreds of bp long and contains the normal sequence is used (Richardson et al. 2002). This method has been utilized with mutations in the CFTR (Kunzelmann et al. 1996,; Goncz et al. 1998, 2001) and dystrophin (Kapsa et al. 2001) genes, and partial gene corrections were obtained. The current SFHR method yields a low correction efficiency, and further improvements are needed.

To overcome the limitations of the current SFHR method, we have developed new DNA fragments for gene correction. The first type is a ds DNA fragment prepared by restriction enzyme digestion of plasmid DNA isolated from *Escherichia coli*. The other type is a single-stranded (ss) DNA fragment prepared by restriction enzyme digestion of ss phagemid DNAs. SFHR-mediated gene correction efficiency was enhanced using these ds and ss DNA fragments.

This chapter focuses on the preparation of these DNA fragments for use in SFHR-mediated gene correction. The correction of an inactivated Hyg-EGFP gene (fusion gene of hygromycin resistance and enhanced green fluorescence protein genes) serves as a model system (Fig. 1). In this system, codon 34 of the inactive Hyg-EGFP gene (TGA, termination codon) is the target, and its conversion to the TCA sequence (Ser)

Graduate School of Pharmaceutical Sciences, Hokkaido University, Kita 12, Nishi 6, Sapporo 060-0812, Japan

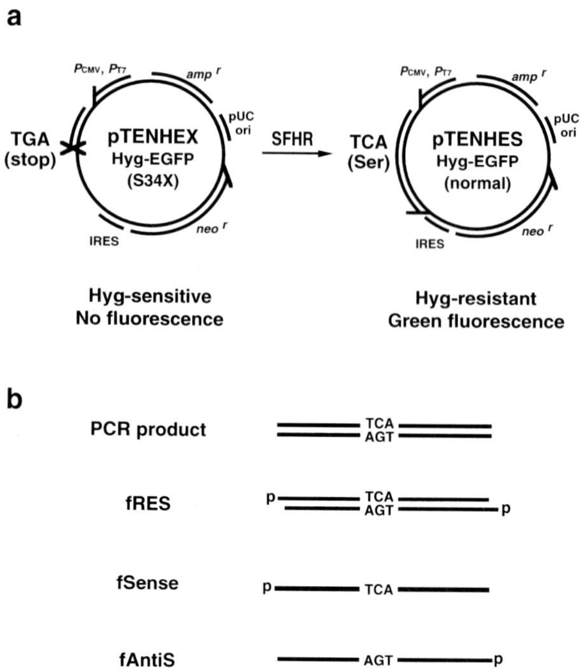

FIG. 1. **a** The gene correction assay using the Hyg-EGFP gene. The plasmid pTENHES harbors the normal Hyg-EGFP gene under the control of the CMV (P_{CMV}) and T7 (P_{T7}) promoters. The plasmid pTENHEX has the same nucleotide sequence as pTENHES, except that the 34th codon is a termination codon. This plasmid was used as the target for the gene correction reaction. **b** The DNA fragments for gene correction described in this chapter

makes the cells EGFP-positive and hygromycin-resistant. Preparation of the 606-bp ds DNA fragment (fRES) by *Xho*I digestion of pTENHES containing the normal (TCA) sequence, as well as the 606-nt ss DNA fragments by *Xho*I digestions of ss phagemid DNAs containing sense and antisense sequences, is described.

2 Materials

2.1 Preparation of DNA Fragments for Gene Correction

1. The pTENHES plasmid, carrying the normal (codon 34 = TCA) Hyg-EGFP gene, and the pTENHEX plasmid, bearing an inactivated (codon 34 = TGA) Hyg-EGFP gene, were constructed in our laboratory.
2. The phagemids pBSHES/Sense and pBSHES/AntiSense contain a portion of the normal Hyg-EGFP gene, and each ss form carries the sense and antisense sequence of the gene, respectively (Fig. 1a). These phagemids were constructed by inserting the *Xho*I fragment of pTENHES, containing the normal Hyg-EGFP gene, into the *Xho*I site of the pBluescript II SK+ phagemid (Stratagene, La Jolla, CA, USA).

3. Single-stranded scaffold oligodeoxyribonucleotides are designed to be complementary to the phagemid DNA near the target restriction enzyme site. A *Tm* value of ~60°C is recommended for use with enzymes whose incubation temperature is 37°C, and the recognition site should be located in the center of the oligodeoxyribonucleotide.

3 Methods

3.1 Preparation of DNA Fragments for Gene Correction

3.1.1 Preparation of Plasmid DNAs

Isolate and purify plasmid DNA with a commercially available kit (e.g., EndoFree Plasmid Mega kit, Qiagen, Hilden, Germany).

3.1.2 Preparation of ss Phagemid DNAs

1. Culture *E. coli* JM105 strain harboring phagemid DNA containing the f1 origin (pBSHES/Sense or pBSHES/AntiSense).
2. Infect the *E. coli* with helper phage (e.g., VCSM13, Stratagene) to produce a ss circular phagemid, according to the supplier's instructions.
3. Separate the phages from the bacterial cells by centrifugation (6,760 × *g*, 15 min).
4. Add 20% PEG 8000/2.5 M NaCl to the phage-containing supernatant (final concentrations: 3% PEG and 375 mM NaCl), and incubate the mixture at 4°C overnight.
5. Precipitate the phages by centrifugation (13,000 × *g*, 20 min) and resuspend them in 0.3 M sodium acetate/1 mM EDTA.
6. Recover the ss phagemid by standard phenol/chloroform extraction. Mix the DNA solution with 2.5 volumes of ethanol or an equal volume of isopropanol, and place at room temperature for 15 to 60 min.
7. Precipitate the ss DNA by centrifugation (13,000 × *g*, 20 min) and dissolved it in H_2O.

3.1.3 Preparation of a ds DNA Fragment by Restriction Enzyme Digestion

The 606-bp ds DNA fragment (fRES) is prepared by *Xho*I digestion of pTENHES, according to the supplier's instructions. When other restriction enzymes are used, their amounts should be altered appropriately.

1. Incubate 200 µg pTENHES with 500 U *Xho*I at 37°C overnight.
2. Collect the DNA by ethanol precipitation.
3. Separate the DNA by low-melting-point agarose gel electrophoresis.
4. Cut out the slice of agarose containing the band and transfer it to a clean tube.
5. Add ~5 volumes of buffer solution containing EDTA (e.g., TE, 10 mM Tris-HCl, 1 mM EDTA, pH 8.0), and incubate at 65°C for 5 min.
6. Add an equal volume of TE-saturated phenol and vortex the mixture. Recover the aqueous phase by centrifugation.
7. Extract the aqueous phase once with phenol-chloroform and once with chloroform.

8. Collect the DNA by ethanol precipitation.
9. Purify the desired fragment by gel filtration chromatography. Disposable gel filtration columns (e.g. NAP-5, GE Healthcare, Waukesha, WI, USA) may be used.
10. Measure the UV spectrum to confirm the fragment's purity and to calculate the yield. The concentration is determined by the molar absorption coefficient of DNA (1.0 OD_{260} = 50 μg ds DNA).

3.1.4 Preparation of ss Linear DNA Fragments

The 606-nt ss DNA fragments are obtained by *Xho*I digestion of ss pBSHES/Sense and ss pBSHES/AntiSense, after annealing with scaffold oligodeoxyribonucleotides complementary to the two *Xho*I sites within the phagemid (fSense and fAntiS). The scaffold oligodeoxyribonucleotides are designed to introduce the recognition sequence of the target enzyme in the center, and the extra sequences are necessary to ensure digestion efficiency.

1. Anneal scaffold oligodeoxyribonucleotides to ss phagemid DNA (5:1 molar ratio).
2. Incubate 111 μg phagemid DNA with 500 U *Xho*I at 37°C overnight.
3. Collect the DNA by ethanol precipitation.
4. Purify the desired fragment by low-melting-point agarose gel electrophoresis and by gel filtration chromatography, as in Sect. 3.1.3.
5. Measure the UV spectrum to confirm the fragment's purity and to calculate the yield. The concentration is determined by the molar absorption coefficient of DNA (1.0 OD_{260} = 40 μg ss DNA).

3.2 *Introduction of DNA Fragments*

3.2.1 Culture of CHO-K1 Cells

1. Culture CHO-K1 cells in DMEM/F12 medium supplemented with 10% fetal bovine serum and antibiotics, in a 5% CO_2 atmosphere at 37°C. Culture 1×10^5 cells in a 10-cm dish every 3 days.
2. One day before transfection, suspend 3×10^5 cells in 4 ml culture medium, and place in a 6-cm dish.

3.2.2 Preparation of DNA for Transfection

1. Denature the DNA fragment by heating at 98°C for 5 min, and chill immediately on ice for at least 5 min until transfection.
2. Mix 10 pmol DNA fragment with 125 ng (25 fmol) pTENHEX carrying the target gene.
3. Add an appropriate amount of unrelated plasmid DNA to keep the total amount of DNA constant.

3.2.3 Transfection and Recovery of DNA

1. Transfect the target plasmid (pTENHEX) and the heat-denatured DNA fragment into CHO-K1 cells with the Lipofectamine Plus Reagent (Invitrogen, Carlsbad, CA, USA) or other available reagents, according to the supplier's instructions.

2. Harvest the cells at 48-h posttransfection and resuspend them in 100 μl TEG (25 mM Tris-HCl, 10 mM EDTA, 50 mM glucose, pH 8.0).
3. Mix slowly with 200 μl 0.2 N NaOH/1% SDS and incubate for 5 min at room temperature (cell lysis).
4. Add 150 μl 8 M $AcONH_4$ and chill on ice for 15 min.
5. Centrifuge at 13,000 × *g* for 15 min.
6. Transfer the supernatant into a new tube.
7. Recover the plasmid DNA by isopropanol precipitation and dissolve in 20 μl H_2O.

3.3 Determination of Gene Correction Frequency

There are many assay systems to calculate gene correction efficiencies. In the following, a method is described that is based on the assumption that *E. coli* cells become EGFP-positive and hygromycin-resistant if gene correction occurs. Due to the high transformation efficiency, the recovered plasmid DNA is first electroporated into the DH5α strain, which does not express the Hyg-EGFP gene. The plasmid amplified in this bacterial strain is then electroporated into the BL21(DE3) *E. coli* strain, in which T7 RNA polymerase can be inducibly expressed with isopropyl-β-D-thiogalactopyranoside (IPTG). When the amount of DNA recovered from the transfected CHO-K1 cells is enough to transform the BL21(DE3) *E. coli* strain, this amplification step can be omitted.

3.3.1 Introduction of DNA into Bacterial Cells

1. Inoculate 125 μl of the overnight *E. coli* culture into 12.5 ml L-broth and incubate at 37°C for 3 h.
2. Chill in ice for 10 min.
3. Collect the *E. coli* by centrifugation (2,220 × *g* for 15 min).
4. Resuspend the pellet in 7.5 ml ice-cold water.
5. Collect the *E. coli* by centrifugation (2,220 × *g* for 15 min).
6. Resuspend the pellet in 1 ml ice-cold water and transfer the contents into a microtube.
7. Collect the *E. coli* by centrifugation (13,000 × *g* for 3 min).
9. Resuspend in 200 μl ice-cold water.
10. Collect the *E. coli* by centrifugation (13,000 × *g* for 3 min).
11. Resuspend the pellet in 75 μl ice-cold water. The final volume should be ~ 100 μl.
12. Place microtubes and 0.1-cm cuvettes on ice.
13. In a cold microtube, mix 100 μl cell suspension with 2 μl DNA, and leave in ice for 5 min.
14. Transfer the contents to a chilled cuvette.
15. Introduce the DNA into the competent *E. coli* cells by electroporation, according to the supplier's instructions.
16. Quickly add 1 ml of pre-warmed SOC medium and incubate at 37°C for 1 h.
17. Proceed to Sect. 3.3.2 or Sect. 3.3.3.

3.3.2 Recovery of DNA from Bacterial Cells

When DNA amplification is necessary, the plasmid DNA is recovered from the transformed DH5α cells. Other antibiotics may be used when a gene other than *amp*^{r} is located on the plasmid DNA.

1. Spread a portion (usually 5–10 μl after dilution) of the transformed DH5α *E. coli* culture (step 16 of Sect. 3.3.1) onto an agar plate containing 50 μg ampicillin/ml and incubate at 37°C overnight. Transfer the residual culture into 5–10 ml fresh L-broth containing 50 μg ampicillin/ml and incubate at 37°C overnight.
2. Count the number of colonies on the ampicillin plate to check the transformation efficiency.
3. Pour 1.8 ml *E. coli* culture into a 2-ml tube.
4. Collect the *E. coli* by centrifugation (13,000 × *g* for 3 min at 4°C).
5. Isolate plasmid DNA by the standard alkaline lysis method.
6. Extract the DNA with phenol-chloroform.
7. Collect the DNA by the ethanol precipitation.
8. Dissolve the DNA in sterilized water and use for transformation of BL21(DE3) (step 13 of Sect. 3.3.1).

3.3.3 Determination of Gene Correction Efficiency

1. Following an incubation in 1 ml of SOC medium at 37°C for 1 h (step 16 of Sect. 3.3.1), dilute a 50-μl aliquot of bacterial cells into 1 ml LB medium containing 50 μg ampicillin /ml and 10 μM IPTG, and incubate further for 3 h.
2. Dilute the cell suspension by 1,000-to 10,000-fold and seed onto LB agar plates containing 50 μg ampicillin/ml and 10 μM IPTG (Hyg-0 plates). Incubate at 37°C for 12–24 h.
3. Dilute the cell suspension by 10- to 100-fold and seed onto LB agar plates containing 50 μg ampicillin/ml, 10 μM IPTG, and 75 μg hygromycin/ml (Hyg-75 plates). Incubate at 37°C for 36–48 h.
4. Count the number of colonies on the Hyg-0 plates.
5. Count the number of EGFP-positive colonies on the Hyg-75 plates with an image analyzer (e.g., Fuji FLA2000G, Tokyo, Japan).
6. Calculate the gene correction frequencies by dividing the number of EGFP-positive colonies on the Hyg-75 plates by the number of colonies on the Hyg-0 plates.

Figure 2 shows the EGFP-expressing colonies (seen as black dots) on a Hyg-75 plate. False-positive colonies also emerged on the plate, but the actual positive colonies were easily distinguished by their green fluorescence.

Using this assay system, the activities of the newly designed DNA fragments for SFHR were quantitatively evaluated: (1) a ds DNA fragment prepared from plasmid DNA (fRES) and (2) actual ss DNA fragments with sense and antisense sequences (fSense and fAntiS). When each heat-denatured DNA fragment was used in a 400-fold molar excess relative to the target plasmid, the ds fragment derived from the plasmid showed two-fold higher gene correction activity, than obtained with the conventional PCR fragment (fRES: ~0.3%; PCR fragment: ~0.15%). The ss DNA fragment with the sense sequence (fSense) dramatically improved the gene correction efficiency, with a

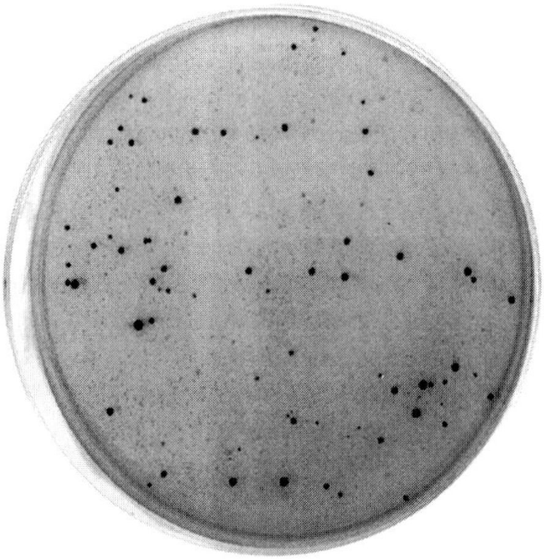

FIG. 2. EGFP-expressing colonies on a hygromycin plate. The plasmid was recovered from CHO-K1 cells cotransfected with fSense. The recovered plasmid was amplified in the DH5α *E. coli* strain and then electroporated into the BL21(DE3) *E. coli* strain. Colonies expressing EGFP were analyzed with an image analyzer. The *black dots* on the plate indicate the EGFP-positive colonies

12-fold enhancement over the conventional PCR fragment (~2%). By contrast, the correction with fAntiS was comparable to that of fRES.

3.4 Comments

The gene correction efficiency was improved by using a ds fragment derived from plasmid and a ss DNA fragment containing the sense sequence. To our knowledge, this is the first report of gene correction activity enhancement by modifying the DNA. These newly designed ss and dsDNA fragments possess an additional advantage, in that they have a highly accurate nucleotide sequence. Since replication in living cells is extremely precise (Drake 1969), the possibility that the ss DNA fragment contained an unexpected mutation within the fragment was quite low. Indeed, no unexpected sequence alterations existed within the sequenced region. By contrast, deoxyribonucleotide misincorporation occurs at a high frequency during in vitro amplification, even when a "high fidelity" PCR enzyme is used (Takagi et al. 1997). Thus, SFHR with a PCR fragment might introduce a mutation at another site when the targeted position is corrected.

In conclusion, we have successfully developed safer and more efficient DNA fragments for SFHR gene correction. With further improvements, these fragments will be an attractive strategy for personalized medicine based on an individual's genetic information.

References

Drake JW (1969) Comparative rates of spontaneous mutation. Nature 221:1132

Goncz KK, Kunzelmann K, Xu Z, Gruenert DC (1998) Targeted replacement of normal and mutant CFTR sequences in human airway epithelial cells using DNA fragments. Hum Mol Genet 7:1913–1919

Goncz KK, Colosimo A, Dallapiccola B, Gagne L, Hong K, Novelli G, Papahadjopoulos D, Sawa T, Schreier H, Wiener-Kronish J, Xu Z, Gruenert DC (2001) Expression of ΔF508 CFTR in normal mouse lung after site-specific modification of CFTR sequences by SFHR. Gene Ther 8:961–965

Kapsa R, Quigley A, Lynch GS, Steeper K, Kornberg AJ, Gregorevic P, Austin S, Byrne E (2001) In vivo and in vitro correction of the mdx dystrophin gene nonsense mutation by short-fragment homologous replacement. Hum Gene Ther 12:629–642

Kunzelmann K, Legendre JY, Knoell DL, Escobar LC, Xu Z, Gruenert DC (1996) Gene targeting of CFTR DNA in CF epithelial cells. Gene Ther 3:859–867

Richardson PD, Augustin LB, Kren BT, Steer CJ (2002) Gene repair and transposon-mediated gene therapy. Stem Cells 20:105–118.

Takagi M, Nishioka M, Kakihara H, Kitabayashi M, Inoue H, Kawakami B, Oka M, Imanaka T (1997) Characterization of DNA polymerase from Pyrococcus sp. strain KOD1 and its application to PCR. Appl Environ Microbiol 63:4504–4510.

Evaluation of Gene Expression In Vivo After Intravenous and Intraportal Administration of Lipoplexes

Tatsuya Okuda, Shigeru Kawakami, and Mitsuru Hashida

1 Introduction

Numerous kinds of synthetic compounds, such as cationic lipids and cationic polymers, have been developed as candidates for non-viral gene carrier molecules for use in gene therapy. Cationic compounds generally form positively charged complexes with DNA, and many of them have shown good results in in-vitro gene transfection experiments. However, in vivo, there are considerable effects of inactivation through non-specific interactions with plasma proteins (Dash et al. 1999, 2000; Tandia et al. 2003) and of rapid clearance by the reticuloendothelial system. Indeed, compounds that can be efficiently transfected into cultivated cells are not always necessarily applicable in vivo. Thus, new gene carriers, applicable to gene delivery in vivo, are needed.

Gene carrier molecules are required to have not only transfection ability but also therapeutic effects mediated by a transgene encoding therapeutic proteins, such as cytokines. A simple assay of in vivo transgene expression would allow the screening of effective gene carriers from a number of candidates. At present, the reporter gene assay is generally used to evaluate the transfection ability of gene carriers. In this system, firefly luciferase, β-galactosidase and green fluorescence protein are commonly used as reporter genes. The luciferase system is superior to the because it is easy to carry and does not require complicated techniques or expensive equipment. In addition, its simplicity enables the screening of a large number of samples, from various organs and a wide range of sample preparation methods.

This chapter focuses on the evaluation of transgene expression in vivo using firefly luciferase as the reporter gene after intravenous and intraportal injection. In addition, a method to separate parenchymal cells (PC) and non-parenchymal cells (NPC) from liver that facilitates the hepatic cellular localization of luciferase activity is introduced.

Department of Drug Delivery Research, Graduate School of Pharmaceutical Sciences, Kyoto University, 46-29 Yoshida, Shimoadachi-cho, Sakyo-ku, Kyoto 606-8501, Japan

2 Materials

2.1 Chemicals

N-2-hydroxyethylpiperazine-*N*′-2-ethanesulfonic acid (HEPES), cholesteryl chloroformate, and bovine serum albumin (BSA) were obtained from Sigma (St. Louis, MO, USA). Diphosphatidyl-ethanolamine (DOPE) was purchased from Avanti Polar Lipids (Alabaster, AL, USA). *N*-(4-aminobutyl)carbamic acid *tert*-butyl ester and *N*-[1-(2,3-dioleyloxy)propyl]-*N,N,N*-trimethylammonium chloride (DOTMA) were obtained from Tokyo Chemical Industry (Tokyo, Japan). The cationic cholesterol derivative, 3β[*N*-(*N*′,*N*′-dimethylaminoethane)carbamoyl] cholesterol (DC-Chol) was synthesized as previously reported (Gao and Huang 1991). All other chemicals were obtained commercially as reagent-grade products.

2.2 Preparation of Plasmid DNA (pCMV-Luc)

pCMV-Luc was constructed by subcloning the *Hin*dIII/*Xba*I firefly luciferase cDNA fragment from the pGL3-control vector (Promega, Madison, WI, USA) into the polylinker of pcDNA3 vector (Invitrogen, Carlsbad, CA, USA). Plasmid DNA was amplified in the *Escherichia coli* strain DH5α, isolated, and purified using a QIAGEN Plasmid Giga Kit (QIAGEN, Hilden, Germany). Purity was confirmed by 1% agarose gel electrophoresis followed by ethidium bromide staining. The DNA concentration was measured by UV absorption at 260 nm.

2.3 Preparation of Liposomes

1. A round-bottomed flask of appropriate size
2. Rotary evaporator
3. Vacuum desiccator
4. Bath-type and probe-type sonicator
5. Extruder (Liplex Biomembrane, Vancouver, Canada) or 0.45-μm pore size filter

2.4 Evaluation of Transgene Expression in Mice

1. Lysis buffer; 0.1 M Tris-HCl, 0.05% Triton X-100, 2 mM EDTA, pH 7.8
2. Homogenizer; Omni-Mixer TH (Omni International., Marietta, GA, USA)
3. Pica gene luminescence kit (TOYO B-Net, Tokyo, Japan)
4. Luminometer; Lumat LB 9507 (EG & G Berthold, Bad Wildbad, Germany)
5. Protein Quantification Kit (DOJINDO LABOLATORIES, Kumamoto, Japan)

2.5 Separation of Parenchymal and Non-parenchymal Cells from Liver

1. Ca^{2+}-, Mg^{2+}-free perfusion buffer; 10 mM HEPES, 137 mM NaCl, 5 mM KCl, 0.5 mM NaH_2PO_4, and 0.4 mM Na_2HPO_4, pH 7.2
2. Ca^{2+}- and collagenase-supplemented perfusion buffer; Ca^{2+}-, Mg^{2+}-free perfusion buffer supplemented with 5 mM $CaCl_2$ and 0.05% (w/v) collagenase (type I), pH 7.5

3. Hank's-HEPES buffer; 10 mM HEPES, 137 mM NaCl, 0.4 mM NaH_2PO_4, 0.4 mM KH_2PO_4, 0.9 mM $CaCl_2$, 0.8 mM $MgSO_4$, 3 mM $NaHCO_3$, 5.6 mM glucose, pH 7.3
4. Peristaltic pump SJ-1211 (ATTO, Tokyo, Japan)

3 Methods

3.1 Synthesis of Sugar-Modified Cholesterol Derivatives

Galactosylated and mannosylated cholesterol derivatives, cholesten-5-yloxy-*N*-(4-((1-imino-2-β-D-thiogalactosyl-ethyl)amino)butyl) formamide (Gal-C4-chol) and cholesten-5-yloxy-*N*-(4-((1-imino-2-β-D-thiomannosyl-ethyl)amino)butyl) formamide (Man-C4-Chol), respectively, were synthesized as previously described (Kawakami et al. 1998, 2000c). Briefly, cholesteryl chloroformate and *N*-(4-aminobutyl)carbamic acid *tert*-butyl ester were reacted in chloroform for 24 h at room temperature. A solution of trifluoroacetic acid and chloroform was added dropwise and the mixture was stirred for 4 h at 4°C. The C4-Chol was reacted with five equivalents of 2-imino-2-methoxyethyl-1-thiogalactoside or 2-imino-2-methoxyethyl-1-thiomannoside, which were synthesized as previously described (Lee et al. 1976), in pyridine containing 1.1 equivalents of triethylamine for 24 h at room temperature. After evaporation of the reaction mixture in vacuo, the resultant material was suspended in water and dialyzed against water for 48 h using a dialysis membrane with a 12-kDa cut-off). After the dialyzate was lyophilized, the crude product was purified by re-crystallization three times in ethyl acetate.

3.2 Preparation of Liposomes

1. Dissolve lipids and/or cholesterol derivatives in chloroform at appropriate ratio.
2. Evaporate the mixture to dryness in a round-bottomed flask by rotary evaporator. The lipid film should be as thin and uniform as possible.
3. Vacuum-desiccate the resultant lipid film for overnight. Chloroform should be removed completely, since residual chloroform affects hydration of the lipid film and the formation of liposomes.
4. Hydrate the lipid film with sterile 5% dextrose solution.
5. Sonicate the dispersion for 5 min in a bath-type sonicator to form liposome.
6. Sonicate the liposome for 3 min on ice by using a probe-type sonicator.
7. Extrude the liposome ten times through a double-stacked 100-nm pore size polycarbonate membrane filter at 60°C by using an extruder. Alternatively, filter the liposomes with a 0.45-μm pore size filter.

3.3 Preparation of Lipoplexes

The liposome/DNA complexes (lipoplexes) for in vivo experiments were prepared as described by Templeton et al (1997). Briefly, equal volumes of DNA and stock liposome solution were diluted with 5% dextrose to produce various ratios of DNA/liposomes and then mixed in a microtube at room temperature.

1. Dilute DNA and liposome stock solution with 5% dextrose.
2. Add DNA solution rapidly to the surface of the liposome solution using a Pipetman pipette.

3. Agitate the mixture rapidly by pumping it up and down twice in the pipette tip.
4. Allow to stand for 30 min at room temperature.

3.4 Evaluation of Transgene Expression in the Mouse

1. Administer 300 µl of lipoplexes containing appropriate amounts of DNA into a mouse via the appropriate route.
2. Six hours later, kill the mouse by cervical dislocation or perfusion with saline through the portal vein.
3. Excise the organs (spleen, liver, kidney, heart and lung) and wash them twice with ice-cold saline.
4. Weigh the excised organs and add an appropriate volume (5 µl/mg for liver and 4 µl/mg for other organs) of lysis buffer.
5. Homogenize the organs using a homogenizer.
6. Transfer 400 µl of homogenate into a microtube and perform three cycles of freezing (liquid N_2 for 3 min) and thawing (37°C for 3 min).
7. Centrifuge the homogenate at 10,000 × *g* for 10 min at 4°C.
8. Analyze the luciferase activity in 20 µl supernatant using a luminometer.
9. Determine the protein concentration of lysate using a Protein Quantification Kit.

3.5 Cellular Localization of Luciferase Activity in Liver

In order to reduce unexpected side effects, tissue- or site-specific targeted delivery of the exogenous therapeutic gene using a gene carrier is necessary for gene therapy. For this reason, many researchers select liver for the target tissue, because asialoglycoprotein and mannose receptors are specifically expressed on the surface of hepatocytes (liver parenchymal cell) and macrophages, such as Kupffer's cells (liver non-parenchymal cell), respectively. Therefore, it is important to evaluate whether the sugar modified gene carrier system is working correctly. In the following, a method to evaluate the hepatic cellular localization of luciferase activity by separating parenchymal and non-parenchymal cells is described.

3.5.1 Separation of Parenchymal and Non-parenchymal Cells from Liver

1. Anesthetize a mouse with an intraperitoneal injection of pentobarbital sodium (50 mg/kg) 6 h after the administration of lipoplexes.
2. Perfuse the liver with Ca^{2+}, Mg^{2+}-free perfusion buffer for 10 min at a perfusion rate of 3–4 ml/min. As soon as perfusion is started, cut the vena cava and aorta.
3. Perfuse the liver with perfusion buffer supplemented with Ca^{2+} and type I collagenase for a further 10 min. Again, a perfusion rate of 3–4 ml/min should be maintained.
4. Excise the liver and remove its capsular membrane.
5. Disperse the liver cells into ice-cold Hank's-HEPES buffer containing 0.1% BSA by gentle stirring.
6. Centrifuge the cell suspension at 50 × *g* for 1 min.
7. Wash the pellet, containing parenchymal cells (PC), with Hank's-HEPES buffer twice by centrifuging at 50 × *g* for 1 min.

8. Centrifuge the supernatant, containing non-parenchymal cells (NPC), twice at $50 \times g$ for 1 min.
9. Centrifuge the resultant supernatant twice at $200 \times g$ for 2 min.
10. Resuspend PC and NPC into 2 ml of ice-cold Hank's-HEPES buffer separately.
11. Determine the cell number and cell viability using the Trypan blue exclusion method.
12. Continue with step 5 of Sect. 3.4.

3.6 Comments

DOTMA/Chol (1:1) liposome was prepared and mixed with 50 μg pCMV-Luc to form lipoplexes (see Sects. 3.2 and 3.3). It has been reported that mixing complexes at low ionic strength prevents aggregation, although large complexes resulting from aggregation showed high transfection efficiency in vitro in the case of DNA/transferrin-PEI complexes (Ogris et al. 1998). Based on this report, the complexes in this study were prepared with 5% dextrose solution. After administering the lipoplexes by intravenous injection of the tail vein, the in vivo transfection ability of DOTMA/Chol liposomes was evaluated (See Sect. 3.4) (Kawakami et al. 2000a). The results showed that DOTMA/Chol liposome had high transfection ability in the lungs (Fig. 1). Similar results were reported with 1,2-bis(oleoyloxy)-3-(trimethylammonio) propane (DOTAP)/protamine/DNA complexes (Li and Huang 1997). The authors of that study named this phenomenon "first-passage effect" because the lung is the first capillary bed that the lipoplexes encounter after intravenous injection.

In our laboratory, novel galactosylated and mannosylated cholesterol derivatives (Gal-C4-Chol and Man-C4-Chol) were synthesized (Fig. 2; see Sect. 3.1) and two liposome formulations containing these sugar-modified cholesterol derivatives, DOTMA/Chol/Gal-C4-Chol (1:0.5:0.5) for hepatocyte targeting and Man-C4-Chol/DOPE (6:4) for macrophage targeting, were prepared for receptor-mediated

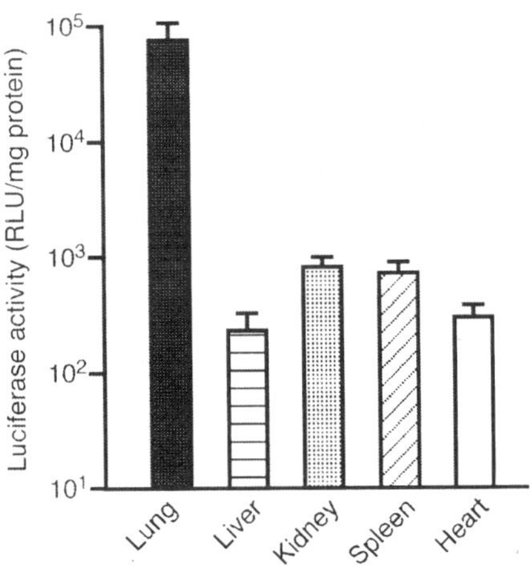

FIG. 1. Transfection activity of DOTMA/Chol (1:1) liposome/DNA complexes after intravenous injection in mice. Plasmid DNA (pCMV-Luc, 30 μg) was complexed with cationic DOTMA/Chol (1:1) liposome at a charge ratio of 2.3. Luciferase activity was measured 6 h post-injection in each organ. Each value represents the mean ± S.D. ($n = 3$)

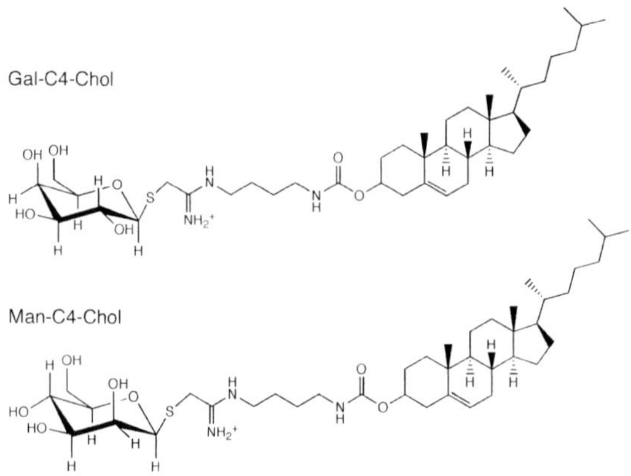

FIG. 2. Structures of novel galactosylated and mannosylated cholesterol derivatives, cholesten-5-yloxy-*N*-(4-((1-imino-2-β-D-thiogalactosyl-ethyl)amino)butyl) formamide (Gal-C4-chol) and cholesten-5-yloxy-*N*-(4-((1-imino-2-β-D-thiomannosyl-ethyl)amino)butyl) formamide (Man-C4-Chol)

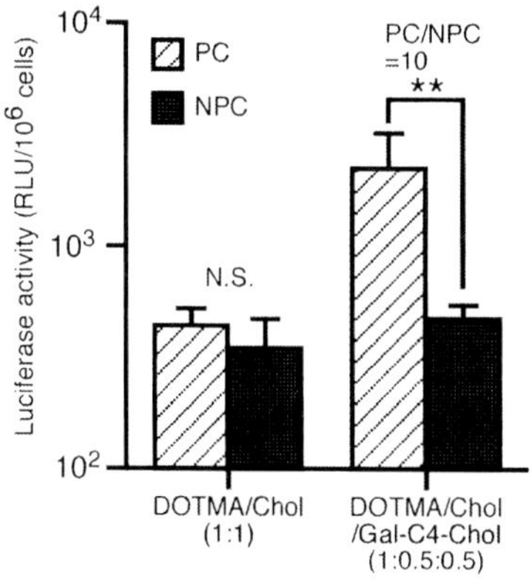

FIG. 3. Intrahepatic gene expression of liposome/DNA complexes after intraportal injection in mice. Plasmid DNA (pCMV-Luc, 50 μg) was complexed with DOTMA/Chol (1:1) and DOTMA/Chol/Gal-C4-chol (1:0.5:0.5) at a charge ratio of 2.3. Luciferase activity was measured 6 h post-injection in parenchymal cells (PC) and non-parenchymal cells (NPC). Statistical analysis was performed by analysis of variance ($^{**}P < 0.01$). *N.S.* Not significant. Each value represents the mean ± S.D. ($n = 3$)

gene delivery to the liver. After formation of the DNA complexes as DOTMA/Chol/Gal-C4-Chol and Man-C4-Chol/DOPE liposomes, these liposome formulations were administrated into mice by intraportal and intravenous injection, respectively. Six hours later, PC and NPC were separated from the liver (see Sect. 3.5), and the hepatic cellular localization of luciferase activity was subsequently evaluated (Kawakami et al. 2000a, b). In the case of DOTMA/Chol/Gal-C4-Chol liposomes, a 10-fold higher luciferase activity was observed in PC than in NPC, while the control liposome, DOTMA/Chol (1:1), showed no significant difference in luciferase activity between PC and NPC (Fig. 3). By contrast, injection of Man-C4-Chol/DOPE liposome yielded significantly higher transgene expression in NPC than in PC, whereas gene expression using the control liposomes, DC-Chol/DOPE (6:4), was slightly higher in NPC (Fig. 4). In addition, galactosylated and mannosylated BSA effectively inhibited gene expression from, respectively, DOTMA/Chol/Gal-C4-Chol and Man-C4-Chol/DOPE liposomes. These results suggesting that hepatic cellular targeting can be achieved by incorporating Gal-C4-Chol or Man-C4-Chol into liposomes at the appropriate ratio of lipids.

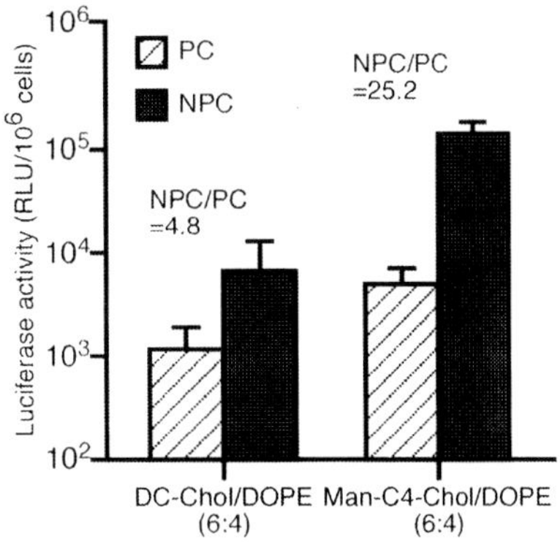

FIG. 4. Intrahepatic gene expression of liposome/DNA complexes after intravenous injection in mice. Plasmid DNA (pCMV-Luc, 50 μg) was complexed with DC-Chol/DOPE (6:4) and Man-C4-chol/DOPE (6:4) at a ratio of 1:7.0 (μg/nmol). Luciferase activity was measured 6 h post-injection in PC and NPC. Each value represents the mean ± S.D. ($n = 3$)

In summary, the luciferase reporter gene assay described in this chapter is a simple and convenient method for monitoring in vivo transgene expression. This system can also be used to evaluate the in vivo transfection ability of other types of gene carrier candidates as well as transgene expression in tumor-bearing animal models.

References

Dash PR, Read ML, Barrett LB, Wolfert MA, Seymour LW (1999) Factors affecting blood clearance and in vivo distribution of polyelectrolyte complexes for gene delivery. Gene Ther 6:643–650

Dash PR, Read ML, Fisher KD, Howard KA, Wolfert M, Oupicky D, Subr V, Strohalm J, Ulbrich K, Seymour LW (2000) Decreased binding to proteins and cells of polymeric gene delivery vectors surface modified with a multivalent hydrophilic polymer and retargeting through attachment of transferrin. J Biol Chem 275:3793–3802

Gao X, Huang L (1991) A novel cationic liposome reagent for efficient transfection of mammalian cells. Biochem Biophys Res Commun 179:280–285

Kawakami S, Yamashita F, Nishikawa M, Takakura Y, Hashida M (1998) Asialoglycoprotein receptor-mediated gene transfer using novel galactosylated cationic liposomes. Biochem Biophys Res Commun 252:78–83

Kawakami S, Fumoto S, Nishikawa M, Yamashita F, Hashida M (2000a) In vivo gene delivery to the liver using novel galactosylated cationic liposome. Pharm Res 17:306–313

Kawakami S, Sato A, Nishikawa M, Yamashita F, Hashida M (2000b) Mannose receptor-mediated gene transfer into macrophages using novel mannosylated cationic liposomes. Gene Ther 7:292–299

Kawakami S, Wong J, Sato A, Hattori Y, Yamashita F, Hashida M (2000c) Biodistribution characteristics of mannosylated, fucosylated, and galactosylated liposomes in mice. Biochim Biophys Acta 1524:258–265

Lee YC, Stowell CP, Krantz MJ (1976) 2-Imino-2-methoxyethyl 1-thioglycosides: new reagents for attaching sugars to proteins. Biochemistry 15:3956–3963

Li S, Huang L (1997) In vivo gene transfer via intravenous administration of cationic lipid-protamine-DNA (LPD) complexes. Gene Ther 4:891–900

Ogris M, Steinlein P, Kursa M, Mechtler K, Kircheis R, Wagner E (1998) The size of DNA/transferring-PEI complexes is an important factor for gene expression in cultured cells. Gene Ther 5:1425–1433

Tandia BM, Vandenbranden M, Wattiez R, Lakhdar Z, Ruysschaert JM, Elouahabi A (2003) Identification of human plasma proteins that bind to cationic lipid/DNA complex and analysis of their sffects on transfection efficiency: implications for intravenous gene transfer. Mol Ther 8:264–273

Templeton NS, Lasic DD, Frederik PM, Strey HH, Roberts DD, Pavlakis GN (1997) Improved DNA:liposome complexes for increased systemic delivery and gene expression. Nature Biotechnol 15:647–652

Evaluation by Southern Blot Hybridization of DNA Administered with a Gene Carrier to Organs

Takahito Kawano and Takuro Niidome

1 Introduction

The biodistribution of plasmid DNA administered with gene carriers is an important parameter in evaluating the characteristics of gene carriers in vivo and the expected therapeutic effects of transgenes. Radioisotopic labeling of DNA is the most common method for determining biodistribution, and the amount of labeled DNA present in each organ can be measured directly by measuring the radioactivity of the organ. This technique has several advantages, such as the high sensitivity of detection; a wide dynamic range which allows highly accurate quantification; and the simplicity of the procedure, which facilitates the handling of many samples at a time. However, radioisotopic labeling cannot distinguish between intact and degraded DNA in the organs. In addition, radioisotopes have to be used in an authorized location and, due to the half-life of the radioactivity, within a limited time frame. Alternatively, fluorescently labeled DNA can also be used to assess biodistribution. Although fluorescence detection is advantageous for observing cross-sections of organs under the microscope, it is unsuitable for determining the total amount and quality of the DNA in each organ. In order to solve these problems, Southern blotting and non-radio isotope (RI) probe hybridization techniques can be utilized. The technique of Southern blotting, named after its inventor, Edward Southern, allows the size of the particular DNA of interest within the smear of genomic genes in the organs to be identified. Using this technique, information about the quality (intact or degraded) and quantity of the DNA in organs is obtained. Of course, procedures for non-RI Southern blot hybridization are more complicated, and the accuracy and dynamic range of quantification is lower than with the direct RI system. However, a highly sensitive cooled CCD camera (e.g., NightOWL, Berthold Technologies, Bad Wildbad, Germany) that detects the chemical luminescence from hybridized fragments over a wide dynamic range has been developed. This technique allows both the quality and the quantity of DNA to be evaluated and is expected to be a powerful tool for studying gene delivery in vivo.

Department of Applied Chemistry, Faculty of Engineering, Kyushu University, Fukuoka 812-8581, Japan

This chapter provides an example of Southern blot hybridization as applied to detecting genes delivered to organs by intravenous administration with gene carriers into mice. Two kinds of gene carriers, a cationic lipid (DOTAP) and a sixth-generation dendrimer (dendritic poly(L-lysine), are compared with respect to differences in the pharmacokinetics in tumor-bearing mice.

2 Materials

2.1 Plasmid DNA

Plasmid DNA (pCMV-Luc) was constructed by subcloning the *Hin*dIII/*Xba*I firefly luciferase cDNA fragment from the pGL3-control vector (Promega Madison, WI, USA) into the polylinker of the pcDNA3 vector (Invitrogen, Carlsbad, CA, USA). The plasmid DNA was amplified in *Escherichia coli* strain DH5α, and then isolated and purified using a QIAGEN Plasmid Mega Kit (Qiagen, Hilden, Germany).

2.2 Gene-Carrier Molecules

As a general cationic lipid gene-carrier molecule, DOTAP (dioleoyl trimethylammonium propane) was employed (Fig. 1A). DOTAP/Chol (1/1) liposomes were prepared as previously described by Li et al. (1999). Dendritic sixth-generation poly(L-lysine) (KG6), which previously showed high transfection ability without significant cytotoxicity in vitro was synthesized as previously described by Ohsaki et al. (2002) (Fig. 1B).

2.3 Animals

Male ddY mice (Seac Yoshitomi, Fukuoka, Japan) were used in all experiments. Tumor-bearing mice were established by subcutaneous injection of $1 \times x10^6$ B-16 melanoma cells in 200 μl HBSS (Hanks' balanced salt solution). When tumors had reached approximately 300 mm^3, DNA complexes were injected into the tail vein of the mouse. Animal experiments were carried out according to the Guidelines for Animal Experimentation at Kyushu University.

3 Methods

3.1 Preparation of DNA Complexes

Ten microliters of plasmid DNA at a concentration of 2.5 mg/ml of was added to 100 μl of 10% dextrose. After addition of 65 μl of H_2O, 25 μl of 25.8 mM KG6 solution was added to the mixture to yield a C/A ratio (molar ratio of cation to anion) of 8. In the case of DOTAP/Chol (1/1) liposome, 10 μl of plasmid DNA (2.5 mg/ml) was added to 100 μl of 10% dextrose. After addition of 36 μl of H_2O, 64 μl of 10 mg DOTAP/Chol (1/1) liposome/ml was added to the mixture to yield a C/A ratio of 11. After a 5-min incubation at room temperature, 200 μl of the complex solution was intravenously injected into the mouse.

3.2 Intravenous Injection of DNA Complexes into Mice and the Collection of Organs

For intravenous administration, 200 μl of DNA complexes in 5% dextrose were injected via the tail vein. After the indicated time points, blood was directly collected

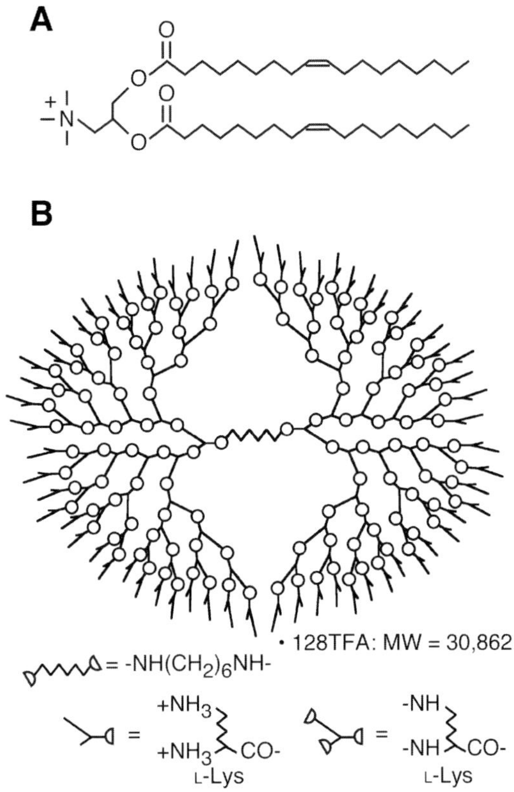

FIG. 1. Structures of **A** dioleoyl trimethylammonium propane (DOTAP) and **B** sixth-generation dendritic poly(L-lysine)(KG6)

from the heart under anesthesia, mixed with EDTA in microfuge tubes and frozen in liquid nitrogen. Tissue samples from other organs, such as the liver, lungs, spleen and kidneys, were collected and frozen in liquid nitrogen. The blood and tissue samples were stored at –80°C until DNA extraction.

3.3 Total DNA Extraction from Organs

DNA was isolated from blood and tissue samples using a DNeasy Tissue Kit (Qiagen), which allows rapid isolation of total DNA from a variety of samples, including fresh and frozen animal tissues and cells. After tissue samples are lysed, the procedure can be completed in as little as 20 min.

Protocol for liver, lung, spleen, and kidney:

1. Chop tissue samples (25 mg except for 10 mg of spleen) into small pieces, and add 180 µl buffer ALT and 20 µl proteinase K.
2. Mix by vortexing and incubate at 55°C until the tissue is completely lysed. Vortex occasionally during incubation.
3. Add 200 µl buffer AL, mix by vortexing, and incubate at 70°C for 10 min.
4. Add 200 µl ethanol, mix by vortexing.
5. Apply the mixture to the DNeasy spin column, and centrifuge at 6,000× *g* for 1 min.
6. Add 500 µl buffer AW1 to the DNeasy spin column, and centrifuge at 6,000× *g* for 1 min.

7. Add 500 μl buffer AW2 to the DNeasy spin column, and centrifuge at full speed for 3 min to dry the DNeasy membrane.
8. Place the DNeasy spin column into a clean microtube.
9. Add 200 μl buffer AE to the DNeasy spin column. Incubate at room temperature for 1 min and centrifuge at 6,000× *g* for 1 min to elute the purified DNA into the microtube.

Protocol for blood:

1. Add 20 μl proteinase K to blood samples (100 μl); adjust volume to 220 μl with PBS.
2. Add 200 μl buffer AL, mix by vortexing, and incubate at 70°C for 10 min.
3. Continue with step 4 of the protocol for animal tissues.

3.4 Blotting

The principal technique is outlined in Fig. 2A. The system has been successfully used not only for Southern blotting but also for Northern and colony blotting applications.

3.4.1 Gel Electrophoresis

1. Prepare a 1% agarose gel in TBE (90 mM Tris-borate buffer, pH 8.0 containing 2 mM EDTA).
2. Apply DNA samples to wells in the gel (0.3 μg of total DNA from the blood, lung, spleen and kidney and 1.0 μg of total DNA from the liver), and electrophoresis in TBE buffer at 100 V.

3.4.2 Processing the Gel

1. Put the gel into a plastic box, cover with depurination solution (250 mM HCl) and begin agitation. Treatment should be stopped when the bromophenol blue dye has turned completely yellow.

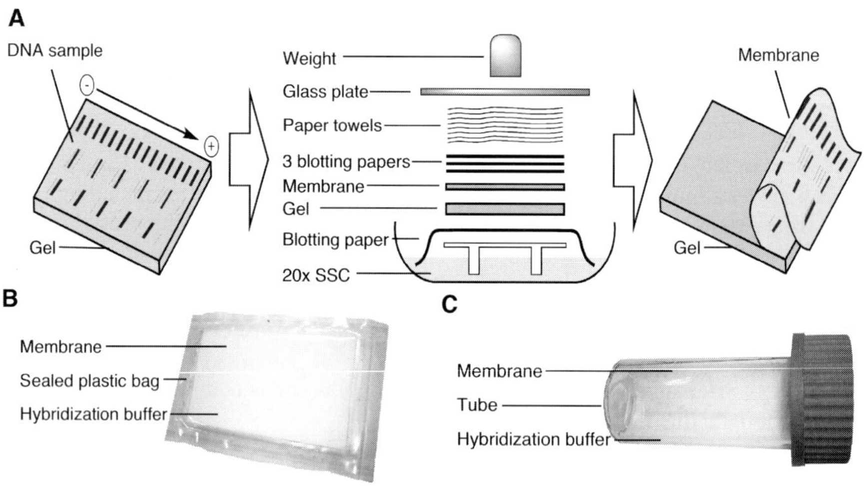

Fig. 2. **A** Outline of capillary blotting technique. **B** The plastic bag used for hybridization. **C** A tube for the rotary hybridization oven

2. Replace with denaturation solution (1.5 M NaCl, 0.5 M NaOH) and begin agitation. Treatment should be stopped 25 min after the bromophenol blue dye turned blue again.
3. Replace with neutralizing solution (1.5 M NaCl, 0.5 M Tris-HCl, pH 7.5), and continue agitation for 30 min.

3.4.3 Capillary Blotting

1. Prepare an appropriate amount of 20× SSC (0.3 M Na_3-citrate, 3 M NaCl, pH 7.0), and make a supporting platform. Cover it with a blotting paper wetted with 20× SSC.
2. Place the gel on the paper, taking care to avoid trapping air bubbles.
3. Cut a sheet of Immobilon-NY+ membrane (Millipore Bedford, MA, USA) to the size of the gel. Pre-wet the membrane by briefly immersing in distilled water, followed by soaking in 20× SSC for 5 min.
4. Place the membrane on the gel, without trapping air bubbles.
5. Place the three blotting papers, cut to size and wetted with 20× SSC, on the membrane, without trapping air bubbles.
6. Place a 5- to 7-cm stack of absorbent paper towels on the blotting paper.
7. Place the glass plate on the paper towels and apply a weight. Leave overnight or for at least 4 h.

3.4.4 Processing the Blot

1. Remove the membrane and the gel together, and place membrane side down on clean blotting paper. Peel off the gel and discard.
2. Place the membrane in 6× SSC for 1 min with gentle agitation to remove any gel fragments.
3. Bake at 80°C for 2 h.

3.5 Hybridization and Signal Detection

Hybridization and signal detection were carried out using the ECL Direct Nucleic Acid Labeling and Detection System (Amersham Biosciences Piscataway, NJ, USA). For quantification of the plasmid DNA, different amounts of plasmid DNA are usually loaded on the agarose gel as standards.

3.5.1 Preparation of Labeled Probe

As the gene of interest is injected plasmid DNA, the plasmid DNA itself was used as a probe. The following protocol is for labeling 100 ng of probe DNA. However, the amount of probe DNA depends on the size of the blot and the amount of target DNA. In particular, the size of blot determines the hybridization volume, while the length of the probe sequence determines the probe concentration. The protocol included in the system recommends that 10 ng probe/ml is appropriate for many applications.

1. Denature 100 ng of the DNA sample (10 μl) to be labeled by heating for 5 min in boiling water.
2. Immediately cool the DNA sample on ice for 5 min.
3. Add an equivalent volume of the DNA labeling reagent to the cooled DNA sample, and mix gently.

4. Add the glutaraldehyde solution, using a volume equivalent to that of the DNA labeling reagent, and mix thoroughly.
5. Incubate for 10 min at 37°C.
6. Add glycerol to a final concentration of 50% (v/v). Labeled probe may be stored at −30°C for 6 months.

3.5.2 Hybridization and Stringency Washes

Hybridization can be carried out in a plastic bag, as shown in Fig. 2B, without the need for a special rotary hybridization oven. Adequate circulation of the buffer is essential and it is important that the blots are allowed to move freely in the bag.

1. Prepare the hybridization buffer: take the required volume of hybridization buffer (0.25 ml/cm^2 is recommended for a 100-cm^2 blot). Add NaCl and the blocking agent to a final concentration of 0.5 M and 5% (w/v). Mix at room temperature for 1 h, then at 42°C for 0.5–1 h.
2. Insert the blot between two plastic sheets (0.1-mm thickness) and seal around of the blot using a heat sealer. After cutting an edge of the bag, add the buffer into the sealed plastic bag using a pipet. Then, seal the bag, again and prehybridize for 1 h at 42°C.
3. Add the labeled probe into the sealed plastic bag to avoid placing it directly on the membrane and mix gently. Incubate with gentle agitation overnight at 42°C on a shaker.
4. Carefully transfer the blots to the primary wash solution (0.4% SDS, 0.5× SSC) in a plastic box, and wash at 42°C for 20 min with gentle agitation.
5. Wash the blots again in fresh primary wash buffer at 42°C for 20 min.
6. Wash in secondary wash buffer (2× SSC) at room temperature for 5 min with gentle agitation.

3.5.3 Hybridization and Stringency Washes in Special Tubes for a Rotary Hybridization Oven

Hybridization and stringency washing can be conveniently performed in a rotary hybridization oven (Fig. 2C). The oven allows the continuous movement of fluid and therefore only minimal volumes of reagents, especially the probe, are needed. They are particularly economical on probe.

1. Prepare the hybridization buffer (see Sect. 3.5.2, step 1) at 42°C.
2. Pre-wet the blots in 5× SSC. Loosely roll the blot and place inside the tube without air bubbles between the blot and the tube.
3. Add the hybridization buffer (0.0625–0.125 ml/cm^2) to the tube.
4. Prehybridize in the rotary oven for 15 min at 42°C.
5. Add the labeled probe to the prehybridization buffer. Hybridize overnight in the rotary oven at 42°C.
6. Discard the hybridization buffer; add 50–100 ml 5× SSC. Rotate the tube for 5 min.
7. Discard the 5× SSC, and replace with the primary wash solution (0.4% SDS, 0.5× SSC) so that the tube is one-third full. Rotate the tube in the oven and wash the blot in for 20 min at 42°C.
8. Replace the fresh primary wash buffer and rotate for 10 min at 42°C. Repeat this wash twice.

9. Remove the blot from the tube, cover with the secondary wash buffer, and wash for 5 min at room temperature in a plastic box. Repeat this wash twice.

3.5.4 Signal Generation and Detection

For optimum sensitivity, it is necessary to work as quickly as possible. The required equipment consists of an X-ray film cassette, cling-film and autoradiography (X-ray) film. In this protocol, chemical luminescence from the hybridized probes is detected on X-ray film. However, as noted above, an imaging system using a cooled CCD camera is more convenient and allows quantification of the signals on the blot.

1. Mix an equal volume of detection reagent 1 with detection reagent 2 (0.125 ml/cm^2).
2. Add the detection reagent directly to the blot on the side carrying the DNA.
3. Incubate for 1 min at room temperature.
4. Wrap the blot in cling-film. Gently smooth out air bubbles.
5. Place the blot DNA side up in the film cassette and place on the sheet of autoradiography film (Hyperfilm, Amersham Biosciences). Expose for 1–5 min.
6. Remove the film and develop.

4 Biodistribution and Pharmacokinetics of Intravenously Injected DNA with Gene Carriers

The Southern blot hybridization technique was tested by examining the biodistribution of plasmid DNA administered together with either a cationic liposome or a dendrimer into tumor-bearing mice by intravenous injection. Two types of gene carriers were used: (1) DOTAP/Chol (1/1) liposomes and (2) sixth-generation dendritic poly(L-lysine) (KG6) (Fig. 1). Figure 3A shows the biodistribution of the DNA and its pharmacokinetics when delivered with DOTAP/Chol (1/1) liposomes. The DNA was mainly distributed in the lungs, where it was retained for 60 min. The accumulation in the lungs was consistent with the data reported by Song et al. (1998). In this case, no DNA was observed in the tumors. By contrast, DNA delivered with KG6 was still detected in the blood after 180 min, although a gradual decrease in the amount was noted (Fig. 3B). The long-lasting circulation of the KG6-DNA complexes was comparable to that of PEG-modified carrier, made up of a block copolymer of poly-L-lysine and poly(ethylene glycol), which stabilized DNA in the blood for 180 min (Harada-Shiba et al. 2002). The amount of DNA in the lungs was also maintained for 180 min. In the liver, the amount of intact DNA reached a maximum at 60 min, and decreased thereafter. These results suggest that a part of the DNA intravenously injected with KG6 was trapped in the liver, while another part circulated in the blood. DNA trapped in the liver would be expected to slowly degrade, whereas DNA circulating in the blood would eventually become trapped in the liver and spleen, and then degraded. More importantly, DNA was clearly observed in the tumor after 60 min.

The amount of DNA in each organ was roughly calculated from the blot. Figure 3C shows the amount of DNA at 30 min after injection. DNA injected with KG6 was localized mainly in the liver (20%), while 10% of the injected DNA remained in the blood. In the cases of DOTAP/Chol (1/1) liposomes and DNA alone, the amounts of DNA in the blood were less than 1% of the injected DNA, while the DOTAP/Chol (1/1) liposomes were mostly distributed in the lungs (15%). No naked DNA was detected.

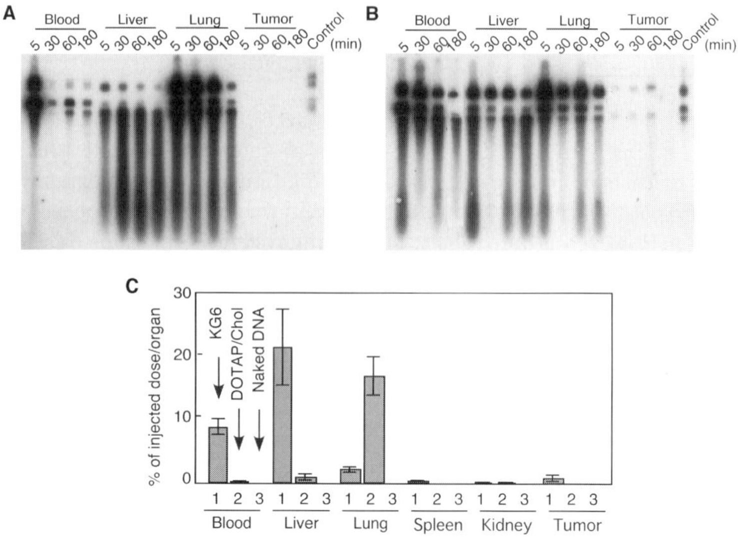

FIG. 3. Time courses of the amounts of plasmid DNA in tissues after intravenous injection of **A** DOTAP/Chol (1/1) liposomes at a C/A ratio of 11 and **B** KG6 complexes at a C/A ratio of 8. **C** The amounts of DNA at 30 min after injection were roughly quantified from the blot. KG6-DNA complexes at a C/A ratio of 8 (1), DOTAP/Chol (1/1) liposomes at a C/A ratio of 11 (2) and naked DNA (3). Data represent the mean value from $n = 3$, and the standard deviation from the mean

Prolonged circulation in the blood and delivery of the plasmid DNA into tumors were achieved by complexing the DNA with KG6. Since the complexes had a weak cationic surface (+3 mV) at a C/A ratio of 8.0 (Ohsaki et al. 2002), non-specific binding of serum components that enhance DNA clearance via the reticuloendothelial system would be lower than that using other simple cationic carriers, such a DOTAP/Chol liposome. The stealth character of the DNA-KG6 complexes in the blood should enhanced permeability and retention (EPR) effect in the tumor. In addition, compared to simple cationic carriers such as DOTAP/Chol liposome, the inert character of the complexes in the blood results in differences in the biodistribution.

References

Harada-Shiba M, Yamauchi K, Harada A, Takamisawa I, Shimokado K, Kataoka K (2002) Polyion complex micelles as vectors in gene therapy-pharmacokinetics and in vivo gene transfer. Gene Ther 9:407–414.

Li S, Wu SP, Whitemore M, Loeffert EJ, Wang L, Watkins SC, Pitt BP, Huang (1999) Effect of immune response on gene transfer to the lung via systemic administration of cationic lipidic vectors. Am J Physiol 276:796–804.

Ohsaki M, Okuda T, Wada A, Hirayama T, Niidome T, Aoyagi H (2002) In vitro gene transfection using dendritic poly(L-lysine). Bioconjugate Chem 13:510–517.

Song YK, Liu F, Chu S, Liu D (1997) Characterization of cationic liposome-mediated gene transfer in vivo by intravenous administration. Human Gnen Ther 8:1585–1594.

Evaluation of the Immune Response After Administration of Plasmid DNA-Non-viral Vector Complexes

Fuminori Sakurai[1,3], Yoshinobu Takakura[2], and Mitsuru Hashida[1]

1 Introduction

The success of gene therapy is largely dependent on the development of gene transfer vector systems. So far, various types of gene delivery vectors, both viral and non-viral, have been developed. Among these, synthetic non-viral vectors, such as cationic lipids and cationic polymers, which are electrostatically complexed with plasmid DNA, are particularly suitable for in vivo gene delivery due to their simplicity of preparation and the absence of a specific immune response. Some of these non-viral vectors have shown promise in in-vivo gene delivery settings (Dow et al. 1999; Hood et al. 2002, Sakurai et al. 2003; Yamada et al. 2003).

Despite the potential utility of synthetic vector systems, the in vivo application of these vectors containing plasmid DNA elicits a range of acute physiological responses, such as transient leukopenia thrombocytopenia, and elevations in serum transaminase and proinflammatory cytokines (TNF-α, IFN-γ, IL-12, etc.) (Dow et al. 1999; Gautam et al. 2001; Li, S. and Huang, L. 1997; Sakurai et al. 2002; Tousignant et al. 2000). In particular, the intravenous administration of plasmid DNA-cationic liposome complexes results in systemic induction of high amounts of proinflammatory cytokines. These immune responses are due to the bacterially derived plasmid DNA, which is recognized as foreign material by vertebrate cells. Bacterial DNA is rich in unmethylated cytosine-phosphate-guanine (CpG) dinucleotides, in contrast to mammalian DNA, which contains a low frequency of CpG dinucleotides that are mostly methylated. Unmethylated CpG in specific sequence contexts, such as 5′-Pur-Pur-CpG-Pry-Pry-3′, is recognized by vertebral cells as a danger signal and thus induces the production of large quantities of proinflammatory cytokines by lymphocytes, natural killer cells, dendritic cells and macrophages (Hemmi et al. 2003; Sivori et al. 2004;

[1] Department of Drug Delivery System, Graduate School of Pharmaceutical Sciences, Kyoto University, Sakyo-ku, Kyoto 606-8501, Japan

[2] Department of Biopharmaceutics and Drug Metabolism, Graduate School of Pharmaceutical Sciences, Kyoto University, Sakyo-ku, Kyoto 606-8501, Japan

[3] Present address: Project III, National Institute of Health Sciences, Osaka Branch, Fundamental Research Laboratories for Development of Medicine, 7-6-8 Asagi Saito, Ibaraki, Osaka 567-0085, Japan

Wagner et al. 2004; Yasuda et al. 2002). Such cytokines can be toxic and limit transgene expression while causing refractory behavior on repeated dosing of the transgene at frequent intervals (Li et al. 1999; Tan et al. 1999). This is a potential problem in gene therapy using plasmid DNA. However, it appears that induction of proinflammatory cytokines activates the immune system and leads to strong antitumor effects in mice (Nomura et al. 1999; Whitmore et al. 1999). For these reasons, it is clearly of great importance to evaluate cellular immune responses after administration of plasmid DNA-non-viral vector complexes.

This chapter focuses on the immune responses, especially proinflammatory cytokine production induced by a plasmid-DNA-based gene delivery system. While there are a large numbers of studies on the development of non-viral vectors and their transfection efficiencies, there is little information about the immune responses induced by plasmid DNA-non-viral vector complexes. However, such studies will aid in the development of safe and efficient gene delivery systems and in the elucidation of the mechanisms of CpG-motif-dependent cytokine production.

2 Materials and Methods

2.1 Preparation of Plasmid DNA

Plasmid DNA is amplified in *Escherichia coli* strain DH5α and purified using a Qiagen Plasmid Giga kit (Qiagen, Valencia, CA, USA). (It is recommended that plasmid DNA be prepared using commercially available kit which can prevent lipopolysaccharide (LPS) contamination). The concentration of plasmid DNA is measured by UV absorption at 260 nm. Plasmid DNA purity is assessed using agarose gel electrophoresis and A_{260}/A_{280} ratios. The LPS concentration in the plasmid DNA preparation is measured using a LAL assay kit (Limus F Single Test Wako, Wako Pure Chemical, Osaka, Japan). (LPS contamination in the plasmid DNA samples should be checked, since it will increase the proinflammatory cytokine responses.)

In the study described in this chapter, pcDNA3 vector (Invitrogen, Carlsbad, CA, USA), which has 27 CpG motifs displaying 5′-Pur-Pur-CpG-Pry-Pry-3′ sequences, was used. The LPS contamination was negligible (less than 22 pg/μg DNA), which corresponds to the manufacturer's information.

2.2 In Vitro Study

Immune cells, such as macrophages and dendritic cells, play a central role in proinflammatory cytokine production induced by plasmid DNA-based gene delivery systems. Here, in vitro proinflammatory cytokine production by mouse peritoneal macrophages is evaluated.

1. Resident macrophages are collected from the peritoneal cavity of the mice with 3–5 ml ice-cold RPMI1640 medium or PBS.
2. The cells are washed, suspended in RPMI1640 medium containing 10% FCS, penicillin G (100 U/ml) and streptomycin (100 μg/ml), and then seeded on 24-well plates at a density of 5×10^5 cells/well.
3. The cells are washed with the medium 4 h after seeding to remove erythrocytes.

4. The next day after seeding, plasmid DNA-non-viral vector complex in 0.5 ml culture medium is added to the cells, which have been washed three times with 0.5 ml culture medium.
5. At the indicated time, the supernatants are collected and centrifuged (6000 *g*, 10 min, 4°C). The supernatants are kept at –80°C until time of ELISA.
6. The amounts of proinflammatory cytokines in the supernatants are determined by ELISA kits.

2.3 In Vivo Study

Systemic induction of high levels of proinflammatory cytokines in blood lead to a massive systemic inflammatory response. It is therefore important to monitor cytokine levels in blood after administration of gene delivery vectors. The authors' protocol is described below. It should be noted that the production levels of many cytokines are dependent on the mouse strain. Expression level of Toll-like receptor 9 (TLR9), which recognizes bacterial DNA and activates NF-κB, are different between mouse strains (Liu et al. 2002).

1. After administration of plasmid DNA-non-viral vector complex into mice, blood samples are collected by retro-orbital or tail vein bleeding. A larger volume of blood

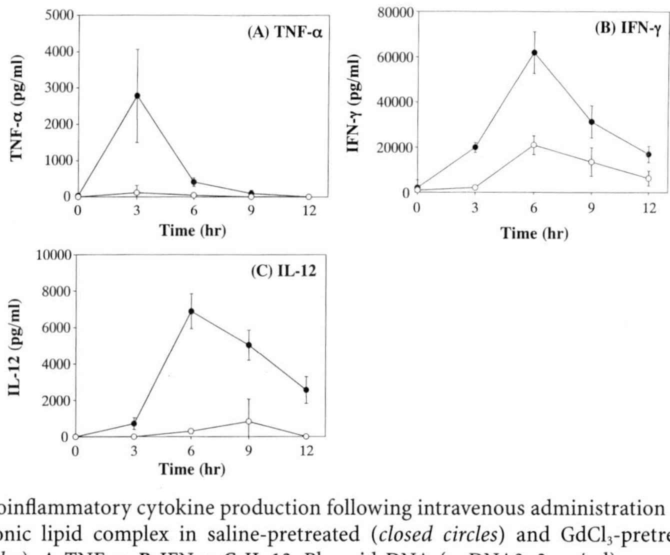

FIG. 1. Proinflammatory cytokine production following intravenous administration of plasmid DNA-cationic lipid complex in saline-pretreated (*closed circles*) and $GdCl_3$-pretreated mice (*open circles*). **A** TNF-α, **B** IFN-γ, **C** IL-12. Plasmid DNA (pcDNA3, 2 mg/ml) was added to a DOTMA/Chol liposome suspension (DOTMA:Chol = 1:1 molar ratio, 1 mg DOTMA/ml) at a charge ratio of +2.24. The mixtures were incubated for 30 min before use. The complex was injected intravenously via the tail vein into CDF1 mice at a dose of 25 μg DNA. Amounts of TNF-α, IFN-γ, and IL-12 in serum after intravenous administration of the complex were analyzed using an ELISA kit. For transient tissue macrophage blockade, gadolinium chloride ($GdCl_3$; Sigma, St Louis, MO, USA) was dissolved in saline, and 45 mg/kg body weight was injected via the tail vein 24 h before plasmid DNA-cationic lipid complex injection, in a total volume of 150 μl. Control mice were injected with 150 μl saline. (From Sakurai et al. 2002, with permission)

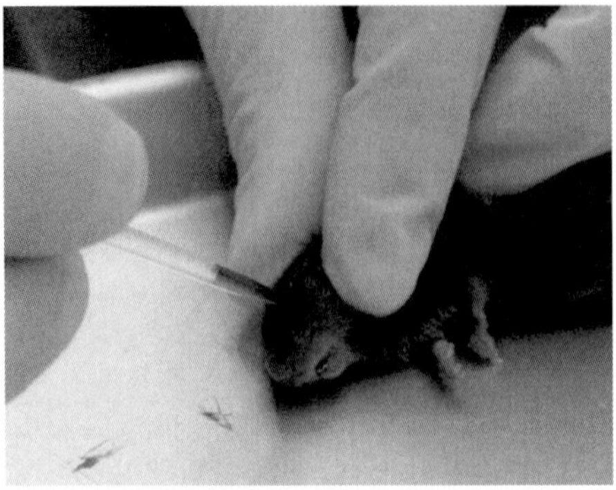

Fig. 2. Blood collection from mice by retro-orbital bleeding

(more than 100 μl) can be easily collected by retro-orbital bleeding (Fig. 2). However, collection of large volume of blood samples several times a day might affect the cytokine concentrations and physiological condition of the mouse. In our study (Fig. 1), blood samples (40–60 μl) were collected by tail-vein bleeding and approximately 20 μl of serum samples were obtained. These were diluted 20- to 1,000 fold for ELISA analysis.

2. The blood samples are incubated at 4°C for 2–3 h to allow clotting and then centrifuged to obtain serum (6000 g, 10 min, 4°C). Serum samples can be stored at −80° until needed.
3. Amounts of proinflammatory cytokines in serum are analyzed using ELISA kits (Genzyme, Cambridge, MA, USA). Hepatotoxicity after plasmid DNA-non-viral vector complex administration is also evaluated by measuring the serum levels of glutamic oxaloacetic transaminase (GOT) and glutamic pyruvic transaminase (GPT) using commercially available reagents (Transaminase CII-Test Wako, Wako Pure Chemical).

As shown in Fig. 1, intravenous administration of the plasmid DNA-cationic liposome (DOTMA/Chol liposome) complex systemically induced large amounts of proinflammatory cytokines (TNF-α, IFN-γ, and IL-12). The highest amount of TNF-α was observed 3 h after i.v. injection of the complex in saline-pretreated mice (2800 pg/ml). In the case of IFN-γ and IL-12, the peak was reached 6 h after injection (IFN-γ; 61900 pg/ml, IL-12; 6800 pg/ml). In $GdCl_3$-pretreated mice, the amounts of these cytokines were significantly lower than those in the saline-pretreated mice. The amounts of TNF-α, IFN-γ and IL-12 at the peak times were reduced 25-fold, 3-fold, and 23-fold in $GdCl_3$-pretreated mice, respectively.

2.4 RT-PCR Analysis of Proinflammatory Cytokines in Mouse Organs

Messenger RNA expression of proinflammatory cytokines, including those expressed at very low levels, in the organs can be easily detected by RT-PCR analysis. While quantitative analysis of mRNA expression can be achieved by real-time RT-PCR, conventional RT-PCR analysis is described here.

1. After administration of plasmid DNA-non-viral vector complex, the animals are bled from the inferior vena cava, while saline is perfused via the portal vein to

remove blood from the organs. Mouse organs are isolated, washed with ice-cold saline, and stored at −80°C until RNA isolation.
2. Total RNA is extracted using Trizol (Invitrogen).
3. RNA (2 μg) is reverse-transcribed to cDNA using oligo(dT) primers and the SUPERSCRIPT First-Strand Synthesis System for RT-PCR (Invitrogen).
4. The cDNA products are amplified by PCR using the primers listed in Table 1.
5. PCR products are electrophoresed on a 2% TBE agarose gel and then stained with ethidium bromide.

Figure 3 shows the RT-PCR analysis of TNF-α expression in selected organs after intravenous administration of the plasmid DNA-cationic lipid complex into mice. TNF-α mRNA expression increased in all of the organs of saline-pretreated mice. In

TABLE 1. Primers for RT-PCR analysis of mouse proinflammatory cytokines

Gene	Expected PCR product Size		Sequence (5′ to 3′)
TNF-α (Yoshidome et al. 1999)	446 bp	Forward	AGCCCACGTAGCAAACCACCAA
		Reverse	ACACCCATTCCCTTCACAGAGCAAT
IFN-γ (Wesselingh et al. 1994)	306 bp	Forward	AGCTCTGAGACAATGAACGC
		Reverse	GGACAATCTCTTCCCCACCC
IL-1β (Wesselingh et al. 1994)	381 bp	Forward	AAGGAGAACCAAGCAACGAC
		Reverse	GAGATTGAGCTGTCTGCTCA
IL-6 (Wesselingh et al. 1994)	308 bp	Forward	TGCTGGTGACAACCACGGCC
		Reverse	GTACTCCAGAAGACCAGAGG
IL-12p40 (Todt et al. 2000)	431 bp	Forward	CTCACCTGTGACACGCCTGA
		Reverse	CAGGACACTGAATACTTCTC
MCP-1 (Muruve et al. 1999)	306 bp	Forward	CACAGTTGCCGGCTGGAGCATC
		Reverse	GGTGCTGAAGACCTTAGGGCAG
IP-10 (Muruve et al. 1999)	548 bp	Forward	GGACGGTCCGCTGCAACTGCATCC
		Reverse	GCAGCCTGGGCATGGCACATGGTG
RANTES (Sonoda et al. 2003)	295 bp	Forward	ACAGGTCAAACTACAACTCCA
		Reverse	TCAGCTCTTAGCAGACATTGG
GAPDH (Control) (Wesselingh et al. 1994)	508 bp	Forward	ACCACCATGGAGAAGGCTGG
		Reverse	CTCAGTGTAGCCCAGGATGC
β-actin (Control) (Yoshidome et al. 1999)	245 bp	Forward	GTGGGCCGCTCTAGGCACCA
		Reverse	CGGTTGGCCTTAGGGTTCAGGGGGG

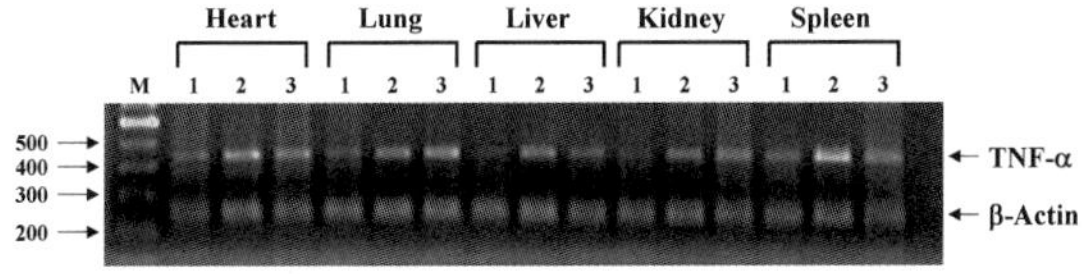

FIG. 3. Detection of TNF-α expression by RT-PCR in various organs following i.v. administration of the complexes. *Lane 1*, Control mice; *lane 2*,mice pretreated with saline 24 h before complex injection; *lane 3*, mice pretreated with $GdCl_3$ 24 h before complex injection. The complexes and the tissue macrophage blockade were carried out as described in Fig. 1. Mice were intravenously injected with the complex at a dose of 25 μg DNA per mouse; 3 h after injection, the organs were collected. For RT-PCR analysis, the following parameters were used: 60 s at 95°C, 90 s at 59°C, and 60 s at 72°C for 30 cycles. (From Sakurai et al. 2002, with permission)

$GdCl_3$-pretreated mice, there was an apparent decrease in the level of TNF-α mRNA expression in the liver and spleen, indicating that these organs play important roles in the production of proinflammatory cytokines induced by the complexes.

3 Comments

In this chapter, proinflammatory cytokine production induced by a plasmid-DNA-based gene delivery was evaluated. Inflammatory immune responses induced by gene delivery vectors, including viral and non-viral vectors, are one of the most serious barriers to successful gene delivery. During the gene therapy trial for ornithine transcarbamylase (OTC) deficiency, carried out at the University of Pennsylvania (USA) in 1999, an escalated dose of adenovirus vector (3.8×10^{13} vector particles) systemically administered through the hepatic artery induced a massive systemic inflammatory response that led to fever, disseminated intravascular coagulation, multiorgan failure and the eventual death of an 18-year-old patient (Raper et al. 2003). It is well known from animal experiments that systemic administration of adenovirus vector causes a rapid induction of proinflammatory cytokines (Muruve et al. 2001), but plasmid DNA-non-viral vector complexes were assumed to be safer than viral vectors, including adenovirus vectors. However, it has been shown that systemic administration of plasmid DNA-non-viral vector complexes induces comparable, or higher amounts of proinflammatory cytokines in blood than adenovirus vectors, suggesting that the risks associated with the induction of inflammatory responses by plasmid DNA-non-viral vector complexes are similar to those caused by the administration of adenovirus vectors.

In a previous study, we demonstrated that intravenous administration of plasmid DNA-cationic lipid complex induces significant proinflammatory cytokine production in blood and that tissue macrophages involving liver Kupffer cells and spleen macrophages are closely involved in TNF-α production (Sakurai et al. 2002). The liver and spleen are well known to play important roles in the removal of foreign particles and invasive microorganisms from the blood stream, mainly via Kupffer cells and spleen macrophages. Several groups have reported that plasmid DNA-cationic lipid complexes accumulate in the liver and spleen, as well as in the lung, after intravenous administration, and that liver Kupffer cells and spleen macrophages are directly involved in uptake of the complexes (Mahato et al. 1995; McLean et al. 1997; Osaka et al. 1996). We demonstrated that tissue macrophage blockade by $GdCl_3$ dramatically reduced serum levels of proinflammatory cytokines (Fig. 1), liver accumulation of the complexes (data not shown) and mRNA expression of TNF-α in organs (Fig. 3). These results indicate that tissue macrophages, including liver Kupffer cells and spleen macrophages, play a central role in TNF-α production.

Proinflammatory cytokines trigger damage and subsequent apoptosis of vascular endothelial cells and liver hepatocytes. The toxicity of these cytokines would limit gene expression and induce refractory behavior on repeated dosing of plasmid DNA-cationic lipid complexes at frequent intervals. Our results demonstrated that suppression of inflammatory cytokine production by $GdCl_3$ resulted in higher transfection efficiency in the lung following repeated injection of plasmid DNA-cationic lipid complexes (Sakurai et al. 2002). Suppression of apoptosis by anti-mouse

TNF-α and anti-mouse IFN-γ antibodies or dexamethasone also led to significantly higher levels of transgene expression in the lung (Li et al. 1999; Tan et al. 1999). Thus, preventing the uptake of plasmid DNA-non-viral vector complexes by tissue macrophages, mainly liver Kupffer cells and spleen macrophages, should lead to a reduction in proinflammatory cytokine production and an improvement in transgene expression.

Immunostimulatory CpG motifs in plasmid DNA are recognized by TLR9 (Hemmi et al. 2000), which has a critical role in the immune responses induced by plasmid DNA-non-viral vector complexes. Zhao et al. demonstrated that intravenous administration of plasmid DNA-cationic lipid (GL-62) complexes into TLR9-knockout (TLR9-KO) mice resulted in a significant decrease of proinflammatory cytokine induction and hematological changes, compared with wild-type mice (Zhao et al. 2004). TLR9 is found mainly on the inner surface of endosomes (Ahmad-Nejad et al. 2002; Lund et al. 2003); therefore plasmid DNA-non-viral vector complex must be endocytosed into cells in order for TLR9 to recognize the CpG motifs in plasmid DNA, inducing the production of proinflammatory cytokines. However, since plasmid DNA is tightly complexed with cationic lipids or polymers, it remains to be elucidated how the plasmid DNA in the complex is exposed to TLR9. In addition, inflammatory responses induced by the plasmid DNA-cationic lipid complexes were not completely diminished in TLR9-KO mice, indicating that other receptors might be involved in immune responses (Zhao et al. 2004). Recently, TLR-knockout mice and myeloid differentiation factor 88 (MyD88)-knockout mice became commercially available (Oriental Yeast, Tokyo, Japan). Plasmid DNA encoding TLR and TLR-related genes, and CpG-free plasmid DNA vectors can be purchased from Invivogen (San Diego, CA, USA). Yew et al. (2002) demonstrated that systemic delivery of CpG-free plasmid DNA-cationic lipid complexes induced less inflammation, fewer hematological changes, and decreased liver toxicity. Further experiments with these knockout mice and modified plasmid DNA will improve our understanding of the mechanism of inflammatory immune response to plasmid-DNA-based gene delivery systems.

Finally, a clear understanding of the proinflammatory cytokine cascade induced by gene delivery vectors is needed in order to monitor and modulate unfavorable inflammatory responses.

References

Ahmad-Nejad P, Hacker H, et al. (2002) Bacterial CpG-DNA and lipopolysaccharides activate Toll-like receptors at distinct cellular compartments. Eur J Immunol 32:1958–1968

Dow SW, Elmslie RE, et al. (1999) Intravenous cytokine gene delivery by lipid-DNA complexes controls the growth of established lung metastases. Hum Gene Ther 10:2961–2972

Dow SW, Fradkin LG, et al. (1999) Lipid-DNA complexes induce potent activation of innate immune responses and antitumor activity when administered intravenously. J Immunol 163:1552–1561

Gautam A, Densmore CL, et al. (2001) Pulmonary cytokine responses associated with PEI-DNA aerosol gene therapy. Gene Ther 8:254–257

Hemmi H, Kaisho T, et al. (2003) The roles of Toll-like receptor 9, MyD88, and DNA-dependent protein kinase catalytic subunit in the effects of two distinct CpG DNAs on dendritic cell subsets. J Immunol 170:3059–3064

Hemmi H, Takeuchi O, et al. (2000) A Toll-like receptor recognizes bacterial DNA. Nature 408:740–745

Hood JD, Bednarski M, et al. (2002) Tumor regression by targeted gene delivery to the neovasculature. Science 296:2404–2407

Li S and Huang L. (1997) In vivo gene transfer via intravenous administration of cationic lipid-protamine-DNA (LPD) complexes. Gene Ther 4:891–900

Li S, Wu SP, et al. (1999) Effect of immune response on gene transfer to the lung via systemic administration of cationic lipidic vectors. Am J Physiol 276:L796–804

Liu T, Matsuguchi T, et al. (2002) Differences in expression of toll-like receptors and their reactivities in dendritic cells in BALB/c and C57BL/6 mice. Infect Immun 70:6638–6645

Lund J, Sato A, et al. (2003) Toll-like receptor 9-mediated recognition of Herpes simplex virus-2 by plasmacytoid dendritic cells. J Exp Med 198:513–520

Mahato RI, Kawabata K, et al. (1995) Physicochemical and pharmacokinetic characteristics of plasmid DNA/cationic liposome complexes. J Pharm Sci 84:1267–1271

McLean JW, Fox EA, et al. (1997) Organ-specific endothelial cell uptake of cationic liposome-DNA complexes in mice. Am J Physiol 273:H387–404

Muruve DA, Barnes MJ, Stillman IE, Libermann TA (1999) Adenoviral gene therapy leads to rapid induction of multiple chemokines and acute neutrophil-dependent hepatic injury in vivo. Hum Gene Ther 10:965–976

Muruve DA, Cotter MJ, et al. (2004) Helper-dependent adenovirus vectors elicit intact innate but attenuated adaptive host immune responses in vivo. J Virol 78:5966–5972

Nomura T, Yasuda K, et al. (1999) Gene expression and antitumor effects following direct interferon (IFN)-gamma gene transfer with naked plasmid DNA and DC-chol liposome complexes in mice. Gene Ther 6:121–129

Osaka G, Carey K, et al. (1996) Pharmacokinetics, tissue distribution, and expression efficiency of plasmid [33P]DNA following intravenous administration of DNA/cationic lipid complexes in mice: use of a novel radionuclide approach. J Pharm Sci 85:612–618

Raper SE, Chirmule N, et al. (2003) Fatal systemic inflammatory response syndrome in a ornithine transcarbamylase deficient patient following adenoviral gene transfer. Mol Genet Metab 80:148–158

Sakurai F, Terada T, et al. (2003) Therapeutic effect of intravenous delivery of lipoplexes containing the interferon-beta gene and poly I: poly C in a murine lung metastasis model. Cancer Gene Ther 10:661–668

Sakurai F, Terada T, et al. (2002) The role of tissue macrophages in the induction of proinflammatory cytokine production following intravenous injection of lipoplexes. Gene Ther 9:1120–1126

Sivori S, Falco M, et al. (2004) CpG and double-stranded RNA trigger human NK cells by Toll-like receptors: Induction of cytokine release and cytotoxicity against tumors and dendritic cells. Proc Natl Acad Sci USA

Sonoda KH, Sasa Y, Qiao H, Tsutsumi C, Hisatomi T, Komiyama S, Kubota T, Sakamoto T, Kawano Y, Ishibashi T (2003) Immunoregulatory role of ocular macrophages: the macrophages produce RANTES to suppress experimental autoimmune uveitis. J Immunol 171:2652–2659

Tan Y, Li S, et al. (1999) The inhibitory role of CpG immunostimulatory motifs in cationic lipid vector-mediated transgene expression in vivo. Hum Gene Ther 10:2153–2161

Todt J, Sonstein J, Polak T, Seitzman GD, Hu B, Curtis JL (2000) Repeated intratracheal challenge with particulate antigen modulates murine lung cytokines. J Immunol 164: 4037–4047

Tousignant JD, Gates AL, et al. (2000) Comprehensive analysis of the acute toxicities induced by systemic administration of cationic lipid:plasmid DNA complexes in mice. Hum Gene Ther 11:2493–2513

Wagner M, Poeck H, et al. (2004) IL-12p70-dependent Th1 induction by human B cells requires combined activation with CD40 ligand and CpG DNA. J Immunol 172:954–963

Wesselingh SL, Levine B, Fox RJ, Choi S, Griffin DE (1994) Intracerebral cytokine mRNA expression during fatal and nonfatal alphavirus encephalitis suggests a predominant type 2 T cell response. J Immunol 152:1289–1297
Whitmore M, Li S, et al. (1999) LPD lipopolyplex initiates a potent cytokine response and inhibits tumor growth. Gene Ther 6:1867–1875
Yamada T, Iwasaki Y, et al. (2003) Nanoparticles for the delivery of genes and drugs to human hepatocytes. Nat Biotechnol 21:885–890
Yasuda K, Ogawa Y, et al. (2002) Plasmid DNA activates murine macrophages to induce inflammatory cytokines in a CpG motif-independent manner by complex formation with cationic liposomes. Biochem Biophys Res Commun 293:344–348
Yew NS, Zhao H, et al. (2002) CpG-depleted plasmid DNA vectors with enhanced safety and long-term gene expression in vivo. Mol Ther 5:731–738
Yoshidome H, Kato A, Edwards MJ, Lentsch AB (1999) Interleukin-10 suppresses hepatic ischemia/reperfusion injury in mice: implications of a central role for nuclear factor kappaB. Hepatology 30:203–208
Zhang Y, Chirmule N, et al. (2001) Acute cytokine response to systemic adenoviral vectors in mice is mediated by dendritic cells and macrophages. Mol Ther 3:697–707
Zhao H, Hemmi H, et al. (2004) Contribution of Toll-like receptor 9 signaling to the acute inflammatory response to nonviral vectors. Mol Ther 9:241–248

Gateway RNAi

Knut R. Madden, Adam N. Harris, Jennifer M. Kilzer,
Heidi L. Welchin, and Peter J. Welch

1 Introduction

1.1 Abstract

Recombinant DNA vector technology developed to express cDNAs also supports the expression of double stranded RNA molecules for RNA interference (RNAi) in loss-of-function genetic experiments. The Gateway (Invitrogen, Carlsbad, CA, USA) recombination system allows the precise, high fidelity transfer of DNA between multiple plasmid vectors in a rapid, simple process, enabling high-throughput procedures. Initially designed to facilitate cDNA cloning and expression, the Gateway (Invitrogen) system has similar benefits for the expression of RNAi transcripts. Gateway (Invitrogen) adapted RNAi expression cassettes can be created in small, stable, easy-to-manipulate plasmids. The relative knockdown activity of multiple RNAi cassettes can be compared in these Gateway (Invitrogen) "entry" vectors. Select expression cassettes can then be Gateway (Invitrogen) transferred to alternative vectors for: optimal delivery (i.e. Lenti or Adenoviral), insertion into a specific genomic site, or to change selectable markers/reporters. Similarly, libraries consisting of numerous different RNAi expression cassettes can be efficiently transfer between vectors. Additionally, Gateway (Invitrogen) technology facilitates recombination of multiple RNAi expression cassettes into a single vector as well as combining knockdown and cDNA expression cassettes onto the same plasmid (Multisite Gateway, Invitrogen).

1.2 Background

The quest to improve the human condition has been bolstered by the capability of cell biologists to suppress expression of specific genes with double stranded RNA (dsRNA)—a process referred to as RNA interference (RNAi). RNAi is directed by short interfering RNA (siRNA) in a cell's cytoplasm, where the antisense strand can guide a multi-protein RNA-Induced Silencing Complex (RISC) to cleave targeted mRNA (1). Synthetic siRNA can be delivered into cells by transfection with cationic lipids; alternatively, RNAi can be induced by production of siRNA from a DNA template within

Invitrogen, 1600 Faraday Ave., Carlsbad, CA 92008, USA

the cell. Intracellular expressed dsRNA can be made as two independent complimentary strands or as a single, short-hairpin RNA (shRNA) that is processed into siRNA (2, 3). Many RNAi expression cassettes use RNA polymerase III (pol III) promoters to drive shRNA expression, such as the type III pol III promoters from the U6 and H1 genes (refs 4–8). These expression cassettes code for RNA with non-complimentary loops between a sense and antisense sequence that are similar or complementary to ~20 bases of target sequence in a mRNA transcript. Termination of these nascent pol III shRNA transcripts is directed by short polythymidine tracts (9). Such loss-of-function RNAi tools will complement over-expression technologies to manipulate protein expression.

The utility of expressing RNAi and cDNA transcripts in a wide variety of applications is limited by the molecular manipulations required to generate DNA expression vectors. To overcome this limitation, cloning technologies that exploit recombinases have been invented to facilitate the manipulation of DNA and support high-throughput (HTP) processes, while maintaining maximum flexibility and compatibility. Two commercially available systems that use *Int* and *Cre* recombinases in vitro are the "Gateway" (Invitrogen) and "Creator" (BD – Clontech, Palo Alto, CA, USA) systems, respectively (10, 11). Alternatively, the Mating-Assisted-Genetically-Integrated-Cloning system (MAGIC) exploits recombination in bacteria (12). The Gateway (Invitrogen) system will be modeled in this article to demonstrate the advantages recombinase cloning systems provide RNAi applications. Similar benefits from other recombination systems are likely, although limitations due to the alternative mechanics of the respective recombination processes may exist.

2 Gateway

Gateway (Invitrogen) Recombination reactions use the lambda bacteriophage DNA recombinase, Integrase (Int), and bacterial DNA phage attachment (att) sites to swap DNA fragments between vectors or genomes (10, 13). Briefly, Int complexes cut two independent pieces of double stranded DNA within att sites creating intermediate single-stranded overhangs. The ends participate in strand exchange and ultimately Int catalyzed rejoining of the recombinant DNA strands. The att sites are positioned in plasmids to orchestrate transfer of a section of DNA from one vector to a similar position between two att sites in another plasmid. Two supporting proteins, Integration Host Factor (IHF) and Excisionase (Xis), allow the strand exchange to occur in a directional, yet reversible, fashion.

The Gateway (Invitrogen) system supports in vitro reactions with att sites flanking the DNA fragment of interest in an Entry plasmid (ENTR), as well as att sites flanking a negative selection marker (*ccdB*) in a second, Destination, plasmid (DEST) (Figure 1). The *ccdB* gene is toxic to most *E. coli* strains but can be propagated in specific mutant strains such as DB31 and *ccdB* Survival (Invitrogen) (14). Recombination between specific att sites is controlled by the sequence identity in a core seven basepair region of the att site. While recombination is efficient between att sites with identical sequences, single base differences within this core prevent recombination. By changing the core att sequence, multiple att sites can be created with unique recombination specificities. Multiple att sites can be brought together in a single reaction, orchestrated by the posi-

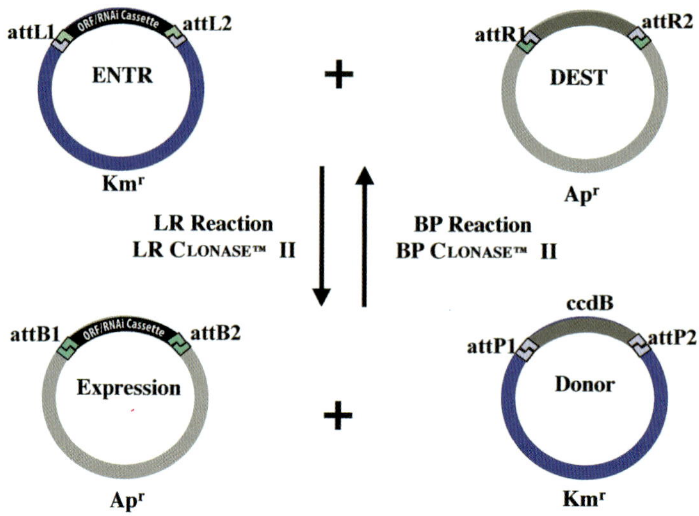

Fig. 1. Gateway recombination reactions. Recombination between ENTR and DEST vectors with LR Clonase II (Int, IHF and Xis) occurs through *att*L and *att*R sites, respectively, resulting in Expression and Donor vectors with *att*B and *att*P sites, respectively. A similar reaction with BP clonase II (Int and IHF) reverses the direction of recombination

tioning of different att sites, to create plasmids that are a compilation of fragments from the original, input vectors (Multisite Gateway, Invitrogen) (15).

In the simplest form, the Gateway (Invitrogen) system has attL1 and attL2 sites flanking a gene in an ENTR vector, and compatible att sites with the same core sequence, attR1 and attR2, flanking a *ccdB* gene in a DEST vector. After incubation in vitro with the appropriate enzymes (LR Clonase II is Int, IHF and Xis in an appropriate buffer, Invitrogen), the gene is transferred into the DEST vector, preserving the orientation. Recombination between attR1 and attL1 sites generate attB1 and attP1 sites; reversing the reaction, attB1 can be recombined with attP1 to form attL1 and attR1 sites (Figure 1). Selecting a specific plasmid from Gateway (Invitrogen) reactions is accomplished by positive selection for the appropriate antibiotic resistance marker and negative selection against the *ccdB* cassette.

The Gateway (Invitrogen) system was initially developed to allow the facile mixing and matching of different regulatory elements and tags for the expression of cDNAs and open reading frames, (ORFs). Today, a growing number of Gateway (Invitrogen) adapted reagents have been created to provide similar flexibility and time saving benefits for the cloning and transfer of RNAi expression cassettes. Of note is that Gateway (Invitrogen) "shuffling" of ORFs into a variety of different expression vector systems has primarily been done by flanking ORFs with att sites. By design, an ORF can be easily recombined with different promoters, in-frame-fusions/tags or transcription termination sequences, as long as the intervening 25 bp att site sequences are tolerated in the final expression construct. In this case, the ORF in an initial ENTR vector does not have the regulatory elements for expression. Once the ORF is recombined into a DEST vector, the desired regulatory elements are brought into position with the ORF for expression. In the Gateway (Invitrogen) RNAi expression system described below, the entire RNAi expression cassette is flanked by att sites (i.e. att1—pol III promoter—shRNA coding sequence—terminator—att2). Therefore, expression of shRNA is possible in an ENTR vector. Subsequent transfer of this functional expression cassette into a DEST vector provides the resulting recombinant construct with the expression cassette and any capabilities provided by the DEST vector (such as viral delivery).

3 Experimental Protocols

3.1 Vector Construction

Oligonucleotides encoding each shRNA were cloned into linear pENTR/U6 and pENTR/H1/TO entry vectors by 5 min room temperature ligations, transformation of TOP10 cells, and selection on LB/kanamycin plates according to the vector manuals (Invitrogen). Typically, a single clone was mid-scale plasmid prepped for each shRNA, and the hairpin sequence was verified using the primers supplied in each kit. To transfer RNAi cassettes to the desired destination vectors, one hour LR Clonase reactions (Invitrogen) were performed, transformed into TOP10 or Stbl3 cells (Invitrogen) for lentiviral vectors, and selected on ampicillin according to the product manuals. Clones were verified by restriction digest (no additional sequencing is necessary due to the high fidelity of the Gateway (Invitrogen) recombination reactions).

3.2 Transient Transfections

Cells were transfected in 24-well plates using Lipofectamine 2000 CD reagent (Invitrogen). In each well, 1.2 μl Lipofectamine 2000, 100 ng each of luciferase and *lacZ* reporter plasmids, and 300 ng H1/TO shRNA ENTR clone or carrier plasmid were transfected in a final volume of 600 μl. Duplicate wells were lysed and assayed for Beta-galactosidase and luciferase activities two days after transfection.

3.3 Transductions

Adenoviral and lentiviral particles were produced according to their respective manuals (Invitrogen). Transductions were carried out in a final volume of 500 μl medium in 12-well plates. The day after transduction, medium was replaced with complete medium. Cells expressing tet Repressor were maintained in medium containing 10 μg/ml Blasticidin. For lentiviral transduction, selection was started on the second day post-transduction with 100 μg/ml Zeocin (Invitrogen) and continued thereafter. Clones were isolated by dilution into 96-well plates and sub-culturing in selection medium.

4 Application of Gateway System

The Gateway (Invitrogen) system supports cloning and testing of RNAi expression cassettes with small (~3 kbp), stable ENTR vectors, and the subsequent transfer of select cassettes into destination vectors. Cloning and testing shRNA knockdown efficacy directly in plasmid vectors designed for delivering DNA to a cell can be problematic. For example, manipulating a 35 kb Adenoviral genome, or maintaining Lentiviral vectors without LTR recombination, can be difficult and complicate data interpretation. Pre-screening multiple shRNA constructs to determine which knockdown cassettes are best is, however, easily done in an ENTR vector. All the elements required for shRNA expression are contained between the Gateway (Invitrogen) att sites and present in the ENTR vector. ENTR vector screening allows non-effective RNAi cassettes to be eliminated. Effective shRNA cassettes can be efficiently transferred into multiple different vector backbones in the Gateway (Invitrogen) DEST format. Libraries of shRNA expression cassettes can be transferred as easily as a single clone, and collections of clones can be transferred in HTP processes.

Following is a demonstration of how the Gateway (Invitrogen) system supports the Invitrogen BLOCK-iT (Invitrogen) RNAi expression vector products: 1) creating shRNA expression cassettes in two different ENTR vectors, 2) testing the expression and regulation of a tet inducible H1 promoter during the development of a BLOCK-iT (Invitrogen) RNAi expression cassette, 3) the utility of the Gateway (Invitrogen) system to transfer select RNAi expression cassettes to a viral DEST, and 4) a demonstration of tet regulated RNAi after genomic integration of Gateway (Invitrogen) shRNA expression cassettes.

4.1 BLOCK-iT RNAi Entry Vectors

The BLOCK-iT (Invitrogen) pENTR (Invitrogen)/U6 and inducible pENTR (Invitrogen)/H1/TO entry vectors utilize compatible dsDNA oligo cloning schemes. Oligos encoding an shRNA of interest and carrying the 5′ 4 nt overhangs, as shown in Figure 2A, were ligated into either vector, transformed into *E. coli*, and selected for kanamycin resistance. Both the U6 and H1/TO promoters are constitutively active in many mammalian cell types (data not shown). In cells expressing tet Repressor (TR) protein, the H1/TO promoter is transcriptionally repressed until a small amount of tetracycline (tet) is added to the culture medium (Figure 2B), which induces high-level expression of the shRNA (16). [The H1/TO promoter was created in the laboratory of Dr. T. Tuschl at the Max-Planck-Institute for Biophysical Chemistry].

4.2 Gateway DEST Vectors

The RNAi cassette (promoter, shRNA sequence, and terminator) formed by insertion of the dsDNA oligo into the ENTR vectors can be tested by transient transfection (both U6 and H1/TO vectors) or stable transfection with Zeocin (Invitrogen) selection (pENTR (Invitrogen)/H1/TO only). Once active shRNA sequences are identified, their RNAi cassettes can be transferred by Gateway (Invitrogen) recombination to any BLOCK-iT (Invitrogen) DEST vector or alternative DEST vector (examples shown in Figure 3). The pBLOCK-iT (Invitrogen) DEST vectors shown include pBLOCK-iT (Invitrogen) 3-DEST and pBLOCK-iT (Invitrogen) 6-DEST for stable transfections with Geneticin (Invitrogen) or Blasticidin selection, respectively; pAd/BLOCK-iT (Invitrogen)-DEST for transient delivery by adenovirus; and pLenti4/BLOCK-iT (Invitrogen)-DEST and pLenti6/BLOCK-iT (Invitrogen)-DEST for lentiviral delivery and stable selection with Zeocin (Invitrogen) or Blasticidin, respectively (Figure 3). Multiple shRNA ENTR vectors targeting the luciferase and *lacZ* gene products were designed, created and tested for knockdown efficacy (data not shown). The most effective pENTR (Invitrogen)H1/TO shRNA clones where additionally tested for tet regulated expression (Figure 4) and recombined with Lenti/BLOCK-iT (Invitrogen)- DEST vectors.

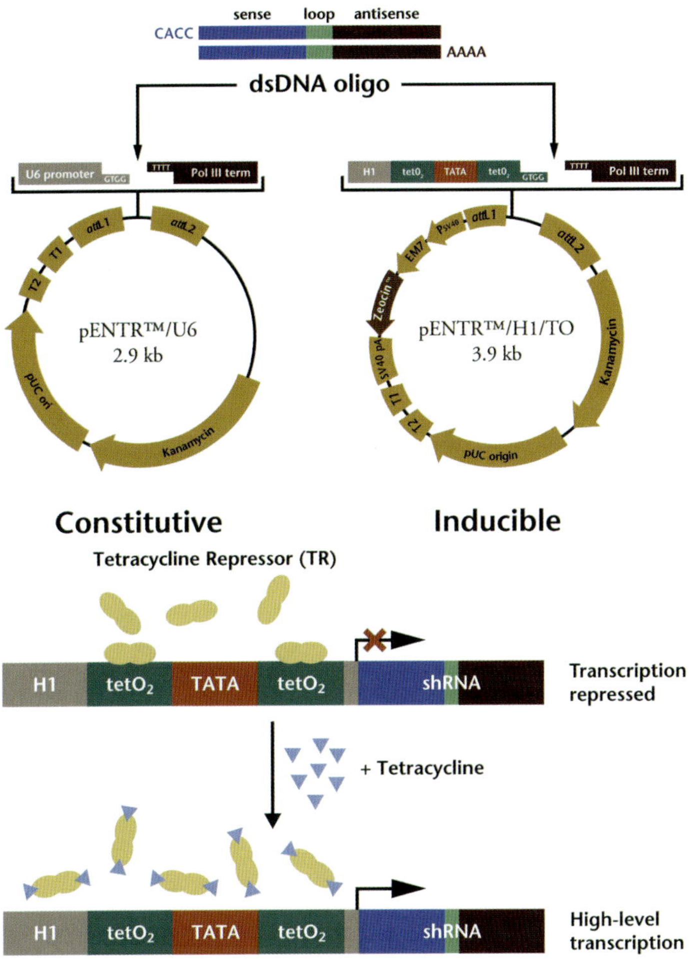

FIG. 2. Scheme for cloning an shRNA-encoding dsDNA oligo into the pENTR (Invitrogen)/U6 and pENTR (Invitrogen)/H1/TO vectors. The dsDNA oligo is designed with 5′ 4 nt overhangs compatible with the linear vectors. The left overhang derives from the promoter; the right one derives from the polyT terminator. Design of the shRNA is flexible but is shown here with the sense target sequence upstream of the loop. Regulation of the H1/TO promoter by tet repressor (TR) and tetracycline. In cells expressing TR, the promoter is repressed; it can be derepressed by addition of tetracycline or its analogs to the culture medium

4.3 Inducible RNAi

To test regulation from the pENTR (Invitrogen)/H1/TO promoter, cell lines expressing the TR protein were cotransfected with BLOCK-iT (Invitrogen) ENTR shRNA clones and both the luciferase and *lacZ* reporter genes. Reporter activity was assayed after two days (Figure 4). Both a clonal cell line commercially available from Invitrogen (T-REx (Invitrogen) CHO) and a population of Blasticidin-selected cells transduced with lentiviral particles expressing the TR gene (HeLa + Lenti6/TR) were tested.

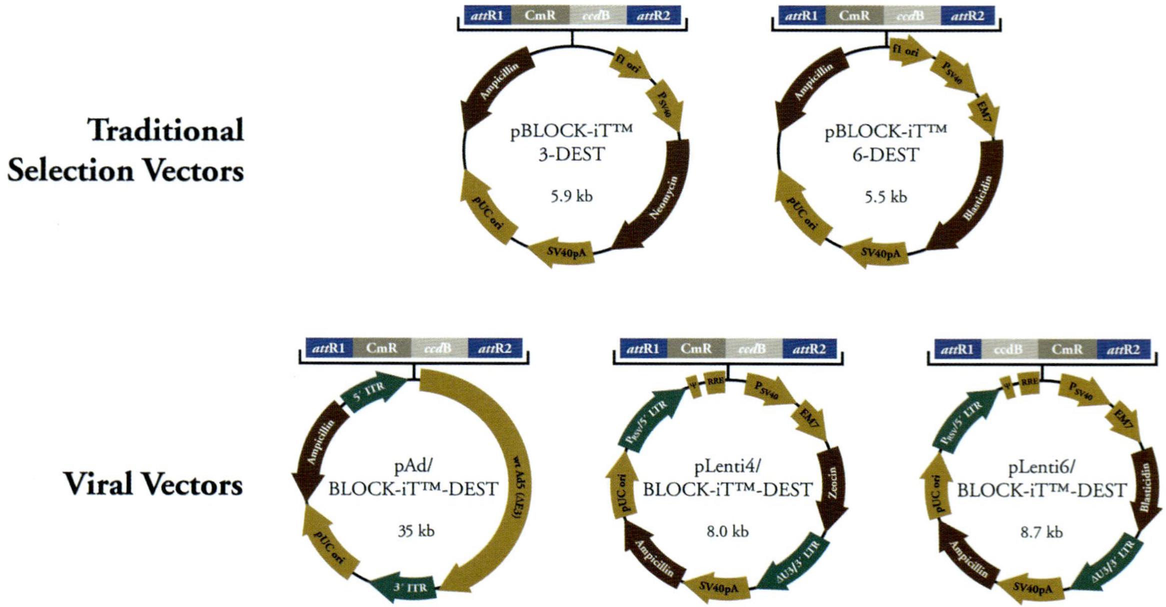

FIG. 3. The *att*L1 and *att*L2 sites in the BLOCK-iT (Invitrogen) entry vectors (Fig. 1) can be easily and quickly recombined with the *att*R1 and *att*R2 sites in the five destination vectors shown above, moving the entire RNAi cassette (promoter-shRNA-terminator) into a new backbone without the need for resequencing. These backbones can provide selection markers for stable transfection (top) or be used to create adenoviruses and lentiviruses carrying the shRNA of interest (bottom)

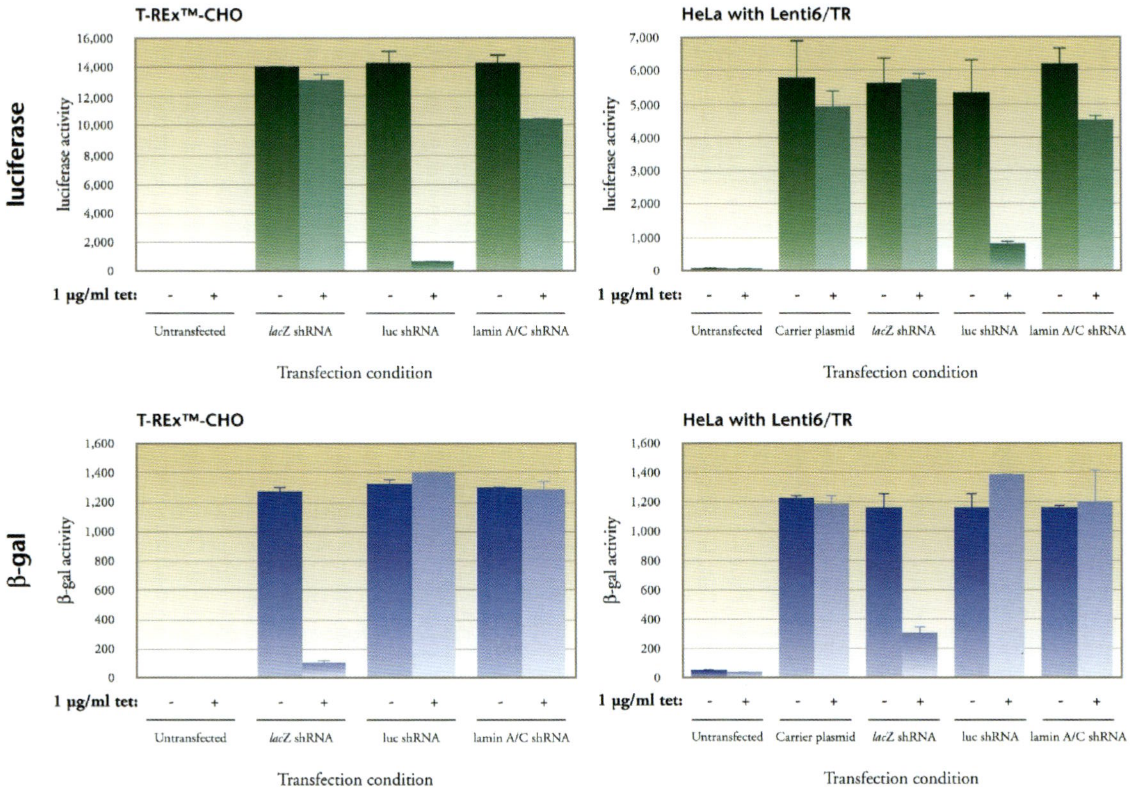

Fig. 4. T-REx (Invitrogen)-CHO cells (left panels) or a polyclonal population of HeLa cells stably transduced with lentiviral particles expressing the TR protein (right panels) were cotransfected with *lacZ* and luciferase (luc) reporter genes plus H1/TO entry clones inducibly expressing *lacZ*-, luc-, or lamin A/C-directed shRNAs. For the HeLa cells, a non-specific carrier plasmid was also included as a control. After three hours, medium was replaced with fresh medium with or without 1 μg/ml tetracycline (tet) as indicated. Cells were lysed 48 hours post-transfection and assayed for luciferase (top) or β-gal (bottom) enzyme activities

Robust knockdown of each reporter gene was only evident upon tet induction of the corresponding shRNA. Moreover, in the absence of tet, no knockdown was observed from the same shRNA, indicating that basal, or "leaky," expression from the H1/TO promoter was not detectable by functional testing. Testing basal expression levels and tet induction of the H1/TO promoter was effective in the ENTR vector without the complicating factors of LTR recombination seen with Lentiviral vectors or the lower transfection efficiencies of larger plasmids.

4.4 Gateway Transfer and Expression Testing in a Lenti DEST Vector

Oligos targeting the Lamin A/C gene were designed, synthesized and tested for inhibition in a similar fashion to the selection of productive knockdown vectors for luciferase and *lacZ* described above. The most effective pENTR (Invitrogen) /H1/TO Lamin A/C shRNA vector was selected and the expression cassette recombined into pLenti6/BLOCK-iT (Invitrogen)-DEST. The resulting plasmid, a Lamin RNAi expression cassette in the lentiviral DEST vector, was tested for it's ability to regulate inhibition of the endogenous lamin A/C gene in another commercially available cell line from Invitrogen, T-REx (Invitrogen) HeLa (Figure 5). After three weeks of Zeocin (Invitrogen) selection, cells were kept uninduced, or induced in medium with tet, for up to 60 days (Figure 5A). Lamin A/C protein levels, as detected by periodic Western blotting, were specifically suppressed by the lamin A/C-directed shRNA only after tet induction. Residual lamin A/C on Day 3 was greatly diminished by Day 6 and presumably resulted from slow protein turnover. The protein was almost completely eliminated by Day 16 and did not return for the duration of the experiment, demonstrating the power of shRNA expression for long-term knockdown. Ten clones derived from the lamin A/C shRNA-transduced population behaved nearly identical to the population itself after eight days of induction (Figure 5B), demonstrating the efficiency of lentiviral delivery. While strong lamin A/C silencing was observed in the stable T-REx (Invitrogen) HeLa clones after induction with tet, limited knockdown was also possible by reducing the amount of tet inducer added to the cell culture medium indicating suppression of lamin A/C in this cell line was tet dose-dependent (data not shown). Additionally, removal of tet resulted in a reversal of the inhibition and de novo expression of Lamin A/C (data not shown). Effective testing, transfer and regulation of Lamin shRNA expression from a lentiviral genome was facilitated by Gateway (Invitrogen), and the H1/TO regulation observed in the context of the ENTR vector was similar to that seen after lentiviral integration of the RNAi expression cassette into the cells genome.

Similar Gateway (Invitrogen) recombination processes also supported the testing and transfer of shRNA knockdown cassettes into an Adenoviral genomes of ~35 kbp (pAd/BLOCK-iT (Invitrogen)-DEST) preventing many of the problems associated with manipulating a genome of this size (data not shown).

5 Conclusions

The Gateway (Invitrogen) recombination cloning system has successfully been combined with the Invitrogen BLOCK-iT (Invitrogen) RNAi expression technologies to

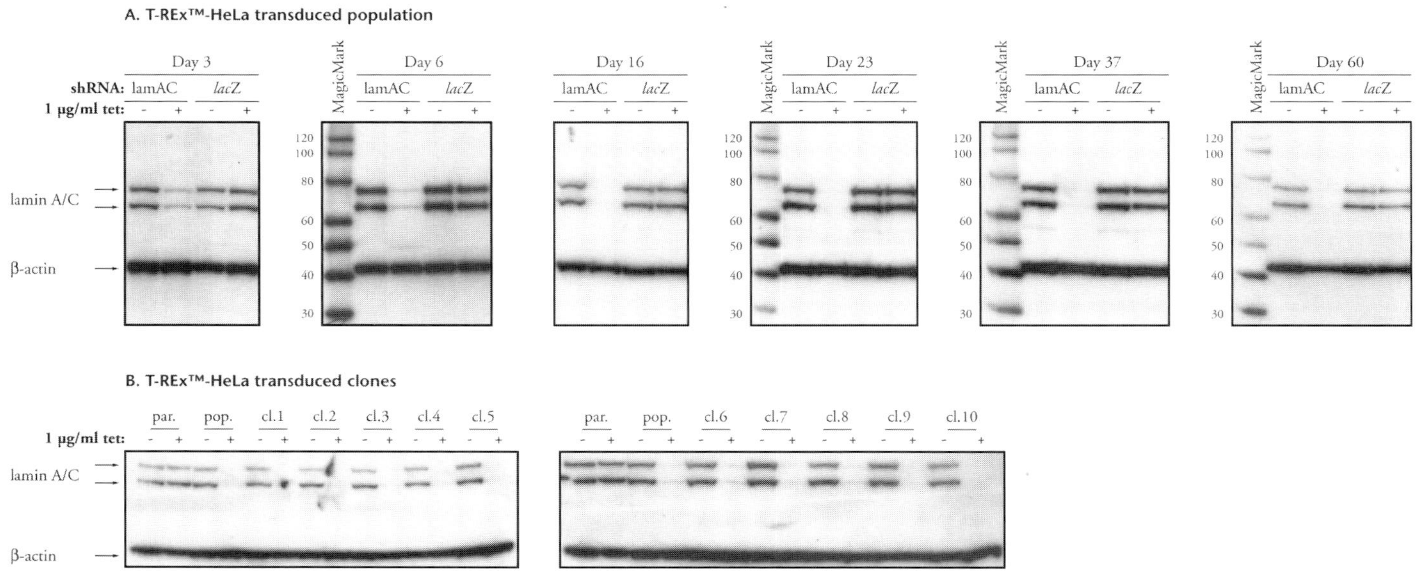

Fig. 5. T-REx (Invitrogen)-HeLa cells were transduced at an effective MOI < 1 with lentiviruses expressing lamin A/C-directed (lamAC) or control *lacZ*-directed shRNAs from the tet inducible H1/TO promoter. **A**) Stable populations of cells were selected with Zeocin (Invitrogen) for three weeks and then plated with or without tet at 1 µg/ml for the number of days indicated. Whole cell lysates were Western blotted and probed with antibodies against lamin A/C and β-actin as a loading control. **B**) The population from A was dilution plated, and ten individual clones were isolated. The clones (cl.1–10), population (pop.), and untransduced T-REx (Invitrogen)-HeLa parental cell line (par.) were incubated with and without 1 µg/ml tet for eight days, then lysed and blotted as in A

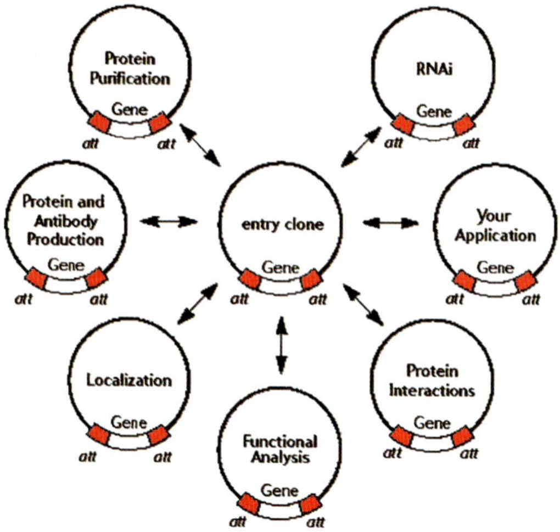

FIG. 6. Gateway flexibility

produce a system through which constitutive and tet inducible shRNA expression cassettes can be efficiently tested and transferred to alternative vector backbones for knockdown experimentation. An example of Gateway (Invitrogen) recombination into a lentiviral DEST was presented, however multiple other DEST vectors are available or can be created by inserting the appropriate attR sites at defined positions in candidate destination vectors (Fig. 6). The transfer of RNAi expression cassettes is fast, reliable and prevents many of the problems associated with manipulating viral genomes, or large plasmids, during the assessment of multiple knockdown constructs. Additionally, the inducible H1/TO promoter allows establishment of stable cell lines with minimal basal shRNA expression. This feature will be critical for the creation of cell lines in which essential genes are to be targeted. Alternatively regulated RNAi expression cassettes, such as with t-RNA promoters (17), modified ribozymes (18) or Cre-lox (19–21) can similarly be Gateway (Invitrogen) adapted as has been demonstrated with Cre-lox (21).

The Gateway (Invitrogen) BLOCK-iT (Invitrogen) RNAi system makes alternative delivery systems amendable to HTP processes. For example, large collections, or libraries, of RNAi cassettes can be delivered with lentiviral vectors facilitating genetic screens, trans-formation of primary and growth-arrested cells, or the creation of transgenic knockdown animals—changing the pace of mammalian genetics. The flexibility provided by the Gateway (Invitrogen) system also enables validated knockdown cassettes to recombined with new delivery vectors/technologies as they are developed.

Finally, the recombination specificity conferred by the core att site sequences allows multiple functional att sites to be created (i.e. att1–7). Simultaneous recombination between three or more att sites is possible and supported by Multisite Gateway (Invitrogen) products and application notes (Invitrogen). Traditionally, Multisite Gateway (Invitrogen) applications have brought together a promoter, ORF and terminator for protein expression. However, similar Multisite Gateway (Invitrogen) reactions can be applied to clone multiple shRNA expression cassettes into a single DEST vector (Figure 7)—shRNA cassettes targeting the same, or different, genes. Multisite Gateway

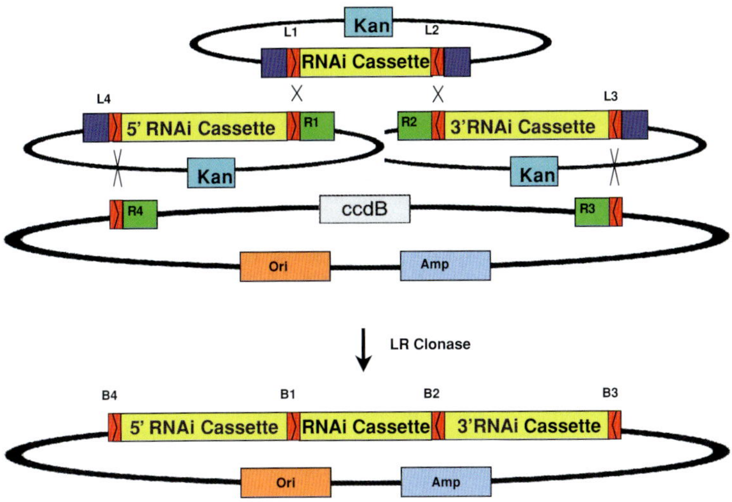

FIG. 7. Multisite Gateway

(Invitrogen) recombination may also be successful at mixing and matching shRNA knockdown cassettes with cDNA expression cassettes to, for example, test if the expression of a mutant cDNA will rescue a RNAi phenotype. Delivering multiple cassettes (shRNA, cDNA or both) linked in cis on the same plasmid is likely to ensure more uniform levels of expression from cell to cell relative to the expression from cassettes delivered on independent plasmids.

References

1. Dykxhoorn DM, Novina CD & Sharp PA (2003), Killing the messenger: short RNAs that silence gene expression. *Nature Reviews Molecular Cell Biology* 4(6):457–467
2. Shi Y (2003), Mammalian RNAi for the masses Trends Genet Jan;19(1):9–12
3. Wadhwa R, Kaul SC, Miyagishi M, Taira K. (2004), Vectors for RNA interference. *Curr Opin Mol Ther* 6(4): 367–372
4. Paddison PJ *et al.* (2002), Short hairpin RNAs (shRNAs) induce sequence-specific silencing in mammalian cells. *Genes & Development* 16(8):948–958
5. Brummelkamp TR, Bernards R & Agami R (2002), A system for stable expression of short interfering RNAs in mammalian cells. *Science* 296(5567):550–553
6. Miyagishi M, Taira K (2002), Development and Application of siRNA expression vector. Nucleic Acid Res Suppl 2:113–114
7. Yu JY, DeRuiter SL & Turner DL (2002), RNA interference by expression of short-interfering RNAs and hairpin RNAs in mammalian. *PNAS* 99(9):6047–6052
8. Miyagishi M, Taira K (2004), RNAi expression vectors in mammalian cells. *Methods Mol Biol* 252:483–491
9. Paule MR, White RJ (2000), Survey and summary: transcription by RNA polymerases I and III. Nucleic Acids Res 15;28(6):1283–1298
10. Hartley J, Temple G, and Brasch M (2000), DNA cloning using in vitro site-specific recombination. *Genome Res* 10(11):1788–1795

11. Liu Q, Li M, Leibham D, Cortez D, and Elledge S (1998). The univector plasmid-fusion system, a method for rapid construction of recombinant DNA without restriction enzymes. *Curr Biol* 8:1300–1309
12. Paddison PJ, Silva JM, Conklin DS, Schlabach M, Li M, Aruleba S, Balija V, O'Shaughnessy A, Gnoj L, Scobie K, Chang K, Westbrook T, Cleary M, Sachidanandam R, McCombie WR, Elledge SJ, Hannon GJ (2004), A resource for large-scale RNA-interference-based screens in mammals. *Nature* 25;428(6981):427–431
13. Landy A (1989), Dynamic, structural, and regulatory aspects of lambda site-specific recombination. Annu Rev Biochem 58:913–949
14. Bernard P, Couturier M 1992. Cell killing by the F plasmid CcdB protein involves poisoning of DNA-topoisomerase II complexes. *J Mol Biol* 226:735–745
15. Cheo DL, Titus SA, Byrd DRN, Hartley JL, Temple GF, Brasch MA 2004. Concerted assembly and cloning of multiple DNA segments using in vitro site-specific recombination: functional analysis of multi-segment expression clones. *Genome Res* 14: 2111–2120
16. Yao F, Svensjo T, Winkler T, Lu M, Eriksson C, Eriksson E (1998) tetracycline repressor, tetR, rather than the tetR-mammalian cell transcription factor fusion derivatives, regulates inducible gene expression in mammalian cells. *Human Gene Therapy* 9(13):1939–1950
17. Kawasaki H, Taira K (2003), Short hairpin type of dsRNAs that are controlled by tRNA(Val) promoter significantly induce RNAi-mediated gene silencing in the cytoplasm of human cells. Nucleic Acids Res 15;31(2):700–777
18. Shinagawa T, Ishii S (2003), Generation of Ski-knockdown mice by expressing a long double-strand RNA from an RNA polymerase II promoter. Genes Dev 1;17(11): 1340–1345
19. Fritsch L, Martinez LA, Sekhri R, Naguibneva I, Gerard M, Vandromme M, Schaeffer L, Harel-Bellan A (2004), Conditional gene knock-down by CRE-dependent short interfering RNAs. EMBO Rep 5(2):178–182
20. Kasim V, Miyagishi M, Taira K 2004), Control of siRNA expression using the Cre-loxP recombination system. Nucleic Acids Res 23;32(7):e66
21. Tiscornia G, Tergaonkar V, Galimi F, Verma IM (2004), CRE recombinase-inducible RNA interference mediated by lentiviral vectors. Proc Natl Acad Sci USA 11;101(19): 7347–7351

4. Design of Genes Based on Current RNA Technology

Design of Intracellularly Active Ribozymes and siRNAs

MAKOTO MIYAGISHI[1,3] and KAZUNARI TAIRA[2,3]

1 Ribozymes

1.1 Introduction

Hammerhead ribozymes are small and versatile catalytic RNA molecules that cleave RNAs at specific sites (Fig. 1A). The rapidly developing field of RNA catalysts is of current interest not only because of their intrinsic catalytic properties but also because of their potential utility as therapeutic agents and specific regulators of gene expression (Rossi, 1995; Rossi and Sarver 1990; Sarver et al. 1990; Sioud 2004). However, despite extensive efforts, the activity of ribozymes in vivo has not usually been high enough to achieve the desired biological effects. Unlike in vitro, conditions in vivo are very complex, and many parameters must be considered. Under certain conditions, the interaction of a ribozyme or its gene with intracellular proteins seems to be significant. Thus, many modifications and improvements are required in the ribozyme expression system as well as methods for introducing ribozymes into cells.

For the successful inactivation of a specific gene by a ribozyme in vivo, we have developed an efficient system for the expression of ribozymes, and we have attempted to apply this system for potential gene therapy and functional analysis of genes of interest. In addition, we have developed two novel ribozymes: an allosterically controllable ribozyme and an RNA-protein hybrid ribozyme. The former, termed a maxizyme, has sensor arms that recognize target mRNA sequences and, in the presence of such sequences, the maxizyme forms a cavity that can capture catalytically indis-

[1] 21st Century Center of Excellence Program, Graduate School of Medicine, The University of Tokyo, 7-3-1 Hongo, Bunkyo-ku, Tokyo 113-8656, Japan
[2] Department of Chemistry and Biotechnology, School of Engineering, The University of Tokyo, 7-3-1 Hongo, Bunkyo-ku, Tokyo 113-8656, Japan
[3] Gene Function Research Center, National Institute of Advanced Industrial Science and Technology (AIST), Central 4, 1-1-1 Higashi, Tsukuba Science City 305-8562, Japan

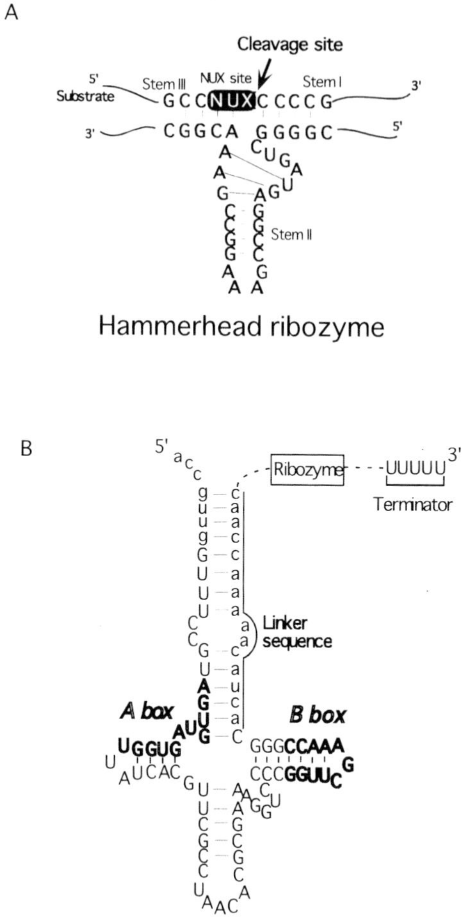

FIG. 1. **A** Secondary structure of a hammerhead ribozyme. The hammerhead ribozyme consists of the substrate-binding region (stems I and III) and a catalytic core with a stem-loop II region. When the catalytic core captures the catalytically indispensable Mg^{2+} ions, cleavage occurs only after the NUX sequence (N, any base; X, A, C, and U). **B** Secondary structure of a $tRNA^{Val}$-driven transcript. In the RNA polymerase-III-based transcription system, the promoter is located within the tRNA sequence (indicated as *A* and *B boxes*). The ribozyme is linked downstream of the partially modified human $tRNA^{Val}$ through a linker (indicated by *underlining*)

pensable Mg^{2+} ions and thereby cleaves the target (Kuwabara et al. 1998a, 1998b, 1999; Tanabe et al. 2000a, 2000b). The latter, a hybrid ribozyme that couples the cleavage activity of the hammerhead ribozyme with the unwinding activity of an RNA helicase, can cleave the target mRNA efficiently, regardless of the secondary structure of the target mRNA (Wadhwa et al. 2003; Warashina et al. 2001).

Many pol III expression systems that were developed for the ribozyme turned out to be very useful for the expression of more recently developed siRNAs, described below (Miyagishi and Taira 2003; Ohkawa and Taira 2000; Shiota et al. 2004). Some parameters that govern ribozyme activity also govern siRNA activity, but other parameters are specific to each system. We will describe those common and/or specific parameters in this chapter.

1.2 Methods for the Introduction of Ribozymes into Cells

In order for ribozymes to work inside cells, they must be internalized into individual cells and access the target mRNA. However, cellular uptake of ribozymes and other naked nucleic acids is usually inefficient, due to their charged composition and large molecular size. To overcome this problem, liposomes and charged lipids are commonly used as delivery systems for ribozymes. Complexes of nucleic acids with cationic lipids are usually internalized into the cells by endocytosis (Brisson and Huang, 1999). Modified liposomes are also used for the delivery of synthetic siRNAs (Hirabayashi et al. 2004).

For application of ribozymes in vivo, chemically synthesized ribozymes can be directly administered, or a plasmid vector encoding ribozyme genes can be introduced into cells, where ribozymes can be transcribed by transcriptional factors in the host. Since naked nucleic acids are rapidly degraded by nucleases in cells, especially in the gastrointestinal tract and blood, ribozymes synthesized in vitro should be protected by chemical modification, such as thio modification or alkylation at the 2′ position of the ribose ring. Such chemically modified ribozymes are being evaluated in clinical trials (Usman and Blatt 2000). However, those modifications sometimes result in increased cell toxicity and higher costs for their preparation. In addition, exogenous administration of the modified ribozymes limits the duration of efficacy, even if the stability is increased. As an alternative approach, vector-based delivery has attracted attention, because the ribozyme gene can be administered as stable DNA, which will then constitutively produce ribozymes inside target cells. It should be mentioned that, in contrast to naked ribozymes, synthetic siRNAs in cells can be stabilized, probably by interacting protein factors and, thus, naked siRNAs are effective in cells (Hirabayashi et al. 2004).

Many viral-based vectors have shown potential as vehicles for in vivo gene delivery, including adenovirus, herpes virus, retrovirus, adeno-associated virus, and lentivirus (Kovesdi et al. 1997; Mountain 2000; Naldini, 1998; Rabinowitz and Samulski, 1998; Sioud 2004; Tiscornia et al. 2004; Wu and Ataai 2000). Rapid advances in viral vector technology have led to not only improved efficiency for introduction of ribozymes into cells but also control of transduction, such as tissue-specific transduction. Although various viral vectors are derived from detoxified viruses, their exploitation has been fraught with problems related to production, immunogenicity and safety (Ferber 2001; Kay et al. 1997; Yang et al. 1995). Thus, the preparation and purification of many viral vectors are still being improved.

At the same time, artificial non-viral vectors are being developed for non-viral gene and oligonucleotide delivery, as described in this book. Non-viral vectors are more cost-effective and less laborious to produce, and safer to use than viral vectors. Polymers and cationic lipids are by far the reagents most widely used to enhance intracellular delivery, and, recently, a wide range of reagents, including liposomes, polypeptides and polymeric dendrimers, have been tested (Alino et al. 1994; Caplen et al. 1995; Cloninger 2002; Miller 2004). However, a critical problem is that the efficiency of introduction by non-viral vectors in specific tissues and organs is lower than that of viral delivery.

Thus, the development of both viral and non-viral vector-based deliveries is advantageous for introducing ribozymes into cells, depending on the aims. Together with the improvement of delivery technologies, the improvement of intracellular ribozymes is necessary, considering the environment inside the cells.

1.3 Ribozyme Expression System

1.3.1 Use of the Pol III System

After introduction of ribozyme-expression vectors into cells, the efficacy of the ribozyme in vivo should be influenced by at least five factors: (1) the amount of ribozyme transcripts, (2) the intracellular stability of ribozymes, (3) the subcellular localization of the ribozymes, (4) the activity of the ribozymes, and (5) the accessibility to ribozymes of their target mRNA. Of these factors, (1)–(3) can be controlled by the choice of an appropriate expression system.

In early studies of ribozyme expression systems, RNA polymerase II (pol II) promoters were widely used as for the expression of ribozymes. In this system, the cap structure and the poly(A) tail are automatically added at the 5′ and 3′ ends of the transcripts, respectively. Thus, the transcripts are protected from degradation by exonucleases and exported from the nucleus to the cytoplasm, just like an mRNA. However, the pol II expression system that is usually suitable for long RNAs (several hundred to several thousand bases) might not be appropriate for the transcription of short RNAs, such as ribozymes and siRNAs. Although the addition of extra sequences around the ribozyme transcript is likely to increase the stability of transcript in the pol II system, these extra sequences might decrease the activity of the ribozyme by perturbing its higher-order structure or by eliciting cellular proteins bound to pol-II-driven transcripts. To avoid this problem, the long pol II transcripts that contain either a *trans*-acting ribozyme or siRNA can be trimmed in *cis* by trimming ribozymes (Kato and Taira 2003; Ohkawa et al. 1993; Taira et al. 1990, 1991).

However, taking all factors together, RNA polymerase III (pol III) systems seem more attractive for the production of ribozymes than pol II systems. Pol III systems are mainly involved in the transcription of short RNAs, such as tRNA (Geiduschek and Tocchini-Valentini 1988) and U6 snRNA, and their level of transcription is two to three orders of magnitude higher than that of pol II systems (Cotten and Birnstiel 1989). In addition, fewer extra sequences are needed for transcription. Therefore, pol III promoters should provide an ideal system for the expression of ribozymes and siRNAs. Indeed, pol III expression systems based on human $tRNA^{met}$ and U6 promoters have been used for the expression of hammerhead and hairpin ribozymes in cells (Good et al. 1997) and the latter is the most widely used promoter for siRNAs (Wadhwa et al. 2004).

Among the pol III promoters, we have chosen the human $tRNA^{Val}$ promoter for the expression of ribozymes. In the tRNA promoter-based system, the promoter regions are located within the tRNA sequence. Thus, it is inevitable that a portion of the tRNA becomes incorporated into the ribozyme. In general, initially transcribed tRNAs are processed at the 5′ and 3′ ends, and matured tRNAs are then exported to the cytoplasm from the nucleus. In our $tRNA^{Val}$-based system, in order to avoid processing at the 3′-end of tRNA, the 3′-end stem sequence was modified to block the release of the ribozyme from the $tRNA^{Val}$ portion, so that the transcript has an extended long sequence, including the ribozyme sequence, at the 3′ end of $tRNA^{Val}$ (Fig. 1B) (hereafter this transcript is referred to as $tRNA^{Val}$-attached ribozyme). As described below, $tRNA^{Val}$-attached ribozymes were engineered so that they could be exported to the cytoplasm in mammalian cells while maintaining a high level of intracellular activities. Moreover, the attached tRNA sequence does not reduce the ribozyme activity, but rather the extra tRNA sequence appears to have a favorable effect on the intracellular activity of ribozymes, most probably by conferring higher resistance to RNases (Kawasaki et al. 1996, 1998; Koszul et al. 2003). This $tRNA^{Val}$-based system can also be used for the expression of siRNAs, since the $tRNA^{Val}$-attached shRNA transcripts can be processed by Dicer to produce siRNAs in the cytoplasm (Kawasaki and Taira 2003).

Another group has also reported a unique pol-III-based system for the expression of ribozymes (Chang et al. 2002; Yu et al. 1993). The dual U6/tRNA^{Lys3} promoter, with U6 and tRNA promoters in tandem, afforded higher levels of transcription than the single tRNA promoter. The dual U6/tRNA^{Lys3}-driven ribozyme exhibited strong activity for the cleavage of cytoplasmic targets (Chang et al. 2002).

1.3.2 Relationship Between Higher-Order Structure and Activity of Ribozymes

In our ribozyme expression system, the hammerhead ribozyme is linked downstream of a partially modified sequence of human tRNA^{Val} through a linker (Fig. 1B). We found that the higher-order structure of ribozymes strongly affects their intracellular stability. In our system, a small stem-loop structure is added at the 3′ end of the ribozyme to enhance its stability in cells. In addition, the transcript should contain a bulge structure at the double-stranded region corresponding to the linker, in order to block processing of the pre-tRNA-like transcript and degradation by nucleases that specifically recognize long double-stranded regions.

In our system, the higher-order structure of the transcript is greatly affected by the length of the linker (Koseki et al. 1999). Thus, the relationship between the higher-order structure of ribozyme transcripts and their activity and stability in cells was investigated (Koseki et al. 1999). First, three different expression systems were constructed in which the sequence of the ribozyme targeted against the conserved region in HIV-1 was the same, but the length of linkers was different (Fig. 2A). When the secondary structures of transcripts were predicted by the Zuker method (Zuker, 1989), the 5′ half of the transcripts, including the tRNA portion, but not the ribozyme region (indicated by boldface in Fig. 2A), showed secondary structures closely similar to those of tRNAs.

Owing to the catalytic properties of ribozymes, the ease with which they bind to their substrates is an important determinant of their activity. Thus, the degree of freedom and the accessibility of the substrate-binding regions of a ribozyme should substantially determine its activity. In Fig. 2A, ribozymes 1, 2, and 3 (Rz1-Rz3) show different degrees of freedom of the substrate-binding regions (indicated by underlines). As expected, these three ribozymes showed different levels of in vitro activity toward a short substrate: Rz3, with the most exposed binding arms, had the highest activity (Koseki et al. 1999). Next, the intracellular stability of these ribozymes was examined. After transfection of plasmids that encoded each ribozyme, the amounts of ribozymes in the cells were measured by Northern hybridization. Rz2 was 26-fold more stable than Rz1 and five-fold more stable than Rz3. It remains unclear why these structures dramatically affected the stability of transcripts, because their overall structures are closely similar.

Next, the activity of each ribozyme in cells was evaluated by using the luciferase reporter assay (Koseki et al. 1998). The luciferase chosen was a firefly enzyme that catalyzes the chemiluminescent reaction. With the addition of luciferin and the ATP energy source, the levels of expression of the luciferase could be easily and accurately determined by measuring the light emission. In our assay, the ribozyme targeted against the 5′ untranslated (5′ UTR) region of the HIV-1 gene was fused with the luciferase gene. If the ribozyme cleaves its target efficiently, the level of light emission should be decreased. This experiment confirmed that Rz2, which is the most stable one in cells, showed the highest activity of the three ribozymes. The results strongly

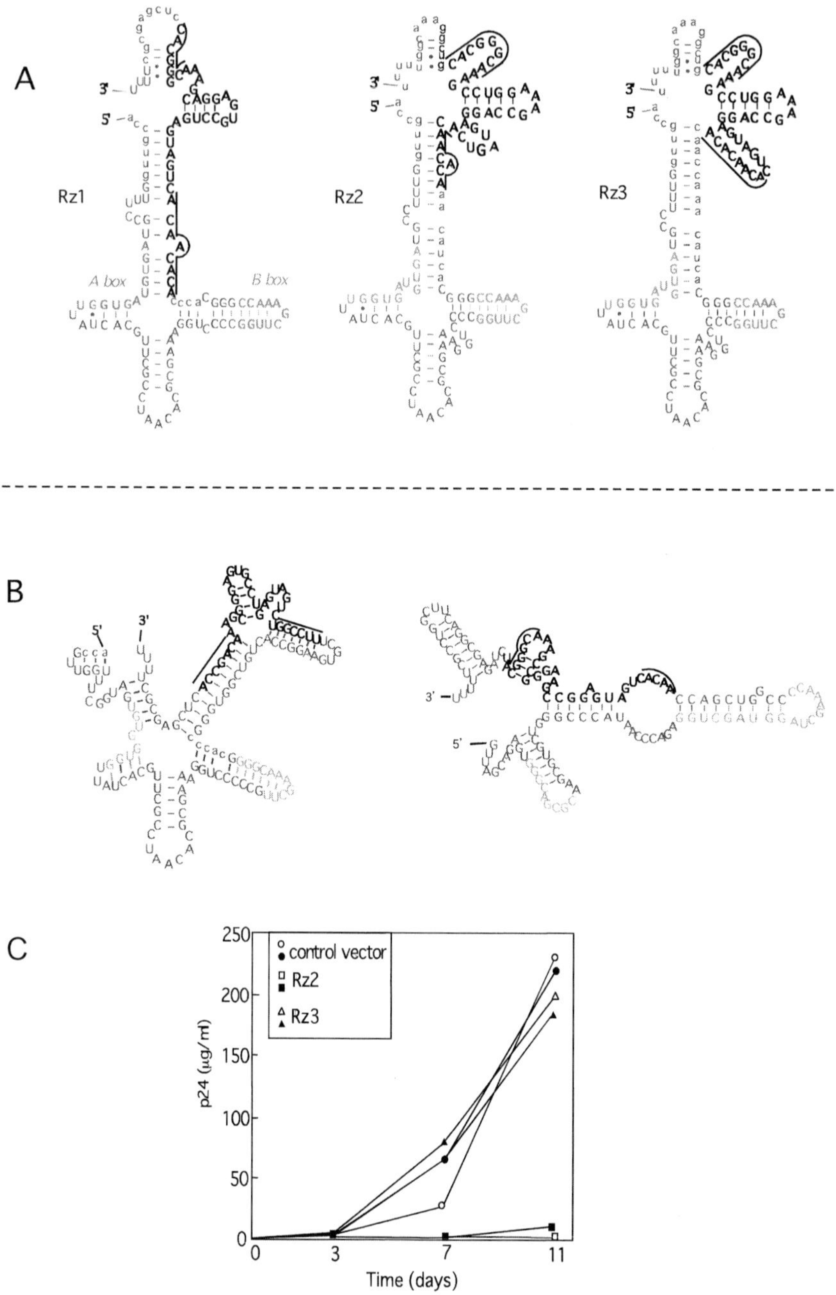

FIG. 2. **A** Secondary structures of tRNA-attached ribozymes (tRNA-Rzs) that are efficiently exported to the cytoplasm. Artificial linker sequences are indicated by *lower-case letters*. The ribozyme sequence is shown in *boldface* and the substrate-binding sites are *underlined*. **B** Secondary structures of tRNA-Rzs that accumulate in the nucleus. **C** Inhibitory effects of the tRNA-attached ribozymes on the expression of p24

indicated a correlation between the activity of ribozymes and their intracellular stability. The activity of ribozymes toward HIV-1 was also investigated in tissue culture cells (Fig. 2C). Cells carrying each ribozyme expression vectors were infected with HIV-1, and viral protein synthesis was then measured as an index of viral replication. The result was consistent with the data of the luciferase assay. Thus, these results indicate that intracellular stability is one of the most important determinants of ribozyme efficacy (Koseki et al. 1999).

1.3.3 Effects of Subcellular Localization on Ribozyme Activity

The subcellular localization of the ribozyme after transcription is also an important factor that determine the activity of ribozymes (Bertrand et al. 1997; Good et al. 1997; Kato et al. 2001). The target mRNA of the ribozyme is initially transcribed in the nucleus and undergoes maturation via splicing, followed by transport into the cytoplasm, where it is translated into a protein. Because nuclear pre-mRNAs form complexes with heterogeneous nuclear proteins and small nuclear ribonuclear proteins, and interact with various RNA-binding proteins involved in splicing or transport, nuclear pre-mRNAs might be less accessible to ribozymes than cytoplasmic mature mRNAs. Moreover, it is also likely that the stable structure of mRNAs is disrupted by various RNA helicases in the cytoplasm (Wadhwa et al. 2003; Warashina et al. 2001). Thus, in order to enhance the efficacy of ribozymes, it is preferable that ribozymes should colocalize with their target mRNA in the cytoplasm.

Our earlier data indicated that tRNA-attached ribozymes with a high level of activity were efficiently exported to the cytoplasm, whereas those with a low level of activity accumulated in the nucleus (Kato et al. 2001; Koseki et al. 1999). Thus, we systematically attempted to identify the cellular compartment in which a ribozyme could act effectively. To examine the correlation between the intracellular localization of ribozymes and their activity in vivo, several types of ribozymes that targeted the same substrate were designed by using two kinds of promoter. One was the promoter of the gene for $tRNA^{Val}$ described above, and the other was a U6 promoter that transcribes only the ribozyme portions of the hammerhead ribozyme (Das et al. 1988; Ohkawa and Taira 2000). Thus, the intracellular localization of transcripts in the cytoplasm or in the nucleus could be controlled using $tRNA^{Val}$ or U6 promoter, respectively.

When the localizations of tRNA-driven ribozymes (tRNA-attached ribozymes) and U6-driven ribozymes were determined by Northern blotting analysis after the fractionation of cells, approximately the same levels of transcript expression from both promoters were observed. The tRNA-driven ribozymes localized predominantly in the cytoplasm, whereas the U6-driven ribozymes localized in the nucleus. Furthermore, the activity of these ribozymes in cultured cells was estimated. Although both types of ribozyme targeted the identical site and had similar activities in vitro, the tRNA-attached ribozymes that were exported to the cytoplasm had strong inhibitory effects, whereas U6-driven ribozymes that remained in the nucleus were completely ineffective. From these results, it became clear that the cytoplasmic localization of ribozymes is a critical determinant of their activity in cells (Kato et al. 2001).

1.3.4 Mechanism of the Export of tRNA-Attached Ribozymes

Although all our $tRNA^{Val}$-attached ribozymes were localized in the cytoplasm, other groups reported that $tRNA^{Met}$-driven ribozymes accumulated in the nucleus (Bertrand et al. 1997; Good et al. 1997). The tRNA^{Met}-driven ribozyme (Fig. 2B, right) targeted against HIV-1 could not efficiently inhibit the expression of the target (Bertrand et al. 1997). Comparison of the secondary structures of the tRNA^{Met}-driven ribozyme and our tRNA^{Val}-driven ribozyme revealed that they were quite different (Fig. 2A, B, right). We further designed another tRNA-driven ribozyme (tRNA^{Val}-Rz; Fig. 2B, left), whose secondary structure did not maintain the cloverleaf structure, and examined its localization. As expected, this tRNA-attached ribozyme remained in the nucleus (Koseki et al. 1999).

From these experiments, we clarified that the secondary structure of the tRNA-attached ribozymes is an important determinant of their localization in cells. As shown in Fig. 2A, all tRNA^{Val}-attached ribozymes, which are transported to the cytoplasm, form a tRNA-like cloverleaf structure in the 5′ half portion, including the tRNA. In contrast, ribozymes remaining in the nucleus do not maintain such cloverleaf structures (Fig. 2B). Thus, it appears that tRNA^{Val}-attached ribozymes with a cloverleaf structure are exported to the cytoplasm via an export pathway for tRNAs.

In recent years, rapid progress has been made in efforts to understand the mechanism involved in the export of tRNAs to the cytoplasm (Arts et al. 1998a, 1998b; Grosshans et al. 2000; Kutay et al. 1998; Lipowsky et al. 1999; Lund and Dahlberg 1998). The transport of tRNAs requires a tRNA-binding protein, called exportin-t (Xpo-t), and Ran GTPase, and rapid transport is accomplished by an interesting mechanism involving GTP hydrolysis. According to studies in *Xenopus* oocytes and in yeast, before leaving the nucleus, transcribed pre-tRNAs must be accurately trimmed at both 5′ and 3′ ends and; moreover, CCA must be attached at the processed 3′ end. Only such processed tRNAs can be recognized by Xpo-t (Arts et al. 1998a, 1998b; Kutay et al. 1998; Lipowsky et al. 1999; Lund and Dahlberg, 1998). In addition, aminoacylation of the matured tRNAs appears to be critical for their transport from the nucleus to the cytoplasm in *Xenopus* oocytes (Lund and Dahlberg 1998) and in yeast (Grosshans et al. 2000). In *Xenopus* oocytes, immature tRNAs with extra nucleotides at the 3′ end are not recognized by Xpo-t, and thus they cannot be exported to the cytoplasm. This phenomenon suggests the existence of a proofreading mechanism in cells, whereby only tRNAs that are usable in the cytoplasm can be exported to the cytoplasm. However, our tRNA^{Val}-attached ribozymes were efficiently exported to the cytoplasm in mammalian cells (Kato et al. 2001; Kawasaki et al. 1998; Kuwabara et al. 1998a, 1998b, 1999, 2000a, 2000b, 2001a; Warashina et al. 2000a), even though they could be considered equivalent in form to certain immature tRNAs because of the extra sequences at their 3′ ends. Thus, we investigated the discrepancy between the reported observations in *Xenopus* oocytes that led to the proposal of the existence of a proofreading mechanism and our own observations of the efficient export of tRNA^{Val}-attached ribozymes (Kuwabara et al. 2001b).

We first considered the possibility of an alternative pathway for the export of tRNAs. In yeast, the fact that genetic mutation of los-1p, a yeast homologue of Xpo-t, did not affect cell survival suggested the existence of additional pathways for tRNA export. However, contrary to our expectation, we found that Xpo-t could recognize tRNA^{Val}-attached ribozymes in a similar manner to the tRNA, and Xpo-t was likely to be involved in the transport of tRNA^{Val}-attached ribozymes in somatic cells (Kuwabara et al. 2001a, 2001b). In contrast, a tRNA-attached ribozyme was not exported to the cytoplasm in *Xenopus* oocytes, as predicted by the proofreading hypothesis. Further

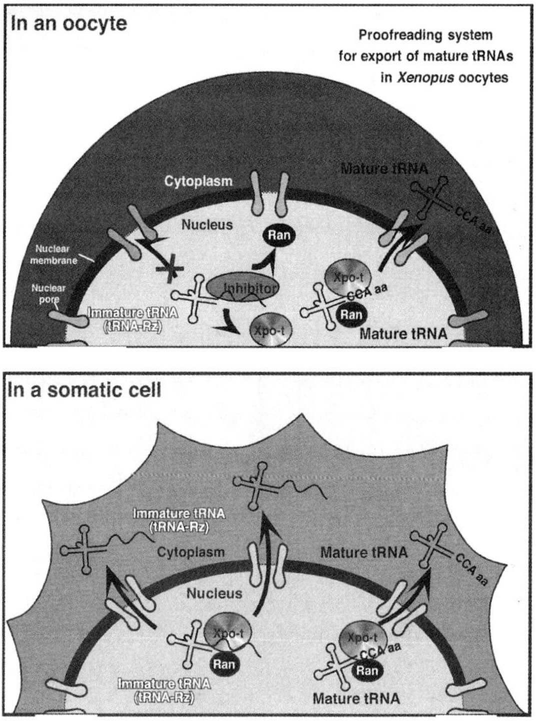

FIG. 3. Schematic representation of transport to the cytoplasm of mature tRNAs and of tRNA-attached ribozymes. The original proofreading mechanism has been modified to include the proposed inhibitor-mediated proofreading of tRNA transcripts in *Xenopus* oocytes (*upper panel*). A more general pathway for the export of mature and tRNA-attached ribozymes in somatic cells is presented in the *bottom panel*

investigation revealed that the Xpo-t/RanGTP complex did not interact with tRNA-attached ribozymes in oocytes, even though such an interaction could be observed in vitro and, more importantly, in several lines of somatic cells. These findings hinted at the presence of inhibitor(s) in *Xenopus* oocytes rather than the involvement of an alternative pathway in somatic cells. Indeed, a nuclear extract from *Xenopus* oocytes strongly inhibited the export of tRNA-attached ribozymes in somatic cells, suggesting the presence of a strong inhibitor(s) in oocytes (Kuwabara et al. 2001b).

It seems likely that the export of tRNAs in oocytes is subject to a special kind of regulation. Moreover, the proofreading mechanism in *Xenopus* oocytes seems to be involved with a specific inhibitor(s) that might recognize immature tRNAs, including tRNA-attached ribozyme, specifically. In somatic cells, when the linker and the ribozyme sequence were appropriately adjusted, a tRNA-attached ribozyme seems to be recognized by Xpo-t and then exported to the cytoplasm (Fig. 3).

Similarly, $tRNA^{Val}$-attached shRNAs can be transported by Exp-t to the cytoplasm by the same mechanism, whereas U6-driven shRNAs are recognized by Expotin-5 and transported to the cytoplasm (Kawasaki and Taira 2003; Lee et al. 2002b).

1.4 Maxizymes, Allosterically Controllable Ribozymes

In the following, a new type of ribozyme that has been developed through studies of shortening the hammerhead ribozyme is described. We succeeded in creating an allosteric ribozyme, termed a maxizyme, that functions as a dimer with significant specificity and activity both in vitro and in vivo (Kuwabara et al. 1998a, 1998b, 1999; Tanabe et al. 2000a, 2000b). This allosteric ribozyme could be defined as a biosensor in vitro and in vivo (see below).

First, a hammerhead ribozyme without the stem II region was created and its activity in vitro was evaluated. Initially, the shortened (minimized) ribozyme was termed a "minizyme", but this term has the negative connotation of a ribozyme with minimal activity. Our minizymes can function as dimers, and they have extremely high activity in vitro and in vivo (Kuwabara et al. 1998a, 1998b, 1999). Thus, we renamed the new dimeric minizyme "maxizyme" [*m*inimized, *a*ctive, *x*-shaped (functions as a dimer), and *i*ntelligent (allosterically controllable) ribo*zyme*].

1.4.1 Shortened Hammerhead Ribozymes That Function as Dimers

The cleavage of substrate RNA by the ribozyme usually requires a divalent metal ion, such as magnesium (Takagi and Taira 2002; Takagi et al. 2004; Tanaka et al. 2004; Suzumura et al. 2004; Warashina et al. 2004). A catalytic domain of the hammerhead ribozyme captures the catalytically indispensable Mg^{2+} ions. Precise control of the conformational change of the ribozyme, in the presence of Mg^{2+} ions, is needed to create an allosteric ribozyme.

Since downsizing of the hammerhead ribozyme is desirable for medical applications, researchers have generated a compact size by replacing the stem-loop II region of the ribozyme with a short linker (Amontov et al. 1996; Fu 1994; McCall et al. 1992; Tuschl and Eckstein 1993). Whereas the activity of minizymes was much lower than that of the parental ribozymes, we were able to create a shortened ribozyme lacking the stem-loop II region that had high activity, equivalent to the parental one (Amontov et al. 1996). Further kinetic and NMR analyses revealed that this shortened ribozyme exhibited extremely high activity as a dimer (Fig. 4A) (Kuwabara et al. 1996, 1998a).

Although we initially called this shortened ribozyme a "dimeric minizyme", we later renamed this molecule "maxizyme" (Kuwabara et al. 1998a, 1998b, 1999; Tanabe et al. 2000a; Tanabe et al. 2000b). As shown in Fig. 4B, we also designed a heterodimeric maxizyme using two different monomers, maxizyme left (MzL) and maxizyme right (MzR) (Kuwabara et al. 1996; 1998b, 1999). Since this maxizyme has two substrate-binding arms, it can recognize two sites on the same target RNA simultaneously. This unique feature of the maxizyme led us to create an allosteric enzyme with sensor function, as described below.

1.4.2 Design of an Allosterically Controllable Maxizyme

As stated above, in order for ribozymes to target RNAs at specific sites, certain minimum sequences are required for efficient cleavage. The cleavage of a substrate occurs immediately after the NUX sequence (where N is any base and X is A, C, or U) within the target (Shimayama et al. 1995; Zhou and Taira, 1998). However, in some cases, the cleavable triplet sequences may not be located at a suitable position on the target. For example, this problem occurs with certain abnormal chimeric mRNAs. Since chimeric mRNA is usually generated via chromosomal translocation, the two halves of the fused mRNA are derived from different genes. Such chromosomal translocation is often involved in the pathogenesis of disease. A well-known example is the Philadelphia chromosome, which causes chronic myelogenous leukemia (CML) (Muller et al. 1991; Nowell and Hungerford 1960). The Philadelphia chromosome

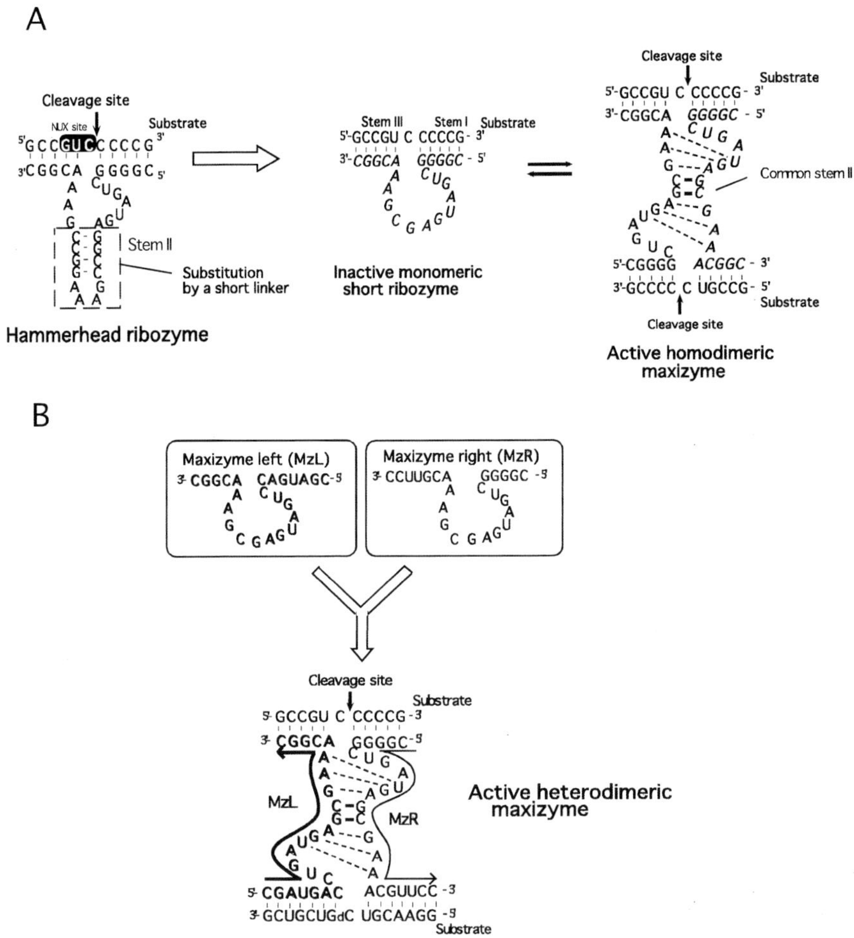

FIG. 4. **A** Development of an allosterically controllable maxizyme. Secondary structures of a parental hammerhead ribozyme (*left*), an inactive monomeric short ribozyme (*center*) and an active (homo) dimeric maxizyme (*right*). A monomeric short ribozyme, termed a minizyme, was designed by deleting a stem-loop II region from a hammerhead ribozyme. **B** The heterodimer (MzL and MzR) can generate two different binding sites: one is complementary to the sequence of interest (activator or inhibitor), and the other is complementary to a cleavable sequence

occurs as a result of reciprocal translocation involving the *BCR* and *ABL* genes, resulting in a fused mRNA. Being tumor-specific and pathogenetically important, fused mRNAs are ideal targets for nucleic-acid-based therapeutics as a paradigm (Shtivelman et al. 1986). Nevertheless, because of the absence of NUX sequences near the chimeric junction, a wild-type hammerhead ribozyme could not exclusively target the chimeric mRNA (Kuwabara et al. 1997, 1998b; Warashina et al. 2000b). Since both *BCR* and *ABL* genes are normally important for cell survival, it is essential to destroy only the abnormal chimeric mRNA, without affecting the normal mRNA.

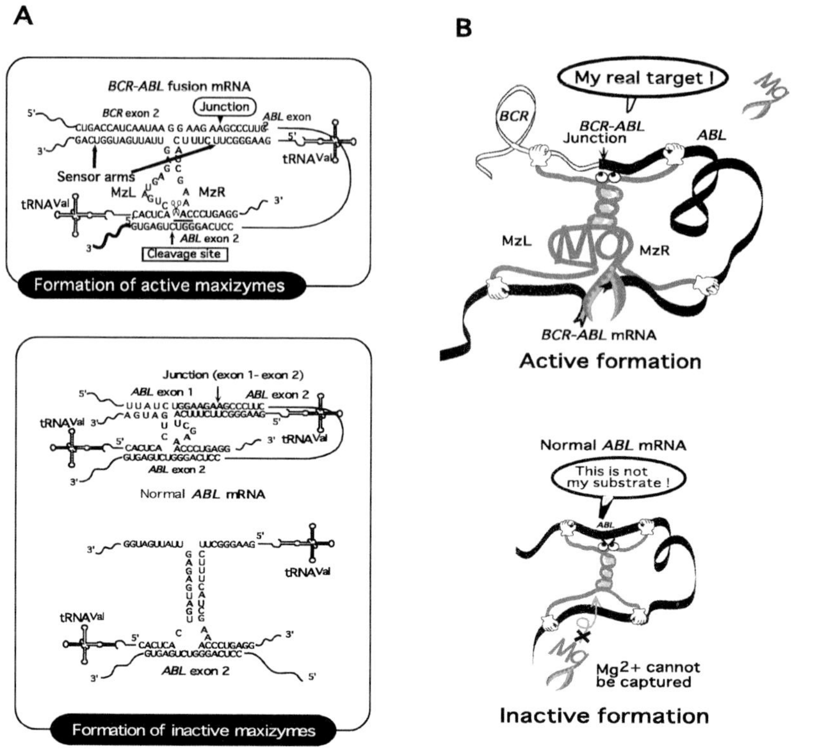

FIG. 5. **A** Formation of active or inactive maxizymes regulated by specific effector sequences. The heterodimer (MzL and MzR) can generate two different binding sites: one is complementary to the sequence of interest (activator or inhibitor), and the other is complementary to a cleavable sequence. To exhibit strict substrate -specificity, maxizymes should take an active conformation only in the presence of the abnormal *BCR-ABL* junction (upper), but not in the presence of only normal *ABL* mRNA or in the absence of the *BCR-ABL* junction (bottom). **B** Maxizymes can discriminate limited differences among various transcripts, and specifically cleave the correct target in each case

Although attempts were made to cleave only chimeric mRNAs using wild-type hammerhead ribozymes, these were unsuccessful (James et al. 1996; Pachuk et al. 1994). As mentioned above, the maxizyme can bind to two different target sites. Thus, if the maxizyme can simultaneously recognize a junction site by one of its arms and a distant NUX site by the other, it should be possible for the maxizyme to cleave only the chimeric mRNA. Through repeated trial and error, we established an ideal maxizyme that has two independently functional arms. One arm can only bind the substrate as the "eye (sensor)", to discriminate chimeric mRNA from normal mRNA, and the other arm serves as the "scissor" that actually cleaves the target (Fig. 5B, top).

In designing an allosteric maxizyme, it is important that the dimer can act as a molecular switch. In other words, only when the junction sequence is present should the maxizyme take an active conformation by binding to the junction, followed by cleaving the NUX sequence. Precise switching should be controlled through the pairing of bases between the MzL and MzR (Fig. 5A). If the base pairing is too stable, the dimer

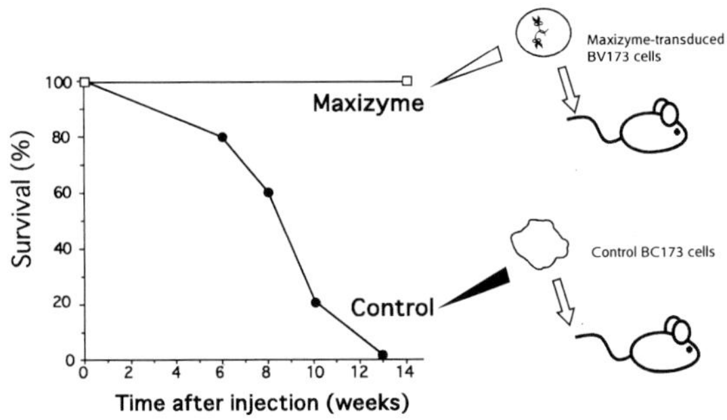

FIG. 6. Anti-tumor effects of the maxizyme in a murine model of chronic myelogenous leukemia (CML). The survival of animals was monitored daily for more than 20 weeks after inoculation. All control mice died within 13 weeks, whereas maxizyme-treated mice remained disease-free for the entire period of the investigation

will take the active conformation even in the absence of a substrate. In contrast, if the base paring is less stable, the dimer will not take the active conformation even if substrate RNA is present. Thus, the precise control of the stability of the base paring is a key point in generating the allosterically controllable maxizyme.

Figure 5A shows a schematic representation of the maxizyme with active and inactive conformations. The active conformation as the dimer can be formed only when the two arms of the maxizyme bind correctly to the two sites on the chimeric mRNA (Fig. 5A, top, and Fig. 5B, top). In contrast, in the presence of only normal *ABL* mRNA, the maxizyme takes an inactive conformation by changing the structure of the central core (Fig. 5A, bottom). The maxizyme with the inactive conformation cannot cleave its substrates, because it cannot capture the magnesium ions that are essential for its cleavage activity. When chemically synthesized maxizymes were evaluated in vitro for specificity to their substrates, they were found to specifically cleave only the chimeric *BCR*/*ABL* mRNA, without damaging the normal *ABL* mRNA.

The efficacy of the maxizyme was also evaluated in cultured mammalian cells. For producing large amounts of maxizyme in cells, the tRNA^{Val}-promoter was again employed for its expression. The high level of expression ensures efficient dimerization. When maxizimes were introduced into cells derived from patients with CML, they showed extremely strong activity compared to parental wild-type ribozymes (Kuwabara et al. 1998b). Moreover, as expected, maxizymes inhibited the expression of the *BCR-ABL* mRNA specifically, without affecting that of normal *ABL* mRNA.

Thus, we succeeded in creating an artificial allosteric enzyme, a maxizyme, that exhibits high specificity and high activity only in the presence of the oncogenic chimeric mRNA (Kuwabara et al. 1998b; Tanabe et al. 2000a).

1.4.3 Oncogene Inactivation in a Mouse Model

The anti-tumor effect of the maxizyme was further examined in animal models (Tanabe et al. 2000a). The maxizyme was introduced into leukemic cells by using a retroviral vector that encoded the two components of the maxizyme under the control of the tRNA^{Val} promoter in tandem. CML cell lines (BV173) that had been transduced with either the maxizyme gene or a negative control gene were independently injected into the tail veins of mice. All of the mice injected with control BV173 cells died of diffuse leukemia, confirmed at necropsy, 6–13 weeks afterwards (median survival time, 9 weeks), whereas mice injected with maxizyme transduced BV173 cells remained healthy (Fig. 6). The results indicated that the maxizyme introduced by the retroviral vector could be produced at sufficient concentrations to support the dimerization in vivo. Moreover, the maxizyme successfully functioned in animals.

At present, kinase inhibitor (Druker et al. 1996) or allogeneic transplantation is employed for the effective therapy of this type of leukemia. Only half of the patients on average are eligible for the transplantation because of limited donor availability and age restrictions. Our results raise the possibility that the maxizyme could be useful for purging bone marrow in patients with CML who are treated by autologous transplantation, and that it would presumably reduce the incidence of relapse by decreasing the tumorigenicity of contaminating CML cells in the transplant (Tanabe et al. 2000a).

1.4.4 Generality of the Maxizyme Technology

The maxizyme technology is not limited to the disruption of the abnormal chimeric gene in CML. Abnormal chimeric genes generated from reciprocal chromosomal translocation have frequently been observed in several leukemia diseases. Maxizymes have successfully cleaved only abnormal targets that lack a NUX cleavage site near the junction in cases of acute lymphoblastic leukemia (ALL) and acute promyelocytic leukemia (APL), without damaging normal genes (Tanabe et al. 2000b). Abnormal chimeric mRNAs are also generated as results of mis-splicings. A maxizyme designed against each splicing variant could discriminate the target specifically by its allosteric function. Indeed, we have already constructed various maxizymes that target different chimeric genes, and these maxizymes showed significant activity and high specificity (Kanamori et al. 2003; Kuwabara et al. 1998b; Tanabe et al. 2000a). Thus, maxizymes should be considered as powerful gene-inactivating agents with allosteric functions for cleaving any type of chimeric mRNA. Importantly, to our knowledge, the maxizyme is the first artificial allosteric enzyme whose activity was demonstrated at the animal level, extending its potential utility into the medical area (Kanamori et al. 2003; Tanabe et al. 2000a).

In contrast to the requirement of the NUX cleavage site for ribozymes, siRNAs can practically target any sequence of interest. Therefore, chimeric genes could also be destroyed by siRNAs (Oshima et al. 2003; Scherr et al. 2003).

Other types of nucleic-acid-based allosteric sensors are being developed (Famulok 2004). Cells can silence certain genes at the stages of transcription (Kawasaki and Taira 2004; Morris et al. 2004) and translation (Bartel 2004) by a mechanism that involves short RNAs, known as microRNAs (miRNAs). This process allows protein expression levels to be altered very quickly. The detection of miRNAs in live cells and tissues will help to address questions about their genetics, biogenesis, trafficking and function. A recent report described a design, based on an RNA enzyme, to develop fluorescent probes for specific miRNAs (Hartig et al. 2004).

1.5 Accessibility of Target mRNA to Ribozymes

Significant activity of a ribozyme inside cells depends on its high-level expression, intracellular stability, and efficient export to the cytoplasm. In addition to these parameters, the accessibility of the ribozyme's target should be also considered, because the rate-limiting step for the ribozyme-mediated cleavage of substrates in vivo seems to be the association/annealing of the ribozyme with its target (Kato et al. 2001).

To enhance such accessibility, computer-generated predictions of secondary structure are typically used to identify target sites that are most likely to have an open conformation (Zuker, 1989). However, these predictions are often inaccurate because of unpredictable RNA-protein interactions that change the structure of RNA in cells. This limitation can be circumvented using a systematic approach to search for huge numbers of candidate sites for antisense molecules (Milner et al. 1997; Patzel and Sczakiel 1998). Another approach, involving oligonucleotide-directed RNase H cleavage of target mRNA in cell extracts, has also been used. It was demonstrated that the efficacy of RNase H cleavage elicited by oligonucleotides well correlated with that of ribozyme-induced cleavage. Thus, accessible sites for ribozymes have often been selected using random oligonucleotide libraries (Scherr and Rossi, 1998; Scherr et al. 2000, 2001). A random ribozyme library has also been employed to identify ribozymes that efficiently cleave the target mRNA (Pan et al. 2003). Although these approaches seemed to be sophisticated and useful, we sought a simpler strategy, one in which a unique ribozyme would be able to access any chosen target site regardless of its secondary structure (Kawasaki and Taira 2002a,b; Wadhwa et al. 2003; Warashina et al. 2001).

1.5.1 RNA-Protein Hybrid Ribozymes

Our goal was to design a ribozyme that would recruit a protein able to unwind any interfering secondary structures, thereby making any site accessible to the ribozyme. To create such a ribozyme, a ribozyme was linked to a RNA helicase, a protein defined as having nonspecific RNA binding, sliding, and unwinding activities (de la Cruz et al. 1999; Jankowsky et al. 2000; Lee and Hurwitz 1993; Wagner et al. 1998). For recruiting the RNA helicase to the ribozyme, the constitutive transport element (CTE), an RNA motif that interacts with an RNA helicase A in vitro and in vivo, was employed

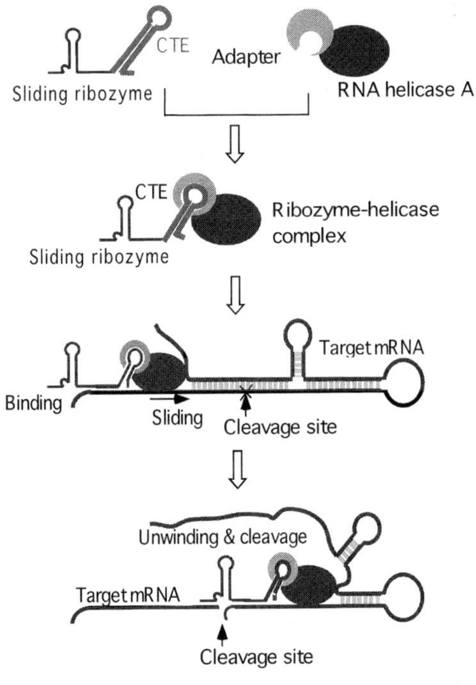

FIG. 7. Schematic model of how a ribozyme with a CTE motif (CTE-Rz) coupled to an RNA helicase cleaves a hidden cleavage site by unwinding the local secondary structure of target mRNA

(Braun et al. 1999; Gruter et al. 1998; Hodge et al. 1999; Kang and Cullen 1999; Li et al. 1999; Schmitt et al. 1999; Tang and Wong-Staal 2000; Tang et al. 1997; Westberg et al. 2000). We hypothesized that an RNA helicase coupled to a ribozyme would efficiently guide the ribozyme to its target site by resolving any inhibitory mRNA structures, thereby leading to efficient substrate cleavage (Kawasaki and Taira 2002a; Wadhwa et al. 2003).

Figure 7 is a schematic representation of how a CTE-ribozyme coupled to an RNA helicase can cleave a hidden target site by unwinding the local secondary structure. To evaluate the expected enhanced ribozyme effect, a vector that expresses the ribozyme with the CTE from the $tRNA^{Val}$ promoter was designed. As the target of the ribozyme, the TAR region within the long terminal repeat (LTR) of HIV-1 was chosen (Kuwabara et al. 1999). Since the conserved TAR region is essential for replication of HIV-1, this region might be an ideal target for gene therapy against HIV-1. Because of its stable stem form, however, it seems difficult for ribozymes to access the TAR region efficiently. If ribozyme attached to the RNA helicase through the CTE could disrupt the stem structure and efficiently cleave the TAR sequence, the ribozyme might be an effective drug against HIV-1 regardless of rapid mutations of HIV-1.

Thus, CTE-connected and unconnected ribozymes targeted against the TAR region were designed. At the same time, we established a HeLa cell line that stably expresses the LTR of HIV-1 fused with a luciferase gene for evaluating the inhibitory effect of ribozymes. As expected, TAR ribozymes without the CTE (non-CTE ribozymes) were unable to inhibit the activity of the luciferase reporter, because they could not access their target sites efficiently. Surprisingly, however, when the CTE was attached to these ribozymes, their inhibitory effects were significantly enhanced, resulting in about 80% reduction of the reporter activity (Warashina et al. 2001). Furthermore, these CTE-coupled ribozymes showed higher activity than non-CTE ribozyme designed to target "open" sites. These results suggested that the attachment of CTE enables all ribozymes to access and cleave their target sites efficiently (Warashina et al. 2001).

To examine the general applicability of the CTE-ribozyme, other CTE-ribozymes targeted against several endogenous genes, such as mouse procaspase-3 (CPP 3), were tested. The cleavage activity of these ribozymes, as expected, was significantly enhanced compared to that of parental non-CTE ribozymes (Warashina et al. 2001). Thus, we demonstrated that CTE-ribozymes have specificity, significant activity, and general utility. All of the CTE-ribozymes showed robust activity in cell culture, even if the parental ribozymes without CTE showed weaker activity (Nawrot et al. 2003; Wadhwa et al. 2003). Therefore, it seems that such hybrid-ribozyme-based technology is broadly applicable due to the ease of design and significant activity in cells. In addition, RNA helicase-recruiting-ribozymes are suitable for a wide range of applications, as described below. As also described below, an alternative motif, the poly(A) sequence, which can recruit a different kind of RNA helicase, eIF4AI, in mammalian cells was also proven useful (Kawasaki and Taira 2002a, b).

1.5.2 Identification of Functional Genes by Use of Hybrid Ribozymes

Based on the improved intracellular activities of the hybrid ribozymes, we next attempted to identify genes associated with specific phenotypes in cells by using the highly active ribozymes with randomized binding arms. Independently, the Wong-Staal, Barber and Taira laboratories developed novel gene discovery systems using randomized ribozyme libraries that were introduced into cells and produced a phenotypic change (Beger et al. 2001; Chatterton 2004; Kawasaki and Taira 2002a, 2002b; Kuwabara et al. 2004; Li et al. 1999; Nelson 2003; Onuki et al. 2002, 2004; Suyama et al. 2003a, 2003b; Taira, 1999; Welch et al. 2000). As reported with hairpin ribozymes (Beger et al. 2001; Chatterton 2004; Li et al. 1999; Welch et al. 2000), randomization of a ribozyme's binding arms allows easy identification of genes of interest. Screening by use of randomized ribozymes that promote anchorage-independent cell growth identified *hPPAN*, the human homolog of the *Drosophila ppan* gene, and the murine telomerase reverse transcriptase (*mTERT*) gene, as tumor suppressors (Li et al. 1999; Welch et al. 2000). Another approach using random ribozyme libraries identified ID4 as a regulator of BRCA1 expression (Beger et al. 2001).

In our case, functional genes were identified by using the hybrid hammerhead ribozyme, which coupled cleavage activity with the unwinding activity of an endogenous RNA helicase (Nelson 2003; Taira 1999). In this study, we created novel hybrid ribozymes by attaching a poly(A) tail (a stretch of 60 As) to the 3′ end of a tRNA-attached ribozyme (Rz-A60) instead of the CTE (Warashina et al. 2001). The poly(A) sequence is known to interact with an endogenous RNA helicase, elF4AI, via interactions with poly(A)-binding protein (PABP) and PABP-interacting protein-1 (PAIP). It was expected that the poly(A)-mediated complex would unwind the RNA duplex of the substrate efficiently, and then cleave inaccessible target sites. By introducing the library of hybrid ribozymes with randomized binding arms into cells, any cell might show phenotypic changes. Then, the target genes involved in such phenotypic changes could be identified by sequencing of the specific ribozyme clones (Nelson 2003; Taira 1999).

For example, using randomized Rz-A60-expression libraries, we established a novel system for screening the genes involved in a signal pathway for Fas-induced apoptosis in HeLa-Fas cells (Fig. 8). In this system, retroviral vectors carrying Rz-A60 with 10-nt randomized sequences in each substrate-binding arm were employed to introduce the libraries into HeLa-Fas cells. Cells were then treated with the Fas-specific antibodies, and surviving cells were harvested for the subsequent isolation of ribozymes. Sequencing of the randomized region of the isolated Rz-A60 enabled genes that are responsible for the Fas-induced apoptotic pathway to be rapidly identified. By this screening, many interesting genes, such as FADD, caspase 8, caspase 9 and caspase 3, were detected. When ribozymes without the poly(A) tail were used for the screening, genes such as FADD and caspase 8 were not identified. This result strongly suggested that hybrid Rz enhanced successful screening in our gene discovery system. Indeed, with this system, we identified many genes (Nelson 2003) involved in other apoptotic pathways (Kawasaki and Taira 2002a, 2002b; Kawasaki et al. 2002), tumor invasion (Suyama et al. 2003a, 2003b, 2004), and Alzheimer's disease (Onuki et al. 2002, 2004). Importantly, novel non-coding RNAs was also identified by use of our gene discovery system (Kuwabara et al. 2004). With the revelation of the human genome, this unique technology becomes extremely valuable for the identification of important genes (Shiota et al. 2004).

The success with the ribozyme library has encouraged us to establish potentially more effective siRNA library, as described below (Miyagishi and Taira 2003).

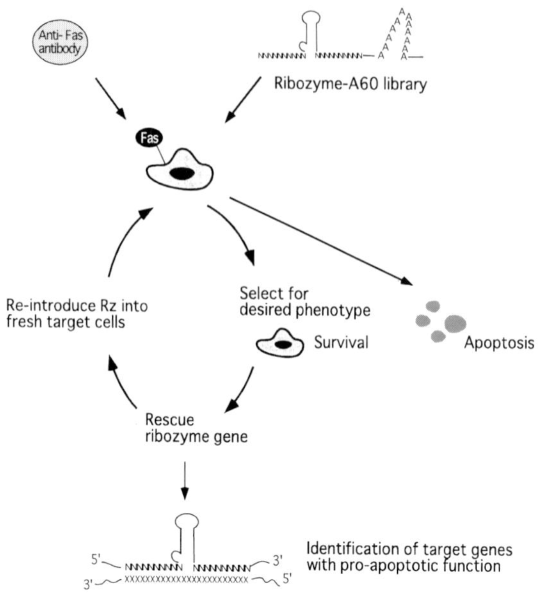

FIG. 8. Schematic diagram of the gene discovery system for identification of genes involved in the Fas-induced apoptosis by use of a poly(A)-connected hybrid-Rz library

2 siRNAs, siRNA-Expression Vectors, and siRNA-Expression Library

2.1 Introduction

RNA interference (RNAi) is an evolutionarily conserved process in plants and animals by which double-stranded RNA (dsRNA) induces sequence-specific degradation of homologous RNA (Fire et al. 1998; reviewed in Novina and Sharp 2004). Although RNAi works well in various organisms, its silencing of specific genes proved difficult to detect in mammalian systems because of dsRNA-dependent nonspecific inhibition of protein synthesis by the PKR pathway and nonspecific RNA degradation by activation of RNase L (Elbashir et al. 2001). However, recent works by Tuschl and colleagues demonstrated that 21- or 22-nt RNAs with two-nt 3′-overhangs (siRNA) can induce gene silencing without nonspecific inhibition of gene expression in cultured mammalian cells (Caplen et al. 2001; Elbashir et al. 2001). Furthermore, in the past year, various groups, including our own, have developed systems for vector-mediated specific RNAi in mammalian cells (Brummelkamp et al. 2002; Kawasaki and Taira 2003; Lee et al. 2002a; Miyagishi and Taira 2002; Paddison et al. 2002; Paul et al. 2002; Sui et al. 2002; Yu et al. 2002; reviewed in McManus and Sharp 2002; Mittal 2004; Tuschl 2002). These vector systems use a pol III promoter, such as the U6, H1, and tRNA promoters, and the systems have been classified into two groups depending on whether the expressed RNAs are tandem-type or hairpin-type (Fig. 9). Although U6 and H1 promoters are not appropriate for ribozyme expression, because the transcripts tend to localize in the nucleus (Kato et al. 2001), they are widely used for siRNA expres-

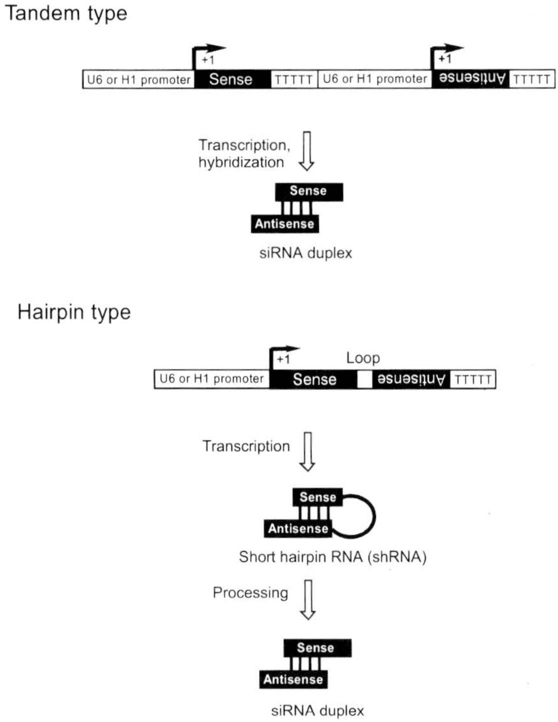

FIG. 9. Schematic representation of tandem-type and hairpin-type siRNA-expression vectors

sion, most probably because the U6- and H1-driven shRNAs (Fig. 9) can be transported to the cytoplasm by Exportin-5, as described above (Lee et al. 2002b). The regulation of pol III promoters by small molecules, such as tetracycline and the Cre-lox P system, also became possible (Coumoul et al. 2004; Fritsch et al. 2004; Kasim et al. 2004; Miyagishi et al. 2004b; van de Wetering et al. 2003), leading to spatial and temporal regulation of gene expression in transgenic animals.

Since gene silencing by siRNA is highly specific and has extremely high activity, genome-wide and comprehensive gene analysis in various organisms can be expected. Systematic functional analysis of a number of genes of *C. elegans* by RNAi has been performed, and many functional genes responsible for a specific mutant phenotype were identified (Fraser et al. 2000; Gonczy et al. 2000). A similar genome-wide RNAi library approach could be extended to mammalian cells by use of an siRNA oligonucleotide library and/or an siRNA expression library that complements the randomized ribozyme library, as described above, for the identification of novel genes (Beger et al. 2001; Chatterton 2004; Kawasaki and Taira 2002a, 2002b; Kuwabara et al. 2004; Li et al. 1999; Nelson 2003; Onuki et al. 2002, 2004; Suyama et al. 2003a, 2003b; Taira, 1999; Welch et al. 2000) In fact, several groups, including ours, have already started to generate siRNA libraries against the entire human genome (Berns et al. 2004; Miyagishi and Taira 2003; Paddison et al. 2004). In this section, we discuss the optimization of siRNA-expression systems as well as practical approaches for solving some of the difficulties that are faced in constructing siRNA-expression vectors and the siRNA-expression library.

2.2 Optimization of the siRNA Expression System: Tandem-Type or Hairpin-Type?

First, optimization of the siRNA expression system, in terms of high suppressive activity and high genetic stability, was addressed. Since the level of transcription by pol III promoters is significantly higher than that of pol II promoters, we describe pol III promoter-based siRNA expression systems (Koseki et al. 1999).

The major pol III expression systems have been classified into two groups, tandem-type and hairpin-type, depending on the type of expressed RNAs (Fig. 9). In the case of the tandem-type siRNA expression vector, both sense and antisense strands are driven separately by their own pol III promoter, followed by annealing in the cells and the formation of siRNA duplexes with about 4-nt overhangs at each 3′-end (Lee et al. 2002a; Miyagishi and Taira 2002). In the hairpin-type, sense and antisense nucleotides are connected by a loop and expressed as a single chain (Brummelkamp et al. 2002; Kawasaki and Taira 2003; Paddison et al. 2002; Paul et al. 2002; Sui et al. 2002; Yu et al. 2002), leading to the rapid formation of hairpin structures with a stem and loop. The hairpin RNA seems to be exported to the cytoplasm by Exportin-5, and to be processed to siRNA by Dicer, as is miRNA (Kawasaki and Taira 2003; Ketting et al. 2001; Lee et al. 2002b). The suppressive activities of the two vector systems were examined. Both types of siRNA expression vector targeted against the same site of firefly luciferase gene were constructed and their suppressive activities were compared. Although at a high concentration both siRNA expression vectors had high levels of suppressive activities, at low concentrations the hairpin-type siRNA-expression vector (21-nt stem length) had significantly higher suppressive activity than the tandem-type siRNA-expression vector (Fig. 10A).

2.3 Optimized Hairpin-Type siRNA Expression Vectors

Constructing hairpin-type siRNA-expression vectors posed two serious technical problems. First, it was difficult to sequence constructs that contained a hairpin region, probably because of the tight palindromic structure. Second, approximately 20–40% of constructs were mutated within the hairpin region upon introduction into *E. coli* cells. Sequencing of mutated constructs revealed that those with two or more point mutations or insertions/deletions in the sense or antisense region could be sequenced without any problems. Therefore, we examined whether constructs with such mutations or insertions might retain silencing activity.

As many as five point mutations from C to T, which generated G:U base-pairing, in the sense strand did not affect the silencing effect (Fig. 10A). As many as four bulge insertions in the sense strand also did not affect the activity. By contrast, constructs with five G-to-A mutations or more than four A-to-C mutations had significantly reduced silencing activity. In particular, insertion of a single G-to-A mutation in the antisense strand [G > A1(AS)], which recognizes the target transcript, resulted in reduced suppression, and insertions of two or five G-to-A mutations in the antisense strand resulted in almost no silencing of the target gene. Therefore, it appears that, in practice, it is not possible to introduce mutations and bulges into the antisense strand and retain activity.

It should also be emphasized that, while generating these constructs with mutated hairpin structures, we found that the rates of mutation in *E. coli* of the hairpin region

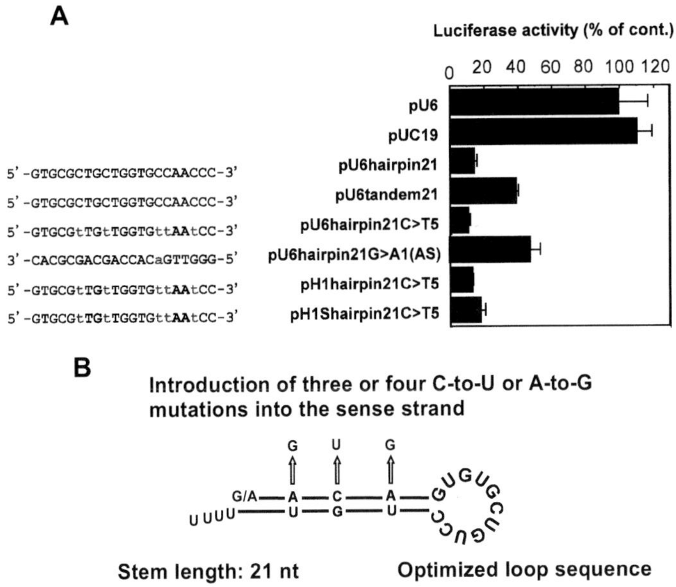

FIG. 10. **A** Comparative analysis of the effects of the U6 promoter-driven 21-nt tandem-type and 21-nt shRNA vectors, the H1 promoter-driven 21-nt shRNA vector and the U6 promoter-driven 21-nt shRNA with various mutations in the hairpin region. HeLa S3 cells were cotransfected with 10 ng pRL-RSV, 30 ng pGL3, and 30 ng of each siRNA expression vector targeted to the firefly gene for luciferase. The activities of firefly luciferase were normalized by reference to those of *Renilla* luciferase. Each *bar* indicates an average value and *vertical bars* indicate standard errors of triplicate assays. **B** Schematic representation of the proposed siRNA expression system

of the constructs were markedly reduced when more than three mutations were inserted in the sense or antisense strand. This property of such constructs is obviously important for the maintenance of corresponding plasmids in prokaryotic host cells and possibly so in mammalian cells as well. Collectively, our data suggested that the way to create a useful hairpin-type siRNA-expression vector was to introduce multiple C-to-T (or A-to-G) mutations (Fig. 10B), which not only prevent unwanted mutations but also enhance the silencing activity (compare results for pU6hairpin21 with pU6hairpin21C > T5 in Figs. 10A and 11 for increased activities due to mutations), and/or to introduce multiple bulges into the sense region. These mutations, especially near the 5′-antisense terminal base pairing, enhance unwinding of double-stranded siRNAs and result in high levels of siRNA activities (Khvorova et al. 2003; Schwarz et al. 2003).

2.4 Optimization of the Loop Sequence

We next examined the effects of the loop sequence on the silencing activity of the hairpin-type siRNA-expression vectors. When a 19-bp stem sequence was used, the hairpin RNA with a 9-nt loop sequence had greater silencing activity than the corresponding RNA with a shorter loop sequence (Brummelkamp et al. 2002), although a short 4-nt, 5′-UUCG-3′, tetraloop could support significant siRNA activi-

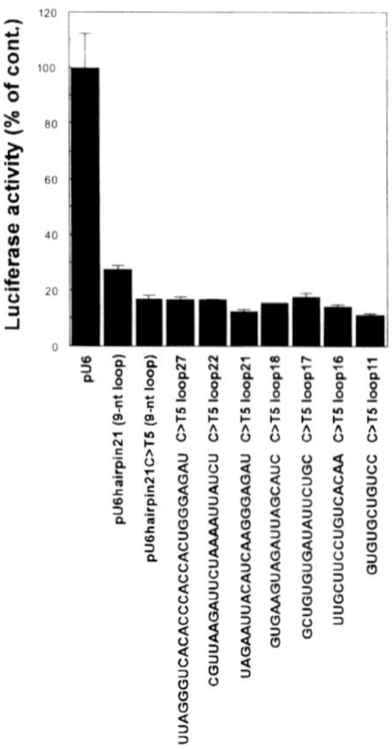

FIG. 11. The effects of various loop sequences on suppressive activity. Details are the same as those in the legend to Fig. 10A

ties (Paul et al. 2002). We wondered whether the natural loop sequence of microRNAs might be a preferable sequence for the hairpin RNA produced from hairpin-type siRNA-expression vectors; therefore, seven loop sequences derived from human microRNAs were examined (Miyagishi and Taira 2003; Miyagishi et al. 2004a, 2004b).

In transfection experiments using siRNA-expression vectors targeted to the firefly gene for luciferase, each of which included a microRNA-derived loop sequence, it was found that all the siRNA-expression vectors tested had similar or even greater silencing activity than the corresponding vectors with a 9-nt loop sequence (Fig. 11). In particular, loop21 and loop11 had slightly higher activities. Although it appeared that the loop sequence affected silencing activity only marginally (Fig. 11), in practice, severalfold higher concentrations of C > T5 with the original 9-nt loop were required to achieve the suppressive activity of C > T5 loop11. Therefore, inclusion of a natural microRNA-derived loop is advantageous, especially when the copy number of the template is low.

2.5 Selection of Favorable Target Sites

One of the most important and critical problems with siRNAs and siRNA-expression vectors is the selection of target sites. It is well known that the effectiveness of siRNA is highly dependent on the target site of a message, and the possibility that siRNA acting at an arbitrarily selected target site has high suppressive activity seems to be empirically about 20 to 40% or less. In the case of ribozyme target sites, the activity strongly depends on the structure of the target mRNAs, which cannot easily be predicted. However, in the case of siRNA target sites, the activity appeared to depend on interactions of shRNA or siRNA with endogenous protein factors, affecting recognition, processing, unwinding, and so on. If so, it should be possible to formulate an algorithm that can predict highly effective sites for siRNAs. To this end, we collaborated with Suzuki and generated more than 800 siRNAs and more than 300 shRNA-expres-

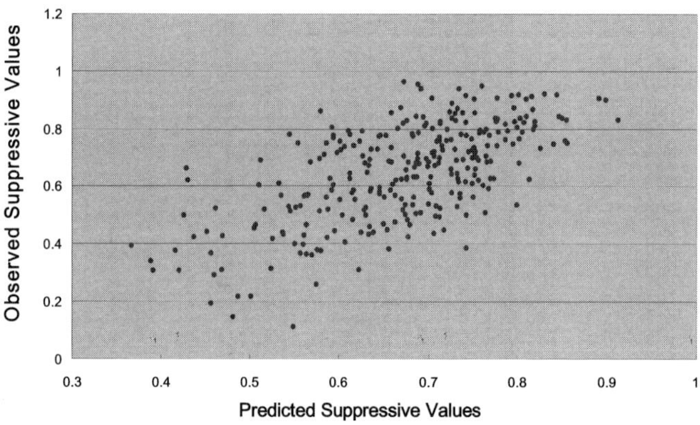

FIG. 12. Development of the algorithm for prediction of effective target sites. The graph shows the leave-one-out cross-validation result between values predicted by application of the algorithm and experimental results using reporter genes

sion vectors against genes for GFP and luciferases (Katoh et al. 2003; Taira and Miyagishi 2003). Examination of the relationship between the activity and the sequence enabled us to extract dozens of correlated factors; consequently an algorithm was successfully generated by nonlinear regression methods (Miyagishi and Taira 2003).

Statistical analysis based on our accumulated data indicated that some nucleotides at specific positions are positively or negatively correlated with the efficacy of siRNAs. siRNAs with an A or U residue at the first nucleotide position from the 5′ end of the antisense strand tended to have relatively strong suppressive activity, whereas those with a G or C residue at the same position tended to be less effective (Miyagishi and Taira 2003; Yoshinari et al. 2004), as also reported by others (Khvorova et al. 2003; Schwarz et al. 2003). In addition, in our analysis, an A residue at the tenth position from the 5′-end within the 21-nt antisense strand of siRNAs with a 2-nt overhang tended to be effective. Moreover, there was also a significant negative correlation between the GC content of the 5′ half of siRNAs (especially, from the 1st to the 8th nucleotide in the antisense strand) and a positive correlation for a U residue at 14th nucleotide.

Figure 12 shows the predicted versus measured results for siRNAs targeted against firefly luciferase, which were calculated by a cross-validation method (Miyagishi and Taira 2003). The correlation coefficient was 0.7. Notably, all siRNAs with a predicted score of more than 0.85 showed highly suppressive activity, indicating that, at high values, the predicted scores are accurate.

2.6 Confirmation of the Specificity of the Target Sequence

In selecting a target site in a specific gene, it is necessary to check the specificity of its sequence, that is, to confirm that there are no highly homologous sequences in other genes. If very stringent conditions are applied in the selection, for example, if more than three base mismatches are needed to discriminate between the correct and homologous targets, a significant portion of favorable target sites that were predicted to be highly active could be eliminated by the BLAST search program. However, if less stringent conditions are applied, for example, if only one base mismatch is allowed, the designed siRNA might have the potential to disrupt not only the correct target but also other homologous genes, although the cleavage efficiency is reduced for the latter. Under these circumstances, we decided to use relatively low stringency for examining target site specificity, because multiple positive candidates can be further analyzed by creating new siRNAs targeted at different sites within the candidate gene.

Alternative splicing and the existence of a highly homologous gene family are also major problems in target site selection. Recent genome-wide analysis of transcripts reveals that a significant portion of genes has alternatively spliced forms. If it is desirable to disrupt a specific alternative variant, an siRNA expression vector targeted against the specific variant can be created that will not damage other, alternative variants originating from the same pre-spliced form. However, for the generation of a first draft library of the common sequence of all alternative transcripts, genes that have multiple alternative splicing forms (similarly, the common sequence of a homologous gene family) can be set as the target site. In creating an siRNA-expression library, this is more economical, and additional siRNAs can easily be made once the targeted gene turns out to be an attractive candidate.

2.7 Improved Method for Library Construction

In constructing an siRNA-expression library against all genes in the genome, some practical problems, such as how to construct the library efficiently, speedily, and cheaply, have to be solved. We have established a large-scale method for constructing such a library, as shown in Fig. 13. Specifically, 96 sets of oligonucleotides that include hairpin sequences corresponding to 96 different target sequences were synthesized and annealed separately. Then, the annealed oligonucleotides were mixed and ligated into the Bsp MI site of the U6-driven siRNA-expression vector. After transformation of *E. coli* cells, 384 clones were picked up and identified by sequencing. Analysis revealed that more than 90% of the initial oligonucleotide sequences were correctly

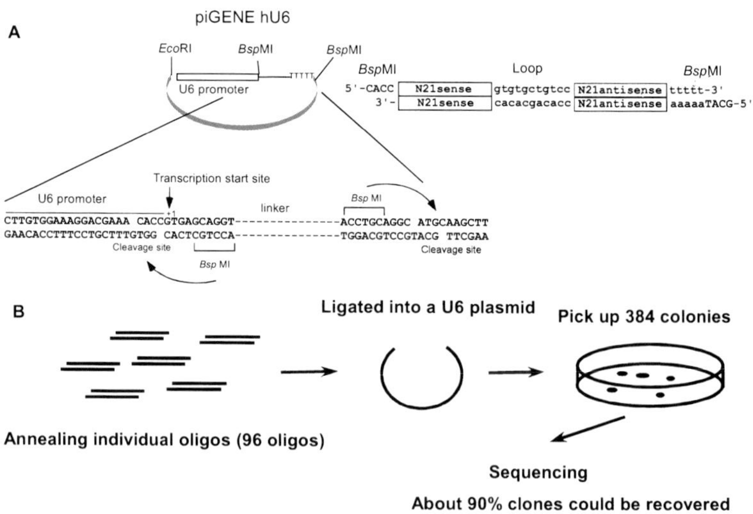

FIG. 13. **A** Cloning site and the insert sequence of siRNA expression vector. **B** Our strategy for the construction of siRNA expression library

inserted into the designed plasmids in a single procedure. Those that were not correctly inserted were added to the next round of the construction. This bulk procedure allowed siRNA-expression vectors to be made much more cheaply and faster than by the traditional procedure, in which each clone is made separately. Using this procedure, 3,000 to 5,000 siRNA-expression vectors can be synthesized per month (Taira and Miyagishi 2001).

2.8 Merits and Demerits of an siRNA-Expression Library Compared to an siRNA Oligonucleotide Library

The plasmid-based siRNA-expression library and the alternative oligonucleotide-based siRNA library each have merits and demerits. The transfection efficiency of siRNA into cells using lipofection depends on the cell type, and the RNAi effect seems to be sustained for only a limited period of time. The advantage of the plasmid-based siRNA is the capability of removing those cells that were not transfected with the plasmids by selecting only transfected cells with antibiotics-resistance genes. Moreover, the RNAi effects last much longer when plasmid-based siRNAs are used. Additionally, virus vectors allow delivery of the siRNA expression cassette into cells with high transfection efficiency and, in the case of lentivirus and retrovirus, it is easy to make stable knockdown cells by integration of the virus vector into the genome. Since stably knocked down cells can be generated only by the use of siRNA-expression vectors, and since such a bulk knockdown library can easily be generated by the use of virus vectors, virus-based construction would be most suitable for bulk library screening.

The advantage of oligonucleotide-based siRNA is that, in some favorable cells, transfection efficiency is greater than 90%, and thus higher than that of plasmid-based siRNA. However, once plasmid-based siRNAs are generated, they can be amplified without limit, especially with our construct (Fig. 10B), which does not undergo unwanted mutation (unlike conventional ones) during amplification in *E. coli* cells, resulting in the more economical construction of a library.

Probably the most significant advantage of using an siRNA-expression system over synthetic siRNA is the reduction of interferon responses in the case of vector-derived siRNAs (Taira et al. 2004). Figure 14 shows the suppressive effects of synthetic (Fig. 14A) and vector-derived (Fig. 14B) siRNAs. In this analysis, 50-bp siRNAs with appropriate C-to-U (or A-to-G) mutations only within the sense strand were used in order to sensitize interferon responses, as shown in Fig. 10B. The antisense strand is completely complementary to a part of the firefly luciferase gene, but it does not bind to the control *Renilla* luciferase mRNA. On appearance, synthetic (in vitro transcribed) dsRNAs showed suppressive effects on firefly luciferase activity in a dose-dependent manner, as can be seen from the relative luciferase activity (firefly luciferase activity divided by the control *Renilla* luciferase activity; Fig. 14A, left). However, when individual activities were compared, as in Fig. 14A right, non-specific effects can clearly be detected: namely, when the relative luciferase activity was 10% of the control, not only firefly luciferase but also *Renilla* luciferase activities were reduced by over 90%. Therefore, apparently sequence-specific gene silencing (Fig. 14A, left) must have occurred, accompanying significant non-specific effects on expressions of both firefly and *Renilla* luciferase genes by in-vitro-transcribed long dsRNA.

In contrast, the U6-based 50-bp-long hairpin RNA expression vector, the stem region of which is the same as the synthetic dsRNA in Fig. 14A, effectively suppressed

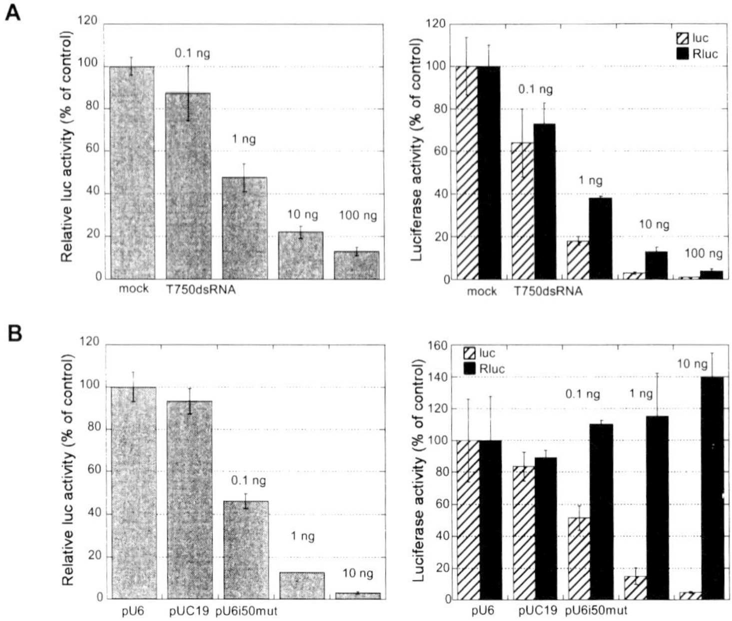

FIG. 14A, B. Comparison of the RNAi effect of 50-bp hairpin RNA expression vector on firefly luciferase activity with the effects on this activity of in vitro transcribed 50-bp RNA in HeLa S3 cells. **A** Effects of long dsRNA transcribed by T7 promoter that is targeted against luciferase gene. HeLa S3 cells were co-transfected with 25 ng luciferase expression plasmid, 2.5 ng *Renilla* luciferase expression plasmid, and the indicated amount of duplex RNA. *Left panel*, relative luciferase activity normalized to *Renilla* luciferase activity; *right panel*, firefly and *Renilla* luciferase activities. Triplicate transfection experiments were performed for each concentration of nucleic acid. **B** Effects of U6-driven long hairpin RNA expression vector targeted against the firefly luciferase gene. Transient luciferase assay was performed as in Fig. 5A. The amounts of dsRNA expression plasmids are indicated inside the histogram. *Left panel*, relative luciferase activity normalized to *Renilla* luciferase activity; *right panel*, firefly and *Renilla* luciferase activities, shown individually

expression of firefly luciferase gene with no observable side effects on expression of *Renilla* luciferase gene (Fig. 14B). The 50-bp-long hairpin RNA vectors in amounts corresponding to 1% of co-transfected luciferase expression vector were sufficient for about 80% reduction of relative luciferase activity (Fig. 14B). Considering that hundreds of copies of plasmids would be introduced into cells in transient transfection experiments, these data indicate that a few copies of shRNAs or long hairpin RNA expression vectors could suppress almost completely the expression of their target gene, importantly, without inducing interferon responses. Since some short siRNAs are known to induce interferon responses, depending on their sequences (Bridge et al. 2003; Sledz et al. 2003), it is much safer to use siRNA-expression vectors that can avoid such responses (Fig. 10B) for the analysis of a specific function of a gene of interest. It is especially advantageous to use our long hairpin RNA expression vectors in targeting viruses, such as HIV and HCV, that are known to undergo mutations, as a result of which mutated viruses might escape from siRNA-mediated cleavage (Lee

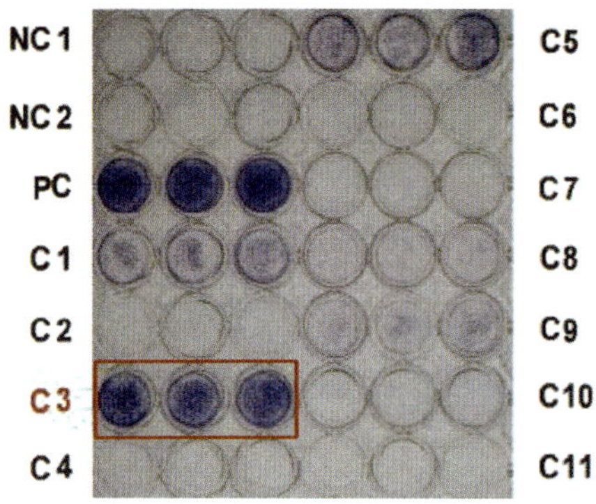

FIG. 15. Identification of genes involved in apoptosis by using siRNA expression library. An example of the screening using siRNA expression library. *PC* Positive control (siRNA expression vector against PKR), *NC1*, *NC2* negative controls, *C1-C11* siRNA expression vectors targeted against specific genes, for example kinases, transcription factors, or apoptosis-related genes. The shRNAi vector in C7 inhibits the dsRNA-dependent apoptosis

and Rossi 2004; Yokota et al. 2003). For the suppression of genes by siRNA-induced methylation of a promoter, methylation over a wider range of the promoter by the present strategy should be significantly more effective than methylation within a short ~21-bp region (Kawasaki and Taira 2004).

3 Concluding Remarks

We have described methods to improve rebozyme activity in vivo, an easy-to-use and highly active siRNA expression vector, and a system for very efficiently constructing a high-quality and genome-wide siRNA-expression library. Recently, we have started to screen functional genes by using the constructed siRNA-expression library. Figure 15 shows an example of screening of functional genes in an apoptosis pathway. Positive controls (siRNA-expression vectors targeted against PKR) could block dsRNA-induced apoptosis (Fig. 15; PC), and in experiments using siRNA-expression vectors targeted against unknown genes (Fig. 15; C1-C11), one positive clone could be identified from 11 candidates (Fig. 15; C7). Within the next few years, the genome-wide siRNA library should enable us to identify a number of known and novel functional genes. In the past, we have identified a miRNA sequence by use of our ribozyme library (Kuwabara et al. 2004). Since miRNA plays an important role in development and in other biological functions of mammalian cells (Kawasaki and Taira 2004; Kuwabara et al. 2004), the siRNA-expression library should also cover miRNA and undefined noncoding RNAs as targets.

Two versatile RNAs, ribozymes and siRNAs, are now available for use as gene knockdown tools. Since siRNAs are involved in natural cellular mechanisms, there exist many proteins that assist the cleavage of the target RNA, whereas the ribozyme cannot depend on intracellular factors. Thus, naturally, siRNAs are more effective than ribozymes. However, ribozymes have the advantage of not inducing interferon response; thus, we can clearly detect only the specific effects, despite their lower activity compared to siRNA (Bridge et al. 2003; Sledz et al. 2003; Taira et al. 2004). In conclusion, for functional analysis of genes of interest, it is advisable to use both siRNA and ribozyme technologies in parallel. Indeed, such specifically genetated siRNA libraries, directed against individual genes, enabled us to identify essential participants in RNAi, including genes that encode a slicer, elF2C2, and a helicase (unpublished data).

In view of the strong specificity of ribozymes and the strong activities of siRNAs, it is clear that the two technologies will play complementary and important roles in the identification of novel genes and their functions.

References

Alino SF, Crespo J, Bobadilla M, Lejarreta M, Blaya C, Crespo A (1994) Expression of human alpha 1-antitrypsin in mouse after in vivo gene transfer to hepatocytes by small liposomes. Biochem Biophys Res Commun 204:1023–1030

Amontov S, Nishikawa S, Taira K (1996) Dependence on Mg2+ ions of the activities of dimeric hammerhead minizymes. FEBS Lett 386:99–102

Arts GJ, Fornerod M, Mattaj IW (1998a) Identification of a nuclear export receptor for tRNA. Curr Biol 8:305–314

Arts GJ, Kuersten S, Romby P, Ehresmann B, Mattaj IW (1998b) The role of exportin-t in selective nuclear export of mature tRNAs. Embo J 17:7430–7441

Bartel DP (2004) MicroRNAs: genomics, biogenesis, mechanism, and function. Cell 116: 281–297

Beger C, Pierce LN, Kruger M, Marcusson EG, Robbins JM, Welcsh P, Welch PJ, Welte K, King MC, Barber JR, Wong-Staal F (2001) Identification of Id4 as a regulator of BRCA1 expression by using a ribozyme-library-based inverse genomics approach. Proc Natl Acad Sci USA 98:130–135

Berns K, Hijmans EM, Mullenders J, Brummelkamp TR, Velds A, Heimerikx M, Kerkhoven RM, Madiredjo M, Nijkamp W, Weigelt B, Agami R, Ge W, Cavet G, Linsley PS, Beijersbergen RL, Bernards R (2004) A large-scale RNAi screen in human cells identifies new components of the p53 pathway. Nature 428:431–437

Bertrand E, Castanotto D, Zhou C, Carbonnelle C, Lee NS, Good P, Chatterjee S, Grange T, Pictet R, Kohn D, Engelke D, Rossi JJ (1997) The expression cassette determines the functional activity of ribozymes in mammalian cells by controlling their intracellular localization. Rna 3:75–88

Braun IC, Rohrbach E, Schmitt C, Izaurralde E (1999) TAP binds to the constitutive transport element (CTE) through a novel RNA-binding motif that is sufficient to promote CTE-dependent RNA export from the nucleus. Embo J 18:1953–1965

Bridge AJ, Pebernard S, Ducraux A, Nicoulaz AL, Iggo R (2003) Induction of an interferon response by RNAi vectors in mammalian cells. Nat Genet 34:263–264

Brisson M, Huang L (1999) Liposomes: conquering the nuclear barrier. Curr Opin Mol Ther 1:140–146

Brummelkamp TR, Bernards R, Agami R (2002) A system for stable expression of short interfering RNAs in mammalian cells. Science 296:550–553

Caplen NJ, Parrish S, Imani F, Fire A, Morgan RA (2001) Specific inhibition of gene expression by small double-stranded RNAs in invertebrate and vertebrate systems. Proc Natl Acad Sci USA 98:9742–9747

Caplen NJ, Alton EW, Middleton PG, Dorin JR, Stevenson BJ, Gao X, Durham SR, Jeffery PK, Hodson ME, Coutelle C et al. (1995) Liposome-mediated CFTR gene transfer to the nasal epithelium of patients with cystic fibrosis. Nat Med 1:39–46

Chang Z, Westaway S, Li S, Zaia JA, Rossi JJ, Scherer LJ (2002) Enhanced expression and HIV-1 inhibition of chimeric tRNA(Lys3)-ribozymes under dual U6 snRNA and tRNA promoters. Mol Ther 6:481–489

Chatterton JE, Hu X, Wong-Staal F (2004) Ribozymes in gene identification, target validation, and drug discovery. Target 3:10–17

Cloninger MJ (2002) Biological applications of dendrimers. Curr Opin Chem Biol 6: 742–748

Cotten M, Birnstiel ML (1989) Ribozyme mediated destruction of RNA in vivo. Embo J 8: 3861–3866

Coumoul X, Li W, Wang RH, Deng C (2004) Inducible suppression of Fgfr2 and Survivin in ES cells using a combination of the RNA interference (RNAi) and the Cre-LoxP system. Nucleic Acids Res 32:e85

Das G, Henning D, Wright D, Reddy R (1988) Upstream regulatory elements are necessary and sufficient for transcription of a U6 RNA gene by RNA polymerase III. Embo J 7: 503–512

de la Cruz J, Kressler D, Linder P (1999) Unwinding RNA in Saccharomyces cerevisiae: DEAD-box proteins and related families. Trends Biochem Sci 24:192–198

Druker BJ, Tamura S, Buchdunger E, Ohno S, Segal GM, Fanning S, Zimmermann J, Lydon NB (1996) Effects of a selective inhibitor of the Abl tyrosine kinase on the growth of Bcr-Abl positive cells. Nat Med 2:561–566

Elbashir SM, Harborth J, Lendeckel W, Yalcin A, Weber K, Tuschl T (2001) Duplexes of 21-nucleotide RNAs mediate RNA interference in cultured mammalian cells. Nature 411:494–498

Famulok M (2004) Green fluorescent RNA. Nature 430:976–977

Ferber D (2001) Gene therapy. Safer and virus-free? Science 294:1638–1642

Fire A, Xu S, Montgomery MK, Kostas SA, Driver SE, Mello CC (1998) Potent and specific genetic interference by double-stranded RNA in Caenorhabditis elegans. Nature 391: 806–811

Fraser AG, Kamath RS, Zipperlen P, Martinez-Campos M, Sohrmann M, Ahringer J (2000) Functional genomic analysis of C. elegans chromosome I by systematic RNA interference. Nature 408:325–330

Fritsch L, Martinez LA, Sekhri R, Naguibneva I, Gerard M, Vandromme M, Schaeffer L, Harel-Bellan A (2004) Conditional gene knock-down by CRE-dependent short interfering RNAs. EMBO Rep 5:178–182

Fu DJ, Benseler F, McLaughlin LW (1994) Hammerhead ribozymes containing non-nucleoside linkers are active RNA catalysts. J. Am. Chem. Soc. 116:4591–4598

Geiduschek EP, Tocchini-Valentini GP (1988) Transcription by RNA polymerase III. Annu Rev Biochem 57:873–914

Gonczy P, Echeverri C, Oegema K, Coulson A, Jones SJ, Copley RR, Duperon J, Oegema J, Brehm M, Cassin E, Hannak E, Kirkham M, Pichler S, Flohrs K, Goessen A, Leidel S, Alleaume AM, Martin C, Ozlu N, Bork P, Hyman AA (2000) Functional genomic analysis of cell division in C. elegans using RNAi of genes on chromosome III. Nature 408:331–336

Good PD, Krikos AJ, Li SX, Bertrand E, Lee NS, Giver L, Ellington A, Zaia JA, Rossi JJ, Engelke DR (1997) Expression of small, therapeutic RNAs in human cell nuclei. Gene Ther 4:45–54

Grosshans H, Hurt E, Simos G (2000) An aminoacylation-dependent nuclear tRNA export pathway in yeast. Genes Dev 14:830–840

Gruter P, Tabernero C, von Kobbe C, Schmitt C, Saavedra C, Bachi A, Wilm M, Felber BK, Izaurralde E (1998) TAP, the human homolog of Mex67p, mediates CTE-dependent RNA export from the nucleus. Mol Cell 1:649–659

Hartig JS, Grune I, Najafi-Shoushtari SH, Famulok M (2004) Sequence-specific detection of MicroRNAs by signal-amplifying ribozymes. J Am Chem Soc 126:722–723

Hirabayashi K, Nakagawa S, Yamaguchi T, Nogawa M, Kobayashi I, Naito H, Kitagawa H, Ishiyama K, Ohgi T, Yano J (2004) Antitumour activity of siRNA/cationic liposome complex in a mouse model of liver metastasis. Clinical Cancer Research 10:7721–7726

Hodge CA, Colot HV, Stafford P, Cole CN (1999) Rat8p/Dbp5p is a shuttling transport factor that interacts with Rat7p/Nup159p and Gle1p and suppresses the mRNA export defect of xpo1-1 cells. Embo J 18:5778–5788

James H, Mills K, Gibson I (1996) Investigating and improving the specificity of ribozymes directed against the bcr-abl translocation. Leukemia 10:1676

Jankowsky E, Gross CH, Shuman S, Pyle AM (2000) The DExH protein NPH-II is a processive and directional motor for unwinding RNA. Nature 403:447–451

Kanamori T, Nishimaki K, Asoh S, Ishibashi Y, Takata I, Kuwabara T, Taira K, Yamaguchi H, Sugihara S, Yamazaki T, Ihara Y, Nakano K, Matuda S, Ohta S (2003) Truncated product of the bifunctional DLST gene involved in biogenesis of the respiratory chain. Embo J 22:2913–2923

Kang Y, Cullen BR (1999) The human Tap protein is a nuclear mRNA export factor that contains novel RNA-binding and nucleocytoplasmic transport sequences. Genes Dev 13: 1126–1139

Kasim V, Miyagishi M, Taira K (2004) Control of siRNA expression using the Cre-loxP recombination system. Nucleic Acids Res 32:e66

Kato Y, Taira K (2003) Expression of siRNA from a single transcript that includes multiple ribozymes in mammalian cells. Oligonucleotides 13:335–343

Kato Y, Kuwabara T, Warashina M, Toda H, Taira K (2001) Relationships between the activities in vitro and in vivo of various kinds of ribozyme and their intracellular localization in mammalian cells. J Biol Chem 276:15378–15385

Katoh T, Susa M, Suzuki T, Umeda N, Watanabe K (2003) Simple and rapid synthesis of siRNA derived from in vitro transcribed shRNA. Nucleic Acids Res Suppl:249–250

Kawasaki H, Taira K (2002a) Identification of genes by hybrid ribozymes that couple cleavage activity with the unwinding activity of an endogenous RNA helicase. EMBO Rep 3:443–450

Kawasaki H, Taira K (2002b) A functional gene discovery in the Fas-mediated pathway to apoptosis by analysis of transiently expressed randomized hybrid-ribozyme libraries. Nucleic Acids Res 30:3609–3614

Kawasaki H, Taira K (2003) Short hairpin type of dsRNAs that are controlled by tRNA(Val) promoter significantly induce RNAi-mediated gene silencing in the cytoplasm of human cells. Nucleic Acids Res 31:700–707

Kawasaki H, Taira K (2004) Induction of DNA methylation and gene silencing by short interfering RNAs in human cells. Nature 431:211–217

Kawasaki H, Onuki R, Suyama E, Taira K (2002) Identification of genes that function in the TNF-alpha-mediated apoptotic pathway using randomized hybrid ribozyme libraries. Nat Biotechnol 20:376–380

Kawasaki H, Ohkawa J, Tanishige N, Yoshinari K, Murata T, Yokoyama KK, Taira K (1996) Selection of the best target site for ribozyme-mediated cleavage within a fusion gene for adenovirus E1A-associated 300 kDa protein (p300) and luciferase. Nucleic Acids Res 24:3010–3016

Kawasaki H, Eckner R, Yao TP, Taira K, Chiu R, Livingston DM, Yokoyama KK (1998) Distinct roles of the co-activators p300 and CBP in retinoic-acid-induced F9-cell differentiation. Nature 393:284–289

Kay MA, Meuse L, Gown AM, Linsley P, Hollenbaugh D, Aruffo A, Ochs HD, Wilson CB (1997) Transient immunomodulation with anti-CD40 ligand antibody and CTLA4Ig enhances persistence and secondary adenovirus-mediated gene transfer into mouse liver. Proc Natl Acad Sci U S A 94:4686–4691

Ketting RF, Fischer SE, Bernstein E, Sijen T, Hannon GJ, Plasterk RH (2001) Dicer functions in RNA interference and in synthesis of small RNA involved in developmental timing in C. elegans. Genes Dev 15:2654–2659

Khvorova A, Reynolds A, Jayasena SD (2003) Functional siRNAs and miRNAs exhibit strand bias. Cell 115:209–216

Koseki S, Ohkawa J, Yamamoto R, Takebe Y, Taira K (1998) A simple assay system for examination of the inhibitory potential in vivo of decoy RNAs, ribozymes and other drugs by measuring the Tat-mediated transcription of a fusion gene composed of the long terminal repeat of HIV-1 and a gene for luciferase. J Control Release 53:159–173

Koseki S, Tanabe T, Tani K, Asano S, Shioda T, Nagai Y, Shimada T, Ohkawa J, Taira K (1999) Factors governing the activity in vivo of ribozymes transcribed by RNA polymerase III. J Virol 73:1868–1877

Koszul R, Malpertuy A, Frangeul L, Bouchier C, Wincker P, Thierry A, Duthoy S, Ferris S, Hennequin C, Dujon B (2003) The complete mitochondrial genome sequence of the pathogenic yeast Candida (Torulopsis) glabrata. FEBS Lett 534:39–48

Kovesdi I, Brough DE, Bruder JT, Wickham TJ (1997) Adenoviral vectors for gene transfer. Curr Opin Biotechnol 8:583–589

Kutay U, Lipowsky G, Izaurralde E, Bischoff FR, Schwarzmaier P, Hartmann E, Gorlich D (1998) Identification of a tRNA-specific nuclear export receptor. Mol Cell 1:359–369

Kuwabara T, Warashina M, Taira K (2000a) Allosterically controllable ribozymes with biosensor functions. Curr Opin Chem Biol 4:669–677

Kuwabara T, Warashina M, Taira K (2000b) Allosterically controllable maxizymes cleave mRNA with high efficiency and specificity. Trends Biotechnol 18:462–468

Kuwabara T, Amontov SV, Warashina M, Ohkawa J, Taira K (1996) Characterization of several kinds of dimer minizyme: simultaneous cleavage at two sites in HIV-1 tat mRNA by dimer minizymes. Nucleic Acids Res 24:2302–2310

Kuwabara T, Warashina M, Nakayama A, Ohkawa J, Taira K (1999) tRNAVal-heterodimeric maxizymes with high potential as geneinactivating agents: simultaneous cleavage at two sites in HIV-1 Tat mRNA in cultured cells. Proc Natl Acad Sci USA 96:1886–1891

Kuwabara T, Hsieh J, Nakashima K, Taira K, Gage FH (2004) A small modulatory dsRNA specifies the fate of adult neural stem cells. Cell 116:779–793

Kuwabara T, Warashina M, Tanabe T, Tani K, Asano S, Taira K (1997) Comparison of the specificities and catalytic activities of hammerhead ribozymes and DNA enzymes with respect to the cleavage of BCR-ABL chimeric L6 (b2a2) mRNA. Nucleic Acids Res 25: 3074–3081

Kuwabara T, Warashina M, Orita M, Koseki S, Ohkawa J, Taira K (1998a) Formation of a catalytically active dimer by tRNA(Val)-driven short ribozymes. Nat Biotechnol 16: 961–965

Kuwabara T, Warashina M, Tanabe T, Tani K, Asano S, Taira K (1998b) A novel allosterically trans-activated ribozyme, the maxizyme, with exceptional specificity in vitro and in vivo. Mol Cell 2:617–627

Kuwabara T, Warashina M, Koseki S, Sano M, Ohkawa J, Nakayama K, Taira K (2001a) Significantly higher activity of a cytoplasmic hammerhead ribozyme than a corresponding nuclear counterpart: engineered tRNAs with an extended 3′ end can be exported efficiently and specifically to the cytoplasm in mammalian cells. Nucleic Acids Res 29: 2780–2788

Kuwabara T, Warashina M, Sano M, Tang H, Wong-Staal F, Munekata E, Taira K (2001b) Recognition of engineered tRNAs with an extended 3′ end by Exportin-t (Xpo-t) and transport of tRNA-attached ribozymes to the cytoplasm in somatic cells. Biomacromolecules 2:1229–1242

Lee CG, Hurwitz J (1993) Human RNA helicase A is homologous to the maleless protein of Drosophila. J Biol Chem 268:16822–16830

Lee NS, Rossi JJ (2004) Control of HIV-1 replication by RNA interference. Virus Res 102: 53–58

Lee NS, Dohjima T, Bauer G, Li H, Li MJ, Ehsani A, Salvaterra P, Rossi J (2002a) Expression of small interfering RNAs targeted against HIV-1 rev transcripts in human cells. Nat Biotechnol 20:500–505

Lee Y, Jeon K, Lee JT, Kim S, Kim VN (2002b) MicroRNA maturation: stepwise processing and subcellular localization. Embo J 21:4663–4670

Li J, Tang H, Mullen TM, Westberg C, Reddy TR, Rose DW, Wong-Staal F (1999) A role for RNA helicase A in post-transcriptional regulation of HIV type 1. Proc Natl Acad Sci USA 96:709–714

Lipowsky G, Bischoff FR, Izaurralde E, Kutay U, Schafer S, Gross HJ, Beier H, Gorlich D (1999) Coordination of tRNA nuclear export with processing of tRNA. Rna 5:539–549

Lund E, Dahlberg JE (1998) Proofreading and aminoacylation of tRNAs before export from the nucleus. Science 282:2082–2085

McCall MJ, Hendry P, Jennings PA (1992) Minimal sequence requirements for ribozyme activity. Proc Natl Acad Sci U S A 89:5710–5714

McManus MT, Sharp PA (2002) Gene silencing in mammals by small interfering RNAs. Nat Rev Genet 3:737–747

Miller AD (2004) Nonviral liposomes. Methods Mol Med 90:107–137

Milner N, Mir KU, Southern EM (1997) Selecting effective antisense reagents on combinatorial oligonucleotide arrays. Nat Biotechnol 15:537–541

Mittal V (2004) Improving the efficiency of RNA interference in mammals. Nat Rev Genet 5:355–365

Miyagishi M, Taira K (2002) U6 promoter-driven siRNAs with four uridine 3′ overhangs efficiently suppress targeted gene expression in mammalian cells. Nat Biotechnol 20: 497–500

Miyagishi M, Taira K (2003) Strategies for generation of an siRNA expression library directed against the human genome. Oligonucleotides 13:325–333

Miyagishi M, Matsumoto S, Taira K (2004a) Generation of an shRNAi expression library against the whole human transcripts. Virus Res 102:117–124

Miyagishi M, Sumimoto H, Miyoshi H, Kawakami Y, Taira K (2004b) Optimization of an siRNA-expression system with an improved hairpin and its significant suppressive effects in mammalian cells. J Gene Med 6:715–723

Morris KV, Chan SW, Jacobsen SE, Looney DJ (2004) Small interfering RNA-induced transcriptional gene silencing in human cells. Science 305:1289–1292

Mountain A (2000) Gene therapy: the first decade. Trends Biotechnol 18:119–128

Muller AJ, Young JC, Pendergast AM, Pondel M, Landau NR, Littman DR, Witte ON (1991) BCR first exon sequences specifically activate the BCR/ABL tyrosine kinase oncogene of Philadelphia chromosome-positive human leukemias. Mol Cell Biol 11:1785–1792

Naldini L (1998) Lentiviruses as gene transfer agents for delivery to non-dividing cells. Curr Opin Biotechnol 9:457–463

Nawrot B, Antoszczyk S, Maszewska M, Kuwabara T, Warashina M, Taira K, Stec WJ (2003) Efficient inhibition of beta-secretase gene expression in HEK293 cells by tRNAVal-driven and CTE-helicase associated hammerhead ribozymes. Eur J Biochem 270:3962–3970

Nelson DL, Suyama E, Kawasaki H, Taira K (2003) Use of random ribozyme libraries for the rapid screening of apoptosis- and metastasis-related genes. Targets 2:191–200

Novina CD, Sharp PA (2004) The RNAi revolution. Nature 430:161–164

Nowell PC, Hungerford DA (1960) A minute chromosome in human chronic granulocytic leukemia. Science 132:1497–1499

Ohkawa J, Taira K (2000) Control of the functional activity of an antisense RNA by a tetracycline-responsive derivative of the human U6 snRNA promoter. Hum Gene Ther 11:577–585

Ohkawa J, Yuyama N, Takebe Y, Nishikawa S, Taira K (1993) Importance of independence in ribozyme reactions: kinetic behavior of trimmed and of simply connected multiple ribozymes with potential activity against human immunodeficiency virus. Proc Natl Acad Sci USA 90:11302–11306

Onuki R, Nagasaki A, Kawasaki H, Baba T, Uyeda TQ, Taira K (2002) Confirmation by FRET in individual living cells of the absence of significant amyloid beta -mediated caspase 8 activation. Proc Natl Acad Sci USA 99:14716–14721

Onuki R, Bando Y, Suyama E, Katayama T, Kawasaki H, Baba T, Tohyama M, Taira K (2004) An RNA-dependent protein kinase is involved in tunicamycin-induced apoptosis and Alzheimer's disease. Embo J 23:959–968

Oshima K, Kawasaki H, Soda Y, Tani K, Asano S, Taira K (2003) Maxizymes and small hairpin-type RNAs that are driven by a tRNA promoter specifically cleave a chimeric gene associated with leukemia in vitro and in vivo. Cancer Res 63:6809–6814

Pachuk CJ, Yoon K, Moelling K, Coney LR (1994) Selective cleavage of bcr-abl chimeric RNAs by a ribozyme targeted to non-contiguous sequences. Nucleic Acids Res 22: 301–307

Paddison PJ, Caudy AA, Bernstein E, Hannon GJ, Conklin DS (2002) Short hairpin RNAs (shRNAs) induce sequence-specific silencing in mammalian cells. Genes Dev 16:948–958

Paddison PJ, Silva JM, Conklin DS, Schlabach M, Li M, Aruleba S, Balija V, O'Shaughnessy A, Gnoj L, Scobie K, Chang K, Westbrook T, Cleary M, Sachidanandam R, McCombie WR, Elledge SJ, Hannon GJ (2004) A resource for large-scale RNA-interference-based screens in mammals. Nature 428:427–431

Pan WH, Xin P, Bui V, Clawson GA (2003) Rapid identification of efficient target cleavage sites using a hammerhead ribozyme library in an iterative manner. Mol Ther 7:129–139

Patzel V, Sczakiel G (1998) Theoretical design of antisense RNA structures sub-stantially improves annealing kinetics and efficacy in human cells. Nat Biotechnol 16:64–68

Paul CP, Good PD, Winer I, Engelke DR (2002) Effective expression of small interfering RNA in human cells. Nat Biotechnol 20:505–508

Rabinowitz JE, Samulski J (1998) Adeno-associated virus expression systems for gene transfer. Curr Opin Biotechnol 9:470–475

Rossi JJ (1995) Controlled, targeted, intracellular expression of ribozymes: progress and problems. Trends Biotechnol 13:301–306

Rossi JJ, Sarver N (1990) RNA enzymes (ribozymes) as antiviral therapeutic agents. Trends Biotechnol 8:179–183

Sarver N, Cantin EM, Chang PS, Zaia JA, Ladne PA, Stephens DA, Rossi JJ (1990) Ribozymes as potential anti-HIV-1 therapeutic agents. Science 247:1222–1225

Scherr M, Rossi JJ (1998) Rapid determination and quantitation of the accessibility to native RNAs by antisense oligodeoxynucleotides in murine cell extracts. Nucleic Acids Res 26:5079–5085

Scherr M, Reed M, Huang CF, Riggs AD, Rossi JJ (2000) Oligonucleotide scanning of native mRNAs in extracts predicts intracellular ribozyme efficiency: ribozyme-mediated reduction of the murine DNA methyltransferase. Mol Ther 2:26–38

Scherr M, Battmer K, Winkler T, Heidenreich O, Ganser A, Eder M (2003) Specific inhibition of bcr-abl gene expression by small interfering RNA. Blood 101:1566–1569

Scherr M, LeBon J, Castanotto D, Cunliffe HE, Meltzer PS, Ganser A, Riggs AD, Rossi JJ (2001) Detection of antisense and ribozyme accessible sites on native mRNAs: application to NCOA3 mRNA. Mol Ther 4:454–460

Schmitt C, von Kobbe C, Bachi A, Pante N, Rodrigues JP, Boscheron C, Rigaut G, Wilm M, Seraphin B, Carmo-Fonseca M, Izaurralde E (1999) Dbp5, a DEAD-box protein required for mRNA export, is recruited to the cytoplasmic fibrils of nuclear pore complex via a conserved interaction with CAN/Nup159p. Embo J 18:4332–4347

Schwarz DS, Hutvagner G, Du T, Xu Z, Aronin N, Zamore PD (2003) Asymmetry in the assembly of the RNAi enzyme complex. Cell 115:199–208

Shimayama T, Nishikawa S, Taira K (1995) Generality of the NUX rule: kinetic analysis of the results of systematic mutations in the trinucleotide at the cleavage site of hammerhead ribozymes. Biochemistry 34:3649–3654

Shiota M, Sano M, Miyagishi M, Taira K (2004) Ribozymes: applications to functional analysis and gene discovery. J Biochem (Tokyo) 136:133–147

Shtivelman E, Lifshitz B, Gale RP, Roe BA, Canaani E (1986) Alternative splicing of RNAs transcribed from the human abl gene and from the bcr-abl fused gene. Cell 47: 277–284

Sioud M, ed. (2004) Methods in Molecular Biology: Ribozymes and siRNA Protocols (Totowa, NJ, Humana Press)

Sledz CA, Holko M, de Veer MJ, Silverman RH, Williams BR (2003) Activation of the interferon system by short-interfering RNAs. Nat Cell Biol 5:834–839

Sui G, Soohoo C, Affar el B, Gay F, Shi Y, Forrester WC (2002) A DNA vector-based RNAi technology to suppress gene expression in mammalian cells. Proc Natl Acad Sci USA 99:5515–5520

Suyama E, Kawasaki H, Kasaoka T, Taira K (2003a) Identification of genes responsible for cell migration by a library of randomized ribozymes. Cancer Res 63:119–124
Suyama E, Kawasaki H, Nakajima M, Taira K (2003b) Identification of genes involved in cell invasion by using a library of randomized hybrid ribozymes. Proc Natl Acad Sci USA 100:5616–5621
Suyama E, Wadhwa R, Kaur K, Miyagishi M, Kaul SC, Kawasaki H, Taira K (2004) Identification of metastasis-related genes in a mouse model using a library of randomized ribozymes. J Biol Chem 279:38083–38086
Suzumura K, Takagi Y, Orita M, Taira K (2004) NMR-based reappraisal of the coordination of a metal ion at the pro-Rp oxygen of the A9/G10.1 site in a hammerhead ribozyme. J Am Chem Soc 126:15504–15511
Taira K, Miyagishi M (2001) siRNA expression system and method for producing functional gene knock-down cell using the system. Japanese Patent Application Heisei-13-363385 (PCT/JP02/12447)
Taira K, Miyagishi M (2003) Method and apparatus for predicting RNAi effect of siRNA. Japanese Patent Application PCT/JP2004/014758
Taira K, Miyagishi M, Akashi H (2004) Escape from the interferon response associated with RNA interference using vectors that encode long modified hairpin-RNA. Japanease Patent Application 2003–414591
Taira K, Nakagawa K, Nishikawa S, Furukawa K (1991) Construction of a novel RNA-transcript-trimming plasmid which can be used both in vitro in place of run-off and (G)-free transcriptions and in vivo as multi-sequences transcription vectors. Nucleic Acids Res 19:5125–5130
Taira K, Oda M, Shinshi H, Maeda H, Furukawa K (1990) Construction of a novel artificial-ribozyme-releasing plasmid. Protein Eng 3:733–737
Taira K, Warashina M, Kuwabara T, Kawasaki H (1999) Functional hybrid molecules with sliding ability. Japanese Patent Application
Takagi Y, Taira K (2002) Detection of a proton-transfer process by kinetic solvent isotope effects in NH(4)(+)-mediated reactions catalyzed by a hammerhead ribozyme. J Am Chem Soc 124:3850–3852
Takagi Y, Inoue A, Taira K (2004) Analysis on a cooperative pathway involving multiple cations in hammerhead reactions. J Am Chem Soc 126:12856–12864
Tanabe T, Kuwabara T, Warashina M, Tani K, Taira K, Asano S (2000a) Oncogene inactivation in a mouse model. Nature 406:473–474
Tanabe T, Takata I, Kuwabara T, Warashina M, Kawasaki H, Tani K, Ohta S, Asano S, Taira K (2000b) Maxizymes, novel allosterically controllable ribozymes, can be designed to cleave various substrates. Biomacromolecules 1:108–117
Tanaka Y, Kasai Y, Mochizuki S, Wakisaka A, Morita EH, Kojima C, Toyozawa A, Kondo Y, Taki M, Takagi Y, Inoue A, Yamasaki K, Taira K (2004) Nature of the chemical bond formed with the structural metal ion at the A9/G10.1 motif derived from hammerhead ribozymes. J Am Chem Soc 126:744–752
Tang H, Wong-Staal F (2000) Specific interaction between RNA helicase A and Tap, two cellular proteins that bind to the constitutive transport element of type D retrovirus. J Biol Chem 275:32694–32700
Tang H, Gaietta GM, Fischer WH, Ellisman MH, Wong-Staal F (1997) A cellular cofactor for the constitutive transport element of type D retrovirus. Science 276:1412–1415
Tiscornia G, Tergaonkar V, Galimi F, Verma IM (2004) CRE recombinase-inducible RNA interference mediated by lentiviral vectors. Proc Natl Acad Sci USA 101:7347–7351
Tuschl T (2002) Expanding small RNA interference. Nat Biotechnol 20:446–448
Tuschl T, Eckstein F (1993) Hammerhead ribozymes: importance of stem-loop II for activity. Proc Natl Acad Sci USA 90:6991–6994
Usman N, Blatt LM (2000) Nuclease-resistant synthetic ribozymes: developing a new class of therapeutics. J Clin Invest 106:1197–1202

van de Wetering M, Oving I, Muncan V, Pon Fong MT, Brantjes H, van Leenen D, Holstege FC, Brummelkamp TR, Agami R, Clevers H (2003) Specific inhibition of gene expression using a stably integrated, inducible small-interfering-RNA vector. EMBO Rep 4: 609–615

Wadhwa R, Kaul SC, Miyagishi M, Taira K (2004) Vectors for RNA interference. Curr Opin Mol Ther 6:367–372

Wadhwa R, Ando H, Kawasaki H, Taira K, Kaul SC (2003) Targeting mortalin using conventional and RNA-helicase-coupled hammerhead ribozymes. EMBO Rep 4:595–601

Wagner JD, Jankowsky E, Company M, Pyle AM, Abelson JN (1998) The DEAH-box protein PRP22 is an ATPase that mediates ATP-dependent mRNA release from the spliceosome and unwinds RNA duplexes. Embo J 17:2926–2937

Warashina M, Kuwabara T, Nakamatsu Y, Takagi Y, Kato Y, Taira K (2004) Analysis of the conserved P9-G10.1 metal-binding motif in hammerhead ribozymes with an extra nucleotide inserted between A9 and G10.1 residues. J Am Chem Soc 126:12291–12297

Warashina M, Kuwabara T, Taira K (2000a) Working at the cutting edge: the creation of allosteric ribozymes. Structure Fold Des 8:R207–212

Warashina M, Takagi Y, Stec WJ, Taira K (2000b) Differences among mechanisms of ribozyme-catalyzed reactions. Curr Opin Biotechnol 11:354–362

Warashina M, Kuwabara T, Kato Y, Sano M, Taira K (2001) RNA-protein hybrid ribozymes that efficiently cleave any mRNA independently of the structure of the target RNA. Proc Natl Acad Sci USA 98:5572–5577

Welch PJ, Marcusson EG, Li QX, Beger C, Kruger M, Zhou C, Leavitt M, Wong-Staal F, Barber JR (2000) Identification and validation of a gene involved in anchorage-independent cell growth control using a library of randomized hairpin ribozymes. Genomics 66:274–283

Westberg C, Yang JP, Tang H, Reddy TR, Wong-Staal F (2000) A novel shuttle protein binds to RNA helicase A and activates the retroviral constitutive transport element. J Biol Chem 275:21396–21401

Wu N, Ataai MM (2000) Production of viral vectors for gene therapy applications. Curr Opin Biotechnol 11:205–208

Yang Y, Xiang Z, Ertl HC, Wilson JM (1995) Upregulation of class I major histocompatibility complex antigens by interferon gamma is necessary for T-cell-mediated elimination of recombinant adenovirus-infected hepatocytes in vivo. Proc Natl Acad Sci USA 92:7257–7261

Yokota T, Sakamoto N, Enomoto N, Tanabe Y, Miyagishi M, Maekawa S, Yi L, Kurosaki M, Taira K, Watanabe M, Mizusawa H (2003) Inhibition of intracellular hepatitis C virus replication by synthetic and vector-derived small interfering RNAs. EMBO Rep 4:602–608

Yoshinari K, Miyagishi M, Taira K (2004) Effects on RNAi of the tight structure, sequence and position of the targeted region. Nucleic Acids Res 32:691–699

Yu JY, DeRuiter SL, Turner DL (2002) RNA interference by expression of short-interfering RNAs and hairpin RNAs in mammalian cells. Proc Natl Acad Sci USA 99:6047–6052

Yu M, Ojwang J, Yamada O, Hampel A, Rapapport J, Looney D, Wong-Staal F (1993) A hairpin ribozyme inhibits expression of diverse strains of human immunodeficiency virus type 1. Proc Natl Acad Sci USA 90:6340–6344

Zhou DM, Taira K (1998) The Hydrolysis of RNA: From Theoretical Calculations to the Hammerhead Ribozyme-Mediated Cleavage of RNA. Chem Rev 98:991–1026

Zuker M (1989) Computer prediction of RNA structure. Methods Enzymol 180:262–288

RNAi-Based Inhibition Specific for Mutant Alleles in Autosomal Dominant Diseases: Sequence-Dependent and -Independent Discrimination of Mutant and Wild-Type Alleles by siRNA

Takanori Yokota

1 Introduction

In autosomal recessive diseases, the protein products of the mutant genes are either not expressed (null mutation) or lose their function. In contrast, in autosomal dominant diseases the protein encoded by the mutant gene usually induces disease due to its toxic property. This "gain of toxic function" of a mutant protein is predicted to cause cell death in autosomal-dominant neurodegenerative diseases characterized by a missense point mutation, such as occurs in familial amyotrophic lateral sclerosis, familial Alzheimer disease, prion-induced disease, familial Parkinson's disease, and familial amyloid neuropathy. For all these familial diseases, a rational approach to therapy is to develop a method to specifically eliminate the aberrant protein by RNA interference (RNAi), even if the various mechanisms of cell death are unknown.

RNA interference is the process of sequence-specific, post-transcriptional gene silencing, initiated by double-stranded RNA (dsRNA). This is a multi-step process that involves the generation of 21- to 23-nt small interfering RNA (siRNA) and results in the degradation of the homologous RNA (Elbashir et al. 2001a). The siRNA is long enough to mediate gene-specific suppression, but short enough to evade the adverse fates of long dsRNA in mammalian cells (Elbashir et al. 2001a), and is expected to be a powerful tool for gene therapy of human diseases. In autosomal dominant disease, a most effective therapeutic approach requires effective reduction or elimination of the aberrant mutant protein while leaving the wild-type protein intact. In this chapter, we describe a method to identify the point mutation in familial amyotrophic lateral sclerosis (ALS) and the expanded CAG repeat in Machado-Joseph disease.

2 Optimal Location of a Mutation for Discriminating Among Point Mutations in Familial ALS

In order to search for the best position for discriminating a point mutation, we examined the suppression effects of several siRNAs targeted to G93A SOD1, the mutation

Department of Neurology and Neurological Science, Graduate School, Tokyo Medical and Dental University, 1-5-45 Yushima, Bunkyo-ku, Tokyo 113-8519, Tokyo

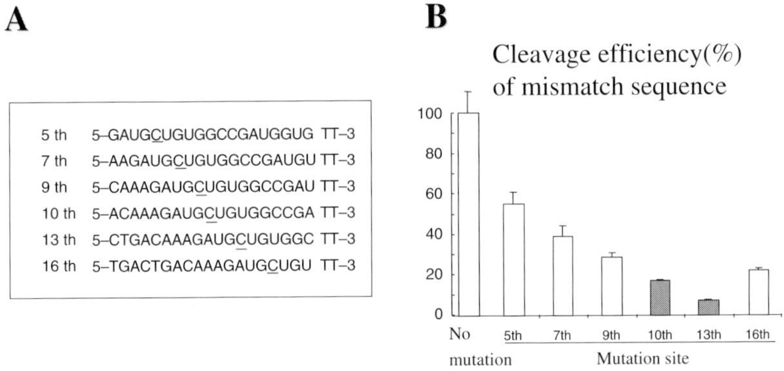

FIG. 1. Effect of a mutation caused by single-nucleotide mismatch in the sense strand on cleavage activity of siRNA. *Underlined* C is the mismatched RNA

causing familial ALS, on the expression of wild-type SOD1 (Fig. 1A). SOD1 expression was monitored by subcloning the coding sequence of wild-type SOD1 cDNA into the 5′ end of the luciferese gene (pGL3-control, Pronega). By changing the mismatch within the region between the fifth and the sixteenth nucleotides from the 5′ end, the mismatch in siRNA at the thirteenth position was found to cause the greatest decrease in the cleavage activity of the siRNA (Fig. 1B).

3 siRNA Specific for G93A SOD1 with Point Mutation in Familial ALS

The siRNAs were demonstrated to cleave the G93A SOD1 mutant, leaving the wild-type intact. The results of two siRNAs targeting G93A SOD1 and corresponding to regions 277–297 (siRNA G93A.1) and 275–294 (siRNA G93A.2) are shown in Fig. 2. Western blot analysis revealed that the siRNA G93A.2 reduced the expression of G93A SOD1 protein by about 90% when expression efficiency was adjusted to that of co-transfected GFP (Fig. 2A). The suppression of G93A SOD1 protein by siRNA G93A.2 increased in a dose-dependent manner when the amount of siRNA was increased from 2.5 to 125 nM (data not shown). The siRNA was able to recognize an alteration of only one nucleotide, because it wild-type SOD1 protein was suppressed to a much lesser extent than G93A SOD1. This was especially the case for siRNAG93A.2, which reduced expression of the wild-type by only 1.8% (Fig. 2A, B). The suppression effect was confirmed by the reduction in GFP fluorescence when siRNA was co-transfected with a GFP-fused SOD1-expressing plasmid (Fig. 2C), using DsRed fluorescence as a control for transfection efficiency (data not shown).

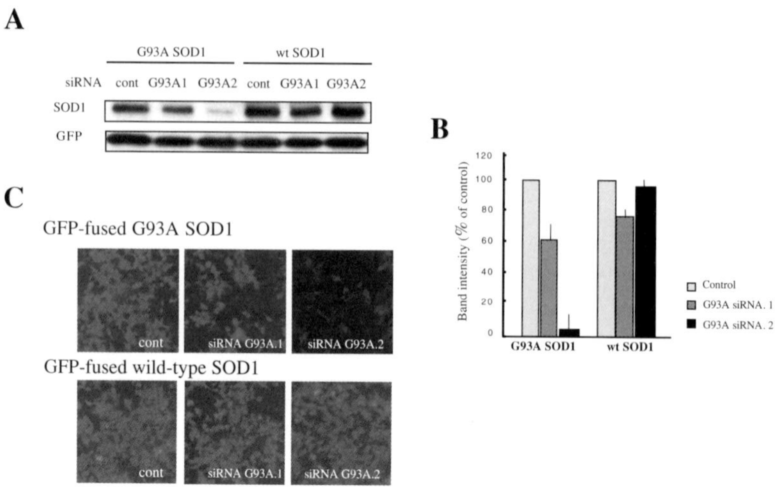

Fig. 2. Effect of siRNA G93A.1 and .2 on G93A and wild-type SOD1 proteins expressed in 293T cells as detected by Western blotting (**A**), and GFP fluorescence (**C**). **B** Percentages of band intensity with siRNAG93A.1 compared to that with each mock transfection. siRNA G93A.2 is more specific for cleaving G93A SOD1 RNA. Values are the mean and SEM. GFP cDNA was fused with G93A SOD1 gene in the expression plasmids (**C**). The expression level of the target protein was adjusted according to the expression level of co-transfected GFP (**B**) and DsRed fluorescence (data not shown). Data obtained at 48 h after transfections

4 siRNA Specific for the Mutant Allele in Machado-Joseph Disease Using a Polymorphorism Related to CAG Repeat Length

There is a C/G polymorphism just after the CAG repeat tract in the Machado-Joseph (MJD)1 gene. Mutant alleles are always (CAG)nC, whereas normal alleles have both (CAG)nG and (CAG)nC at similar frequencies (Fig. 3). Several siRNAs were designed to target the MJD1 RNA sequence, including this C/G polymorphism (data not shown). The siRNA with a mismatch at the eleventh nucleotide from the 5′ end of the sense sequence was best able to discriminate between mutant and wild-type MJD1 RNAs.

The mutant MJD1 gene product (ataxin-3), Q79C, was most effectively suppressed by 25 nM of the MJD siRNA MJD. Western blotting showed a 96% decrease in signal intensity compared with the control (Fig. 4). In contrast, expression of the wild-type ataxin-3, Q22G, was hardly suppressed by the siRNA.

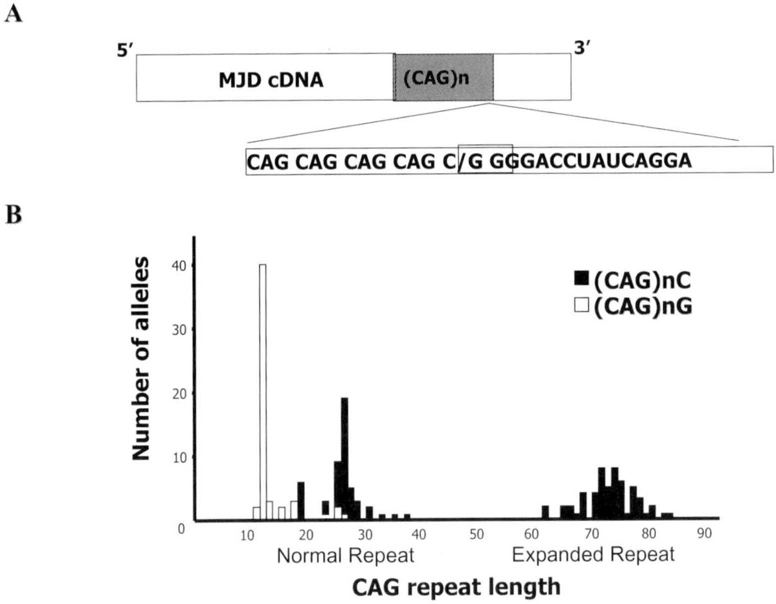

FIG. 3. C/G polymorphorism following the CAG repeat of the MJD1 gene (**A**) and its bias in frequency in mutant and wild-type alleles in 56 Japanese MJD patients (**B**)

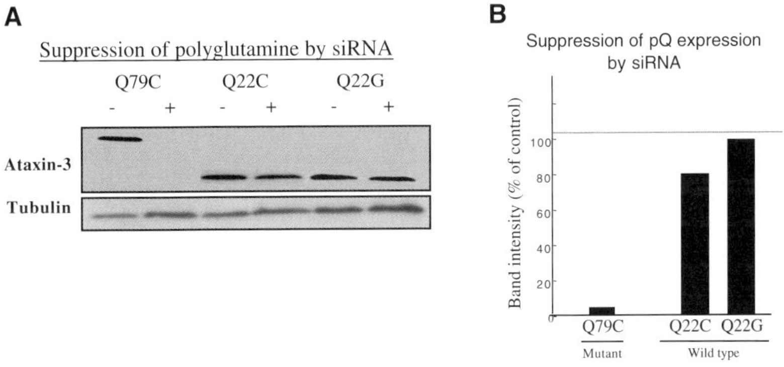

FIG. 4A,B. Specific effect of siRNA on the expression of mutant MJD1. **A** Comparison of the effect of siRNA MJD3 on Q79C, Q22G, and Q22C expression. Western blot analysis of the effect of siRNA MJD3 at 25 nM on Q79C, Q22G, and Q22C expression. The results of siRNA(−) were obtained with siRNAs with randomized sequences. **B** Quantitation of signal intensities. Percentage suppression was determined by comparison with the band intensity of the control. Values are the mean and SEM. Almost no suppression of Q22G expression (5.9%), and mild suppression of Q22C expression (22.5%) were noted using siRNAMJD at 25 nM siRNA, in contrast to robust suppression of Q79C expression (96.0%)

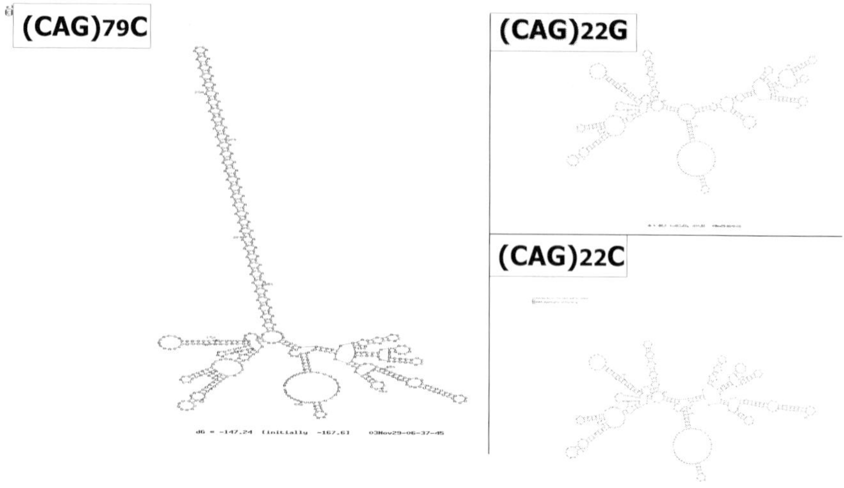

Fig. 5. Secondary structures of (CAG)79C, (CAG)79G, and (CAG)79C mRNA

5 Sequence-Independent Discriminations of Mutant and Wild-Type Alleles by siRNA

Unexpectedly, MJD siRNA decreased the expression of another wild-type allele, Q22C, only slightly, although the target sequence in (CAG)79C and (CAG)22C was the same (Fig. 4, arrowhead). Figure 5 shows the secondary structure of (CAG)79C (CAG)22G and (CAG)22C RNAs. The difference in cleavage activity by the MJD siRNA between (CAG)79C and (CAG)22C might have been due to a difference in the secondary structures of the target RNAs.

6 Discussion

Amyotrophic lateral sclerosis is characterized by the degeneration of lower motor neurons in the spinal cord and brainstem, and degeneration of the descending motor pathway in the corticospinal tracts. Although most cases of ALS are sporadic and have an unknown etiology, 5–10% of ALS cases are familial; of these, approximately 20% are due to mutations in the gene encoding Cu,Zn superoxide dismutase (SOD1) (Rosen et al. 1993). Most of the changes in the SOD1 gene in patients with familial ALS involve a single nucleotide alternation (Radunivic et al. 1996). Recent studies with transgenic mice and cell-culture models of ALS with SOD1 mutations indicated that the mutations induce the disease by the toxic properties of the mutant proteins, and not by a loss of SOD1 activity (Cleveland et al. 1999). Since SOD1 is an important radical scavenger, discrimination of a single nucleotide mismatch is necessary to suppress expression of the mutant allele only, leaving the wild-type protein intact.

Machado-Joseph disease is an autosomal dominant neurodegenerative disorder characterized clinically by cerebellar ataxia, pyramidal and extrapyramidal signs, peripheral neuropathy, and ophthalmoplegia. The number of CAG repeats in the normal MJD1 gene is between 13 and 36, whereas in the defective gene this range is between 62 and 84 (Kawaguchi et al. 1994). The pathogenosis of MJD is also considered to be due to a "gain of toxic function" (Ross 1995; Ikeda et al. 1996). Therefore, an effective and simple gene therapeutic approach for MJD requires reducing the amount of the aberrant mutant protein. Furthermore, it might be necessary to reduce mutant ataxin-3 selectively, leaving the wild-type protein intact, because the wild type MJD1 gene product has an important role in cell survival, such as quality control of processes in the endoplasmic reticulum (Kobayashi et al. 2002) and DNA repair (Wang et al. 2000). We found that the CAG repeat tract in the MJD1 gene is followed by a C or G, and there is extreme bias for this C/G polymorphism between mutant and normal MJD1 alleles; mutant alleles have exclusively (CAG)nC, whereas normal alleles have both (CAG)nG and (CAG)nC at similar frequencies (Matsumura et al. 1996; Gaspar et al. 2001) (Fig. 3). We engineered siRNAs to selectively cleave mutant MJD RNA by targeting the sequence including this C/G polymorphism.

Diverse effects of mismatches between siRNA and target sequences have been reported. A central single mutation, namely, a mismatch in siRNA at the tenth and eleventh positions, produces a marked decrease in siRNA cleavage activity (Elbashir et al. 2001b). Mismatches in the 5′ end have a more negligible effect on siRNA cleavage activity than those more centrally located. This bias might be linked to the proposed existence of a "ruler" region in the siRNA that is used by the RISC complex to "measure" the target RNA for cleavage (Holen et al. 2002). Of the six siRNAs examined in our study, those with mismatches at positions 10 and 13 from the 5′ end of the sense sequence were better able to discriminate mutations than those with outside mismatches, especially at the 5′ end. Our results are consistent with those previously reported. siRNAs designed for familial ALS and MJD suppressed mutant allele expression by recognizing the point mutation in SOD1 mRNA and the C/G polymorphism in MJD1 mRNA.

Surprisingly, there were significant differences in siRNA effects on Q22C and Q79C, although the target sequences of their siRNAs were identical. This is the first report of sequence-independent discrimination of mutant and wild-type alleles by siRNA. One possible reason for this difference is that not all RNA sequences are equally accessible to siRNAs: some sequences might be buried within the secondary structure of target RNAs, especially when they are highly folded (Yoshinari et al. 2004). We also observed that the best target site of siRNA for highly folded RNA is almost the same as that for ribozyme, whose cleavage efficiency is greatly influenced by the secondary structure of the target RNA (Yokota et al. 2003). The target site of the C/G polymorphism is just downstream of the CAG repeat, the secondary structure of which, according to computer prediction, is a tight stem form. Although we have no data indicating siRNA MJD3 is more accessible to (CAG)79C than (CAG)22C, a change in the secondary structure of the MJD RNA due to a large difference of the CAG repeat length might affect the efficiency of siRNA MJD3 (Fig. 5). Another possible explanation is that there is a RNA-binding protein which preferentially binds (CAG)22C over (CAG)79C and thus interferes with the access of siRNA MJD3 to (CAG)22C RNA.

As demonstrated above, design of an siRNA specific for the mutant allele responsible for autosomal dominant disease is possible even for single-nucleotide alternations and change in the secondary structure of the mRNA. However, since these specificities are not complete, the selection of a highly effective siRNA sequence is necessary for effective discrimination between mutant and wild-type alleles. When the efficient and safe delivery of siRNA in vivo and the long-term expression of siRNA are achieved, gene therapy for autosomal dominant diseases will no longer be a dream.

References

Cleveland DW (1999) From Charcot to SOD1: mechanisms of selective motor neuron death in ALS. Neuron 24:515–520

Elbashir SM, Harborth J, Lendeckel W, Yalcin A, Weber K, Tuschl T (2001a) Duplexes of 21 nucleotide RNAs mediate RNA interference in cultured mammalian cells. Nature 411:494–498

Elbashir SM, Martinez J, Patkaniowska A, Lendeckel W, Tuschl T (2001b) Functional anatomy of siRNAs for mediating efficient RNAi in Drosophila melanogaster embryo lysate. EMBO J 20:6877–6888

Gaspar C, Lopes-Cendes I, Hayes S (2001) Ancestral origins of the Machado-Joseph disease mutation: a worldwide haplotype study. Am J Hum Genet 68:523–528

Holen T, Amarzguioui M, Wiiger MT, Babaie E, Prydz H (2002) Positional effects of short interfering RNAs targeting the human coagulation trigger tissue factor. Nucleic Acids Res 30:1757–1766

Ikeda H, Yamaguchi M, Satoshi S, Aze Y, Narumiya S, Kakizuka A (1996) Expanded polyglutamine in the Machado-Joseph disease protein induced cell death in vitro and in vivo. Nat Genet 13:196–202

Kawaguchi Y, Okamoto T, Taniwaki M (1994) CAG expansions in a novel gene for Machado-Joseph disease at chromosome 14q32.1. Nat Genet 8:221–228

Kobayashi T, Tanaka K, Inoue K, Kakizuka A (2002) Functional ATPase activity of p97/valosin-containing protein (VCP) is required for the quality control of endoplasmic reticulum in neuronally differentiated mammalian PC12 cells. J Biol Chem 277: 47358–47365

Matsumura R, Takayanagi T, Murata K, Futamura N, Hirano M, Ueno S (1996) Relationship of (CAG)nC configuration to repeat instability of the Machado-Joseph disease gene. Hum Genet 98:643–645

Radunivic T, Leigh PN (1996) Cu/Zn superoxide dismutase gene mutations in amyotrophic lateral sclerosis; correlation between genotype and clinical features. J Neurol Neurosurg Psychistry 61:565–572

Rosen DR, Siddique T, Patterson D, Figlewicz DA, Sapp P, Hentati A, Donaldson D, Goto J, O'Regan JP and Deng H (1993) Mutations in Cu, Zn superoxide dismutase are associated with familial amyotrophic lateral sclerosis. Nature 362:59–62

Ross CA (1995) When more is less: pathogenesis of glutamine repeat neurodegenerative diseases. Neuron 15:493–496

Wang G, Sawai N, Kotliarova S, Kanazawa I, Nukina N (2000) Ataxin-3, the MJD1 gene product, interacts with the two human homologs of yeast DNA repair protein RAD23, HHR23A and HHR23B. Hum Mol Genet 9:1795–1803

Yokota T, Sakamoto N, Enomoto Y, TanabeY, Maekawa, Makoto Miyagishi M, Taira K, Watanabe M, Mizusawa H (2003) Inhibition of intracellular hepatitis C virus replication by synthetic and vector-derived small interfering RNAs. EMBO Rep 4:602–608

Yoshinari K, Miyagishi M, Taira K (2004) Effect on RNAi of the tight structure, sequence and position of the targeted region. Nucleic Acids Res 32:691–699

In Vivo RNA Interference: Another Tool in the Box?

Silvère Petit, Emmanuel Valentin, and Kader Thiam

1 Introduction

In classical forward genetics, genes were defined by the resulting phenotype. Later, analysis of individual genes at the molecular level became possible. The develop-ment of high-throughput methods for the sequencing and annotation of cDNA and whole genomes has led to the expansion of reverse genetics, in which access to gene sequences makes it possible to elucidate their biological function. Although this approach has greatly extended our knowledge of simple organisms (Hannon 2004), it is of limited use for investigating the more complex mammalian biological systems.

The generation of hypomorphic mutants using mouse knockout technology by homologous recombination provides a powerful means of elucidating gene function in vivo. To date, published knockouts exist for about 10% of mouse genes. These knockouts often display a complicated phenotype as the gene is inactivated at the single-cell stage. Better spatiotemporal control of knockouts has overcome this major limitation of in vivo gene inactivation.

The recent discovery of RNA interference (RNAi) in cultured mammalian cells (Hammond 2001) and the demonstration that RNAi-mediated gene silencing can occur in rodents have led to new opportunities in mouse functional genetics (Lewis 2002).

This chapter reviews the recent findings of in vivo RNAi and focuses on the advantages and limitations of the different approaches. New methods fulfilling major requirements for successful use of RNAi in vivo are also discussed. These methods combine the lessons learned from in vitro and in vivo studies that used RNAi-mediated gene knockdown and knowledge of the advantages and limitations of transgenic models.

RNAi is a sequence-specific post-transcriptional gene silencing (PTGS) mechanism. It is triggered by double-stranded RNA (dsRNA), an important signaling molecule in eukaryotic cells in that it breaks down mRNAs that are homologous in sequence to itself. This elegant defense system evolved to eliminate non-self RNA molecules, such as cytoplasmic replicating RNA viruses, transcripts originating from transgenes, transposable elements, and viruses randomly integrated into the host genome (Hannon 2004; Mello 2004).

GenOway, Immeuble Chateaubriand, 181 avenue, Jean Jaures, 69362 Lyon, Cedex 07, France

The mechanism for RNA interference is not completely known but has been shown to involve the cleavage activity of the ribonuclease III (RNase III)-type protein Dicer and the formation of the RISC complex (RNA-induced silencing complex) (Elbashir 2001; Mello 2004).

Mammalian cells were, until recently, not amenable to RNAi since the use of in vitro transcribed, long dsRNA (>30 bp) led to a sequence-unspecific response. It has been reported that chemically synthesized duplexes of 21-nucleotide (nt) RNAs with 2-nt 3′ overhangs, known as siRNA (small interfering RNA), specifically interferes with gene expression and bypasses the sequence-independent response of mammalian cells to long dsRNAs (Tuschl 2001; Elbashir 2001). Recently, it was demonstrated that RNAi-mediated gene silencing can also be obtained in cultured mammalian cells by endogenous expression of short hairpin RNA (shRNA) harboring a fold-back stem-loop structure (Paddison 2002; Miyagishi 2002a,b, 2004; Brummelkamp 2002).

Cells activate a non-specific innate immune response (i.e., interferon-mediated defense) when exposed to dsRNA longer than 30 bp, which prohibits the application of longer dsRNA. For this reason, si/shRNA-mediated gene silencing is very useful in somatic mammalian cells (Harborth 2001; Tuschl 2001; Elbashir 2001).

Among the others technologies available for in vitro functional genomics (ribozyme technique or antisense techniques), RNAi-mediated gene silencing has been demonstrated to be the technology of choice. Its high gene-knockdown efficiency and high specificity make siRNA attractive, as these characteristics are difficult to assess in other in vitro methods. However, both in vitro and in vivo experiments require caution, as it is uncertain whether long-term expression of si/shRNA may trigger side effects, including the induction of off-target knockdown silencing by cross-hybridization of si/shRNA partially homologous to others transcripts (Hannon 2004). In this case, the observed phenotype may be difficult to interpret. Expression of long-term si/shRNA could also have negative effects by competing with endogenously expressed miRNA. Another major concern with in vivo RNAi-mediated gene silencing is the possibility of activating the interferon defense system (Sledz 2003; Persengiev 2004; Pebernard 2004). Given the major concern about the specificity of RNAi-mediated silencing, working at the lowest possible concentration of si/shRNA has become one of the major rules for effective RNAi assays (Hannon 2004). All of the limitations associated with in vitro siRNA experiments might also apply to mouse experiments.

The availability of well-characterized and validated physical and chemical transfection methods has made the delivery of siRNA into cell-culture systems quite effective. In contrast, the proper delivery of siRNA into the more much more complex mammalian models has been and remains a subject of intense investigation. Table 1 presents some examples from the published studies discussed below.

2 First In Vivo RNAi Proof-of-Concept Experiment: Inoculation of siRNA

The first observations of efficient in vivo RNAi knockdown were reported in 2002 by several groups. Repeated tail-vein or peritoneal injections of very large amounts of naked siRNA molecules (up to 40 μg) or plasmid expressing shRNA targeting viral

TABLE 1. Examples of published RNA; transgenic models.

Delivery or expression methods	Targeted gene	In vivo/in vitro validated	Reference
Inoculation of siRNA			
Tail vein injection of siRNA molecules	EGFP	In vivo	Lewis et al 2002
Tail vein injection of siRNA and shRNA expression vectors	Luciferase, HBV	In vivo	MacCaffrey et al 2002
Tail vein injection	GFP, TNF-α	In vivo	Sorensen et al 2002
Tail vein injection of siRNA	HBV	In vivo	Klein et al 2003
DNA pronucleus microinjection			
Oocytes pronuclear microinjection with shRNA expression vector	mos	In vivo	Svoboda et al. 2001
Oocytes pronuclear microinjection	EGFP	In vivo	Hasuwa et al 2002
Oocytes pronuclear microinjection	mos	In vivo	Stein et al. 2002, 2003
Viral delivery systems			
Lentiviral shRNA expression vector	p53, CD8a, CD25	In vivo	Rubinson et al 2003
Retroviral shRNA expression vector	p53	In vivo	Hemann et al 2003
ES cell stable transfe**ctant**			
Electroporation of ES cells with shRNA expression vector	Rasa I	In vivo	Kunath et al 2003
Electroporation of ES cells with shRNA expression vector	Neil I	In vivo	Carmell et al 2003
Conditional expression system			
PolII shRNA expression vector	GFP, β-glucoronidase	in vitro	Xia et al 2002
PolII shRNA expression vector	β-gal, ski	in vitro	Shinagawa et al 2003
shRNA expression vector using tetracycline system	β-catenin	in vitro	Van de Wetering et al 2003
shRNA expression vector using tetracycline system	Dnmt1	in vitro	Matsukura et al 2003
shRNA expression vector using tetracycline system	p110ßa, p110ß	in vitro	Czauderna et al 2003
shRNA expression vector using Cre-LoxP system	GFP	in vitro	Kasim et al 2003
shRNA expression vector using Cre-LoxP system	Fgfr2, Survivin	in vitro	Coumoul et al 2004
shRNA expression vector using Cre-LoxP system	GFP, lacZ, p53	in vivo	Fristch et al 2004
Lentiviral shRNA expression vector using Cre-LoxP system	p53, Npm, Dnmt1, CD8	in vivo	Ventura et al 2004
Lentiviral shRNA expression vector using Cre-LoxP system	GFP, p53, p65	in vitro	Tiscornia et al 2004

antigens or viruses, reporter genes, or endogenous genes led to a potent and specific reduction in target gene expression (McCaffrey 2002, 2003; Song 2003; Klein 2003; Sorensen 2003; Lewis 2002). The level of the knockdown effects reported in these studies was in the range of 70%–80%.

Even if this type of strategy might be promising to develop innovative siRNA-based therapeutics, it is greatly impaired by several major limitations. First, the gene silencing effect is only short-term, possibly due to the rapid clearance of naked siRNA molecules or shRNA-expressing plasmid in vivo. Second, only repeated or systemic inoculations lead to prolonged knockdown effects. Third, the knockdown is observed in a limited spectrum of organs due to the delivery system used (McCaffrey 2002; Lewis 2003). Finally, the immunological consequences of repeated inoculation of siRNA molecules or shRNA-expressing plasmids have often not been investigated. While several pioneering studies paved the way for in vivo RNAi-mediated knockdown, induction of the interferon defense pathway was observed (Klein 2003).

3 Transgenic: Random Integration of shRNA-Expressing Constructs

One obvious way of inducing more prolonged knockdown effects and of overcoming the IFN response is to transgenically supply shRNA-expressing constructs into the genome. Most of these constructs are made up of constitutive H1 or U6 RNA polymerase III promoters driving the expression of particular shRNA sequences.

Three main transgenic RNAi strategies, leading to the random integration of shRNA-expressing constructs into the genome, have been tested: oocytes pronucleus microinjection, recombinant viral delivery methods, and embryonic stem (ES) cell approaches.

3.1 Random Integration Approach: DNA Pronucleus Microinjection

Microinjection of transgenic constructs into fertilized oocytes is the method of choice for the rapid and cost-effective generation of transgenic rodents. It is not surprising that this approach was used for the first study of transgenic gene silencing through RNAi in mice and rats. Hasuwa and co-workers demonstrated that transgenically supplied siRNA induced strong inhibition of the EGFP fluorescent reporter gene in mice and rats (Hasuwa 2002). Immunoblot quantification indicated that EGFP level of expression was reduced by about 80% in day 10.5 whole embryos. This strong reduction in fluorescence was further confirmed in newborn mice in a large panel of organs, including brain, heart, liver, pancreas, kidney and skin, suggesting that the silencing effect was widespread throughout the body. Moreover, the knockdown phenotype observed in the founder's embryos or adult mice was efficiently transmitted to the germline. Finally, a similar gene-silencing effect was observed after microinjection of EGFP-expressing rat oocytes. Germline transmission of the gene-knockdown effect was also reported. This important proof-of-concept study indicated that transgenic RNAi could constitute a novel method to create gene-knockdown animals. Furthermore, this method could be applied to other species, such as rabbits, in which DNA pronuclear microinjection has been proven to be efficient. Nevertheless, several groups failed in reproducing these data.

The power but also the limitations of random integration of transgenes into the genome are now well understood and characterized since the first transgenic mice were generated two decades ago. The application of this technique to generate transgenic shRNA-expressing animals presents intrinsic limitations similar to those that exist for the generation of "classical" transgenic animals. First, the number of integrated shRNA-expressing constructs cannot be pre-selected. In the case of integration of a large number of copies, the F1-generation phenotype may be highly variable, due to segregation of the mutant alleles followed by alteration of the final level of expressed shRNA molecules. The well-known positional variegation effect, namely the integration of the shRNA-expressing construct into chromatin regions with different transcriptional activities, can also lead to great variation in the level of expression of the shRNA of interest. These important drawbacks make it necessary to study a large number of founders to validate the siRNA effect and/or target specificity. Another consequence of the random integration of the shRNA-expressing construct is that comparative studies between different shRNA designs cannot be performed. In order to perform significant in vivo analysis, several independent transgenic lines (3–5) must be analyzed in order to overcome the positional variegation effect. The limits of these technologies have restricted the use of DNA pronuclear microinjection for generating RNAi transgenic rodents. Finally, the global efficiency of pronucleus microinjection is low, with not more than 10% of reimplanted oocytes leading to viable transgenic animals.

3.2 Random Integration Approach: Recombinant Viral Delivery Systems

Lentiviral vectors enable the generation of transgenic animals by in vitro transduction of fertilized eggs at different preimplantation stages (Pfeifer 2001). This method has been successfully applied to transduce shRNA-expressing construct in rodents (Tiscornia 2003). After the transduction of fertilized eggs from GFP-positive transgenic mice with lentivirus expressing shRNA targeted against GFP, markedly reduced fluorescence was observed in blastocysts. Pups from F1 progeny also had a sharp decrease in fluorescence, indicating germline transmission of the silencing effect. Lentivirus expressing shRNA was also used to infect mouse ES cells and single-cell embryos (Rubinson 2003). When the lentivirus randomly integrated in the mouse genome expressed shRNA targeted against endogenous genes, nearly 90% knockdown was observed in adult mice.

Retroviral infection was also used to transduce hematopoietic stem cells with shRNA directed against the *p53* tumor suppressor gene (Hemann 2003). The injection of these genetically modified stem cells into lethally irradiated recipient mice led to the generation of tumors with varying degrees of aggressiveness. Interestingly, the severity of the observed disease correlated well with the ability of each shRNA knockdown construct to down-regulate *p53* expression (Hemann 2003). This ability to create graded phenotypes holds great promise for studying complex systems, such as gene-dosage-related questions.

The lentivirus approach could potentially be applied to transduce ES and stem cells, embryos, and organs. This would enable the generation of transgenic mice in a variety of genetic backgrounds, including inbred lines, in which DNA microinjection has a low efficiency. Ultimately, gene knockdown models could be generated in species in which classical genome engineering technology is inefficient, including important model species such as rats and non-human primates. However, the lentiviral-associated delivery system also presents important limitations. As with DNA pronuclear microinjection, the number of integrated copies and the site of insertion cannot be predicted. The multiplicity of infection (MOI), which is a critical factor for the successful generation of transgenic animals, must be carefully adjusted in shRNA-expressing constructs. In fact, high MOI will increase the probability of obtaining transgenic animals but will also lead to a high number of integrated copies. Random integration of the shRNA-expressing lentivirus induces high phenotype variability at the F1 generation. Furthermore, mosaicism due to transcriptional shut-off mechanism can be observed. High numbers of integrated viral constructs at different loci can also lead to artifacts in the observed phenotypes, due to insertion mutagenesis. Finally, the consequences of lentiviral vector infection still needs to be investigated.

3.3 Random Integration Approach: ES Cell Stable Transfectant

Another illustration that RNAi can be used to create germline transgenic mice came from the work of Carmell and co-workers (Carmell 2003). Three different shRNA constructs specific to different endogenous genes were first microinjected into fertilized eggs. No significant gene-knockdown effect was observed in the transgenic lines generated. In contrast, 80% gene knockdown was observed in stable integrants after ES cell transfection with the shRNA-expressing constructs targeted toward *Neil1*, a gene involved in DNA repair following ionizing radiation. Two independent ES cell clones were injected into blastocysts, and germline transmission of the shRNA-expressing construct was observed in about half the F1 progeny of both generated transgenic cell lines—Most importantly, a *Neil1* gene-silencing effect similar to the one detected in the ES stable transfectants was observed in the liver of the transgenic mice. This report corroborates our internal data presented at several meetings (Expediting Target Identification and Validation through RNAi, London Nov 17–18, 2003; Second international RNAi 2004 Boston Meeting, May 4–7 2004; In vivo and transgenic models conference, Berlin, Nov 16–17, 2004).

Another study reported the generation of stable ES cell lines by electroporation of a construct expressing an shRNA targeting the gene *RasGAP* under the control of an H1 RNA pol III promoter (Kunath 2003). Different ES cell lines, exhibiting different levels of *RasGAP* inhibition, were used to construct completely ES-cell-derived embryos generated by the tetraploid aggregation method. The severity of the observed phenotypes correlated roughly with the level of *RasGAP* inhibition. Interestingly, this study suggested that different levels of shRNA-mediated gene silencing may be useful in generating hypomorphic allelic series for dissecting gene function.

Moreover, several interesting features are associated with the ES-cell electroporation method. When the target gene is expressed in ES cells, it is possible to monitor the efficiency of the knockdown effect of shRNA constructs. ES cell lines can then be pre-selected, based on either the knockdown effect or the copy number of shRNA constructs, and a high number of founders, with controlled shRNA copy number, can be generated. Nevertheless, it is important to highlight that gene expression into undifferentiated ES cell is often difficult to study, and that differentiation of ES cells into predefined cell fates requires painstaking experimental set-ups.

Furthermore, these reports as well as our unpublished data suggested that one of the key parameters for successful in vivo RNAi effects seems to be to limit the number of integrated copies of shRNA-expressing constructs. In fact, the integration of a high copy number is often associated with loss of RNAi specificity, off-target gene inhibition, and interference with embryo development. Interestingly, the deleterious consequences of shRNA positional variegation effect on shRNA expression can be overcome by the easy and rapid generation of a large number of founders. Finally, the development time of shRNA transgenic mice through ES cell electroporation is compatible with the use of shRNA for gene validation, as founders are available within 3 months.

Taken together, the proof-of-concept studies reviewed here demonstrate that RNAi can provide an alternative for the study of gene loss-of-function phenotypes in mice. However, and as described above, important drawbacks are associated with random integration of shRNA-expressing constructs into the genome of transgenic animals. In vivo analysis must be performed on several independent lines to overcome the position effect. The analysis of the phenotype of several transgenic lines can be labor-intensive, in contrast to the short development time often associated with RNAi projects.

4 Transgenic Models: Targeted Insertion Approach

4.1 *Targeted Insertion Enables Control of the Site and the Number of Inserted Copies*

The limitations described above are found in any transgenic model based on transgene or shRNA random integration. Gene targeting through homologous recombination allows for modifying the genome in a localized and precise manner. Knock-in gene targeting refers to the replacement of one particular genome sequence with another. This approach constitutes the best solution for overcoming the drawbacks linked to random integration of the shRNA-expressing construct. It is based on the insertion through homologous recombination of the shRNA-expressing construct in a ubiquitously expressed, permissive locus, such as *HPRT* (hypoxanthine phosphoribosyl transferase) and *ROSA26* (Zambrowicz 1997). These "neutral loci", with no enhancer activity, are positioned in constantly active regions of the genome that have an open chromatin structure and allow permanent access to transcription factors.

4.2 *Targeted Insertion Approach: Use of a Neutral Locus to Secure shRNA Expression*

Based on our own experience in generating knock-in models, we selected the *HPRT* locus, which has been extensively studied in the literature. *HPRT* encodes a constitutively expressed housekeeping enzyme involved in nucleotide metabolism. It has been shown that the particular properties of the *HPRT* locus protect exogenous sequences inserted in that region against gene-silencing, positional, or methylation effects. In addition, tissue-specific promoters inserted in the *HPRT* locus conserve their expression properties (Jinnah 1991; Bronson 1996; Chatterjee 1996; Guillot 1999; Engle 1996; Hatada 1999; Cvetkovic 2000; Evans 2000).

Moreover *HPRT*-deficient mice appear phenotypically normal, with the exception of minor changes in brain dopamine levels (Finger 1998; Jinnah 1999; Hyland 2004), and do not display any neurobehavioral defects. Based on these observations, *HPRT* became one of the best candidate loci for targeted insertion into a neutral locus. Targeting the *HPRT* locus enables the rapid development of a knock-in line and secured expression of the shRNA-expressing construct. The main drawback of *HPRT* gene knock-in arises from its localization on the X chromosome.

The development of knock-in models through homologous recombination in ES cells is often thought to be a time-consuming and labor-intensive process. But development time is mainly related to the cloning and mapping of the gene to be targeted in order to guarantee precise targeting. This is why the design of ready-to-use, well-characterized *HPRT*-targeting vectors greatly shortens the development time of knock-in models. We have designed and validated vectors targeting the *HPRT* locus with high efficiency (above 10%). This approach has a development time of 4 months for shRNA knock-in models (Fig. 1).

Inserting shRNA-expressing constructs in the *HPRT* locus offers several major advantages. Firstly, the site of integration as well as the number of copies of the shRNA construct is mastered and can be easily preselected by screening ES cell clones. The level of expression of the shRNA-expressing construct is controlled, as the *HPRT* locus exhibits a high accessibility to transcriptional machinery. This approach enables in vivo analysis on a single transgenic lineage, and transgenic animals can be obtained directly with inbred genetic backgrounds (129/SvPas). Most importantly, comparative analysis of the in vivo efficiency of different shRNA constructs becomes possible. However, it is preferable to analyze the expression of the shRNA in male mice (hemizygous) or homozygous females due to the fact that the *HPRT* locus is located on the X chromosome.

5 Future Development of In Vivo RNAi

The conventional knockdown strategies described above show that a specific gene can be knocked-down with high efficiency in rodent models. One important improvement to these approaches would be the development of inducible and tissue-specific shRNA expression systems.

The stable loss-of-function of genes involved in cell survival may cause compensatory responses or an embryonic lethal phenotype. Moreover, constitutive expression of shRNA molecules may induce non-physiological responses and pleiotropic effects in multiple tissues. Thus, phenotypic analysis can be complex and difficult to interpret. Spatiotemporal control of shRNA expression will make it possible to study essential gene function and analyze target gene under physiopathological conditions.

To date, most shRNA expression vectors have been designed to drive shRNA transcription from U6 or H1 RNA polymerase III promoters. These promoters are suitable for efficient constitutive expression of small hairpin RNA, as they generate shRNA harboring no polyA tail, a feature that would abolish their function. However, no tissue specificity is associated with these ubiquitous promoters.

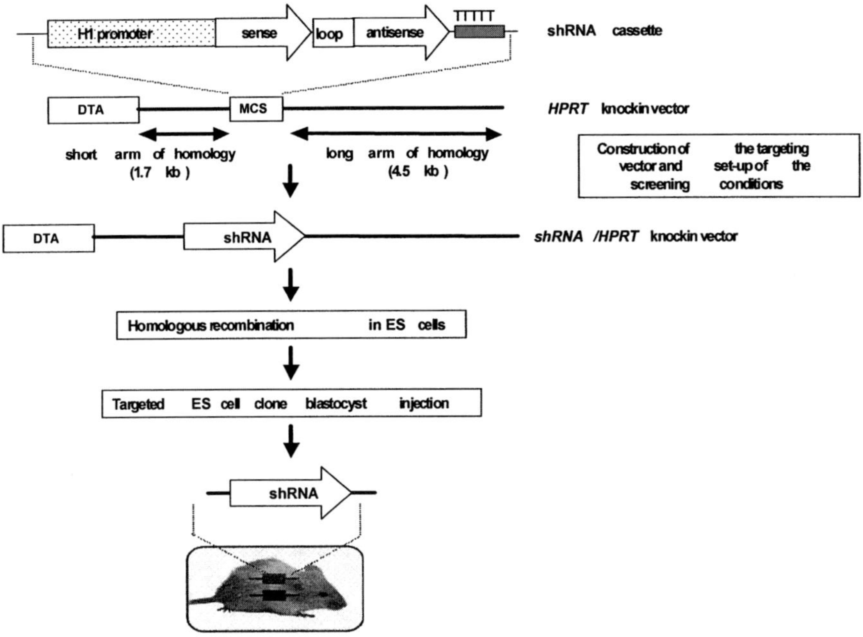

FIG. 1. Use of the *HPRT* neutral locus to secure shRNA expression. *MCS*, Multiple cloning sites; *DTA*, diphtheria toxin A negative selection cassette. The polyT *gray box* represents the pol III promoter stop of the transcription signal

5.1 Control of shRNA Expression: Use of Pol II Promoters

Polymerase II promoters are rarely used to drive shRNA transcription because they generate shRNA harboring a polyA tail. Xia and co-workers reported that viral-mediated shRNA expression driven by the pol II CMV promoter and a minimal poly(A) cassette induced strong inhibition of reporter genes, such as EGFP and luciferase, in mammalian cells. Northern blot analysis and Western blot quantification showed that the level of reporter gene expression was reduced to baseline 72 h after viral infection. The efficiency of the pol II CMV system for silencing genes was further confirmed in vivo by brain injection of adenoviral shRNA expression vectors in transgenic mice expressing EGFP or by tail injection of adenoviral shRNA expression vectors that target endogenous genes (Xia 2003).

The expression of long dsRNA driven by the polymerase II CMV promoter was also used to generate endogenous *ski* knockdown mice by microinjection in fertilized mouse oocytes (Shinagawa 2003). Interestingly, the phenotype of *ski* knockdown embryos was similar to that of *ski* knock-out embryos. Nevertheless no germline transmission was reported in any of the articles listed above.

Studies demonstrating that polymerase II promoters can drive the expression of functional shRNA in vivo could pave the way for the expression of shRNA from tissue-specific polymerase II promoters. Nevertheless, most of the in vivo work performed using shRNA-mediated gene silencing is based on the use of pol III promoters, in order to prevent any risk linked to poly(A) tailing.

Thus, several approaches based on the generation of different inducible/tissue-specific promoters and on the application of site-specific recombinase systems have been investigated. Most of this work has been performed in vitro, and some of these methods have yet to be validated in vivo (Fritsch 2004; Kasim 2004; Coumoul 2004; Tiscornia 2004; Ventura 2004).

5.2 Control of shRNA Expression: Use of the Tet System

Another important improvement would be to specifically induce shRNA expression. The developments of conditional vectors that use the tetracycline operator/repressor interaction system were the first descriptions of an efficient inducible polymerase III promoter driving shRNA transcription (Van de Wetering 2003; Matsukura 2003; Czauderna 2003). This system is characterised by the use of polymerases III promoter H1 or U6, harboring the operator region of the bacterial tetracycline operon sequence (tetO) (Yao 1998) close to the initiation site of transcription (Fig. 2). The tet repressor tetR binds to the tetO sequence and represses any expression of shRNA. In contrast, addition of tetracycline or doxycycline inhibits tetR binding to the tetO sequence, therefore allowing for shRNA transcription.

Using the above approach, the group of van de Wetering first generated stable cell lines expressing the Tet repressor (van de Wetering 2003). Tet repressor cell lines were then stably transfected by shRNA expression vectors harboring the tetO sequence and targeting different endogenous genes. The level of induced knockdown effects at protein level was above 90% after doxycycline treatment. The feasibility of tetracycline induced knockdown effects was further confirmed in other studies (Matsukura 2003; Czauderna 2003).

These findings were consistent with the literature reports on the in vitro use of tetracycline- or doxycycline-inducible systems. Nevertheless, in vivo Tet control systems have already been described as leaky in many transgenic approaches. Not surprisingly, the report of Czauderna et al. (2003) described the same leakiness in vivo. These findings corroborate the data accumulated on the Tet-based control of gene expression in vivo. Finally, it is worth noting that this methodology requires the

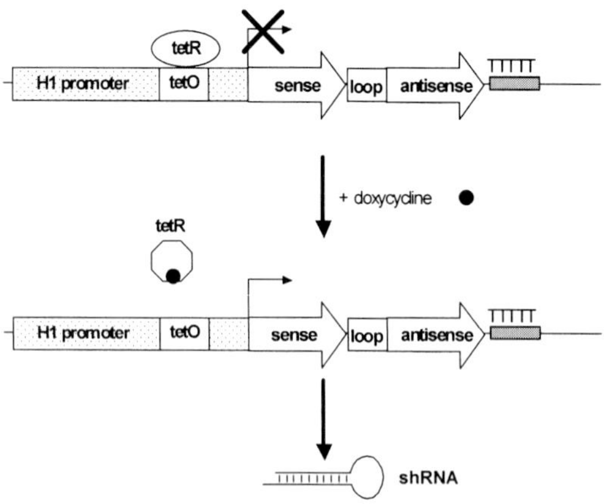

Fig. 2. Doxycycline-based strategy for regulation of shRNA expression. *Black arrows*, LoxP sites. The polyT *gray box* represents the pol III promoter stop of the transcription signal. *TetR*, Tetracycline repressor; *tetO*, tetracycline operator sequence

generation of mouse strains harboring two functional transgenes: the *tet* repressor gene and the shRNA cassette with a promoter containing the tetO sequences.

5.3 Control of shRNA Expression: Use of the Cre-LoxP System

Over the last decade, the manipulation of the mouse genome using site-specific recombinase strategies has proven to be a very useful tool for activating or inactivating specific genes in spatially and temporally restricted patterns. Currently, two main recombinases systems are used in gene targeting techniques: the Cre recombinase, which specifically recognizes loxP consensus sites, and the Flp recombinase, which specifically recognizes FRT consensus sites. Depending on the orientation and localization of the two sites, the recombinase protein leads to DNA modifications, such as deletion, insertion, inversion or translocation (Kwan 2002).

The efficiency and potential of recombinase-based systems suggest that they offer a new alternative for the control of gene expression. Indeed, gene expression could be tightly controlled by a loxP-flanked transcription STOP cassette. This approach presents the following advantages: (1) tight control of gene expression could be achieved as no leakiness is observed. (2) gene expression is generally observed either 24 to 48 h following Cre-recombinase delivery when Cre is delivered locally using intracranial or intraperitoneal injections of lentiviral Cre-expressing vectors (Pfeifer 2001; Ahmed 2004) (3) or at the time the promoter-driving Cre-recombinase is expressed when Cre is delivered using Cre-expressing lines.

Thus, it is not surprising that this approach was investigated for conditional gene knockdown in mice, as the Cre-LoxP system offers all the desired characteristics for tight in vivo control of shRNA expression.

We and several groups focused on two major ways to create robust Cre-dependent shRNA expression cassette that overcome the difficulties linked to the intrinsic compact nature of RNA polymerase III.

The first system (Fritsch 2004; Kasim 2004) is based on the insertion of a loxP-STOP-loxP cassette in the loop of shRNA the expression of which is driven by polymerase III H1 or U6 (Fig. 3A). The loxP-STOP-loxP cassette contains a termination site recognized by the polymerase III enzyme. In the absence of Cre recombinase, incomplete, and hence non-functional shRNA is transcribed. In the presence of Cre, the floxed STOP cassette is removed, enabling the correct transcription of functional shRNA. This strategy was tested in vitro for reporter gene knockdown. The authors demonstrated Cre-dependent production of shRNA and Cre-dependent strong inhibition of GFP in transiently transfected cells. Western blot quantification indicated that the GFP level of expression dropped to a minimum level 60 h after transfection. The strong silencing effect was further monitored in vivo in TA muscles of GFP transgenic mice. Electroporated muscles were examined 12 days later and had a significant decrease in GFP expression (Fritsch 2004).

The second system for generating a Cre-dependent shRNA expression cassette is based on insertion of the loxP-STOP-loxP cassette in the polymerase III promoter (Tiscornia 2004; Coumoul 2004; Ventura 2004; Expediting Target Identification and Validation through RNAi London, Nov 17–18, 2003; Second international RNAi 2004 Boston Meeting, May 4–7, 2004; In vivo and transgenic models conference, Berlin Nov 16–17, 2004). This strategy allows investigation of shRNA function in a tissue-specific

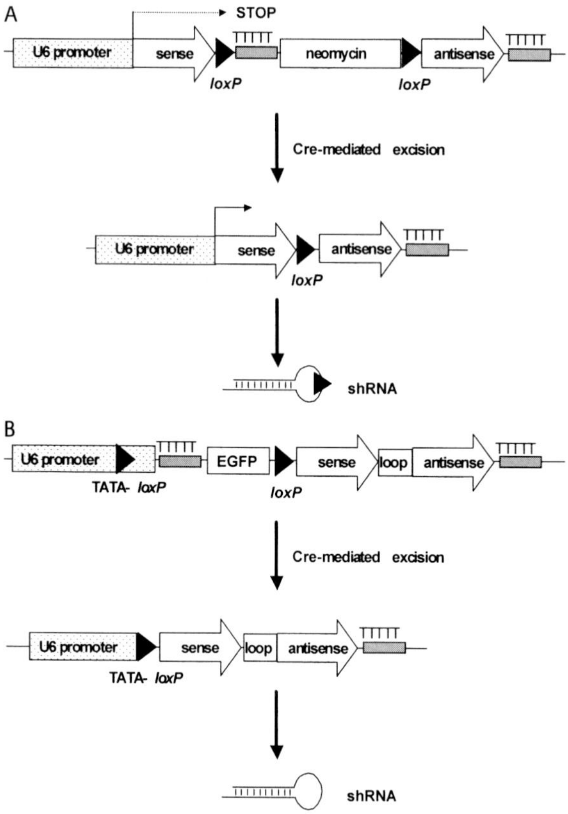

FIG. 3. Cre-Lox system based strategy for regulation of shRNA expression. **A** LoxP site located in the shRNA loop. **B** LoxP site located in the TATA box of the pol III promoter. *Black arrows*, LoxP sites. The polyT *gray box* represents the pol III promoter stop of the transcription signal. *Neomycin* stands for the positive selection cassette

manner, as the target will be knocked down only in tissues where the Cre recombinase is expressed. Coumoul and co-workers developed an shRNA cassette in which a loxP-flanked neomycin cassette is inserted into the promoter. Tiscornia's and Ventura's groups developed a strategy consisting of the construction of lentivirus-based vectors harboring a STOP cassette flanked by a TATA box containing mutated LoxP sites (Fig. 3B).

Both of these approaches have been shown to be efficient for in vitro Cre-mediated control of shRNA expression and thus for Cre-mediated silencing of reporter and endogenous genes. In these co-transfection experiments (transfection of both shRNA-expressing vectors and Cre-expressing vectors), the level of silencing effects on endogenous genes 92 h after vectors transfection was more than 90%.

The same efficiency of the Cre-mediated knockdown mechanism has been demonstrated in mouse ES cells type by Coumoul's group.

More importantly, Ventura and co-workers have demonstrated the feasibility of this system to achieve tissue-specific, conditional RNAi transgenic mice. In the generated model, ES cells were infected with a lentiviral shRNA expressing vector targeted against the T-lymphocyte cell-surface marker CD8. Chimeric mice were obtained and characterized using EGFP visualization for transgene transmission. Then, chimeras were crossed with Cre-transgenic mice expressing Cre recombinase under the control of a *mx2* or *lek* promoter. Fluorescence-activated cell sorter analysis (FACS) demonstrated a tissue-specific knockdown effect of CD8.

These reports, combined with our own data on Cre-LoxP-dependent shRNA expression vectors, have demonstrated that this system could be very promising for efficient and tight control of shRNA expression in vivo. We believe that this approach provides the highest security with regards to shRNA expression in vivo.

5.4 Recombinase-Mediated Cassette Exchange Method

Strategies enabling the reproducible, rapid and cost- and time-effective development of several shRNA knock-in models are a pre-requisite for in vivo comparative studies of shRNA activity. Combining the advantages of the knock-in at the *HPRT* locus using ready-to-use vectors with the power of the recombinase-mediated cassette exchange methodology (RMCE) (Feng 1999; Lauth 2002) meets these requirements. RMCE is a two-step procedure enabling the exchange of a specific DNA sequence with another one provided by a plasmid *in trans*, with each cassette being flanked by nonidentical LoxP or FRT sites (Fig. 4). RMCE has proven to be a highly efficient procedure in ES cell lines. As described in Fig. 4, recognition sites for a site-specific recombinase are inserted at the *HPRT* locus by homologous recombination. Then, site-specific recombination inserts a replacement sequence into this pre-tagged locus. We are currently developing ES cell clones harboring a tagged *HPRT* locus in which the insertion of shRNA-expressing constructs will be highly efficient. Once the constructs are inserted, this procedure only requires a single straightforward cloning step to generate shRNA-expressing plasmid flanked by the mutated recombinase-recognition sites. shRNA cassette exchange is then performed and the cells are further processed for blastocyst injection. This approach enables a very rapid and cost-effective translation toward in vivo analysis of shRNA.

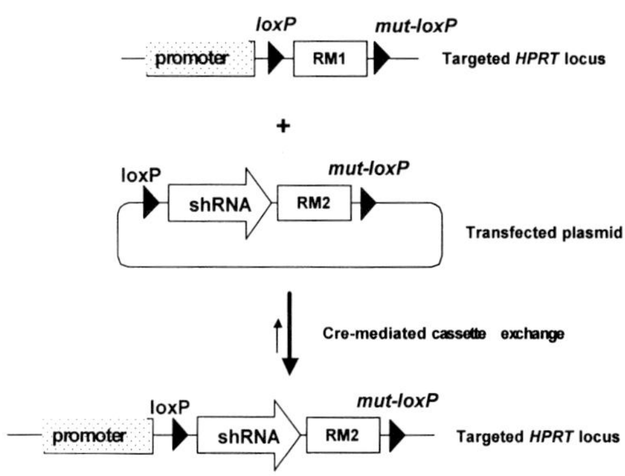

FIG. 4. Cre-mediated cassette exchange strategy for the rapid generation of shRNA knock-in mice. *Black arrows*, wild-type and mutated LoxP sites. The polyT *gray box* represents the pol III promoter stop of the transcription signal. *RM1*, *RM2*, Positive/negative selection cassettes

6 Conclusion

The proof-of-concept of RNAi technology for rapid and efficient gene silencing has been well established in vitro. The technology also holds great promise in vivo and should be considered a new tool for in vivo gene function analysis.

The complementarity existing between classical knockout mice and RNAi gene silencing mice is well illustrated by therapeutic gene-target validation. Inhibition of a protein target in vivo using small therapeutic molecules is never achieved with total efficiency. Knockout models are useful to address the question of compound specificity but might not be the optimal models to mimic the effect of inhibitory therapeutic agents. However, in conditional knockout mice models, the excision of the floxed gene of interest through Cre recombinase action is rarely complete, and may result in mosaicism, i.e., tissues displaying knockout in only a proportion of their cells. However, in vivo RNAi gene silencing effects are never complete but in the range of 30–90%. But the knockdown effect should be observed at the single cell level, and in all cells composing a tissue. Depending on the nature of the therapeutic target, and in particular its cellular localization, one approach could be more appropriate than the other.

Another important novel application of RNAi in addition to conditional knock out is the ability to induce different levels of gene silencing in individual knockdown mice (Hemann 2003). The possibility of controlling a graduated knockdown effect, and then of creating graded phenotypes, open great perspectives for the study of complex pathological situations and gene-dosage-related questions.

Finally, shRNA knockdown allows for targeting a broader spectrum of genes. For instance, particular splice variants could be targeted more easily (Zhang 2003). Multicopy genes that are functionally redundant can theoretically all be knocked down by a single shRNA transgene construct. Furthermore, an entire gene family could be knocked down by using a siRNA construct targeting a highly conserved domain.

One key parameter for obtaining functional RNAi in the mouse is to control the shRNA copy number. In order to secure the use of RNAi in vivo, novel knock-in approaches based on ready-to-use tools have been developed. These knock-in vectors target permissive chromatin environments, such as the *HPRT* locus, secure shRNA expression, and allow for the generation of animal models displaying a single copy of the shRNA-expressing construct. This approach, based on validated and optimized tools, enables the development of shRNA knock-in models within 4 months. shRNA expression could be controlled by combining this technology with the Cre-recombinase-based approach. We and others have confirmed that a loxP-sites-flanked STOP cassette provide tight control of RNAi-mediated gene knockdown. The gene knockdown could then be tissue- and time-specific based on the Cre delivery system used. Nevertheless, careful study of the specificity and efficiency of a given construct remains necessary before any project related to in vivo RNAi can be undertaken, and the above timing might become a reality when starting from available in-vitro-validated shRNA constructs.

The ability to generate several RNAi transgenic animal models and the development of efficient inducible and tissue-specific shRNA expression systems should also enable the generation of larger numbers of knock-down strains, opening new possibilities for large-scale target-gene validation projects.

References

Ahmed BY, Chakravarthy S, Eggers R, Hermens WT, Zhang JY, Niclou SP, Levelt C, Sablitzky F, Anderson PN, Lieberman AR, Verhaagen J (2004) Efficient delivery of Cre-recombinase to neurons in vivo and stable transduction of neurons using adeno-associated and lentiviral vectors. BMC Neurosci Jan 30;5(1):4

Bronson SK, Plaehn EG, Kluckman KD, Hagaman JR, Maeda N, Smithies O (1996) Single-copy transgenic mice with chosen-site integration. Proc Natl Acad Sci U S A Aug 20;93(17):9067–9072

Brummelkamp TR, Bernards R, Agami R (2002) A system for stable expression of short interfering RNAs in mammalian cells. Science Apr 19;296(5567):550–553. Epub 2002 Mar 21

Carmell MA, Zhang L, Conklin DS, Hannon GJ, Rosenquist TA (2003) Germline transmission of RNAi in mice. Nat Struct Biol Feb;10(2):91–92

Chatterjee B, Song CS, Jung MH, Chen S, Walter CA, Herbert DC, Weaker FJ, Mancini MA, Roy AK (1996) Targeted overexpression of androgen receptor with a liver-specific promoter in transgenic mice. Proc Natl Acad Sci U S A Jan 23;93(2):728–722

Coumoul X, Li W, Wang RH, Deng C (2004) Inducible suppression of Fgfr2 and Survivin in ES cells using a combination of the RNA interference (RNAi) and the Cre-LoxP system. Nucleic Acids Res Jun 15;32(10):e85

Cvetkovic B, Yang B, Williamson RA, Sigmund CD (2000) Appropriate tissue- and cell-specific expression of a single copy human angiotensinogen transgene specifically targeted upstream of the HPRT locus by homologous recombination. J Biol Chem Jan 14;275(2):1073–1078

Czauderna F, Santel A, Hinz M, Fechtner M, Durieux B, Fisch G, Leenders F, Arnold W, Giese K, Klippel A, Kaufmann J (2003) Inducible shRNA expression for application in a prostate cancer mouse model. Nucleic Acids Res Nov 1;31(21):e127

Elbashir SM, Harborth J, Lendeckel W, Yalcin A, Weber K, Tuschl T (2001) Duplexes of 21-nucleotide RNAs mediate RNA interference in cultured mammalian cells. Nature 411(6836):494–498

Engle SJ, Womer DE, Davies PM, Boivin G, Sahota A, Simmonds HA, Stambrook PJ, Tischfield JA (1996) HPRT-APRT-deficient mice are not a model for Lesch-Nyhan syndrome. Hum Mol Genet Oct;5(10):1607–1610

Evans V, Hatzopoulos A, Aird WC, Rayburn HB, Rosenberg RD, Kuivenhoven JA (2000) Targeting the HPRT locus in mice reveals differential regulation of tie2 gene expression in the endothelium. Physiol Genomics Mar 13;2(2):67–75

Feng YQ, Seibler J, Alami R, Eisen A, Westerman KA, Leboulch P, Fiering S, Bouhassira EE (1999) Site-specific chromosomal integration in mammalian cells: highly efficient CRE recombinase-mediated cassette exchange. Mol Biol Oct 1;292(4):779–785

Finger S, Heavens RP, Sirinathsinghji DJ, Kuehn MR, Dunnett SB (1988) Behavioral and neurochemical evaluation of a transgenic mouse model of Lesch-Nyhan syndrome. J Neurol Sci Sep;86(2–3):203–213

Fritsch L, Martinez LA, Sekhri R, Naguibneva I, Gerard M, Vandromme M, Schaeffer L, Harel-Bellan A (2004) Conditional gene knock-down by CRE-dependent short interfering RNAs. EMBO Rep Feb;5(2):178–182. Epub 2004 Jan 09

Guillot PV, Liu L, Kuivenhoven JA, Guan J, Rosenberg RD, Aird WC (2000) Targeting of human eNOS promoter to the HPRT locus of mice leads to tissue-restricted transgene expression. Physiol Genomics Mar 13;2(2):77–83

Hammond SM, Boettcher S, Caudy AA, Kobayashi R, Hannon GJ (2001) Argonaute2, a link between genetic and biochemical analyses of RNAi. Science Aug 10; 293(5532):1146–1150

Hannon GJ, Rossi JJ (2004) Unlocking the potential of the human genome with RNA interference. Nature Sep 16;431(7006):371–378

Harborth J, Elbashir SM, Bechert K, Tuschl T, Weber K (2001) Identification of essential genes in cultured mammalian cells using small interfering RNAs. J Cell Sci 114(Pt 24): 4557–4565

Hasuwa H, Kaseda K, Einarsdottir T, Okabe M (2002) Small interfering RNA and gene silencing in transgenic mice and rats. FEBS Lett Dec 4;532(1–2):227–230

Hatada S, Kuziel W, Smithies O, Maeda N (1999) The influence of chromosomal location on the expression of two transgenes in mice. J Biol Chem Jan 8;274(2):948–955

Hemann MT, Fridman JS, Zilfou JT, Hernando E, Paddison PJ, Cordon-Cardo C, Hannon GJ, Lowe SW (2003) An epi-allelic series of p53 hypomorphs created by stable RNAi produces distinct tumor phenotypes in vivo. Nat Genet Mar;33(3):396–400

Hyland K, Kasim S, Egami K, Arnold LA, Jinnah HA (2004) Tetrahydrobiopterin deficiency and dopamine loss in a genetic mouse model of Lesch-Nyhan disease. J Inherit Metab Dis 27(2):165–178

Jinnah HA, Jones MD, Wojcik BE, Rothstein JD, Hess EJ, Friedmann T, Breese GR (1999) Influence of age and strain on striatal dopamine loss in a genetic mouse model of Lesch-Nyhan disease. J Neurochem Jan;72(1):225–229

Jinnah HA, Gage FH, Friedmann T (1991) Amphetamine-induced behavioral phenotype in a hypoxanthine-guanine phosphoribosyltransferase-deficient mouse model of Lesch-Nyhan syndrome. Behav Neurosci Dec;105(6):1004–1012

Kasim V, Miyagishi M, Taira K (2004) Control of siRNA expression using the Cre-loxP recombination system. Nucleic Acids Res Apr 23;32(7):e66

Katome T, Obata T, Matsushima R, Masuyama N, Cantley LC, Gotoh Y, Kishi K, Shiota H, Ebina Y (2003) Use of RNA interference-mediated gene silencing and adenoviral overexpression to elucidate the roles of AKT/protein kinase B isoforms in insulin actions. J Biol Chem Jul 25;278(30):28312–28323. Epub 2003 May 06

Klein C, Bock CT, Wedemeyer H, Wustefeld T, Locarnini S, Dienes HP, Kubicka S, Manns MP, Trautwein C (2003) Inhibition of hepatitis B virus replication in vivo by nucleoside analogues and siRNA. Gastroenterology Jul;125(1):9–18

Kunath T, Gish G, Lickert H, Jones N, Pawson T, Rossant J (2003) Transgenic RNA interference in ES cell-derived embryos recapitulates a genetic null phenotype. Nat Biotechnol May;21(5):559–561

Kwan KM (2002) Conditional alleles in mice: practical considerations for tissue-specific knockouts. Genesis Feb;32(2):49–62

Lauth M, Spreafico F, Dethleffsen K, Meyer M (2002) Stable and efficient cassette exchange under non-selectable conditions by combined use of two site-specific recombinases. Nucleic Acids Res Nov 1;30(21):e115

Lewis DL, Hagstrom JE, Loomis AG, Wolff JA, Herweijer H (2002) Efficient delivery of siRNA for inhibition of gene expression in postnatal mice. Nat Genet Sep;32(1):107–108

Matsukura S, Jones PA, Takai D (2003) Establishment of conditional vectors for hairpin siRNA knockdowns. Nucleic Acids Res Aug 1;31(15):e77

McCaffrey AP, Meuse L, Pham TT, Conklin DS, Hannon GJ, Kay MA (2002) RNA interference in adult mice. Nature Jul 4;418(6893):38–39

McCaffrey AP, Nakai H, Pandey K, Huang Z, Salazar FH, Xu H, Wieland SF, Marion PL, Kay MA (2003) Inhibition of hepatitis B virus in mice by RNA interference. Nat Biotechnol Jun;21(6):639–644

Mello CC, Conte D Jr (2004) Revealing the world of RNA interference. Nature Sep 16;431(7006):338–342

Miyagishi M, Taira K (2002) Development and application of siRNA expression vector. Nucleic Acids Res Suppl (2):113–114

Miyagishi M, Taira K (2002) U6 promoter-driven siRNAs with four uridine 3′ overhangs efficiently suppress targeted gene expression in mammalian cells. Nat Biotechnol May;20(5):497–500

Miyagishi M, Taira K (2004) RNAi expression vectors in mammalian cells. Methods Mol Biol 252:483–491

Paddison PJ, Caudy AA, Bernstein E, Hannon GJ, Conklin DS (2002) Short hairpin RNAs (shRNAs) induce sequence-specific silencing in mammalian cells. Genes Dev Apr 15;16(8):948–958

Pebernard S, Iggo RD (2004) Determinants of interferon-stimulated gene induction by RNAi vectors. Differentiation Mar;72(2–3):103–111

Persengiev SP, Zhu X, Green MR (2004) Nonspecific, concentration-dependent stimulation and repression of mammalian gene expression by small interfering RNAs (siRNAs). RNA Jan;10(1):12–18

Pfeifer A, Brandon EP, Kootstra N, Gage FH, Verma IM (2001) Delivery of the Cre recombinase by a self-deleting lentiviral vector: efficient gene targeting in vivo. Proc Natl Acad Sci USA Sep 25;98(20):11450–11455

Rubinson DA, Dillon CP, Kwiatkowski AV, Sievers C, Yang L, Kopinja J, Rooney DL, Ihrig MM, McManus MT, Gertler FB, Scott ML, Van Parijs L (2003) A lentivirus-based system to functionally silence genes in primary mammalian cells, stem cells and transgenic mice by RNA interference. Nat Genet Mar;33(3):401–406

Shinagawa T, Ishii S (2003) Generation of Ski-knockdown mice by expressing a long double-strand RNA from an RNA polymerase II promoter. Genes Dev Jun 1;17(11): 1340–1345

Sledz CA, Holko M, de Veer MJ, Silverman RH, Williams BR (2003) Activation of the interferon system by short-interfering RNAs. Nat Cell Biol Sep;5(9):834–839

Song E, Lee SK, Wang J, Ince N, Ouyang N, Min J, Chen J, Shankar P, Lieberman J (2003) RNA interference targeting Fas protects mice from fulminant hepatitis. Nat Med Mar;9(3):347–351

Sorensen DR, Leirdal M, Sioud M (2003) Gene silencing by systemic delivery of synthetic siRNAs in adult mice. J Mol Biol Apr 4;327(4):761–766

Stein P, Svoboda P, Anger M, Schultz RM (2003) RNAi: mammalian oocytes do it without RNA-dependent RNA polymerase. RNA Feb;9(2):187–192

Stein P, Svoboda P, Schultz RM (2003) Transgenic RNAi in mouse oocytes: a simple and fast approach to study gene function. Dev Biol Apr 1;256(1):187–193

Svoboda P, Stein P, Schultz RM (2001) RNAi in mouse oocytes and preimplantation embryos: effectiveness of hairpin dsRNA. Biochem Biophys Res Commun Oct 12;287(5):1099–1104

Tiscornia G, Singer O, Ikawa M, Verma IM (2003) A general method for gene knockdown in mice by using lentiviral vectors expressing small interfering RNA. Proc Natl Acad Sci U S A Feb 18;100(4):1844–1848. Epub 2003 Jan 27

Tiscornia G, Tergaonkar V, Galimi F, Verma IM (2004) CRE recombinase-inducible RNA interference mediated by lentiviral vectors. Proc Natl Acad Sci USA May 11;101(19):7347–7351. Epub 2004 Apr 30

Tuschl T (2001) RNA interference and small interfering RNAs. Chem Biochem 2(4):239–245

van de Wetering M, Oving I, Muncan V, Pon Fong MT, Brantjes H, van Leenen D, Holstege FC, Brummelkamp TR, Agami R, Clevers H (2003) Specific inhibition of gene expression using a stably integrated, inducible small-interfering-RNA vector. EMBO Rep Jun;4(6):609–615

Ventura A, Meissner A, Dillon CP, McManus M, Sharp PA, Van Parijs L, Jaenisch R, Jacks T (2004) Cre-lox-regulated conditional RNA interference from transgenes. Proc Natl Acad Sci U S A Jul 13;101(28):10380–10385

Xia H, Mao Q, Paulson HL, Davidson BL (2002) siRNA-mediated gene silencing in vitro and in vivo. Nat Biotechnol Oct;20(10):1006–1010

Yao F, Svensjo T, Winkler T, Lu M, Eriksson C, Eriksson E (1998) Tetracycline repressor, tetR, rather than the tetR-mammalian cell transcription factor fusion derivatives, regulates inducible gene expression in mammalian cells. Hum Gene Ther Sep 1;9(13):1939–1950

Zambrowicz BP, Imamoto A, Fiering S, Herzenberg LA, Kerr WG, Soriano P (1997) Disruption of overlapping transcripts in the ROSA beta geo 26 gene trap strain leads to widespread expression of beta-galactosidase in mouse embryos and hematopoietic cells. Proc Natl Acad Sci U S A Apr 15;94(8):3789–3794

Zhang L, Yang N, Mohamed-Hadley A, Rubin SC, Coukos G (2003) Vector-based RNAi, a novel tool for isoform-specific knock-down of VEGF and anti-angiogenesis gene therapy of cancer. Biochem Biophys Res Commun Apr 18;303(4):1169–1178

Suppression of Gene Expression via Chromatin Remodeling and the siRNA-Induced Silencing of Transcription

Hiroaki Kawasaki[1,3], Yoko Fukuda[2,3], Kazunari Taira[1,3], and Kevin V. Morris[4]

1 Epigenetics

Epigenetics is the study of meiotically and mitotically heritable changes in gene expression that are not encoded in the DNA (Jablonka and Lamb 2003; Egger et al. 2004). Three distinct mechanisms appear to be intricately related and implicated in the initiation and maintenance of epigenetic modifications: DNA methylation; RNA-associated silencing; and the modification of histones (Egger et al. 2004). It has recently become clear that, in human cells, RNA plays a far more profound and complex role in regulating gene expression than had previously been imagined. Such regulatory effects are mediated by RNA-associated silencing; they can act at the transcriptional level, and can involve an RNA interference (RNAi)-based mechanism that is directed specifically by short-interfering RNAs (siRNAs). Moreover, recent observations by our groups indicate that siRNAs can silence target genes at the chromatin level in mammalian cells. In this chapter, we will discuss siRNA-mediated silencing of transcription and siRNA-directed methylation of DNA, as well as their putative mechanisms, in human cells.

Evolution has been defined as "the change in genetic composition of a population during successive generations, due to natural selection acting on the genetic variation among individuals, and resulting in the development of new species." An added layer of evolutionary complexity has emerged with the realization that heritable modifications can occur not only when the genome is replicated and passed on to daughter cell(s) but also when the profile of gene expression in a parent cell influences the next generation. This latter phenomenon is referred to as "epigenetic inheritance" (Lavrov and Mavrodiev 2003; McNairn and Gilbert 2003). Thus, epigenetics is the study of

[1] Department of Chemistry and Biotechnology, School of Engineering, [2] Department of Neuroscience, Graduate School of Medicine, The University of Tokyo, 7-3-1 Hongo, Bunkyo-ku, Tokyo 113-8656, Japan

[3] Gene Function Research Center, National Institute of Advanced Industrial Science and Technology (AIST), Central 4, 1-1-1 Higashi, Tsukuba Science City 305-8562, Japan

[4] Division of Molecular Biology, Fox North Rm. 2002, Beckman Research Institute, 1450 East Duarte Road, Duarte, CA 91010-3011, USA

meiotically and mitotically heritable changes in gene expression that are not encoded in the DNA (Jablonka and Lamb 2003; Egger et al. 2004). Such changes involve three distinct mechanisms, namely, DNA methylation, RNA-associated silencing, and the modification of histones (Egger et al. 2004). It is also clear now that, in human cells, RNA-associated silencing can operate at the transcriptional level via an RNAi-based mechanism that is mediated specifically by siRNAs that direct the silencing of target genes at the chromatin level. The ramifications from this paradigm shift with respect to the regulation of gene expression are vast both in a therapeutic context (i.e., via the potential directed control of gene expression) and biological context, as we strive to understand the evolution of cells and higher organisms.

2 Post-transcriptional Gene Silencing

Double-stranded RNA (dsRNA)-induced post-transcriptional gene silencing (PTGS) is known as RNAi in animals and as quelling in the filamentous fungus *Neurospora crassa* and in plants (Fire et al. 1998; Hutvagner and Zamore 2002; Tijsterman et al. 2002). In this system, dsRNAs are processed to siRNAs of 21–25 nucleotides (nt) by the RNase III known as Dicer (Zamore et al. 2000; Bernstein et al. 2001; Elbashir et al. 2001; Kawasaki et al. 2003). These siRNAs are incorporated into the RNAi-induced silencing complex (RISC) where they then promote the degradation of sequence-specific mRNAs in a reaction mediated by Argonaute 2 (Ago 2) in the cytoplasm (Tuschl et al. 1999; Hammond et al. 2000, 2001; Zeng and Cullen 2002; Song et al. 2004; Liu et al. 2004). Recent reports indicate that, in mammalian cells, siRNAs longer than 25 nt, which can be processed by Dicer with transfer of the resultant processed siRNAs to RISC, have suppressive effects up to 100-fold higher than those of shorter, 19-nt siRNAs that cannot be processed by Dicer (Siolas et al. 2005; Kim et al. 2005). In *Caenorhabditis elegans* and in plants, siRNAs are amplified by an RNA-dependent RNA polymerase (RdRP) that uses the target mRNA as a template (Sijen et al. 2001). By contrast, no mammalian system is known that exploits the RdRP-mediated amplification of siRNA (Fig. 1). It is of interest that siRNAs and their ability to silence gene

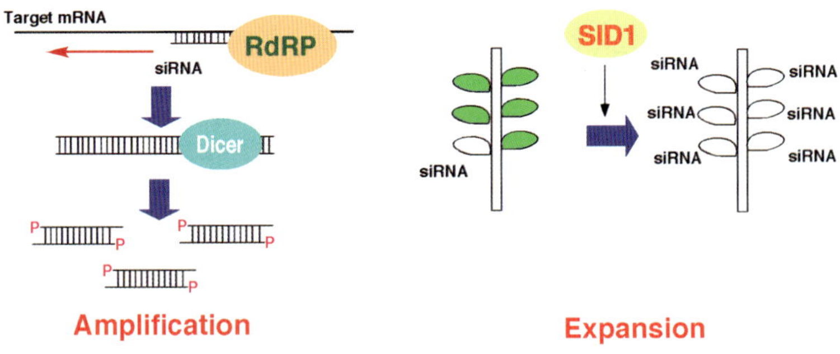

Fig. 1. Amplification and spread of short-interfering RNAs (siRNAs) in *Caenorhabditis elegans* and plants. New double-stranded (ds)RNAs are synthesized by RNA-dependent RNA polymerase (RdRP) with target mRNAs as the template. Then these dsRNAs are processed to siRNAs by Dicer. In addition, siRNAs are transported to distant cells by SID1, resulting in the expansion of siRNA-mediated gene silencing over a wide area

expression can spread to distant cells in *C. elegans* and plants (Winston et al. 2002; Feinberg and Hunter 2003). In mammalian cells, while long dsRNAs induce interferon-mediated non-specific gene silencing (Williams 1997), siRNAs can induce the sequence-specific silencing of gene expression (Elbashir et al. 2001), despite the fact that they are not amplified by an RdRP. The absence of such amplification of siRNA might be linked to the failure to detect any genes for a homolog of RdRP in mammalian genomes. This situation might be advantageous to humans, with respect to potential siRNA-mediated anti-cancer therapy, because oncogenes that originate by point mutation of original functional genes are potential targets of specific siRNAs directed against the mutant but not the wild-type gene. By contrast, if a specific siRNA against the mutant gene were to act as a primer and RdRP-dependent amplification were to occur, the resultant amplified siRNA could interfere with expression not only of the mutant gene but also of the original functional gene.

Silencing by dsRNA was first recognized as an antiviral mechanism that protects organisms from RNA viruses and the random integration of transposable elements (Waterhouse et al. 2001). By contrast, silencing by microRNAs (miRNAs) occurs at the translational level and involves the targeting of partially complementary sequences located within the 3′ untranslated region (UTR) of mRNAs (Fig. 2) that are involved in the differentiation of cells and development in plants and animals (Ambros 2004; Bartel 2004; Kawasaki et al. 2004). In general, miRNAs are first transcribed as a long RNA that is then processed to a pre-miRNA of approximately 70 nt (Lee et al. 2003; Denli et al. 2004; Gregory et al. 2004; Han et al. 2004). This pre-miRNA is transported to the cytoplasm (Lee et al. 2004; Lund et al. 2004) where it is processed by RNase III Dicer to produce the mature miRNA. The mature miRNA is incorporated into ribonucleoprotein complexes (miRNPs), which include eIF2C2 and FMRP, that function in RNAi-mediated gene silencing (Mourelatos et al. 2002; Jun et al. 2004). In both plants and animals, some miRNAs can also function as siRNAs, provided that they are almost perfectly complementary to their target sequences, and can direct the cleavage of complementary mRNAs (Llave et al. 2002; Tang et al. 2003; Mansfield et al. 2004; Yekta et al. 2004). Thus, it is likely that the miRNA and siRNA pathways are very similar and might even overlap to some extent (Fig. 2).

3 Transcriptional Gene Silencing

RNAi-based suppression at the level of gene expression was first observed in viroid-infected plants (Wassenegger et al. 1994) and was shown to be due to RNA-dependent methylation of DNA (RdDM). RdDM requires a dsRNA that is processed to yield short RNAs (Wassenegger et al. 1994; Mette et al. 2000). These short RNAs include sequences that are identical to promoter regions and they are capable of inducing methylation of the homologous promoter and subsequent transcriptional gene silencing (TGS; Fig. 2). In addition, RdDM results in an unusual pattern of methylation of CpG and CpNpG sequences (Wassenegger 2000; Aufsatz et al. 2002). It has been proposed that the short RNAs; one or more DNA methyltransferases (DMTases), such as MET1, DRM1/2 or CMT3 (Jones et al. 2001; Lindroth et al. 2001); and chromatin-modifying factors, such as histone deacetylase HDA6 (Aufsatz et al. 2002), might trigger methylation de novo and the silencing of transcription from the homologous promoter in plants (Fig. 3).

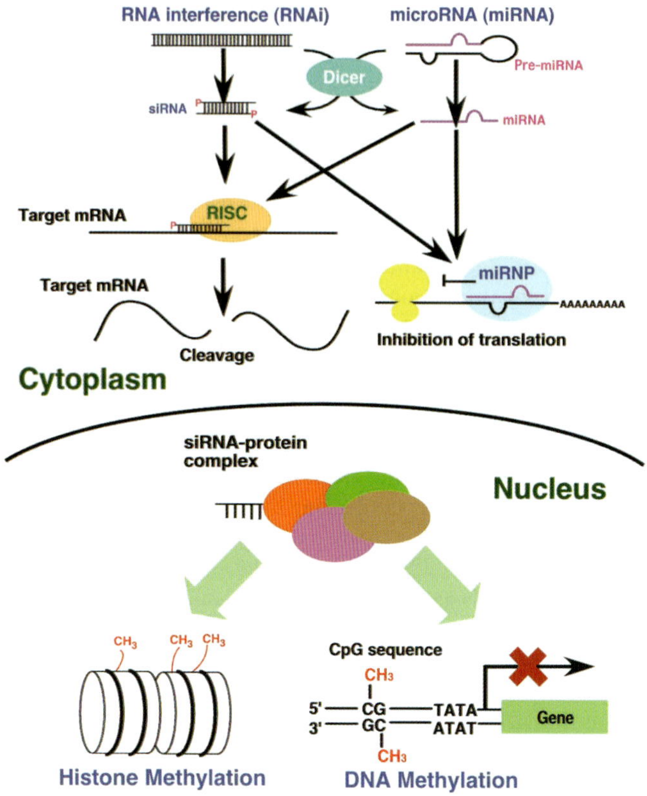

FIG. 2. Pathways of gene silencing by siRNA and miRNA. siRNAs are produced from long dsRNAs by the ribonuclease Dicer. The siRNAs are incorporated into the RNAi-induced silencing complex (RISC) which then cleaves the target RNAs via the action of Argonaute 2. miRNAs are expressed as short hairpin RNAs (pre-miRNAs) and are transported to the cytoplasm. These pre-miRNAs are processed to yield mature miRNAs by Dicer. The miRNAs are incorporated into miRNPs which then inhibit the translation of target mRNAs by forming partial duplexes with target mRNAs. In addition, miRNAs can function as siRNAs when the miRNA is fully complementary to the target mRNAs

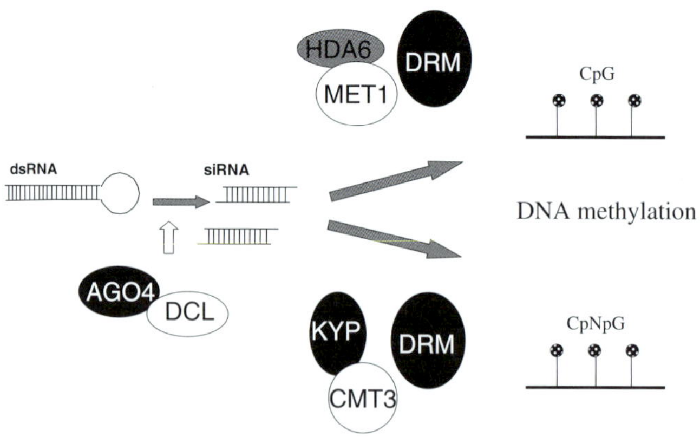

FIG. 3. DNA methylation in response to dsRNA in plants. dsRNAs are processed to siRNAs by members of the DCL family of proteins. These siRNAs can trigger MET1-dependent methylation of CpG or CMT3-dependent methylation of CpNpG

In plants, Northern blotting analysis of siRNAs has consistently revealed the existence of two size classes of siRNA: shorter 21- or 22-nt siRNAs and longer 24- to 26-nt siRNAs (Hamilton et al. 2002; Zilberman et al. 2003). Functional differences between these two size classes were revealed by studies with viral proteins, such as P1, P19, 2b and AC2, that are able to suppress siRNA-mediated gene silencing (Hamilton et al. 2002). These viral proteins can prevent the accumulation of members of one or both size classes of siRNA, with resultant specific inhibition of PTGS and TGS (Fig. 2). In the plant *Nicotiana benthamiana*, prevention of the accumulation of the larger 24- to 26-nt siRNAs by the P1 and P19 proteins inhibited RdDM and the expansion of silencing. These data suggest that longer siRNAs might be involved in RdDM but any relevance of size classes in human cells remains to be demonstrated.

It is possible that the size and functional diversity of siRNAs in plants is due to the multiplicity of DCL enzymes, which are Dicer-homologous enzymes. *Arabidopsis* has four DCL proteins (Schauer et al. 2002). All DCL proteins contain an RNase III domain and a DExH-box RNA helicase domain, but they vary in terms of the presence and number of dsRNA-binding domains, nuclear localization signals (NLSs), and PAZ domains, which are conserved in RNAi-related proteins such as Dicer and Ago2 (Schauer et al. 2002). The PAZ domain would hold the 3′ end of the single-stranded siRNA (Lingel et al. 2004; Ma et al., 2004; Song JJ, et al. 2004). Since DCL1 and DCL4 have one or two predicted NLSs, it is possible that both nuclear and cytoplasmic pathways for the processing of dsRNA might exist in plants. Experiments using the NLS of DCL1 fused with green fluorescent protein (GFP) suggested that DCL1 might be localized in the nuclei of plant cells (Papp et al. 2003). In addition, DCL1 has been shown to be required for the processing of precursors to miRNAs (Park et al. 2002; Reinhart et al. 2002) and such processing might also occur in the nucleus (Papp et al. 2003). Moreover, a deficiency in DCL3 in *Arabidopsis* led to a reduction in levels of longer siRNAs and loss of chromatin-based silencing such as the methylation of DNA and histones, suggesting that DCL3 might be required for siRNA-mediated TGS (Xie et al. 2004).

In *C. elegans*, most transposons are silenced by RdDM. Plants might similarly regulate the expression of transposons by RdDM (Lippman et al. 2003). Most retrotransposons in plants are silenced by the DNA methyltransferase MET1, by the ATPase DDM1, and the histone deacetylase HDA6. In addition, silencing of the retrotransposons *AtCOPIA44* and *AtCOPIA4* requires the histone Lys-9 methyltransferase KYP, the DNA methyltransferase CMT3, and Ago4. CMT3 and KYP are also involved in the transcriptional silencing that is induced by long hairpin RNAs transcribed from regions of the *SUP* and *PAI2* genes that include long inverted repeats. Moreover, it is possible that, in plant, NOSpro and *GYPSY* are both stably silenced by the deacetylation of histone (such as H2A, H2B, and H4) as well as by the methylation of DNA. Further investigations of DNA methylation and the modification of histone are required to define the details of the transcriptional silencing of transposons in plants. In general, the higher the organism, the higher the proportion of junk DNA in its genome, and many higher organisms contain repetitive sequences that include transposons. Is it possible that a mechanism exists that can regulate as well as utilize transposons to elude or alleviate selective pressures subsequently driving the evolution of the cell and higher organisms.

4 RNAi-Induced Methylation of DNA in Human Cells

It became clear some time ago that long dsRNAs can induce the sequence-specific methylation of DNA in plants, regulating gene expression at the transcriptional level. However, it was unclear until recently whether this phenomenon might be relevant to mammalian cells. In studies of mammalian cells, it was demonstrated first that long dsRNAs induce sequence-specific gene silencing in mouse embryonal carcinoma (EC) cells and embryonic stem (ES) cells (Svoboda et al. 2000; Wianny and Zernicka-Goetz 2000), while long dsRNAs (>30 nt) activate a dsRNA-dependent protein kinase (PKR) and a 2′,5′-oligoadenylate synthetase (Stark et al. 1998) in mammalian somatic cells, with a resultant non-specific reduction in levels of mRNAs.

Our groups discovered recently that siRNAs targeted to promoters can induce transcriptional silencing via the methylation of DNA in human cells (Morris et al. 2004; Kawasaki and Taira 2004). A priori, one would assume that the siRNAs must access the nucleus, as well as the targeted promoter, and indeed such has been shown to be the case for synthetic siRNAs (Morris et al. 2004). However, there may be a mechanism that actively transports the promoter-targeted siRNAs to the nucleus since synthetic and tRNA vector-based siRNAs, which are localized predominantly in the cytoplasm, effectively targeted the promoter of the oncogene *erbB2* and induced CpG-specific methylation (Kawasaki and Taira 2004).

The mechanism by which promoter-directed siRNAs are guided to and gain access to genomic DNA remains unknown. It is possible that a small fraction of an uncharacterized siRNA-protein complex might be transported to the nucleus, as has been observed in plants (Waterhouse et al. 2001; Matzke et al. 2001); alternatively, siRNAs might gain access to genomic DNA during cell division, when the nuclear membrane disappears. What we do know about the mechanism is that it involves both DNMT1 and DNMT3b (Kawasaki and Taira 2004). Since the homolog of DNMT1 in plants, MET-1, is necessary for DNA methylation, the mechanism of DNA methylation in plants and animals might involve at least partially similar mechanisms. Moreover, siRNAs can induce not only DNA methylation but also histone Lys-9 methylation in human cells (Kawasaki and Taira 2004). The methylation of histones associated with RNAi has the potential also to lead to DNA methylation. Thus, the mechanism of siRNA-induced DNA methylation might be complicated and might actually involve a chromatin-remodeling complex, at least at some stage of silencing, since 5-azacytidine (5-AzaC) and trichostatin A (TSA) each completely reversed siRNA-mediated TGS (Morris et al. 2004). These data, together with results from a Dicer knockout system (Bernstein et al. 2003; Fukagawa et al. 2004), suggest that RNAi is involved in silencing of the heterochromatin in mammals. Indeed, similar observations have been made in the fission yeast *Schizosacchoromyces pombe* (Noma et al. 2004). While the details remain to be resolved, it is clear that small RNAs play a pivotal role in chromatin remodeling in mammalian cells (Morris et al. 2004; Kawasaki and Taira 2004; Kuwabara et al. 2004).

5 Chromatin-Remodeling Complexes and RNAi

Euchromatin (less condensed and transcriptionally relatively active DNA) is typically associated with acetylated histones and with di-methylation of histone H3 at the lysine residue at position 4 (H3mLys-4), whereas heterochromatin (condensed and transcriptionally relatively inactive DNA) is associated with di-methylation of histone H3 at lysine 9 (H3mLys-9) (Lippman et al. 2004). The acetylation of histone tails by histone acetyltransferase (HAT) results in removal of a positive charge from these histone tails and disruption of the interaction of the positively charged histone tails with DNA, causing relaxation of the chromatin and gene activation (Fig. 4A). Conversely, the deacetylation of histones by histone deacetylases (HDACs) results in condensation of the chromatin and repression of transcription (Fig. 4B) (for review, see Lusser 2004).

There are six well-defined gene-regulatory histone deacetylase complexes that appear to be involved in the suppression of gene expression (for review, see Dobosy and Selker 2001). Of these six complexes, two appear to be of interest in the context of possible mechanisms of siRNA-mediated TGS in mammalian cells. One of those complexes is the nucleosome-remodeling histone deacetylase complex Mi2/NuRD, which contains the retinoblastoma protein Rb, HDAC1 and HDAC2, a histone methyltransferase (SUV39H1) and, possibly, a heterochromatin-associated protein (HP1).

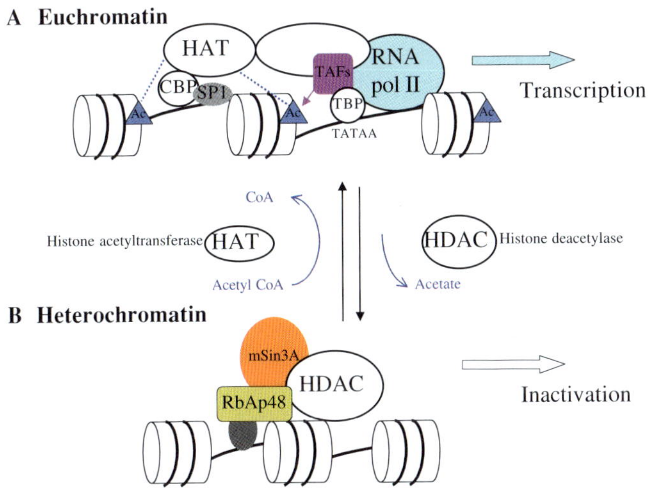

FIG. 4A,B. General regulation of transcription: euchromatin versus heterochromatin. **A** Transcriptionally active euchromatin is shown with nucleosomes, consisting of histones H2A, H2B, H3 and H4, packaged together and blanketed by genomic DNA. Transcription factors, such as Sp1, can bind to the Sp1 site (GGGCGG) and influence gene expression positively by recruiting co-activators such as CREB binding protein (CBP). CBP can associate with a histone acetyltransferase (HAT), which can acetylate histones (H2A, H2B, H3, and/or H4). This acetylation results in the relaxation of the nucleosomes, allowing the binding of the TATA-binding protein (TBP), transcriptional-activating factors (TAFs), and the SWI/SNF complex and, eventually the recruitment of RNA polymerase II, which ultimately catalyzes transcription. **B** The same nucleosomal region can be silenced by directed modifications of chromosome induced by the recruitment of one of six known chromatin-remodeling complexes (reviewed in Dobosy and Selker 2001), which include histone deacetylases (HDACs). Such recruitment results in compaction of the nucleosome and methylation of the DNA

In the MI2/NuRD complex, histone methylation (H3 Lys-9) has been shown to be involved in inactivation of the X chromosome in the mouse (Lee et al. 1999) and in the recruitment of the heterochromatin protein Swi6/HP1 (Hall et al. 2002; Volpe et al. 2002). Methylation of histone H3 Lys-9 creates a binding site for HP1 (Lachner et al. 2001) and H3 Lys-9 was found to be methylated in promoter-specific siRNA-treated cultures of human MCF-7 cells (Kawasaki and Taira 2004). HP1 is an adaptor protein that recognizes methylated lysine residues within histone tails and mediates the repression of transcription by recruiting a histone methyltransferase (Zhang et al. 2002). Moreover, HP1 contains a hinge region, which has been found to have RNA-binding activity, and chromodomains, which are conserved in chromatin-remodeling factors and are able to interact with methylated H3 Lys-9. The repression of transcription by HP1 seems to require both an intact chromodomain and an RNA binding component (Muchardt et al. 2002). The localization of H3m LYs-9 and HP1 to the pericentromeric heterochromatin is abolished by treatment of mouse L929 cells with RNase A (Maison et al. 2002; Bernstein et al. 2003). In addition, treatment of cells with inhibitors of HDAC, such as TSA, results in disruption of the heterochromatin and loss of binding of the HP1 adapter protein (Taddei et al. 2001). HP1 forms a complex with Rb and SUV39HI, and the histone methyltransferase SUV39HI binds not only HP1 but also DNMT1 and DNMT3a (Fuks et al. 2003a,b). Finally, in *Schizosaccharomyces pombe*, there is an association of siRNA with chromodomain-containing proteins that are the homologs of SUV39HI and HP1 (Volpe et al. 2002). This association is an extremely intriguing observation since we can envision that the Mi2/NuRD complex might also bind an siRNA through interactions with HP1 or some as yet to be revealed factor and then might selectively target regions of the chromatin including the promoter that is the target of the siRNA. Such putative regulation of gene expression is shown schematically in Fig. 5.

Alternatively, it is possible that the Sin3-HDAC complex (Figs. 4B, 5), which contains HDACs 1 and 2, and Sin3A and MeCP2, might be involved in siRNA-mediated TGS (Dobosy and Selker 2001). Specifically, the methyl-CpG-binding proteins designated MeCP1 and MeCP2 are good candidates for involvement in siRNA-mediated TGS. MeCP2 binds HDAC and it binds to chromosomes in a methylation-dependant manner; it also contains a transcriptional-repression domain (TRD) (Nan et al. 1998). In mammalian cells, three DNA methyltransferases, namely, DNMT3a, DNMT3b and DNMT1, have been shown to be responsible for DNA methylation (Bestor 2000), and transcriptional gene silencing (TGS) mediated by siRNAs has been shown to involve DNMT1 and DNMT3b (Kawasaki and Taira 2004). Consequently, one mechanism for TGS in mammalian cells might involve siRNA-induced methylation of the target promoter, which recruits the MeCP1 or MeCP2 proteins, which then bind HDAC, with subsequent suppression of transcription.

Ultimately, it may very well turn out that siRNA interacts with one or an amalgamation of the other four gene-regulatory histone deacetylase complexes not discussed in detail above. It is, perhaps, more likely that the siRNA is bound to an as yet undefined complex that contains one or more of the Argonaute proteins (Fig. 5). In plants, Argonaute 4 is actively involved in the methylation of DNA and histones (Zilberman et al. 2003), and Argonaute 2 has been shown to be a functional slicer, that is, a component of the RICS complex with RNase III activity and involved in RNAi in mammals (Liu et al. 2004; Song et al. 2004). The putative siRNA/Argonaute complex might then

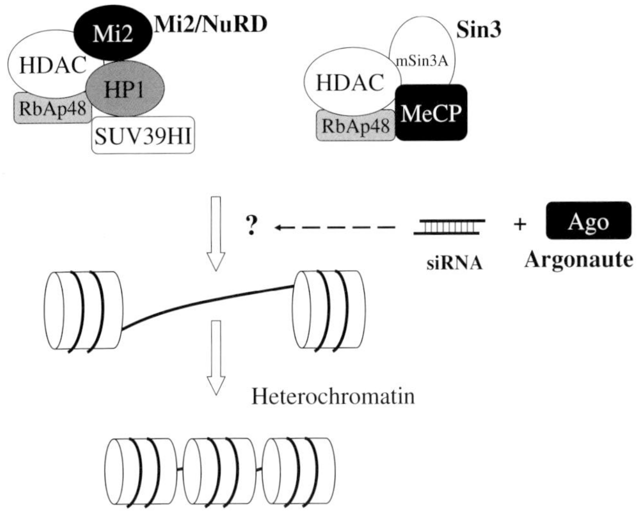

FIG. 5. Putative mechanisms for siRNA-mediated silencing of transcription in mammalian cells. siRNAs might bind directly to a heterochromatin-associated protein (HP1, known to contain an RNA-binding domain) and might subsequently direct HP1 and its cognate complex to the targeted promoter. The siRNA might then function to direct the chromatin-remodeling complex to the targeted promoter, with resultant histone deacetylation, DNA methylation, and silencing of the targeted region. Alternatively, the siRNA somehow associates with the chromatin-remodeling complex Sin3, subsequently facilitating methylation of the targeted promoter and resulting in recruitment of the Sin3 complex and silencing of the promoter. Finally, siRNA might bind to an Argonaute (Ago) or to an Argonaute and an as yet to be described complex that might include an HDAC. The complex might then recruit (or it might already contain) a histone deacetylase that catalyzes the removal of an acetyl group from the modified lysine residues located in the amino-terminal tails of core histones (H2A, H2B, H3 and/or H4) of the nucleosome that contains the targeted promoter region. The modification by HDAC leads ultimately to nucleosomal compaction and a heterochromatic state that might or might not be modified by DNA methylation of the siRNA-targeted promoter

subsequently interact with one or more of the six previously mentioned chromatin-remodeling complexes. Indeed, the observation that siRNA-mediated TGS is associated with CpG methylation in targeted promoters and can be reversed by 5-AzaC or TSA suggests a role for a chromatin-remodeling complex. The story is still unfolding and it is proving to be both exciting and confusing.

6 Conservation of siRNA-Induced TGS and Methylation in Various Species

One potential discrepancy that has yet to be resolved involves reports of the absence of siRNA-induced methylation in mammalian cells. When siRNAs were directed to exons 1 and 4 of the Huntington gene, they were able to reduce the expression of mRNA by more than 50% but were unable to direct methylation of the genomic DNA that corresponded to the targeted exons (Park et al. 2004). By contrast, in plants, siRNAs directed to coding regions have been shown to direct methylation to such targeted regions (Jones et al. 1999; Kamath et al. 2003; and for review, see Agrawal et al. 2003). The suggestion from these data is that the mechanism of siRNA-mediated methylation in plants is fundamentally different from that in animals, or that some regions of animal genomes cannot be targeted by siRNAs for methylation. Such a difference might be the result of differences among DNA methyltransferases, as well as among the signatures of the DNA methylation marks displayed (Jeltsch 2002) in plants, or preferential retention of siRNAs in the cytoplasm of mammalian cells (Zeng and Cullen 2002). Alternatively, these differences might be the result of the evolution of a more complex immune system in animals and of more limited reliance on RNAi for the modulation of unwanted gene expression, as compared with that in plants (Mathieu and Bender 2004). Furthermore, it is possible that the transcriptional and post-transcriptional silencing of gene expression might not be as closely related in mechanistic terms in mammals as it is in plants (Sijen et al. 2001).

An effort to detect sequence-specific siRNA-directed methylation of DNA in mouse oocytes targeted with a ~500-bp short hairpin RNA (shRNA) failed (Svoboda et al. 2004). Such failure supports the hypothesis that not all genomic regions can be targeted by siRNAs; alternatively, that size limitations govern the effectiveness of the shRNA (Svoboda et al. 2004). Localization in the nucleus of the siRNA might be a critical factor in siRNA-mediated TGS, even though $tRNA^{Val}$-driven transcripts are transported efficiently to the cytoplasm (Kuwabara et al. 2001; Kawasaki et al. 2003; Shiota et al. 2004). Indeed, nuclear delivery of synthetic siRNAs directed to the EF1-α promoter appeared essential for detection of a ~40% reduction in the level of expression of EF1-α (Morris et al. 2004).

7 siRNA-Directed Methylation of the Genome

Methylation of DNA is important in gene regulation, imprinting, and inactivation of the X chromosome. In animals, levels and patterns of methylation vary, with low levels in *C. elegans* and *Drosophila melanogaster* and high levels in vertebrate genomes, in which there are large amounts of 5-methyl cytosine. The methylation of the C^5 atom in CpG residues is an important mediator of epigenetic regulation (Bird 2002). CpG islands are regions of unmethylated DNA with a high frequency of the CpG dinucleotide in vertebrate genomes; they are found at the 5′ ends of many housekeeping genes and of some tissue-specific genes. More than 60% of all human genes appear to be associated with CpG islands. Outside such islands, levels of CpG decline to approximately 20% of expected levels (Lander et al. 2001) and such CpGs dinucleotides are partly or completely methylated. Methylation of deoxycytosine occurs at the cytosine of a CpG dinucleotide, resulting in 5-methyl cytosine (Fig. 6).

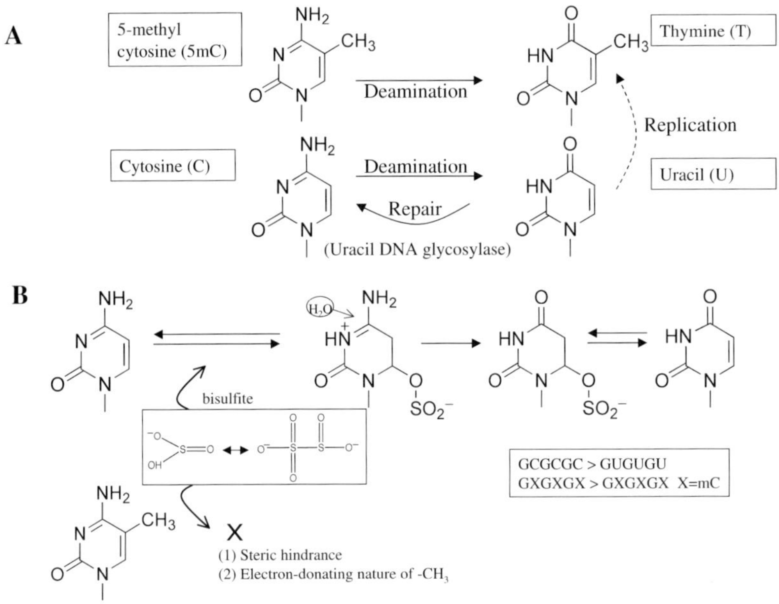

FIG. 6A,B. Deamination of 5-methyl cytosine and cytosine. **A** Deamination of cytosine yields uracil, but the error can be repaired by uracil DNA glycosylase. By contrast, 5-methyl cytosine mutates to thymine at a much higher rate than the rate at which cytosine is deaminated to yield uracil. **B** Bisulfite sequencing. Bisulfite can catalyze deamination of cytosine but not of 5-methyl cytosine because of the electron-donating and bulkier methyl group. This property enables one to detect methylation sites within CpG islands

7.1 siRNA-Directed Methylation and the Evolution of the Genome

It is difficult to resist the temptation to speculate that the methylation of CpG residues impacted on or acted as a naturally selective force in the evolution of the human genome. Specifically, the observation that, in human cells, methylated cytosines can be eliminated over a period of time as the result, most likely, of accidental deamination. Such deamination of a methylated cytosine gives rise to a thymine, which is not recognized by uracil DNA glycosylase. The next result, after cellular division, is that the genome of one of the daughter cells contains a thymine where there has been a cytosine and a genetic change has been induced by what was, initially, an epigenetic modification. In general, this paradigm makes sense if we consider the activity of a particular region of DNA. In the case of promoters, where consistent activity is required for cell fidelity, the CpGs are maintained in a relatively demethylated state that allows any accidental deamination to be appropriately reversed, (i.e., an unmethylated cytosine, after deamination, is recognized as uracil by uracil DNA glycosylase and correctly excised and replaced with a cytosine) (Fig. 6). By contrast, in non-promoter regions, such as Alu repeated regions, which are typically littered with retrotransposons, or mRNA-coding regions actively targeted by lentiviruses, such as

HIV-1 (Wu et al. 2003), one generally finds little of the genome in the form of CpG islands (Cross and Bird 1995). Moreover, the vast majority of non-long terminal repeats (non-LTRs) retrotransposons (LINEs or L1s) in mammals include 5′ truncations, rearrangements, and nonsense mutations (Finnegan 1997), suggesting that application of selective force might have inhibited the activity of these particular genomic regions. There is considerable evidence to indicate that transposable elements are probably the intended targets of RdDM, as noted above. Small RNAs corresponding to transposon sequences have been found in *Arabidopsis* and tobacco (Hamilton et al. 2002; Llave et al. 2002; Lippman et al. 2003; Xie et al. 2004), suggesting that some dsRNAs are generated from transposon sequences. Another piece of evidence is that mutations in RNA-processing factors resulted in the loss of cytosine methylation of the *AtSN1* retrotransposable element in *Arabidopsis* (Lippman et al. 2003; Zilberman et al. 2003; Xie et al. 2004). It is possible that the mechanism driving such selective inactivation might be mediated by RNAi. After integration into the host genome, transposons might be transcribed from endogenous promoters. Subsequently, they might serve as the templates for synthesis of dsRNA by RdRP or the source of both sense and antisense transcripts (Fig. 7) (Sijen and Plasterk 2003). Methylation of transposons and related sequence has two effects on their activity, namely, transcriptional silencing by methylation at promoter regions, as widely recognized and observed in many species and at many developmental stages, and transposon immobilization. Single mutations in either methyltransferase 1 (MET1) or chromomethyltransferase 3 (CMT3) in *Arabidopsis* resulted in the accumulation of transcripts of the *CACTA* transposon but had a much smaller effect on mobilization (Kato et al. 2003). Since MET1 and CMT3 are known to be a CG and a non-CG methyltransferase, respectively, both CG and non-CG methylation might be necessary for full immobilization of a transposon.

The Alu family of retrotransposons, the most abundant short interspersed elements (SINEs), makes up approximately 10% of the human genome. Approximately 20% of Alu elements consist of CpG dinucleotides, accounting for up to 30% of the total 5 mC sites in the human genome. However, fewer than 1% of these repetitive elements have the capacity for transposition despite the fact that many of them still retain a functional promoter (Bird 2002). If siRNA-mediated methylation were to play some role in the immobilization of Alu repetitive element, its contribution would be expanded

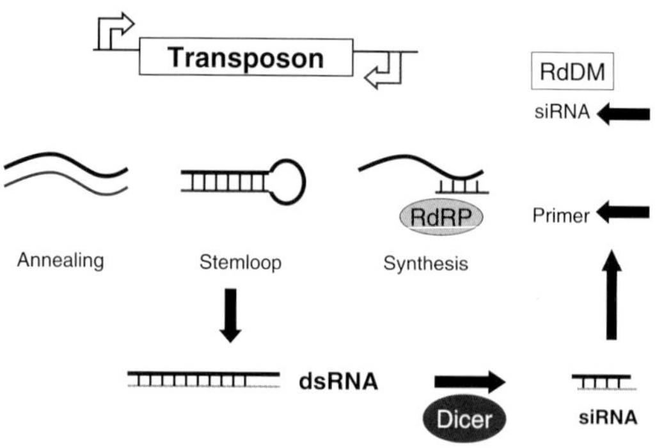

FIG. 7. Generation of dsRNA from transposon sequences. dsRNA can be generated by the annealing of sense and antisense RNAs or by synthesis of novel RNA by RNA-dependent RNA polymerase (RdRP). Sequences generated by processing from dsRNA can serve as priming sequences for RdRP, as well as serving as siRNAs

to include maintenance of the stability of the genome size and genome structure, and would not just include the regulation of gene expression.

7.2 siRNA-Directed Methylation and Its Role in the Protection of Animal Cells

The role of RNAi in protecting animal cells from viral infection remains to be clarified since dsRNAs can induce not only RNAi but also an interferon-mediated non-specific RNA response and other non-specific responses that lead to interruption of protein synthesis and, ultimately, cell death (Abraham et al. 1999). However, if there is indeed an RNAi-based mechanism in mammalian cells that is designed to limit the extent to which transposable elements can move around the genome, such a mechanism might prove to be a double-edged sword, at least with regard to potential future therapeutic applications. Might the establishment of an epigenetic marker, introduced by promoter-directed siRNAs and limiting expression of a particular gene, function ultimately as a natural selective force, driving the evolution of the genome? Such an epigenetic marker could be introduced via the methylation and accidental-deamination scenario described previously. Alternatively, a chromatin-remodeling complex, possibly including components that appeared to be involved in the recently observed TGS, might remain bound to the particular siRNA-targeted region, and/or the newly methylated CpGs of the targeted promoter might be sequestered by methylation-specific proteins and protected from water and accidental deamination (Dobosy and Selker 2001). Whatever the details, it is certain that the most stable cell (i.e., the one least prone to error) is the cell that is in equilibrium internally and with its surroundings. Thus, we can assume that the more control we try to exert over the profile of gene expression in a cell, the greater the potential for untoward effects.

8 Concluding Remarks

When we consider the regulation of gene expression, as described in other chapters of this book, we must consider, in addition to the post-transcriptional suppression of gene expression in the cytoplasm, the regulation of gene expression at the transcriptional level in the nucleus (Fig. 2). The advantage of siRNA technology is that it allows us to easily combine several different siRNAs in a cocktail with which we can target different sites within a transcript and/or exploit both TGS and PTGS simultaneously. We can also use vector-driven siRNAs or shRNAs similarly, since multiple siRNAs can be encoded within a single DNA template. With our recently developed ability to express long shRNAs, of more than 100 bp, we can anticipate even more efficient gene silencing by both TGS and PTGS technologies (Akashi et al. 2005). With each advance, we move one step closer to the creation of therapeutic RNA.

References

Abraham N, Stojdl DF, Duncan PI, Methot N, Ishii T, Dube M, Vanderhyden BC, Atkins HL, Gray DA, McBurney MW, Koromilas AE, Brown EG, Sonenberg N, Bell JC (1999) Char-

acterization of transgenic mice with targeted disruption of the catalytic domain of the double-stranded RNA-dependent protein kinase. PKR. J Biol Chem 274(9):5953–5962
Agrawal N, Dasaradhi PV, Mohmmed A, Malhotra P, Bhatnagar RK, Mukherjee SK (2003) RNA interference: biology, mechanism, and applications. Microbiol Mol Biol Rev 67(4): 657–685
Akashi H, Matsumoto S, Taira K (2005) Gene discovery by ribozyme and siRNA libraries. Nat Rev Mol Cell Biol in press
Ambros V (2004) The functions of animal microRNAs. Nature 431(7006):350–355
Aufsatz W, Mette MF, van der Winden J, Matzke AJ, Matzke M (2002) RNA-directed DNA methylation in Arabidopsis. Proceedings of the National Academy of Sciences 99(4): 16499–16506
Aufsatz W, Mette MF, van der Winden J, Matzke M, Matzke AJ (2002) HDA6, a putative histone deacetylase needed to enhance DNA methylation induced by double-stranded RNA. Embo J 21(24):6832–6841
Bartel DP (2004) MicroRNAs: genomics, biogenesis, mechanism, and function. Cell 116(2): 281–297
Bernstein E, Caudy AA, Hammond SM, Hannon GJ (2001) Role for a bidentate ribonuclease in the initiation step of RNA interference. Nature 409(6818):363–366
Bernstein E, Kim SY, Carmell MA, Murchison EP, Alcorn H, Li MZ, Mills AA, Elledge SJ, Anderson KV, Hannon GJ (2003) Dicer is essential for mouse development. Nat Genet 35(3):215–217
Bestor TH (2000) The DNA methyltransferases of mammals. Hum Mol Genet 9(16): 2395–2402
Bird A (2002) DNA methylation patterns and epigenetic memory. Genes Dev 16(1):6–21
Cross SH, Bird AP (1995) CpG islands and genes. Curr Opin Genet Dev 5(3):309–314
Denli AM, Tops BB, Plasterk RH, Ketting RF, Hannon GJ (2004) Processing of primary microRNAs by the Microprocessor complex. Nature 432(7014):231–235
Dobosy JR, Selker EU (2001) Emerging connections between DNA methylation and histone acetylation. Cell Mol Life Sci 58(5–6):721–727
Dobosy JR, Selker EU (2001) Emerging connections between DNA methylation and histone acetylation. Cellular and Molecular Life Sciences 58:721–727
Egger G, Liang G, Aparicio A, Jones PA (2004) Epigenetics in human disease and prospects for epigenetic therapy. Nature 429(6990):457–463
Elbashir SM, Harborth J, Lendeckel W, Yalcin A, Weber K, Tuschl T (2001) Duplexes of 21-nucleotide RNAs mediate RNA interference in cultured mammalian cells. Nature 411(6836):494–498
Elbashir SM, Lendeckel W, Tuschl T (2001) RNA interference is mediated by 21- and 22-nucleotide RNAs. Genes and Development 15:188–200
Feinberg EH, Hunter CP (2003) Transport of dsRNA into cells by the transmembrane protein SID-1. Science 301(5639):1545–1547
Finnegan DJ (1997) Transposable elements: how non-LTR retrotransposons do it. Curr Biol 7(4):R245–248
Fire A, Xu S, Montgomery MK, Kostas SA, Driver SE, Mello CC (1998) Potent and specific genetic interference by double-stranded RNA in Caenorhabditis elegans. Nature 391(6669):806–811
Fukagawa T, Nogami M, Yoshikawa M, Ikeno M, Okazaki T, Takami Y, Nakayama T, Oshimura M (2004) Dicer is essential for formation of the heterochromatin structure in vertebrate cells. Nat Cell Biol 6(8):784–791
Fuks F, Hurd PJ, Deplus R, Kouzarides T (2003)a The DNA methyltransferases associate with HP1 and the SUV39H1 histone methyltransferase. Nucleic Acids Res 31(9): 2305–2312
Fuks F, Hurd PJ, Wolf D, Nan X, Bird AP, Kouzarides T (2003)b The methyl-CpG-binding protein MeCP2 links DNA methylation to histone methylation. J Biol Chem 278(6): 4035–4040

Gregory RI, Yan KP, Amuthan G, Chendrimada T, Doratotaj B, Cooch N, Shiekhattar R (2004) The Microprocessor complex mediates the genesis of microRNAs. Nature 432(7014):235–240

Hall IM, Shankaranarayana GD, Noma K, Ayoub N, Cohen A, Grewal SI (2002) Establishment and Maintenance of a Heterochromatin Domain. Science 297:2232–2237

Hamilton A, Voinnet O, Chappell L, Baulcombe D (2002) Two classes of short interfering RNA in RNA silencing. Embo J 21(17):4671–4679

Hammond SM, Bernstein E, Beach D, Hannon GJ (2000) An RNA-directed nuclease mediates post-transcriptional gene silencing in Drosophila cells. Nature 404(6775):293–296

Hammond SM, Boettcher S, Caudy AA, Kobayashi R, Hannon GJ (2001) Argonaute2, a link between genetic and biochemical analyses of RNAi. Science 293(5532):1146–1150

Han J, Lee Y, Yeom KH, Kim YK, Jin H, Kim VN (2004) The Drosha-DGCR8 complex in primary microRNA processing. Genes Dev in press

Hutvagner G, Zamore PD (2002) RNAi: nature abhors a double-strand. Curr Opin Genet Dev 12(2):225–232

Jablonka E, Lamb MJ (2003) Epigenetic heredity in evolution. Tsitologiia 45(11):1057–1072

Jeltsch A (2002) Beyond Watson and Crick: DNA methylation and molecular enzymology of DNA methyltransferases. Chembiochem 3(4): 274–293

Jin P, Zarnescu DC, Ceman S, Nakamoto M, Mowrey J, Jongens TA, Nelson DL, Moses K, Warren ST (2004) Biochemical and genetic interaction between the fragile X mental retardation protein and the microRNA pathway. Nat Neurosci 7(2):113–117

Jones L, Hamilton AJ, Voinnet O, Thomas CL, Maule AJ, Baulcombe DC (1999) RNA-DNA interactions and DNA methylation in post-transcriptional gene silencing. Plant Cell 11(12):2291–2301

Jones L, Ratcliff F, Baulcombe DC (2001) RNA-directed transcriptional gene silencing in plants can be inherited independently of the RNA trigger and requires Met1 for maintenance. Current Biology 11:747–757

Kamath RS, Fraser AG, Dong Y, Poulin G, Durbin R, Gotta M, Kanapin A, Le Bot N, Moreno S, Sohrmann M, Welchman DP, Zipperlen P, Ahringer J (2003) Systematic functional analysis of the Caenorhabditis elegans genome using RNAi. Nature 421(6920):231–237

Kato M, Miura A, Bender J, Jacobsen SE, Kakutani T (2003) Role of CG and non-CG methylation in immobilization of transposons in Arabidopsis. Curr Biol 13:421–426

Kawasaki H, Kuwabara T, Miyagishi M, Taira K (2003) Identification of functional genes by libraries of ribozymes and siRNAs. Nucleic Acids Res Suppl (3):331–332

Kawasaki H, Suyama E, Iyo M, Taira K (2003) siRNAs generated by recombinant human Dicer induce specific and significant but target site-independent gene silencing in human cells. Nucleic Acids Res 31(3):981–987

Kawasaki H, Taira K (2004) Induction of DNA methylation and gene silencing by short interfering RNAs in human cells. Nature;431(7005):211–217

Kawasaki H, Wadhwa R, Taira K (2004) World of small RNAs: from ribozymes to siRNA and miRNA. Differentiation 72(2–3):58–64

Kim DH, Behlke MA, Rose SD, Chang MS, Choi S, Rossi JJ (2005) Synthetic dsRNA Dicer substrates enhance RNAi potency and efficiency. Nat Biotechnol 23(2):222–226

Kuwabara T, Hsieh J, Nakashima K, Taira K, Gage FH (2004) A small modulatory dsRNA specifies the fate of adult neural stem cells. Cell 116(6):779–793

Kuwabara T, Warashina M, Sano M, Tang H, Wong-Staal F, Munekata E, Taira K (2001) Recognition of engineered tRNAs with an extended 3′ end by Exportin-t (Xpo-t) and transport of tRNA-attached ribozymes to the cytoplasm in somatic cells. Biomacromolecules 2(4):1229–1242

Lachner M, O'Carroll D, Rea S, Mechtler K, Jenuwein T (2001) Methylation of histone H3 Lysine 9 creates a binding site for HP1 proteins. Nature 410:116–120

Lander ES, Linton LM, Birren B, Nusbaum C, Zody MC, Baldwin J et al. (2001) Initial sequencing and analysis of the human genome. Nature 409:860–921

Lavrov SA, Mavrodiev EV (2003) Epigenetic inheritance and its possible role in the evolution of plant species. Zh Obshch Biol 64(5):403–420

Lee JT, Lu N, Han Y (1999) Genetic analysis of the mouse X inactivation center defines an 80-kb multifunction domain. Proc Natl Acad Sci U S A 96(7):3836–3841

Lee Y, Kim M, Han J, Yeom KH, Lee S, Baek SH, Kim VN (2004) MicroRNA genes are transcribed by RNA polymerase II. EMBO J 23(20):4051–4060

Lee Y, Lee Y, Ahn C, Han J, Choi H, Kim J, Yim J, Lee J, Provost P, Radmark O, Kim S, Kim VN (2003) The nuclear RNase III Drosha initiates microRNA processing. Nature 425(6956):415–419

Lindroth AM, Cao X, Jackson JP, Zilberman D, McCallum CM, Henikoff S, Jacobsen SE (2001) Requirement of CHROMOMETHYLASE3 for maintenance of CpXpG methylation. Science 292(5524):2077–2080

Lingel A, Simon B, Izaurralde E, Sattler M (2004) Nucleic acid 3′-end recognition by the Argonaute2 PAZ domain. Nature Struct Mol Biol 11(6):576–577

Lippman Z, Gendrel AV, Black M, Vaughn MW, Dedhia N, McCombie WR, Lavine K, Mittal V, May B, Kasschau KD, Carrington JC, Doerge RW, Colot V, Martienssen R (2004) Role of transposable elements in heterochromatin and epigenetic control. Nature 430(6998):471–476

Lippman Z, May B, Yordan C, Singer T, Martienssen R (2003) Distinct Mechanisms Determine Transposon Inheritance and Methylation via Small Interfering RNA and Histone Modification. PLoS Biol 1(3):E67.

Liu J, Carmell MA, Rivas FV, Marsden CG, Thomson JM, Song JJ, Hammond SM, Joshua-Tor L, Hannon GJ (2004) Argonaute2 is the catalytic engine of mammalian RNAi. Science 305(5689):1437–1441

Llave C, Kasschau KD, Rector MA, Charrington JC (2002) Endogenous and silencing-associated small RNAs in plants. Plant Cell 14:1605–1619

Llave C, Xie Z, Kasschau KD, Carrington JC (2002) Cleavage of Scarecrow-like mRNA targets directed by a class of Arabidopsis miRNA. Science 297(5589):2053–2056

Lund E, Guttinger S, Calado A, Dahlberg JE, Kutay U (2004) Nuclear export of microRNA precursors. Science 303(5654):95–98

Lusser A (2004) Acetylated, methylated, remodeled: chromatin states for gene regulation. Current Opinion in Plant Biology 5:437–443

Lutsenko E, Bhagwat AS (1999) Principal causes of hot spots for cytosine to thymine mutations at sites of cytosine methylation in growing cells. A model, its experimental support and implications. Mutat Res 437(1):11–20

Ma JB, Ye K, Patel DJ (2004) Structural basis for overhang-specific small interfering RNA recognition by the PAZ domain. Nature 429(6989):318–322

Maison C, Bailly D, Peters AH, Quivy JP, Roche D, Taddei A, Lachner M, Jenuwein T, Almouzni G (2002) Nat Genet 30(3):329–334

Mansfield JH, Harfe BD, Nissen R, Obenauer J, Srineel J, Chaudhuri A, Farzan-Kashani R, Zuker M, Pasquinelli AE, Ruvkun G, Sharp PA, Tabin CJ, McManus MT (2004) MicroRNA-responsive 'sensor' transgenes uncover Hox-like and other developmentally regulated patterns of vertebrate microRNA expression. Nat Genet 36(10):1079–1083

Mathieu O, Bender J (2004) RNA-directed DNA methylation. J Cell Sci 117(Pt 21):4881–4888

Matzke MA, Matzke AJ, Pruss GJ, Vance VB (2001) RNA-based silencing strategies in plants. Curr Opin Genet Dev 11(2):221–227

McNairn AJ, Gilbert DM (2003) Epigenomic replication: linking epigenetics to DNA replication. Bioessays 25(7):647–656

Mette MF, Aufsatz W, Van der Winden J, Matzke AJM Matzke MA (2000) Transcriptional silencing and promoter methylation triggered by double-stranded RNA. EMBO J 19(19): 5194–5201

Morris KV, Chan SW, Jacobsen SE, Looney DJ (2004) Small Interfering RNA-Induced Transcriptional Gene Silencing in Human Cells. Science 305(5688):1289–1292

Mourelatos Z, Dostie J, Paushkin S, Sharma A, Charroux B, Abel L, Rappsilber J, Mann M, Dreyfuss G (2002) miRNPs: a novel class of ribonucleoproteins containing numerous microRNAs. Genes Dev 16(6):720–728

Muchardt C, Guilleme M, Seeler JS, Trouche D, Dejean A, Yaniv M (2002) Coordinated methyl and RNA binding is required for heterochromatin localization of mammalian HP1alpha. EMBO Rep 3(10):975–981

Nan X, Ng HH, Johnson CA, Laherty CD, Turner BM, Eisenman RN, Bird A (1998) Transcriptional repression by the methyl-CpG-binding protein MeCP2 involves histone deacetylase complex. Nature 393:386–389

Noma K, Sugiyama T, Cam H, Verdel A, Zofall M, Jia S, Moazed D, Grewal SI (2004) RITS acts in cis to promote RNA interference-mediated transcriptional and post-transcriptional silencing. Nat Genet 36(11):1174–1180

Papp I, Mette MF, Aufsatz W, Daxinger L, Schauer SE, Ray A, van der Winden J, Matzke M, Matzke AJ (2003) Evidence for nuclear processing of plant micro RNA and short interfering RNA precursors. Plant Physiol 132(3):1382–1390

Park CW, Chen Z, Kren BT, Steer CJ (2004) Double-stranded siRNA targeted to the huntingtin gene does not induce DNA methylation. Biochem Biophys Res Commun 323(1):275–280

Park W, Li J, Song R, Messing J, Chen X (2002) CARPEL FACTORY, a Dicer homolog, and HEN1, a novel protein, act in microRNA metabolism in Arabidopsis thaliana. Curr Biol 12(17):1484–1495

Reinhart BJ, Weinstein EG, Rhoades MW, Bartel B, Bartel DP (2002) MicroRNAs in plants. Genes Dev 16(13):1616–1626

Schauer SE, Jacobsen SE, Meinke DW, Ray A (2002) DICER-LIKE1: blind men and elephants in Arabidopsis development. Trends Plant Sci, 7(11):487–491

Shiota M, Sano M, Miyagishi M, Taira K (2004) Ribozymes: applications to functional analysis and gene discovery. J Biochem (Tokyo) 136(2):133–147

Siolas D, Lerner C, Barchard J, Ge W, Linsley PS, Paddison PJ, Hannon GJ, Cleary MA (2005) Synthetic shRNAs as potent RNAi triggers. Nat Biotechnol 23(2):227–231

Sijen T, Fleenor J, Simmer F, Thijssen KL, Parrish S, Timmons L, Plasterk RH, Fire A (2001) On the role of RNA amplification in dsRNA-triggered gene silencing. Cell 107(4):465–476

Sijen T, Plasterk RHA (2003) Transposon silencing in the Caenorhabditis elegans germ line by natural RNAi. Nature 426:310–314

Sijen T, Vign I, Rebocho A, Blokland R, Roelofs D, Mol J, Kooter J (2001) Transcriptional and posttranscriptional gene silencing are mechansitically related. Current Biology 11:436–440

Song JJ, Smith SK, Hannon GJ, Joshua-Tor L (2004) Crystal structure of Argonaute and its implications for RISC slicer activity. Science 305(5689):1434–1437

Stark GR, Kerr IM, Williams BR, Silverman RH, Schreiber RD (1998) How cells respond to interferons. Annu Rev Biochem. 67:227–264

Svoboda P, Stein P, Filipowicz W, Schultz RM (2004) Lack of homologous sequence-specific DNA methylation in response to stable dsRNA expression in mouse oocytes. Nucleic Acids Res 32(12):3601–3606

Svoboda P, Stein P, Hayashi H, Schultz RM (2000) Selective reduction of dormant maternal mRNAs in mouse oocytes by RNA interference. Development. 127(19):4147–56

Taddei A, Maison C, Roche D, Almouzni G (2001) Reversible disruption of pericentric heterochromatin and centromere function by inhibiting deacetylases. Nat Cell Biol 3(2):114–120

Tang G, Reinhart BJ, Bartel DP, Zamore PD (2003) A biochemical framework for RNA silencing in plants. Genes Dev 17(1):49–63

Tijsterman M, Ketting RF, Plasterk RH (2002) The genetics of RNA silencing. Annu Rev Genet 36:489–519

Tuschl T, Zamore PD, Lehmann R, Bartel DP, Sharp PA (1999) Targeted mRNA degradation by double-stranded RNA in vitro. Genes Dev 13(24): 3191–3197

Volpe TA, Kidner C, Hall IM, Teng G, Grewal SI, Martienssen RA (2002) Regulation of heterchromatic silencing and histone H3 Lysine-9 methylation by RNAi. Science 297: 1833–1837

Wassenegger M (2000) RNA-directed DNA methylation. Plant Molecular Biology 43: 203–220

Wassenegger M, Graham MW, Wang MD (1994) RNA-directed de novo methylation of genomic sequences in plants. Cell 76:567–576

Waterhouse PM, Wang MB, Lough T (2001) Gene silencing as an adaptive defence against viruses. Nature 411(6839):834–842

Wianny F, Zernicka-Goetz M (2000) Specific interference with gene function by double-stranded RNA in early mouse development. Nat Cell Biol 2(2):70–75

Williams BR (1997) Role of the double-stranded RNA-activated protein kinase (PKR) in cell regulation. Biochem Soc Trans 25(2):509–153

Winston WM, Molodowitch C, CP Hunter (2002) Systemic RNAi in C. elegans requires the putative transmembrane protein SID-1. Science 295(5564):2456–2459

Wu X, Li Y, Crise B, Burgess SM (2003) Transcription start regions in the human genome are favored targets for MLV integration. Science 300(5626):1749–1451

Xie Z, Johansen LK, Gustafson AM, Kasschau KD, Lellis AD, Zilberman D, Jacobsen SE, Carrington JC (2004) Genetic and functional diversification of small RNA pathways in plants. PLoS Biology 2:642–652

Yekta S, Shih IH, Bartel DP (2004) MicroRNA-directed cleavage of HOXB8 mRNA. Science 304(5670):594–546

Zamore PD, Tuschl T, Sharp PA, Bartel DP (2000) RNAi: double-stranded RNA directs the ATP-dependent cleavage of mRNA at 21 to 23 nucleotide intervals. Cell 101(1):25–33

Zeng Y, Cullen BR (2002) RNA interference in human cells is restricted to the cytoplasm. RNA 8(7):855–860

Zhang CL, McKinsey TA, Olson EN (2002) Association of class II histone deacetylases with heterochromatin protein 1: potential role for histone methylation in control of muscle differentiation. Molecular and Cellular Biology 20:7302–7312

Zilberman D, Cao X, Jacobsen SE (2003) ARGONAUTE4 control of locus-specific siRNA accumulation and DNA and histone methylation. Science 299(5607):716–719

Intracellular Delivery of Nucleic Acids: Differences Between Transfection and siFection Reflect Differences Between DNA and RNA, and Between Oligodeoxynucleotides and Oligonucleotides

Michael Keller[1] and Andrew D. Miller[1,2]

1 Introduction

1.1 Delivery of DNA—Transfection

Cationic liposomes are widely used in the process of DNA (gene) delivery to cells (*transfection*) in vitro. A variety of cationic liposome formulations are now commercially available that are generally optimized for specific rather than general applications in vitro owing to their different formulations and transfection characteristics (Miller 1998, 2003). Structure-activity correlations remain difficult to establish, with occasional exceptions (Stewart et al. 2001; Keller et al. 2003b). Therefore, differences in cationic liposome-mediated transfection efficacy under different experimental conditions involving different cell lines are quite poorly understood. However, what is clear is that the nature of the lipids and their formulations are pivotal factors in the constitution of a successful DNA delivery vehicle in vitro. For this reason, biophysical studies on cationic liposome-DNA (lipoplex, LD) particle formation provide an important primary evaluation tool to attempt an understanding of structure/activity relationships pertinent to cationic liposome-mediated nucleic acid delivery to cells (Gershon et al. 1993; Zelphati et al. 1998; Hansma et al. 1998; Zuidam and Barenholz 1999; Golan et al. 1999; Stewart et al. 2001; Keller et al. 2003b). Of especial importance seems to be an understanding of the immense structural reorganization that both DNA and cationic liposomes can undergo when LD particle formation takes place directly following initial electrostatic interactions (Pector et al. 2000).

With respect to in vivo applications, extremely rapid developments in molecular biology have been making gene therapy a potentially promising new therapeutic

---

[1] IC-Vec Ltd, Flower Building, Armstrong Rd, London SW7 2AZ, UK

[2] Imperial College Genetic Therapies Centre, Department of Chemistry, Imperial College London, London SW7 2AZ, UK

modality (Lander et al. 2001; Venter et al. 2001). As of July 1, 2004, there were 976 gene-therapy clinical trial protocols logged worldwide. Remarkably, 8.6% of these logged clinical trials involved non-viral, cationic-liposome-based approaches to therapy (*lipofection*). However, clinical trials and other studies have amply demonstrated that acute and chronic instability of LD particles with respect to fusion and aggregation even in low-ionic strength buffer substantially reduces the efficacy, effectiveness, and reliability of cationic liposome-mediated transfection in vivo and in clinical applications (Templeton et al. 1997). Hence, cationic liposome-based approaches need a period of logical and patient development of new lipids, formulations, and other technical innovations prior to the next wave of animal proof-of-principle studies and clinical trials thereafter (Miller 2003). Simple cationic liposome systems alone are unlikely to be clinically effective.

Formulations of LD particles can be improved through the formation of ternary LD particles in which DNA is pre-condensed by a cationic entity prior to combination with cationic liposomes. Cationic entities such as cationic peptides or polyamines condense DNA firstly by neutralizing the negative charges of the phospho-diester backbone of DNA. Following charge neutralization, DNA condenses to a fraction of the volume of its natural state (Schellmann and Gosule 1976; Bloomfield 1996). This is particularly true of plasmid DNA (pDNA), which is the most frequently used form of gene expression system delivered by cationic liposomes. Poly- and oligo-L-lysine, protamine, polyimidazole, and oligo-L-lysine containing peptides are some of the best-known DNA pre-condensing agents (Laemmli, 1975; Sorgi et al. 1997; Schwartz et al. 1999; Zhang et al. 1999; Pack et al. 2000). Even bifunctional cationic peptides containing a receptor-targeting moiety have also been described in conjunction with LD systems (Colin et al. 1998, 2001; Cooper et al. 1999). Alternative non-peptide species that can be potentially used as pre-condensing agents include histone proteins, polyethyleneimine, poly[(2-dimethylamino)ethylmethacrylate], cationic peptoids, and starburst dendrimers (Haensler and Szoka 1993a, b; Fritz et al. 1996; Hagstrom et al. 1996; KukowskaLatallo et al. 1996; Murphy et al. 1998; Remy et al. 1998).

We have recently described a ternary LD system, referred to as liposome:mu:DNA (LMD) (Murray et al. 2001; Tagawa et al. 2002). This synthetic non-viral vector system is based around using the cationic adenoviral core peptide μ (mu or M, MRRAHHRRRR-ASHRRMRGG) that pre-condenses pDNA into MD (mu:DNA) nanoparticles (100 ± 20 nm in diameter) at an optimal ratio of 0.6 equivalents (w/w) peptide per pDNA (Keller et al. 2002). Thereafter, combination of negatively charged MD particles with cationic liposomes results in essentially mono-disperse, uniform LMD particles (120 ± 30 nm in diameter) (Tagawa et al. 2002). However, although LMD particles have real elements of stability in serum and are surprisingly resistant to aggregation in low-ionic strength medium as well as amenable to long-term storage, the intracellular delivery properties of LMD systems are insufficient. Studies on intracellular trafficking events (Keller et al. 2003a), show that LMD particles disintegrate to individual components within minutes of cell entry by endocytosis. Thereafter, the peptide component localizes to the cellular nuclei first (15 min), followed by the pDNA (30–45 min) (Fig. 1) (Keller et al. 2003a). When cells are fully growth arrested (with aphidicolin), pDNA does not enter cellular nuclei at all. This is in spite of the fact that mu peptide comprises a potent nuclear localization sequence (NLS). Nuclear entry

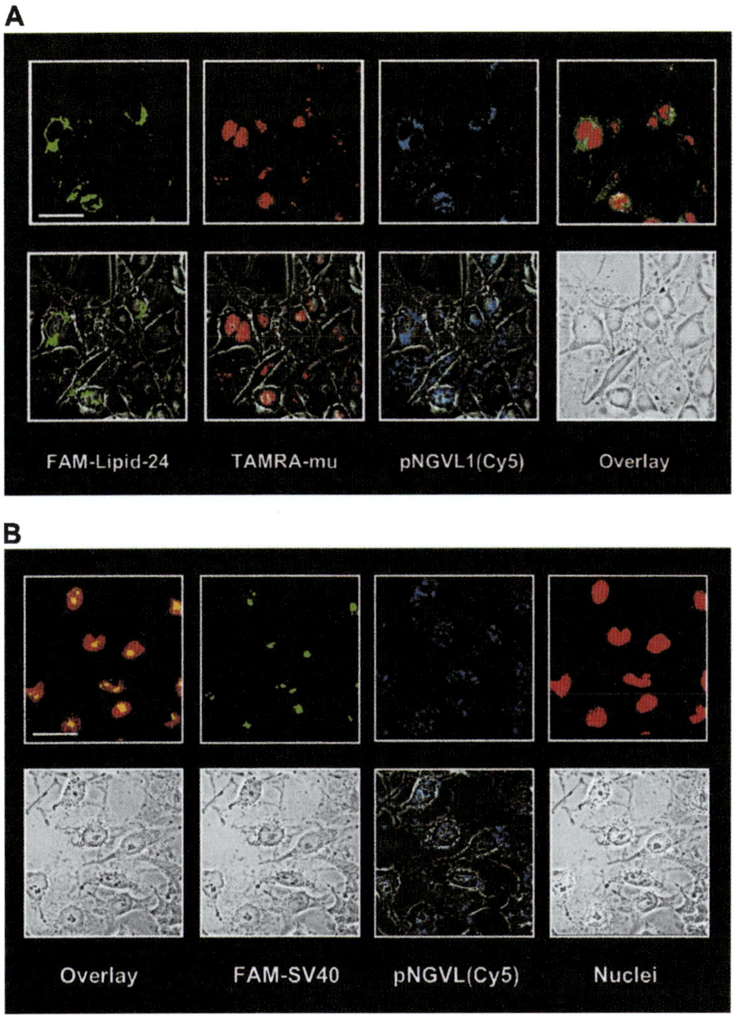

Fig. 1. Confocal microscopy analyses of three-fold labeled liposome : mu : DNA (LMD) (**A**) and a similar two-fold-labeled ternary lipoplex (LD) particle (**B**) after 15 min incubation of vector system with 56 FHTE8o^{-} cells. In the case of three-fold-labeled LMD (**A**), TAMRA-mu is in *red*, the FAM-Lp-24 lipid in *green* and the pDNA(Cy5) in *blue*. For the two-fold-labeled vector system (**B**), FAM-SV40 derived peptide is in *green* and the pDNA(Cy5) in *blue*. The *red* color in **B** is due to staining of the nuclei with propidium iodide. *Bar* 10 μm. The figure is reproduced from Keller et al. (2003a) with permission from J. Wiley & Sons

into the nuclei of growth-arrested (quiescent cells) is an essential prerequisite for gene therapy since most target cells are quiescent at the time of DNA delivery, even in tumours. This nuclear barrier problem and the requirement for long-term expression after delivery to cells represent the two most significant barriers unique to effective DNA delivery for gene therapy today. Efforts to overcome these drawbacks are currently underway in several laboratories.

2 Delivery of Short, Double-Stranded Synthetic RNA—siFection

RNA interference is a new frontier in nucleic acid delivery (Novina and Sharp 2004). Initially, gene suppression phenomena were characterized in the early 1990s, when plant biologists working with petunias were surprised to find that introducing numerous copies of a gene led to petals with an unexpectedly dark purple hue instead of the more usual patchy appearance (Napoli et al. 1990; van der Krol et al. 1990). Moreover, when plants were infected with an RNA virus that had been genetically engineered to contain fragments of a plant gene, the plant's gene itself became inactivated (Wassenegger et al. 1994). Only when Mello and co-workers published their pioneering work in 1998 did both these observation make sense in terms of the phenomenon of "RNA interference" (Fire et al. 1998).

The basis of RNA interference has recently come into view. RNA viruses replicate through double stranded RNA (dsRNA) intermediates that are now known to be cleaved by a cytosolic enzyme (DICER) (Bernstein et al. 2001) into short RNA duplexes typically comprising a central anti-parallel 19- to 21-bp double helical region with 2-base overhangs at each 3′-end. Each 5′-end is also phosphorylated (Tuschl et al. 1999; Zamore et al. 2000; Elbashir et al. 2001a, b). These short stretches of dsRNA have been dubbed short interfering or small (short) interference RNA (siRNA). A multi-protein complex called RISC (RNA-induced silencing complex) is now known to directionally separate sense (S) from antisense (AS) strands of the siRNA duplex and become activated through sequestration of the AS strand (Bernstein et al. 2001). This RISC-bound AS strand presumably becomes a tool for the sequence-selective molecular recognition of other RNA molecules. Hence viral mRNA is captured by activated RISC and becomes degraded catalytically by one of the RISC proteins (Ago2), thus eliminating the source of infection (Meister et al. 2004).

The phenomenon of RNA interference would be little more than a curiosity of viral anti-infectivity were it not for the growing realization that the phenomenon appears to have significant applications and implications way beyond the boundaries of the initial observations and discoveries. Apparently, siRNAs are generated when transposons, viruses, or endogenous genes are transcribed into dsRNA or when dsRNA is introduced experimentally into plant and animal cells to trigger RNA interference (*gene silencing*) (Fire et al. 1998; Hamilton and Baulcombe 1999; Zamore et al. 2000; Elbashir et al. 2001a). Moreover, synthetic siRNAs can also surrogate for siRNA generated in situ, allowing for purpose-designed, specific gene (or, more correctly, mRNA) knock-down that results in specific down-regulation of protein expression in cells such as cultured *Xenopus* oocytes and even in mammalian cells (Abdelrahim et al. 2002; Anantharam et al. 2003). In other words, numbers of siRNAs may be identified with the capacity to complement corresponding regions in a target mRNA of interest, after which these siRNAs may be sifted with defined rule sets so as to identify those siRNAs (approx. 1%–3%/gene) with S- and AS-strand base sequences that are optimal for RISC-mediated destruction of the target mRNA of interest (Ui-Tei et al. 2004). Furthermore, these sequences may be screened at one higher level by means of high-end bioinformatics analyses (such as the siDIRECT analysis), ensuring that they have no likelihood of cross-reactivity with other off-target mRNA sequences and hence have a reduced likelihood of eliciting undesirable cellular toxicities (Naito et al. 2004).

Given such capacities, the gateway is now firmly open in principle for the RNA interference phenomenon and for siRNA technology to transform functional genomics programs and even nucleic-acid-based therapeutics itself. Nevertheless, such statements need to be tempered by the reality that research into applications of RNA interference is generally at a preliminary level. Early research in vivo has demonstrated the potential of synthetic siRNA and DNA-directed siRNA to knock-down both exogenous and endogenous gene expression in adult mice (Hasuwa et al. 2002; Sorensen et al. 2003), although induction of the interferon system appears to be a potential side effect (Bridge et al. 2003; Sledz et al. 2003). Furthermore, research into siRNA delivery (siFection) is itself at quite a preliminary level. For instance, in spite of the widespread use of cationic liposome systems to deliver pDNA and oligodeoxynucleotides (ODNs) to cells (Felgner et al. 1987; Lasic and Papahadjopoulos, 1995; Hansma et al. 1998; Zelphati et al. 1998; Golan et al. 1999; Keller et al. 2003b), little has been reported in the literature concerning the use of cationic liposome systems to delivery siRNA to cells (*siFection*), including even basic studies on methodologies for the formulation of siRNA lipoplex (LsiR) particles from cationic liposomes (L) and siRNA (siR).

One reason for this may be the apparent misconception that all nucleic acids are much alike and should be delivered to cells in comparable ways using comparable delivery systems. Superficially this is true. For instance, both pDNA and siRNA have anionic phosphodiester backbones with identical negative charge/nucleotide (nt) ratios and should therefore interact electrostatically with cationic liposome systems to form cationic liposome-nucleic-acid (lipoplex) particles able to transfer the nucleic acids into cells. However, pDNA and siRNA are otherwise very different from each other in molecular weight and molecular topography, with potentially important consequences. All pDNA condenses into small nanoparticles of 60–100 nm subsequent to neutralization of 70%–90% of its phosphodiester backbone charge with a cationic agent (Bloomfield, 1991, 1996; Boulikas and Martin 1997; Hansma et al. 1998; Golan et al. 1999; Matulis et al. 2000; Stewart et al. 2001; Keller et al. 2002). Mechanistic studies using the mu peptide have revealed how increasing pDNA-peptide interactions lead to progressive base-pair-tilting, generating regions of high and low double helical stability that, in turn, promote super-coiling followed by pDNA hydrophobic collapse (Preuss et al. 2003; Tecle et al. 2003). In kinetic terms, the process of pDNA condensation and the reverse process of pDNA expansion appear to be equivalent to small single-domain protein folding and unfolding, respectively (Tecle et al. 2003).

Cationic-agent-condensed pDNA exists in a variety of different morphologies, such as spherical, toroids, and rods, depending upon the cationic condensing agent (Gershon et al. 1993; Hansma et al. 1998; Golan et al. 1999; Stewart et al. 2001). Irrespective of the agent, there is a minimal size for pDNA condensation corresponding to around 400 bp (Bloomfield 1991). Such behavior ensures that pDNA is almost entirely encapsulated or encased by the cationic agent and protected from enzymatic or physical degradation within nanometric particles (Zelphati et al. 1998; Chesnoy and Huang 1999; Kreiss et al. 1999; Ross and Hui 1999; Zuidam and Barenholz 1999; Zuidam et al. 1999; Simberg et al. 2001; Stewart et al. 2001; Tagawa et al. 2002; Preuss et al. 2003; Keller et al. 2003a). In contrast to pDNA, siRNA cannot condense into particles of nanometric dimensions, being already a small sub-nanometric nucleic acid.

Therefore, electrostatic interactions between siRNA and a cationic liposome system pose two potential problems; firstly, a relatively uncontrolled interaction process leading to LsiR particles of excessive size and poor stability; secondly, incomplete encapsulation of siRNA molecules thereby exposing siRNA to potential enzymatic or physical degradation prior to delivery to cells. Such considerations should make clear that pDNA and siRNA are completely different kinds of nucleic acids and that LsiR particle formulation should be regarded as a distinct and different problem from LD particle formulation. Consequently, it cannot be assumed that what works for pDNA must work for siRNA as well. Therefore, in considering the best approach to study cationic-liposome-mediated siRNA delivery to cells, we recently elected to start from first principles with regard to LsiR formation and then proceed to siFection studies thereafter. Results documented in this chapter suggest that an optimized siRNA formulation procedure is quite different from the pDNA formulation, and that parameters such as toxicity and efficacy can be controlled by the development of specific protocols for LsiR formation.

3 Discussion

In our studies, we focused upon re-tailoring cationic liposomes prepared from the synthetic cationic lipid (cytofectin) N^1-cholesteryloxycarbonyl-3,7-diazanonane-1, 9-diamine (CDAN) (Cooper et al. 1998; Geal et al. 2000; Keller et al. 2003b) and the neutral lipid dioleoyl-L-α-phosphatidylethanol amine (DOPE). CDAN/DOPE cationic liposomes (1 : 1 m/m, Trojene) have proven very effective for pDNA delivery to cells in vitro, with minimal toxicity and minimal handling (Keller et al. 2003b). Therefore, we considered that a reformulation of CDAN/DOPE cationic liposomes could be appropriate for siRNA delivery to cells. This indeed turned out to be the case (Spagnou et al. 2004).

Investigations into LsiR formulation with different CDAN/DOPE cationic liposome formulation mixtures illustrated a number of interesting features, especially the impact of the aqueous medium on the formation of LsiR particles and the extent of dilution during LsiR particle formation. In short, the increased salt concentration of OptiMEM-1 seems to be beneficial for LsiR particle formation and for effective gene knock-down post-siFection. In addition, the extent of dilution during LsiR formation appeared pivotal to the efficiency of gene silencing as well (Fig. 2). This is consistent with the formation of discrete LsiR particles that sediment with a broader surface coverage on cells. In contrast to pDNA delivery, we were unable to observe a difference between the use of small LsiR particles (50–100 nm) and larger aggregates (200–600 nm) for gene knock-down. In the case of pDNA, larger LD particles tend to give better transfection results than smaller particles due in part to a more efficient sedimentation process of the LD particles onto cell surfaces (Ross and Hui 1999; Kreiss et al. 1999; Simberg et al. 2001; Keller et al. 2003b).

However, as interesting as these features of LsiR systems might be, the most important feature of siRNA is gene silencing/knock-down with high specificity (Chi et al. 2003; Semizarov et al. 2003). LsiR siFection must be guaranteed to render specific siRNA gene knock-down, otherwise the value of this process is doubtful. Only cationic liposomes capable of generating LsiR systems that render unambiguous, specific gene

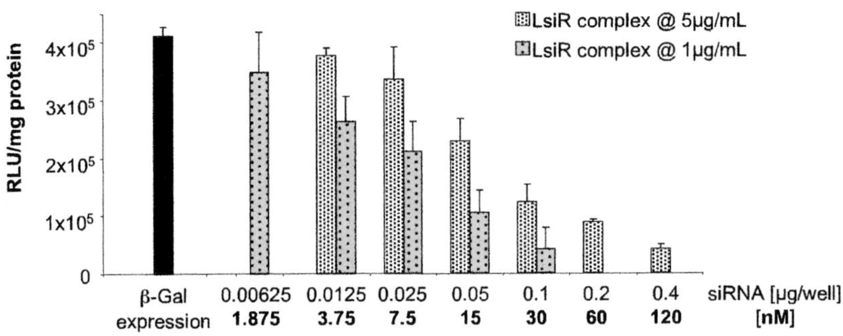

FIG. 2. Influence of the formulation procedure on optimal β-Gal protein down-regulation. LsiR (lipid:siRNA 13:1, *w/w*) systems obtained from CDAN/DOPE (45:55, *m/m*; siFECTamine) cationic liposomes were prepared in a total volume of 100 µl OptiMEM at concentrations of 5 or 1 µg/ml, respectively. These LsiR systems were then pipetted to wells (48-well plate) containing cultured HeLa cells (40,000) in 150 µl complete growth medium (10% serum, antibiotics) and cells were incubated at 37°C/CO_2 for 3 h. β-Gal activity and total protein content were assayed 24 h after pDNA transfection. Note that LsiR systems generated at high dilution yield a consistently better profile of gene knock-down. Figure is reproduced from Spagnou et al. (2004) with permission from the American Chemical Society

knock-down are of value. Using a number of putative cationic liposome/micelle-based siRNA delivery systems, we observed that at siRNA doses of 0.5 µg/well (150 nM/well) on a 48-well plate, non-specific siRNA (control siRNA-NS) was able to mediate significant gene knock-down and down-regulation of protein, a scenario that interferes greatly with functional studies and therefore must be avoided. Hence, we would suggest that siRNA delivery reagents should always be selected that are able to mediate optimal gene silencing at doses significantly lower than 0.5 µg so as to minimize the possibility of non-specific gene knock-down effects in RNA interference in vitro.

In our hands, we determined that CDAN/DOPE (45:55 m/m, siFECTamine) cationic liposomes was the most effective CDAN/DOPE formulation for siFection and was able to mediate effective protein down-regulation routinely at siRNA doses of 0.1 µg/well (30 nM/well) on a 48-well plate (Fig. 2), well below the dose threshold for non-specific gene knock-down. Therefore, we proceeded to perform cellular toxicity tests. In the context of functional genomics studies, minimal to negligible toxicity is self-evidently highly desirable. Hence we examined the effects of both siFection reagents themselves and LsiR particles on cell viability, quantifying cellular mortality (toxicity) by means of a lactate dehydrogenase (LDH) assay. Principle comparison was made with the reagent known as lipofectAMINE2000, which is the siFection reaction most widely used currently. In two cell lines, we observed that CDAN/DOPE (45:55, m/m) cationic-liposome-mediated siFection resulted in equal if not better protein down-regulation than lipofectAMINE2000-mediated siFection (Fig. 3). Moreover, CDAN/DOPE (45:55, m/m) cationic liposomes were found to be neither toxic alone nor in combination with siRNA. By comparison, lipofectAMINE2000 was clearly unreliably toxic at the doses required for optimal siFection reagent even without siRNA (Fig. 3).

Any significant cellular toxicity would certainly jeopardize the outcome of any functional study using any reagent to effect siRNA delivery. We would suggest that the LDH

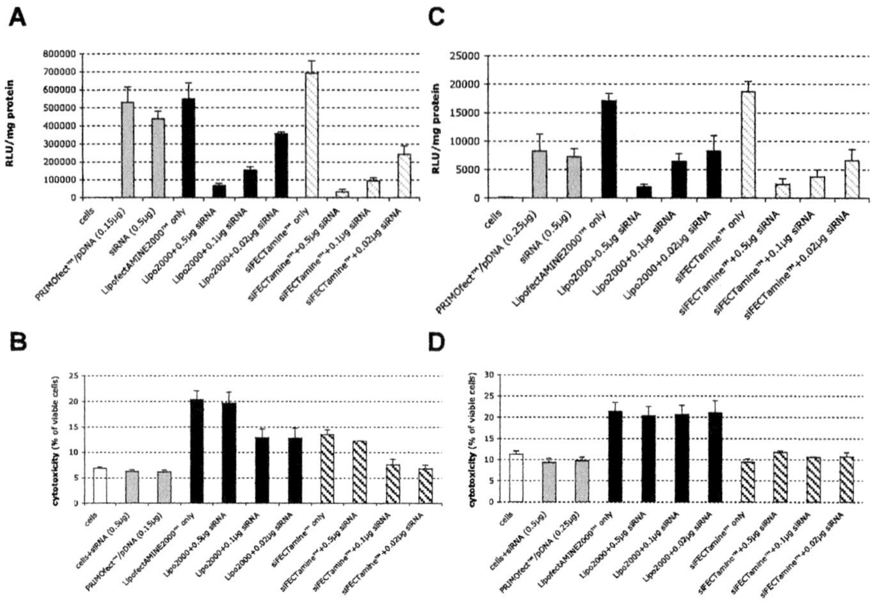

FIG. 3. Comparison of CDAN/DOPE (45:55, *m/m*; siFECTamine) cationic liposomes and lipofectAMINE2000 for specific siRNA-mediated β-Gal gene knock-down in HeLa cells (**A**) and IGROV-1 cells (**C**). PRIMOfect(IC-Vec) was used to transfect HeLa cells for 3 h with pDNA(*lacZ*) (0.15 μg/well) prior to siFection (3 h), β-Gal activity was assayed 24 h post-pDNA transfection; PRIMOfect was then used to transfect IGROV-1 cells for 3 h with pDNA(*lacZ*) (0.25 μg/well) prior to siFection (3 h) and β-Gal activity assays. For both HeLa (**B**) and IGROV-1 (**D**) cells, the cellular mortality induced by toxic effects of siRNA delivery reagents siFECTamine and lipofectAMINE2000 was measured by the lactate dehydrogenase (LDH) assay. Note that the amount of "siFection reagent only" corresponds to a putative dose of 0.5 μg/well (150 nM) siRNA. Figure is reproduced from Spagnou et al. (2004) with permission from the American Chemical Society

cellular toxicity test and others such tests (Spagnou et al. 2004) should be used as a matter of routine for judging the efficacy and suitability of siRNA delivery reagents. The appearance of cellular toxicity in any test should be taken seriously and cast into doubt the suitability of the reagent as a means to deliver siRNA for specific gene knock-down studies, or indeed for any other, more extended, functional genomics studies leading to potential therapeutic applications.

The origins of toxic effects should be determined by detailed mechanistic studies of siRNA delivery. In our own mechanistic studies, we formulated 3′-fluorescein-labeled siRNA with either lipofectAMINE2000 or CDAN/DOPE (45:55, m/m) cationic liposomes. IGROV-1 cells were incubated with LsiR systems for 1 or 3 h, respectively, and fixed after 1, 3, 5 or 24 h with para-formaldehyde according to established procedures (Colin et al. 1998). A striking feature of these experiments was the unexpectedly slow uptake of siRNA mediated by CDAN/DOPE (45:55, m/m) cationic liposomes compared with lipofectAMINE2000 (Fig. 4).

LsiR particles prepared from CDAN/DOPE (45:55, m/m) cationic liposomes were found to require at least 1–3 h before a measurable accumulation of fluorescent siRNA

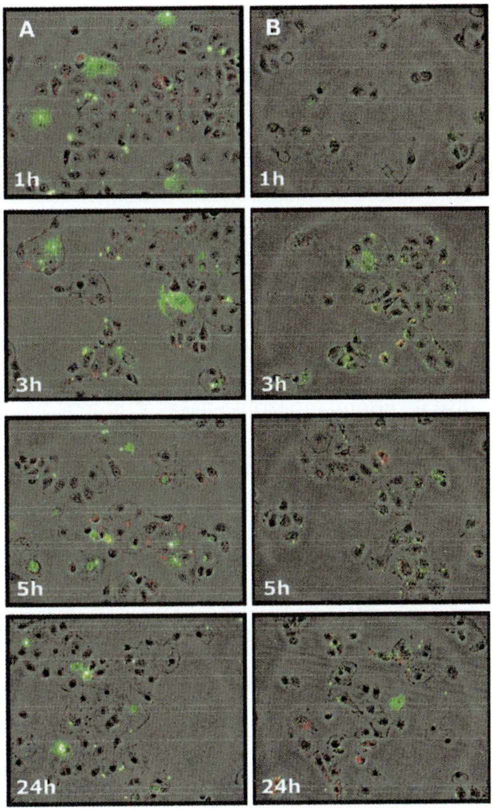

FIG. 4. Mechanistic study on the kinetics of cellular uptake of siRNA mediated by lipofectAMINE2000 (**A**) and CDAN/DOPE (45:55, *m*/*m*; siFECTamine) liposomes (**B**). Cells were incubated with rhodamine-dextran (M_W 3000) (50 μM) for 16 h. Subsequently, 3′-fluorescein-labeled siRNA was formulated with lipofectAMINE2000 or siFECTamine, respectively, and IGROV-1 cells were incubated with either of these LsiR systems for 1, 3, 5, or 24 h before fixing with 4% para-formaldehyde. Fluorescence of the siRNA (*green*) and dextran (*red*) was detected by epi-fluorescence microscopy. Note that in the case of siFECTamine-mediated delivery, significant green fluorescence intensity was observed after only 3 h, whereas for lipofectAMINE, high intensities of green fluorescence were observed after already 1 h incubation of LsiR particles with cells. Figure is reproduced from Spagnou et al. (2004) with permission from the American Chemical Society

could be observed (green) within cells localized in numerous small cytosolic vesicles. These vesicles were found to be different from lysosomes that were otherwise marked with rhodamine-dextrane (red). This result contrasted clearly with data acquired previously, showing that pDNA uptake mediated by a high ratio of CDAN/DOPE liposomes (lipid:pDNA 12:1 w/w) can lead to the intracellular internalization of large amounts of pDNA within minutes (Keller et al. 2003a), hence, demonstrating that siRNA and pDNA can also differ substantially in terms of their respective intracellular uptake rates. Intriguingly, LsiR particles prepared from lipofectAMINE2000 were also found to accumulate substantially inside cells within 1 h, leading to the accumulation of fluorescent siRNA localized in fewer, larger cytosolic vesicles but also in large diffuse patches (Fig. 4). Perhaps the faster speed and more diffuse morphology of siRNA intracellular distribution is linked to toxicity.

4 Summary

Altogether, our study demonstrates that pDNA and siRNA formulate differently with cationic liposomes, and that mechanistically the two species of nucleic acid can vary substantially in both their intracellular up-take rates and accumulation into defined intracellular vesicles. While researchers may be able to tolerate even modest toxicity when it comes to pDNA delivery, such toxicity is completely inappropriate for siRNA delivery especially when functional genomics applications are to be considered. In this respect, CDAN/DOPE (45:55, *m*/*m*; siFECTamine) cationic liposomes appear to be an ideal starting point for functional siRNA delivery given minimal toxicity. In many ways, cationic-liposome-mediated delivery of pDNA is a mature field compared with siRNA delivery. Therefore, as with pDNA delivery before, detailed research into cationic liposome formulations and delivery conditions will be essential in order to try and understand parameters for optimal siRNA delivery involving different cell types, and different growth and handling conditions in vitro. Nonetheless, the nuclear barrier problem and the requirement for long-term expression after delivery to cells represent two enduring hurdles that cationic-liposome-mediated pDNA delivery must solve if cationic-liposome-based synthetic non-viral vector systems are to prevail as the vector systems of choice for gene therapy. By contrast, cationic liposome-mediated siRNA delivery is presented with no such hurdles, implying that cationic-liposome-based synthetic non-viral vectors systems for siRNA therapeutics may forge ahead of applications of the same for gene therapy in spite of the relative infancy of the siRNA field; but such optimism may yet have to be tempered by realism as the siRNA story unfolds!

References

Abdelrahim M, Samudio I, Smith R, Burghardt R, Safe S (2002) Small inhibitory RNA duplexes for Sp1 mRNA block basal and estrogen-induced gene expression and cell cycle progression in MCF-7 breast cancer cells. J Biol Chem 277:28815–28822

Anantharam A, Lewis A, Panaghie G, Gordon E, Mccrossan ZA, Lerner DJ, Abbott GW (2003) RNA interference reveals that endogenous Xenopus MinK-related peptides govern mammalian K+ channel function in oocyte expression studies. J Biol Chem 278:11739–11745

Bernstein E, Caudy AA, Hammond SM, Hannon GJ (2001) Role for a bidentate ribonuclease in the initiation step of RNA interference. Nature 409:363–366

Bloomfield VA (1991) Condensation of DNA by Multivalent Cations: Considerations on Mechanism. Biopolymers 31:1471–1481

Bloomfield VA (1996) DNA Condensation. Curr. Op . Struct. Biol. 6:334–341

Boulikas T, Martin F (1997) Histones, Protamine, and Polylysine but not Poly(E:K) Enhance Transfection Efficiency. Int J Onc 10:317–322

Bridge AJ, Pebernard S, Ducraux A, Nicoulaz AL, Iggo R (2003) Induction of an Interferon Response by RNAi Vectors in Mammalian Cells. Nature Genet. 34:263–264

Chesnoy S, Huang L (1999) DNA Condensed by Polycations and Lipids for Gene Transfer. Stp Pharm Sci 9:5–12

Chi JT, Chang HY, Wang NN, Chang DS, Dunphy N, Brown PO (2003) Genomewide View of Gene Silencing by Small Interfering RNAs. Proc Natl Acad Sci USA 100:6343–6346

Colin M, Harbottle RP, Knight A, Kornprobst M, Cooper RG, Miller AD, Trugnan G, Capeau J, Coutelle C, Brahimi-Horn MC (1998) Liposomes Enhance Delivery and Expression of an RGD-oligolysine Gene Transfer Vector in Human Tracheal Cells. Gene Ther 5: 1488–1498

Colin M, Moritz S, Fontanges P, Kornprobst M, Delouis C, Keller M, Miller AD, Capeau J, Coutelle C, Brahimi-Horn MC (2001) The Nuclear Pore Complex is Involved in Nuclear Transfer of Plasmid DNA Condensed with an Oligolysine-RGD Peptide Containing Nuclear Localisation Properties. Gene Ther 8:1643–1653

Cooper RG, Harbottle RP, Schneider H, Coutelle C, Miller AD (1999) Peptide mini-vectors for gene delivery. Angew Chem Int Ed 38:1949–1952

Cooper RG, Etheridge CJ, Stewart L, Marshall J, Rudginsky S, Cheng SH, Miller AD (1998) Polyamine Analogues of 3-β-[N-(N′,N′-dimethylaminoethane)carbomoyl]cholesterol (DC-Chol) as Agents for Gene Delivery. Chem Eur J 4:137–151

Elbashir SM, Harborth J, Lendeckel W, Yalcin A, Weber K, Tuschl T (2001a) Duplexes of 21-nucleotide RNAs mediate RNA interference in cultured mammalian cells. Nature 411: 494–498

Elbashir SM, Lendeckel W, Tuschl T (2001b) RNA interference is mediated by 21- and 22-nucleotide RNAs. Genes Dev 15:188–200

Felgner PL, Gadek TR, Holm M, Roman R, Chan HW, Wenz M, Northrop JP, Ringold GM, Danielsen M (1987) Lipofection—a Highly Efficient, Lipid-Mediated DNA-Transfection Procedure. Proc Natl Acad Sci USA 84:7413–7417

Fire A, Xu SQ, Montgomery MK, Kostas SA, Driver SE, Mello CC (1998) Potent and specific genetic interference by double-stranded RNA in Caenorhabditis elegans. Nature 391: 806–811

Fritz JD, Herweijer H, Zhang GF, Wolff JA (1996) Gene Transfer Into Mammalian Cells Using Histone-Condensed Plasmid DNA. Hum. Gene Ther. 7:1395–1404

Geall AJ, Taylor RJ, Earll ME, Eaton MAW, Blagbrough IS (2000) Synthesis of Cholesteryl Polyamine Carbamates: pKa Studies and Condensation of Calf Thymus DNA. Bioconjugate Chem 11:314–326

Gershon H, Ghirlando R, Guttman SB, Minsky A (1993) Mode of Formation and Structural Features of DNA Cationic Liposome Complexes Used for Transfection. Biochemistry 32:7143–7151

Golan R, Pietrasanta LI, Hsieh W, Hansma HG (1999) DNA Toroids: Stages in Condensation. Biochemistry 38:14069–14076

Haensler J, Szoka FC (1993a) Polyamidoamine Cascade Polymers Mediate Efficient Transfection of Cells in Culture. Bioconjugate Chem 4:372–379

Haensler J, Szoka FC (1993b) Synthesis and Characterization of a Trigalactosylated Bisacridine Compound to Target DNA to Hepatocytes. Bioconjugate Chem. 4:85–93

Hagstrom JE, Sebestyen MG, Budker V, Ludtke JJ, Fritz JD, Wolff JA (1996) Complexes of Non-Cationic Liposomes and Histone H1 Mediate Efficient Transfection of DNA Without Encapsulation. Biochim Biophys Acta 1284:47–55

Hamilton AJ, Baulcombe DC (1999) A species of small antisense RNA in posttranscriptional gene silencing in plants. Science 286:950–952

Hansma HG, Golan R, Hsieh W, Lollo CP, Mullenley P, Kwoh D (1998) DNA Condensation for Gene Therapy as Monitored by Atomic Force Microscopy. Nucleic Acids Res. 26: 2481–2487

Hasuwa H, Kaseda K, Einarsdottir T, Okabe M (2002) Small Interfering RNA and Gene Silencing in Transgenic Mice and Rats. FEBS Lett 532:227–230

Keller M, Harbottle RP, Perouzel E, Morvane C, Imran S, Ahad R, Vaysse L, Bergau A, Moritz S, Brahimi-Horn MC, Coutelle C, Miller AD (2003a) Nuclear Localization Sequence (NLS)-Templated Non-Viral Gene Delivery Vectors: Investigation of Intracellular Trafficking Events of LMD and LD Vector Systems. ChemBioChem 4:286–298

Keller M, Jorgensen MR, Perouzel E, Miller AD (2003b) Thermodynamic Aspects and Biological Profile of CDAN/DOPE and DC-Chol/DOPE Lipoplexes. Biochemistry 42: 6067–6077

Keller M, Tagawa T, Preuss M, Miller AD (2002) Biophysical Characterization of the DNA Binding and Condensing Properties of Adenoviral Core Peptide μ (mu). Biochemistry 41:652–659

Kreiss P, Cameron B, Rangara R, Mailhe P, Aguerre-Charriol O, Airiau M, Scherman D, Crouzet J, Pitard B (1999) Plasmid DNA Size does not Affect the Physicochemical Properties of Lipoplexes but Modulates Gene Transfer Efficiency. Nucleic Acids Res 27:3792–3798

Kukowskalatallo JF, Bielinska AU, Johnson J, Spindler R, Tomalia DA, Baker JR (1996) Efficient Transfer of Genetic Material into Mammalian Cells Using Starburst Polyamidoamine Dendrimers. Proc Natl Acad Sci USA 93:4897–4902

Laemmli UK (1975) Characterisation of DNA Condensates Induced by Poly(ethylene oxide) and Polylysine. Proc Natl Acad Sci USA 72:4288–4292

Lander ES, Linton LM, Birren B, Nusbaum C, Zody MC, Baldwin, J, Devon K, Dewar K, Doyle M, Fitzhugh W, Funke R, Gage D, Harris K, Heaford A, Howland J, Kann L, Lehoczky J, Levine R, Mcewan P, Mckernan K, Meldrim J, Mesirov JP, Miranda C, Morris W, Naylor J, Raymond C, Rosetti M, Santos R, Sheridan A, Sougnez C, Stange-Thomann N, Stojanovic N, Subramanian A, Wyman D, Rogers J, Sulston J, Ainscough R, Beck S, Bentley D, Burton J, Clee C, Carter N, Coulson A, Deadman R, Deloukas P, Dunham A, Dunham I, Durbin R, French L, Grafham D, Gregory S, Hubbard T, Humphray S, Hunt A, Jones M, Lloyd C, Mcmurray A, Matthews L, Mercer S, Milne S, Mullikin JC, Mungall A, Plumb R, Ross M, Shownkeen R, Sims S, Waterston RH, Wilson RK, Hillier LW, Mcpherson JD, Marra MA, Mardis ER, Fulton LA, Chinwalla AT, Pepin KH, Gish WR, Chissoe SL, Wendl MC, Delehaunty KD, Miner TL, Delehaunty A, Kramer JB, Cook LL, Fulton RS, Johnson DL, Minx PJ, Clifton SW, Hawkins T, Branscomb E, Predki P, Richardson P, Wenning S, Slezak T, Doggett N, Cheng JF, Olsen A, Lucas S, Elkin C, Uberbacher E, Frazier M, Gibbs RA, Muzny DM, Scherer SE, Bouck JB, Sodergren EJ, Worley KC, Rives CM, Gorrell JH, Metzker ML, Naylor SL, Kucherlapati RS, Nelson DL, Weinstock GM, Sakaki Y, Fujiyama A, Hattori M, Yada T, Toyoda A, Itoh T, Kawagoe C, Watanabe H, Totoki Y, Taylor T, Weissenbach J, Heilig R, Saurin W, Artiguenave F, Brottier P, Bruls T, Pelletier E, Robert C, Wincker P, Rosenthal A, Platzer M, Nyakatura G, Taudien S, Rump A, Yang HM, Yu J, Wang J, Huang GY, Gu J, Hood L, Rowen L, Madan A, Qin SZ, Davis RW, Federspiel NA, Abola AP, Proctor MJ, Myers RM, Schmutz J, Dickson M, Grimwood J, Cox DR, Olson MV, Kaul R, Shimizu N, Kawasaki K, Minoshima S, Evans GA, Athanasiou M, Schultz R, Roe BA, Chen F, Pan HQ, Ramser J, Lehrach H, Reinhardt R, Mccombie WR, De La Bastide M, Dedhia N, Blocker H, Hornischer K, Nordsiek G, Agarwala R, Aravind L, Bailey JA, Bateman A, Batzoglou S, Birney E, Bork P, Brown DG, Burge CB, Cerutti L, Chen HC, Church D, Clamp M, Copley RR, Doerks T, Eddy SR, Eichler EE, Furey TS, Galagan J, Gilbert JGR, Harmon C, Hayashizaki Y, Haussler D, Hermjakob H, Hokamp K, Jang WH, Johnson LS, Jones TA, Kasif S, Kaspryzk A, Kennedy S, Kent WJ, Kitts P, Koonin EV, Korf I, Kulp D, Lancet D, Lowe TM, Mclysaght A, Mikkelsen T, Moran JV, Mulder N, Pollara VJ, Ponting CP, Schuler G, Schultz JR, Slater G, Smit AFA, Stupka E, Szustakowki J, Thierry-Mieg D, Thierry-Mieg J, Wagner L, Wallis J, Wheeler R, Williams A, Wolf YI, Wolfe KH, Yang SP, Yeh RF, Collins F, Guyer MS, Peterson J, Felsenfeld A, Wetterstrand KA, Patrinos A, Morgan MJ (2001) Initial Sequencing and Analysis of the Human Genome. Nature 409:860–921

Lasic DD, Papahadjopoulos D (1995) Liposomes Revisited. Science 267:1275–1276

Matulis D, Rouzina I, Bloomfield VA (2000) Thermodynamics of DNA Binding and Condensation: Isothermal Titration Calorimetry and Electrostatic Mechanism. J Mol Biol 296:1053–1063

Meister G, Landthaler M, Patkaniowska A, Dorsett Y, Teng G, Tuschl T (2004) Human Argonaute2 mediates RNA cleavage targeted by miRNAs and siRNAs. Mol Cell 15: 185–197

Miller AD (2003) The problem with cationic liposome/micelle-based non-viral vector systems for gene therapy. Curr. Med. Chem. 10:1195–1211

Miller AD (1998) Cationic Liposomes For Gene Therapy. Angew Chem Int Ed 37:1769–1785

Murphy JE, Uno T, Hamer JD, Cohen FE, Dwarki V, Zuckermann RN (1998) A Combinatorial Approach to the Discovery of Efficient Cationic Peptoid Reagents for Gene Delivery. Proc Natl Acad Sci USA 95:1517–1522

Murray KD, Etheridge CJ, Shah SI, Matthews DA, Russell W, Gurling HMD, Miller AD (2001) Enhanced Cationic Liposome-Mediated Transfection Using the DNA- Binding Peptide μ (mu) From the Adenovirus Core. Gene Ther 8:453–460

Naito Y, Yamada T, Ui-Tei K, Morishita S, Saigo K (2004)siDirect: highly effective, target-specific siRNA design software for mammalian RNA interference. Nucleic Acids Res 32:W124–129

Napoli C, Lemieux C, Jorgensen R (1990) Introduction of a Chimeric Chalcone Synthase Gene into Petunia Results in Reversible Co-Suppression of Homologous Genes in Trans. Plant Cell 2:279–289

Novina CD, Sharp PA (2004) The RNAi revolution. Nature 430:161–164

Pack DW, Putnam D, Langer R (2000) Design of Imidazole-Containing Endosomolytic Biopolymers for Gene Delivery. Biotechnol. Bioeng 67:217–223

Pector V, Backmann J, Maes D, Vandenbranden M, Ruysschaert JM (2000) Biophysical and Structural Properties of DNA Center·diC(14)-Amidine Complexes—Influence of the DNA/Lipid Ratio. J Biol Chem 275:29533–29538

Preuss M, Tecle M, Shah I, Matthews DA, Miller AD (2003) Comparison between the interactions of adenovirus-derived peptides with plasmid DNA and their role in gene delivery mediated by liposome-peptide-DNA virus-like nanoparticles. Org Biomol Chem 1:2430–2438

Remy JS, Abdallah B, Zanta MA, Boussif O, Behr JP, Demeneix B (1998) Gene Transfer with Lipospermines and Polyethyleneimines. Adv. Drug. Del. Rev. 30:85–95

Ross PC, Hui SW (1999) Lipoplex Size is a Major Determinant of In vitro Lipofection Efficiency. Gene Ther 6:651–659

Schellmann JA, Gosule LC (1976) Compact Form of DNA Induced by Spermidine. Nature 259:333–335

Schwartz B, Ivanov MA, Pitard B, Escriou V, Rangara R, Byk G, Wils P, Crouzet J, Scherman D (1999) Synthetic DNA-Compacting Peptides Derived from Human Sequence Enhance Cationic Lipid-Mediated Gene Transfer in vitro and in vivo. Gene Ther 6:282–292

Semizarov D, Frost L, Sarthy A, Kroeger P, Halbert DN, Fesik SW (2003) Specificity of Short Interfering RNA Determined Through Gene Expression Signatures. Proc. Natl. Acad. Sci. U. S. A. 100:6347–6352

Simberg D, Danino D, Talmon Y, Minsky A, Ferrari ME, Wheeler CJ, Barenholz Y (2001) Phase behavior, DNA ordering, and size instability of cationic lipoplexes—Relevance to optimal transfection activity. J. Biol. Chem. 276:47453–47459

Sledz CA, Holko M, De Veer MJ, Silverman RH, Williams BRG (2003) Activation of the Interferon System by Short-Interfering RNAs. Nat. Cell Biol. 5:834–839

Sorensen DR, Leirdal M, Sioud M (2003) Gene Silencing by Systemic Delivery of Synthetic siRNAs in Adult Mice. J. Mol. Biol. 327:761–766

Sorgi FL, Bhattacharya S, Huang L (1997) Protamine Sulfate Enhances Lipid-Mediated Gene Transfer. Gene Ther. 4:961–968

Spagnou S, Miller AD, Keller M (2004) Lipidic Carriers of siRNA: Differences in the Formulation, Cellular Uptake, and Delivery with Plasmid DNA. Biochemistry 43:13348–13356

Stewart L, Manvell M, Hillery E, Etheridge CJ, Cooper RG, Stark H, Van-Heel M, Preuss M, Alton E, Miller AD (2001) Cationic Lipids for Gene Therapy Part 4—Physico-chemical Analysis of Cationic Liposome-DNA Complexes (Lipoplexes) with Respect to in vitro and in vivo Gene Delivery Efficiency. J Chem Soc-Perkin Trans 2:624–632

Tagawa T, Manvell M, Brown N, Keller M, Perouzel E, Murray KD, Harbottle RD, Tecle M, Booy F, Brahimi-Horn MC, Coutelle C, Lemoine NR, Alton EWFW, Miller AD (2002)

Characterisation of LMD Virus-Like Nanoparticles Self-Assembled from Cationic Liposomes, Adenovirus Core Peptide μ (mu) and Plasmid DNA. Gene Ther 9:564–576

Tecle M, Preuss M, Miller AD (2003) Kinetic study of DNA condensation by cationic peptides used in nonviral gene therapy: analogy of DNA condensation to protein folding. Biochemistry 42:10343–10347

Templeton NS, Lasic DD, Frederik PM, Strey HH, Roberts DD, Pavlakis GN (1997) Improved DNA: Liposome Complexes for Increased Systemic Delivery and Gene Expression. Nat Biotechnol 15:647–652

Tuschl T, Zamore PD, Lehmann R, Bartel DP, Sharp PA (1999) Targeted mRNA degradation by double-stranded RNA in vitro. Genes Dev 13:3191–3197

Ui-Tei K, Naito Y, Takahashi F, Haraguchi T, Ohki-Hamazaki H, Juni A, Ueda R, Saigo K (2004) Guidelines for the selection of highly effective siRNA sequences for mammalian and chick RNA interference. Nucleic Acids Res 32:936–948

Van Der Krol AR, LA, M, M, B, JN, M, AR, S (1990) Flavonoid genes in petunia: addition of a limited number of gene copies may lead to a suppression of gene expression. Plant Cell 2:291–299

Venter JC, Adams MD, Myers EW, Li PW, Mural RJ, Sutton GG, Smith HO, Yandell M, Evans CA, Holt RA, Gocayne JD, Amanatides P, Ballew RM, Huson DH, Wortman JR, Zhang Q, Kodira CD, Zheng XQH, Chen L, Skupski M, Subramanian G, Thomas PD, Zhang JH, Miklos GLG, Nelson C, Broder S, Clark AG, Nadeau C, Mckusick VA, Zinder N, Levine AJ, Roberts RJ, Simon M, Slayman C, Hunkapiller M, Bolanos R, Delcher A, Dew I, Fasulo D, Flanigan M, Florea L, Halpern A, Hannenhalli S, Kravitz S, Levy S, Mobarry C, Reinert K, Remington K, Abu-Threideh J, Beasley E, Biddick K, Bonazzi V, Brandon R, Cargill M, Chandramouliswaran I, Charlab R, Chaturvedi K, Deng ZM, Di Francesco V, Dunn P, Eilbeck K, Evangelista C, Gabrielian AE, Gan W, Ge WM, Gong FC, Gu ZP, Guan P, Heiman TJ, Higgins ME, Ji RR, Ke ZX, Ketchum KA, Lai ZW, Lei YD, Li ZY, Li JY, Liang Y, Lin XY, Lu F, Merkulov GV, Milshina N, Moore HM, Naik AK, Narayan VA, Neelam B, Nusskern D, Rusch DB, Salzberg S, Shao W, Shue BX, Sun JT, Wang ZY, Wang AH, Wang X, Wang J, Wei MH, Wides R, Xiao CL, Yan CH, Yao A, Ye J, Zhan M, Zhang WQ, Zhang HY, Zhao Q, Zheng LS, Zhong F, Zhong WY, Zhu SPC, Zhao SY, Gilbert D, Baumhueter S, Spier G, Carter C, Cravchik A, Woodage T, Ali F, An HJ, Awe A, Baldwin D, Baden H, Barnstead M, Barrow I, Beeson K, Busam D, Carver A, Center A, Cheng ML, Curry L, Danaher S, Davenport L, Desilets R, Dietz S, Dodson K, Doup L, Ferriera S, Garg N, Gluecksmann A, Hart B, Haynes J, Haynes C, Heiner C, Hladun S, Hostin D, Houck J, Howland T, Ibegwam C, Johnson J, Kalush F, Kline L, Koduru S, Love A, Mann F, May D, Mccawley S, Mcintosh T, Mcmullen I, Moy M, Moy L, Murphy B, Nelson K, Pfannkoch C, Pratts E, Puri V, Qureshi H, Reardon M, Rodriguez R, Rogers YH, Romblad D, Ruhfel B, Scott R, Sitter C, Smallwood M, Stewart E, Strong R, Suh E, Thomas R, Tint NN, Tse S, Vech C, Wang G, Wetter J, Williams S, Williams M, Windsor S, Winn-Deen E, Wolfe K, Zaveri J, Zaveri K, Abril JF, Guigo R, Campbell MJ, Sjolander KV, Karlak B, Kejariwal A, Mi HY, Lazareva B, Hatton T, Narechania A, Diemer K, Muruganujan A, Guo N, Sato S, Bafna V, Istrail S, Lippert R, Schwartz R, Walenz B, Yooseph S, Allen D, Basu A, Baxendale J, Blick L, Caminha M, Carnes-Stine J, Caulk P, Chiang YH, Coyne M, Dahlke C, Mays AD, Dombroski M, Donnelly M, Ely D, Esparham S, Fosler C, Gire H, Glanowski S, Glasser K, Glodek A, Gorokhov M, Graham K, Gropman B, Harris M, Heil J, Henderson S, Hoover J, Jennings D, Jordan C, Jordan J, Kasha J, Kagan L, Kraft C, Levitsky A, Lewis M, Liu XJ, Lopez J, Ma D, Majoros W, Mcdaniel J, Murphy S, Newman M, Nguyen T, Nguyen N, Nodell M, Pan S, Peck J, Peterson M, Rowe W, Sanders R, Scott J, Simpson M, Smith T, Sprague A, Stockwell T, Turner R, Venter E, Wang M, Wen MY, Wu D, Wu M, Xia A, Zandieh A, Zhu XH (2001) The Sequence of the Human Genome. Science 291:1304–1320

Wassenegger M, Heimes S, Riedel L, Sanger HL (1994) RNA-Directed De-Novo Methylation of Genomic Sequences in Plants. Cell 76:567–576

Zamore PD, Tuschl T, Sharp PA, Bartel DP (2000) RNAi: Double-stranded RNA directs the ATP-dependent cleavage of mRNA at 21 to 23 nucleotide intervals. Cell 101:25–33

Zelphati O, Nguyen C, Ferrari M, Felgner J, Tsai Y, Felgner PL (1998) Stable and Monodisperse Lipoplex Formulations for Gene Delivery. Gene Ther 5:1272–1282

Zhang F, Andreassen P, Fender P, Geissler E, Hernandez JF, Chroboczek J (1999) A Transfecting Peptide Derived from Adenovirus Fiber Protein. Gene Ther 6:171–181

Zuidam NJ, Barenholz Y (1999) Characterization of DNA-lipid complexes commonly used for gene delivery. Int J Pharm 183:43–46

Zuidam NJ, Hirschlerner D, Margulies S, Barenholz Y (1999) Lamellarity of cationic liposomes and mode of preparation of lipoplexes affect transfection efficiency. Biochim Biophys Acta 1419:207–220

In Vivo Antitumor Activity of a New Cationic Liposome siRNA Complex

Junichi Yano, Gerald E. Smyth, Kazuko Hirabayashi, and Tadaaki Ohgi

1 Introduction

Natural viruses readily transfer their own genes into host cells. However, it is extremely difficult to deliver a nucleic acid molecule itself, as a drug, into a target cell because of its susceptibility to degradation and its poor permeation characteristics. Adenovirus vectors are used in gene therapy, but to develop a natural functional nucleic acid, such as a small interfering RNA (siRNA), as a medicine, a drug-delivery system is an essential requirement. Since lipofectin was reported by Felgner et al. (1987), many cationic liposomes have been developed, and they are now widely used as general-purpose research reagents. However, a cationic liposome that can be administered with a high margin of safety to humans has not yet been developed. For the purpose of delivering poly(I):poly(C) (synthetic dsRNA) to human cells, we have developed a new type of cationic liposome. The design of this cationic liposome presented several challenges: the liposome must have low cytotoxicity and readily release dsRNA into the cytosol, while the liposome/nucleic acid complex must be capable of being sterilized by filtration, formulated as a lyophilate, and reconstituted as an aqueous dispersion.

The new cationic liposome we have developed, LIC-101, was optimized for the delivery of dsRNA. Poly(I):poly(C), a well known cytokine inducer, itself has no direct antiproliferative activity against tumor cells, but when combined with LIC-101, the resulting complex, NS-9, shows a strong antitumor effect against various malignant cell lines in vitro and in animal models of cancer due to direct cell killing, mediated by an apoptotic effect (Hirabayashi et al. 1999). NS-9 is currently under clinical trial in the USA.

Meanwhile, Tuschl's group reported that RNA interference (RNAi) can be mediated by short RNA duplexes known as siRNA (Elbashir et al. 2001a, b). Although siRNA is only around 22 base pairs in length, its physicochemical properties are similar to those

Discovery Research Laboratories, Nippon Shinyaku Co, Ltd., 14 Nishinosho-monguchi-cho, Kisshoin, Minami-ku, Kyoto 601-8550, Japan of poly(I):poly(C), which is around 250bp in length. Therefore, if our cationic liposome could also be used as a drug-delivery system for siRNA, the way would be open to therapeutic application of the latter. Because its sequence is the same as that of the target gene, siRNA is suitable for development as a tailor-made medicine with high specificity for its molecular target. Furthermore, the action of RNA interference is extremely strong, so that we can expect molecular target therapy applications for intractable conditions such as viral diseases or cancer. However, while the application of siRNA or siRNA expression vectors to human therapy has attracted the attention of many researchers, the delivery of siRNA molecules to the appropriate cells, tissues, or organs besides the liver remains a major problem and is under development in our laboratories.

The so-called RNA world is a frontier field in the life sciences today, and it is expected that an increasing number of new drug targets for siRNA will be discovered in the near future. For this reason, the field that can best utilize the results of post-genomic research may well be the field of RNA medicines.

2 A New Cationic Liposome, LIC-101

The new cationic liposome that we have developed selectively targets the liver. In recent work, described below, we have succeeded in introducing an siRNA molecule into hepatic parenchymal cells. Therefore, we can expect applications such as the treatment of various hepatitis viruses, severe diabetes mellitus, or disorders of lipid metabolism, in addition to cancer in liver sites. Furthermore, by changing the particle size of the complex, it may be possible to target drug delivery to organs besides the liver, for example to the lungs. In addition, preliminary research suggests that it may be possible to target drugs to other organs by attaching organ-specific markers to the liposome complex.

3 Composition and Formulation of LIC-101

Our cationic liposome, LIC-101, consists of a novel cationic lipid analogue, 2-*O*-(2,2-diethylaminoethyl)-carbamoyl-1,3-*O*-dioleoylglycerol (CLZ-42) (Fig. 1), and a neutral lipid component, egg-yolk lecithin. In addition to maintaining the anti cancer cell activity of dsRNA at a high level, this combination also contributes to lowering the toxicity and the hemolytic activity of the liposome.

In contrast to cell experiments in vitro, in in-vivo administration, particularly systemic intravenous (i.v.) administration, the particle diameter of the complex is an extremely important parameter. Although complexes of cationic liposome and nucleic acid are obtained on simple mixing in solution, at high concentrations of complex components the usual mixing-method often gives aggregates of micron order. However, using high-pressure emulsification, we have developed a high-concentration formulation of a uniform siRNA/LIC-101 complex with particles of nano order. Under such conditions, an oligo RNA such as siRNA does not undergo degradation. The complex has a mean particle diameter of 150nm and is quite stable under storage as an aqueous dispersion. After sterilization by filtration, the complexes can be stored in powder form as a lyophilate.

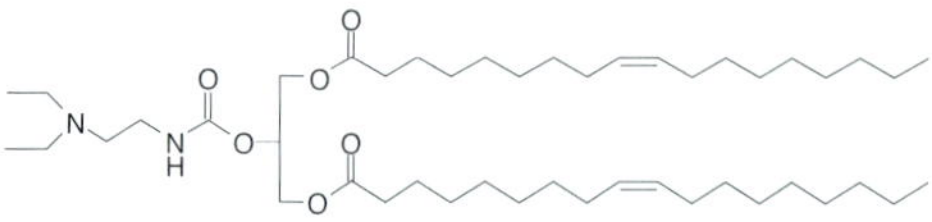

FIG. 1. Chemical structure of CLZ-42, the cationic lipid component of LIC-101

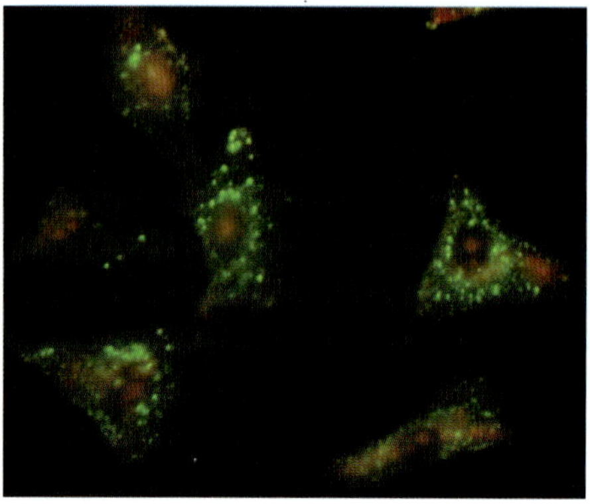

FIG. 2. Localization of fluorescein-labeled siRNA in A549 cells. A549 cells were seeded on a Nunc chamber slide and treated the next day with 100 nM fluorescein-labeled B717 (Dharmacon, Lafayette, Col., USA), a *bcl-2*-specific siRNA, complexed with LIC-101. After incubation for 6 h, the cells were fixed with 4% formaldehyde in phosphate-buffered saline, mounted with glycerol containing propidium iodide (1 μg/ml) to stain the nuclei, and observed under a fluorescence microscope. *Green fluorescence*, fluorescein; *red fluorescence*, propidium iodide

4 Intracellular Delivery of Liposome/Nucleic Acid Complexes

4.1 Localization of Fluorescein-Labeled siRNA/LIC-101 In Vitro

A549 (human lung carcinoma) cells were treated with fluorescein-labeled B717, an siRNA that is sequence-specific for the human *bcl-2* oncogene, complexed with LIC-101, and observed under a fluorescence microscope. After incubation for 6 h, the fluorescence was observed in the cytosol, and not in the nucleus, of A549 cells (Fig. 2).

After adsorption to the cell surface, the liposome/nucleic acid complex is internalized by endocytosis (Labat-Moleur et al. 1996) and the nucleic acid is subsequently released from the endosome to the cytosol. When siRNA released into the cytosol is taken up by the RNAi-induced silencing complex (RISC), it displays its RNAi activity. Various mechanisms of nucleic acid release from the endosome to the cytosol have been proposed, such as the proton-sponge effect (Haensler and Szoka 1993), but the mechanism by which siRNA is released from the LIC-101 complex is still unclear.

4.2 Uptake of Fluorescein-Labeled siRNA/LIC-101 by A549 Cells In Vitro

We compared the intracellular delivery of fluorescein-labeled B717 (*bcl-2*-specific siRNA) into A549 cells mediated by LIC-101 or a commercial cationic lipid transfection reagent (Fig. 3). With LIC-101, more than 90% of the cells were fluorescein-positive 3 h after addition, whereas with the commercial reagent no more than 50% of the cells were fluorescein-positive 6 h after addition. In siRNA delivery mediated by LIC-101, therefore, siRNA is taken up relatively quickly by the majority of cells.

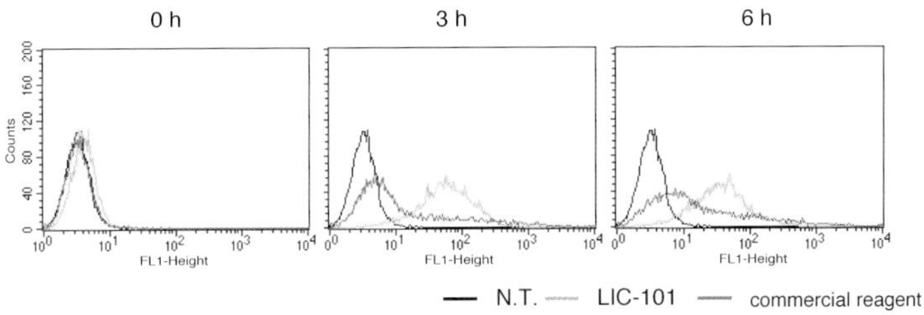

	0 h		3 h		6 h	
	MFI	positive cells (%)	MFI	positive cells (%)	MFI	positive cells (%)
LIC-101	4.19	0.40	75.7	97.1	48.0	92.5
commercial reagent	3.55	0.44	57.1	33.3	58.3	48.2
N.T.	3.44	0.23	-	-	-	-

FIG. 3. LIC-101-mediated intracellular delivery of siRNA in vitro. Fluorescein-labeled B717 siRNA (100 nM) was delivered to A549 cells by LIC-101 or a commercial cationic lipid transfection reagent (Bichko et al., 1994). Cells were withdrawn at 3 and 6 h for flow cytometry analyses. The fluorescence of externally bound complexes was quenched by pretreatment with trypan blue (Van Amersfoort and Van Strijp, 1994) and the fluorescence of complexes taken up into the cell specifically determined. Cells were counted as positive if they had a fluorescence intensity of 10 or greater. *MFI*, mean fluorescence intensity; *NT*, no treatment; *dashes*, no measurements were made

4.3 Uptake of Radiolabeled siRNA/LIC-101 by Hepatocytes In Vivo

^{3}H-labeled B717 siRNA complexed with LIC-101 was administered to mice i.v., and after 1 h the mice were laparotomized, their livers perfused, and parenchymal and non-parenchymal cells isolated and subjected to scintillation counting. The concentration of ^{3}H-B717 was 1.18 μg/10^{7} cells or 0.063 μg/mg protein in parenchymal cells and 0.352 μg/10^{7} cells or 0.349 μg/mg protein in non-parenchymal cells (Fig. 4). The distribution of ^{14}C-CLZ-42 between parenchymal and non-parenchymal cells was similar to that of ^{3}H-B717. This suggests that it is the whole siRNA/LIC-101 complex that is taken up by hepatic cells. While the siRNA appears to be more concentrated in the relatively smaller non-parenchymal cells, more siRNA is taken up per cell by the relatively larger parenchymal cells. The concentrations of ^{3}H-B717 and ^{14}C-CLZ-42 in the parenchymal cells were similar to those measured in A431 (human epidermoid carcinoma) cells under the conditions of an in vitro RNAi assay with 30 nM B717/LIC-101 complex, conditions which gave effective suppression of Bcl-2 protein (Bcl-2) (see Sect. 5.1 below).

5 Biological Activity of siRNA

Bcl-2 regulates the mitochondria-mediated apoptosis pathway; and various cell-death stimuli, including chemotherapeutic agents, activate caspases by this pathway, thereby promoting apoptosis. A high level of expression of *bcl-2* is associated with resistance

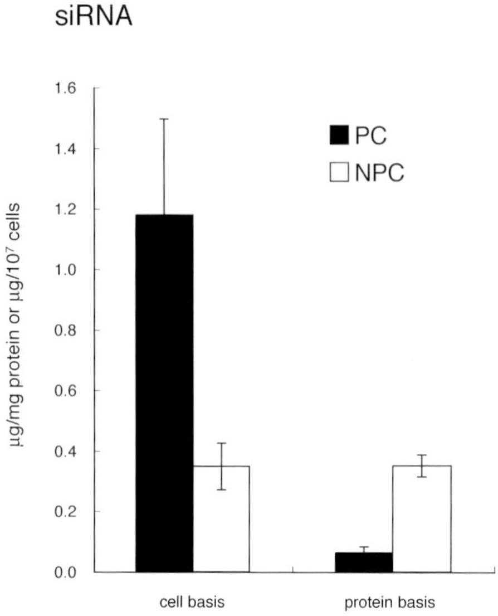

FIG. 4. Hepatic cellular uptake of siRNA/LIC-101 complexes in vivo. ^{3}H-labeled B717 siRNA complexed with LIC-101 was administered intravenously to mice via the tail vein at a dose of 10 mg/kg. After 1 h, mice were laparotomized and the liver was perfused in an open circuit for 10 min with 0.1 M HEPES buffer through the portal vena cava and the aorta and then for 15 min with 0.1 M HEPES buffer containing 0.05% collagenase. The liver was then removed and the non-parenchymal cells (*NPC*) and parenchymal cells (*PC*) isolated by differential centrifugation (Nilsson and Berg 1977). The final 50-g pellet (PC) and the final 600-g pellet (NPC) were resuspended in phosphate-buffered saline and homogenized, after which the radioactivity of the suspensions was determined by scintillation counting. Each *bar* represents the mean ± S.D. of three mice

to chemotherapeutic agents and radiation in a number of tumors, so that a drug to reduce the levels of this protein would promote caspase activation to induce apoptosis and therefore could be considered a promising therapeutic agent (DeVita et al. 2001). The following studies are more fully described in Yano et al. (2004).

5.1 Suppression of bcl-2 in Tumor Cells

5.1.1 Dose-Dependence of RNAi Effect

A431 and MDA-MB-231 (human breast carcinoma) cells were treated with B717/LIC-101 or a negative control, GL3/LIC-101, for 3 days and the expression of Bcl-2 was assessed by Western blotting (Fig. 5). (GL3 is an siRNA that is sequence-specific for firefly luciferase.) B717/LIC-101 inhibited the expression of Bcl-2 in a dose-dependent manner from 3 to 30 nM, whereas GL3/LIC-101 did not inhibit the expression of this protein even at 30 nM.

5.1.2 Time-Course of RNAi and Antisense Effects

B717/LIC-101 and AS-sDNA/LIC-101 were tested on A549 cells and Bcl-2 was assessed at various times (Fig. 6; AS-sDNA is a phosphorothioate-type antisense oligonucleotide that is sequence-specific for human *bcl-2* mRNA). B717/LIC-101 (100 nM) almost completely inhibited the expression of *bcl-2* after 2 or 3 days, and the effect continued for at least 6 days. AS-sDNA/LIC-101 (1 μM) partially inhibited *bcl-2* expression after 2 days, but the effect weakened rapidly.

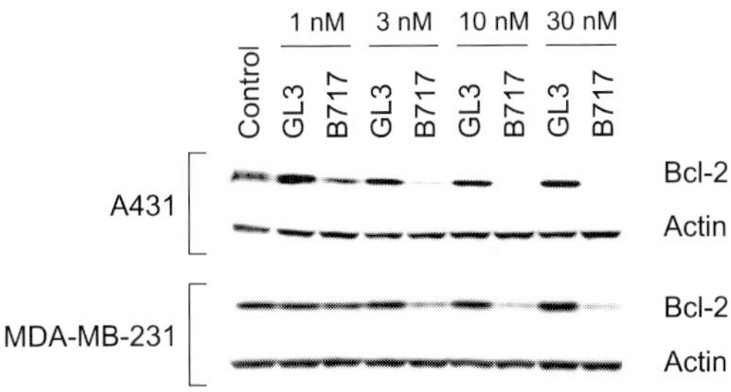

FIG. 5. Inhibition of Bcl-2 protein (Bcl-2) expression by siRNA/LIC-101 (dose dependence). A431 and MDA-MB-231 cells were seeded 15–24h before transfection and treated with 1–30nM B717/LIC-101 or GL3/LIC-101 prepared at an oligonucleotide:LIC-101 ratio of 1:16 (w/w), or saline for control cells. After 3 days, cells were harvested by centrifugation and resuspended in lysis buffer containing 1% Nonidet P-40 and protease inhibitor cocktail. The cell debris was removed by centrifugation and the protein concentration of the cell extracts determined. Samples containing 10–20μg of protein were subjected to SDS-polyacrylamide gel electrophoresis and transferred to polyvinylidene fluoride membrane filters, after which Bcl-2 was detected by Western blotting with anti-Bcl-2 antibody (Dako, Kyoto, Japan) and a peroxidase-conjugated secondary antibody. Antigen-antibody complexes were visualized by enhanced chemiluminescence (ECL Plus, Amersham Biosciences, Buckinghamshire, UK)

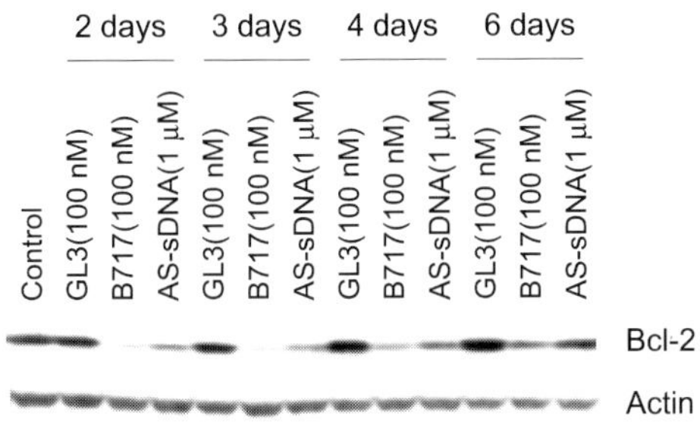

FIG. 6. Inhibition of Bcl-2 expression by siRNA/LIC-101 or antisense/LIC-101 (time course). A549 cells were treated with 100nM B717/LIC-101 (siRNA), 100nM GL3/LIC-101 (negative control), or 1000nM AS-sDNA/LIC-101 (antisense). Saline was added to control cells. The levels of Bcl-2 or actin 2–6 days after transfection were assessed by Western blotting as described in the legend to Fig. 5

5.2 Antiproliferative Activity of siRNA/LIC-101

A431 cells were treated with B717/LIC-101. The complex showed dose-dependent antiproliferative activity leveling off at 10nM (Fig. 7), whereas the negative control GL3/LIC-101 showed no significant activity. B717/LIC-101 also showed dose-dependent activity in A549 and MDA-MB-231 cells, leveling off at 300nM.

5.3 Antitumor Activity of siRNA/LIC-101 in Mice

B717/LIC-101 (10mg/10ml/kg) was administered by i.v. bolus injection to mice once or three times a week for 6 weeks. The average survival of the control group was 42

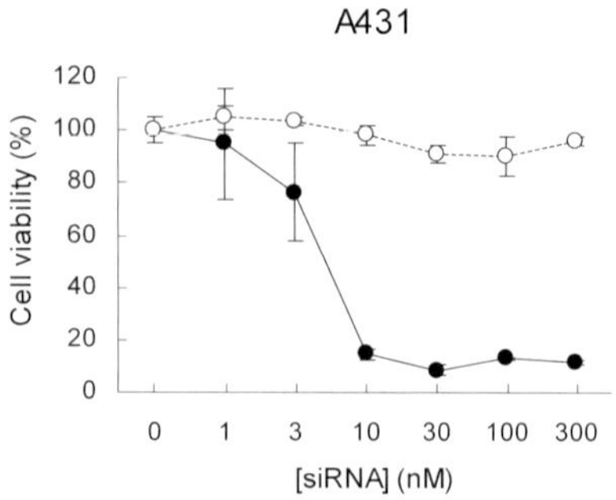

FIG. 7. Antiproliferative activity of siRNA/LIC-101 against A431 cells. Cells were seeded in 96-well plates at a density of 5×10^2 cells/well and treated with B717/LIC-101 (*closed circles*) or GL3/LIC-101 (*open circles*) at the indicated concentrations. The plates were incubated for six days, after which a 5 mg/ml solution of 3-(4,5-dimethylthiazol-2-yl)-2,5-diphenyltetrazolium bromide (MTT) was added to each well and the incubation continued for 4 h. Then 0.04 N HCl in 2-propanol was thoroughly mixed into each well and the absorbance at 595 nm determined. Each assay was performed in triplicate

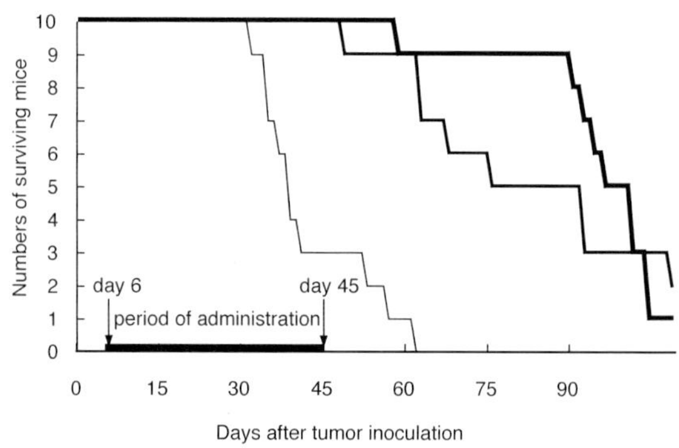

FIG. 8. Antitumor activity of siRNA/LIC-101. BALB/c *nu*/*nu* mice (5-week-old, male; Clea Japan, Tokyo, Japan) were injected with A549 cells by intrasplenic injection on day 0. From day 6 to day 45, B717/LIC-101 (10 mg/kg) was administered intravenously once or three times a week. Control mice were treated with 10% (w/v) maltose solution three times a week. Each group contained ten mice. The survival of the mice was monitored until day 100 and survival was calculated by Kaplan-Meier curves. *Thin line*, control; *medium line*, B717/LIC-101 (once/ week); *thick line*, B717/LIC-101 (3 times/week). **Significantly different from the control by the generalized Wilcoxon test, $p < 0.001$

days, and all of these mice died after 61 days due to liver metastasis (Fig. 8). The average survival of the once-a-week B717/LIC-101 group was over 80 days, and three of 10 mice were still alive on day 100, extending the survival of the group significantly beyond that of the control group. The average survival of the three-times-a-week B717/LIC-101 group was over 94 days, and five of 10 mice were still alive on day 100, extending the survival of the group beyond that of the once-a-week B717/LIC-101 group. GL3/LIC-101 did not show any antitumor activity when it was administered three times a week for 6 weeks. On day 34, the number of tumor nodules in the B717/LIC-101 groups was less than in the control group, and the sizes of the nodules were smaller.

6 Discussion

siRNAs or siRNA expression vectors have previously been administered i.v. to mice by hydrodynamic injection, which accomplishes a rapid infusion of siRNA solution in a volume of 1 ml per 10 g of body weight and is unsuitable for use in humans. Virus-mediated delivery of siRNA has also been reported, but there is concern over the safety of viral vectors. For example, McCaffrey et al. (2002) co-transfected the firefly luciferase gene along with synthetic siRNAs or an shRNA expression vector into mice by hydrodynamic injection and showed that sequence-specific siRNA or shRNA reduces luciferase expression in the liver. Lewis et al. (2002) administered siRNA by hydrodynamic injection to transgenic mice expressing enhanced green fluorescent protein (EGFP) and found that siRNA specific for EGFP reduces the expression of EGFP in the liver, mainly in hepatocytes. Xia et al. (2002) demonstrated that virus-mediated delivery of siRNA specifically reduces the expression of targeted genes in EGFP transgenic mice, both in the striatum after direct injection into the brain and in the liver after intravenous injection through the tail vein. And Song et al. (2003) found that siRNA targeted to the Fas receptor protects mice from liver failure and fibrosis when administered by hydrodynamic injection in two models of autoimmune hepatitis. McCaffrey et al. (2003) also used hydrodynamic transfection to administer plasmids encoding the hepatitis B virus (HBV) genome together with HBV-specific shRNAs to mice and demonstrated that treatment with shRNAs inhibits HBV replication in the liver. Various in vivo effects of siRNA and shRNA are therefore well established. However, the methods of delivery of RNA used, namely, hydrodynamic injection, viral vectors, or direct injection into the brain, are not suitable for use in humans, so that the safe delivery of nucleic acid drugs in human therapy is still an unresolved issue.

7 Concluding Remarks

A cationic liposome such as LIC-101, which can mediate the rapid entry of siRNA into a majority of cells, would appear to be the ideal vector for in vivo delivery of siRNA. Our siRNA/LIC-101 complex has *bcl-2*-inhibitory and antiproliferative activity against various tumor cell lines. Furthermore, siRNA/LIC-101 can be safely administered to animals and it shows strong antitumor activity in mouse models of cancer. This in vivo result takes cancer therapy that includes siRNA one step closer to clinical use, with potential for application to various gene-targeting therapies. Finally, we hope that our cationic liposome can be developed as a uniform drug delivery system for various RNA medicines of the future, including siRNA, siRNA expression vectors, and ribozymes, in non-viral RNA delivery.

References

Bichko V, Netter HJ, Taylor J (1994) Introduction of hepatitis delta virus into animal cell lines via cationic liposomes. J Virol 68:5247–5252

DeVita VT, Hellman S, Rosenberg, SA (2001) Cancer: Principles and Practice of Oncology. Lippincott Williams & Wilkins, Philadelphia, pp 111–122

Elbashir SM, Lendeckel W, Tuschl T (2001a) RNA interference is mediated by 21- and 22-nucleotide RNAs. Genes Dev 15:188–200

Elbashir SM, Harborth J, Lendeckel W, Yalcin A, Weber K, Tuschl T (2001b) Duplexes of 21-nucleotide RNAs mediate RNA interference in cultured mammalian cells. Nature 411:494–498

Felgner PL, Gadek TR, Holm M, Roman R, Chan HW, Wenz M, Northrop JP, Ringold GM, Danielsen M (1987) Lipofection: a highly efficient, lipid-mediated DNA-transfection procedure. Proc Natl Acad Sci USA 84:7413–7417

Haensler J, Szoka FC Jr (1993) Polyamidoamine cascade polymers mediate efficient transfection of cells in culture. Bioconjug Chem 4:372–379

Hirabayashi K, Yano J, Inoue T, Yamaguchi T, Tanigawara K, Smyth GE, Ishiyama K, Ohgi T, Kimura K, Irimura T (1999) Inhibition of cancer cell growth by polyinosinic-polycytidylic acid/cationic liposome complex: a new biological activity. Cancer Res 59:4325–4333

Labat-Moleur F, Steffan AM, Brisson C, Perron H, Feugeas O, Furstenberger P, Oberling F, Brambilla E, Behr JP (1996) An electron microscopy study into the mechanism of gene transfer with lipopolyamines. Gene Ther 3:1010–1017

Lewis DL, Hagstrom JE, Loomis AG, Wolff JA, Herweijer H (2002) Efficient delivery of siRNA for inhibition of gene expression in postnatal mice. Nat Genet 32:107–108

McCaffrey AP, Meuse L, Pham TT, Conklin DS, Hannon GJ, Kay MA (2002) RNA interference in adult mice. Nature 418:38–39

McCaffrey AP, Nakai H, Pandey K, Huang Z, Salazar FH, Xu H, Wieland SF, Marion PL, Kay MA (2003) Inhibition of hepatitis B virus in mice by RNA interference. Nat Biotechnol 21:639–644

Nilsson M, Berg T (1977) Uptake and degradation of formaldehyde-treated ^{125}I-labelled human serum albumin in rat liver cells in vivo and in vitro. Biochim Biophys Acta 497: 171–182

Song E, Lee SK, Wang J, Ince N, Ouyang N, Min J, Chen J, Shankar P, Lieberman J (2003) RNA interference targeting Fas protects mice from fulminant hepatitis. Nat Med 9:347–351

Van Amersfoort ES, Van Strijp JA (1994) Evaluation of a flow cytometric fluorescence quenching assay of phagocytosis of sensitized sheep erythrocytes by polymorphonuclear leukocytes. Cytometry 17:294–301

Xia H, Mao Q, Paulson HL, Davidson BL (2002) siRNA-mediated gene silencing in vitro and in vivo. Nat Biotechnol 20:1006–1010

Yano J, Hirabayashi K, Nakagawa S, Yamaguchi T, Nogawa M, Kashimori I, Naito H, Kitagawa H, Ishiyama K, Ohgi T, Irimura T (2004) Antitumour activity of siRNA/cationic liposome complex in mouse models of cancer: A new clinical application of siRNA. Clin Cancer Res 10:7721–7726

Enhancing RNAi with Synthetic RNA Duplexes

Dong-Ho Kim[1], Mark A. Behlke[2], Scott D. Rose[2], Mi-Sook Chang[3], Sangdun Choi[3], and John J. Rossi[1]

1 Introduction to siRNA-Mediated Silencing

RNA interference (RNAi) is an evolutionarily conserved process by which specific mRNAs are targeted for degradation by complementary small interfering RNAs (siRNAs). It has become the method of choice for mammalian cell genetics and as a potential sequence-specific therapeutic approach (Sharp 1999; Hannon 2002; Hutvagner and Zamore 2002). Long double-stranded (ds) RNAs are degraded by the RNase III class endonuclease Dicer into 21- to 23-nt duplexes that have 2-base 3′-overhangs (Zamore et al. 2000; Bernstein et al. 2001). Dicer's primary role in RNAi is the endonucleolytic processing of long dsRNAs into short 21- to 23-mer effector molecules (siRNAs) (Bernstein et al. 2001; Ketting et al. 2001). Dicer is also involved in the early steps of RNA induced silencing complex (RISC) complex formation and may be required for entry of the siRNA into RISC (Lee et al. 2004; Pham et al. 2004). In *Drosophila*, the protein R2D2 associates with Dicer (specifically Dicer-2) and binds siRNAs prior to entry into RISC (Liu et al. 2003). Although no functional homologue for R2D2 has been identified in mammals, a similar link is assumed to exist between the Dicer cleavage step and entry into RISC in mammalian cells. Human Dicer has been cloned and characterized (Zhang et al. 2002). Interestingly, while the recombinant enzyme can be used to process long dsRNA into functional siRNAs, the process is slow and complete "dicing" of a substrate in vitro can take 24 h incubation (Zhang et al. 2002). Thus, other factors must be involved in promoting rapid Dicer cleavage in vivo.

1.1 Silencing by Synthetic siRNAs

We systematically studied the silencing properties of chemically synthesized duplex RNAs of different lengths and designs. A series of RNA duplexes were synthesized specific to a variety of target genes and are shown in Table 1. Compounds are named

[1] Divison of Molecular Biology, Beckman Research Institute of the City of Hope, Duark, CA 91010, USA
[2] Integrated DNA Technologies, Coralville Iowa, IA 52241, USA
[3] Division of Biology, California Institute of Technology, Pasadena, CA 91125, USA

TABLE 1. Summary of oligonucleotide reagents

	Sequence	Name
5′	**GCAAGCUGACCCUGAAGUUCAUCUGCACCACCGGCAAGC 3′**	**EGFP Site-1**
5′	GCUGACCCUGAAGUUCAUCUU	EGFPS1-21+2
3′	UUCGACUGGGACUUCAAGUAG	
5′	CCUGAAGUUCAUCUGCACCAC	EGFPS1-21+2(1)
3′	UGGGACUUCAAGUAGACGUGG	
5′	CCCUGAAGUUCAUCUGCACCA	EGFPS1-21+2(2)
3′	CUGGGACUUCAAGUAGACGUG	
5′	ACCCUGAAGUUCAUCUGCACC	EGFPS1-21+2(3)
3′	ACUGGGACUUCAAGUAGACGU	
5′	GACCCUGAAGUUCAUCUGCAC	EGFPS1-21+2(4)
3′	GACUGGGACUUCAAGUAGACG	
5′	UGACCCUGAAGUUCAUCUGCA	EGFPS1-21+2(5)
3′	CGACUGGGACUUCAAGUAGAC	
5′	CUGACCCUGAAGUUCAUCUGC	EGFPS1-21+2(6)
3′	UCGACUGGGACUUCAAGUAGA	
5′	GCUGACCCUGAAGUUCAUCUG	EGFPS1-21+2(7)
3′	UUCGACUGGGACUUCAAGUAG	
5′	GCAAGCUGACCCUGAAGUUCAUU	EGFPS1-23-2UU
3′	UUCGACUGGGACUUCAAGUAGAC	
5′	GCUGACCCUGAAGUUCAUCUGUU	EGFPS1-23+2UU
3′	UUCGACUGGGACUUCAAGUAGAC	
5′	GCAAGCUGACCCUGAAGUUCAUUU	EGFPS1-24-2UU
3′	UUCGACUGGGACUUCAAGUAGACG	
5′	GCUGACCCUGAAGUUCAUCUGCUU	EGFPS1-24+2UU
3′	UUCGACUGGGACUUCAAGUAGACG	
5′	GCAAGCUGACCCUGAAGUUCAUCUU	EGFPS1-25-2UU
3′	UUCGACUGGGACUUCAAGUAGACGU	
5′	GCUGACCCUGAAGUUCAUCUGCAUU	EGFPS1-25+2UU
3′	UUCGACUGGGACUUCAAGUAGACGU	
5′	GCAAGCUGACCCUGAAGUUCAUCUUU	EGFPS1-26-2UU
3′	UUCGACUGGGACUUCAAGUAGACGUG	
5′	GCUGACCCUGAAGUUCAUCUGCACUU	EGFPS1-26+2UU
3′	UUCGACUGGGACUUCAAGUAGACGUG	
5′	AAGCUGACCCUGAAGUUCAUCUGCAC	EGFPS1-26+0
3′	UUCGACUGGGACUUCAAGUAGACGUG	
5′	AAGCUGACCCUGAAGUUCAUCUGCACC	EGFPS1-27+0
3′	UUCGACUGGGACUUCAAGUAGACGUGG	
5′	F-AAGCUGACCCUGAAGUUCAUCUGCACC	EGFPS1-27+0
3′	UUCGACUGGGACUUCAAGUAGACGUGG-F	FAM #1
5′	AAGCUGACCCUGAAGUUCAUCUGCACC-F	EGFPS1-27+0
3′	UUCGACUGGGACUUCAAGUAGACGUGG-F	FAM #2
5′	F-AAGCUGACCCUGAAGUUCAUCUGCACC	EGFPS1-27+0
3′	F-UUCGACUGGGACUUCAAGUAGACGUGG	FAM #4
5′	AAGCUGACCCUGAAGUUCAUCUGCACC-F	EGFPS1-27+0
3′	F-UUCGACUGGGACUUCAAGUAGACGUGG	FAM #5
5′	AAGCUGACCCUGAAGUUCAUCUGCAUU	EGFPS1-27+0UU
3′	UUCGACUGGGACUUCAAGUAGACGUGG	
5′	GCAAGCUGACCCUGAAGUUCAUCUGUU	EGFPS1-27-2UU
3′	UUCGACUGGGACUUCAAGUAGACGUGG	
5′	GCUGACCCUGAAGUUCAUCUGCACAUU	EGFPS1-27+2UU
3′	UUCGACUGGGACUUCAAGUAGACGUGG	

TABLE 1. *Continued*

	Sequence	Name
5′	AAGCUGACCCUGAAG**A**UCAUCUGCAUU	EGFPS1-27+0UU/16
3′	UUCGACUGGGACUUC**U**AGUAGACGUGG	
5′	AAGCUGACCCUGAAG**AA**CAUCUGCAUU	EGFPS1-27+0UU/16,17
3′	UUCGACUGGGACUUC**UU**GUAGACGUGG	
5′	AAGCUGACCCUGAA**CAA**CAUCUGCAUU	EGFPS1-27+0UU/15,16,17
3′	UUCGACUGGGACUU**GUU**GUAGACGUGG	
5′	AAGCUGACCCUG**UUCA**UCAUCUGCACC	EGFPS1-27+0/mut
3′	UUCGACUGGGAC**AAGU**AGUAGACGUGG	
5′	AAGCUGACCCUGAAGUUCAUCUGCACCA	EGFPS1-28+0
3′	UUCGACUGGGACUUCAAGUAGACGUGGU	
5′	AAGCUGACCCUGAAGUUCAUCUGCACCAC	EGFPS1-29+0
3′	UUCGACUGGGACUUCAAGUAGACGUGGUG	
5′	**GTTGAACTTGAATCAGAAGATGAAGTCAAATTGGC 3′**	**hnRNPH Site-1**
5′	CUUGAAUCAGAAGAUGAUU	hnRNPH-21+2
3′	UUGAACUUAGUCUUCUACU	
5′	AACUUGAAUCAGAAGAUGAAGUCAAAU	hnRNPH-27+0
3′	UUGAACUUAGUCUUCUACUUCAGUUUA	
5′	**ATAAAACTGGATGAAGGCTGGGTACCTTTGGAGAT 3′**	**La Site-1**
5′	CUGGAUGAAGGCUGGGUACUU	La-21+2
3′	UUGACCUACUUCCGACCCAUG	
5′	AACUGGAUGAAGGCUGGGUACCUUUUU	La-27+0UU
3′	UUGACCUACUUCCGACCCAUGGAAACC	

according to target, length, and overhang. For example, "EGFPS1" refers to an RNA duplex targeting "Site-1" in enhanced green fluorescent protein (EGFP); 27+2 denotes a 27-nt duplex with a 2-base 3′-overhang, 27+0 denotes a 27-nt duplex with no overhang (blunt), and 27–2 denotes a 27-nt duplex with a 2-base 5′-overhang. An expanded set of synthetic RNA duplexes of varying length containing 3′-overhangs, 5′-overhangs, or blunt ends, were tested for their relative potency in a model system targeting Site-1 in EGFP (Kim and Rossi 2003). HEK293 cells were co-transfected with the EGFP expression plasmid EGFP-C1 plus RNA duplexes at varying concentrations. Initial transfections were carried out using 50-nM amounts of the various RNA duplexes (Fig. 1A). The real potency of the longer duplexes was only revealed when transfections were done at subnanomolar concentrations. Using duplex RNA concentrations of 200 pM (Fig. 1B) and 50 pM (Fig. 1C), we observed that potency increased with length. Blunt, 5′-overhang, and 3′-overhanging ends were of similar potency. Maximal inhibitory activity was seen at a duplex length of 27 bp.

1.2 In Vitro Cleavage Reactions with Human Recombinant Dicer

Duplex RNA oligonucleotides ranging from 21 to 27 base pairs incubated with recombinant human Dicer resulted in cleavage of the 23-mer, 25-mer, and 27-mer duplexes but not the 21-mer duplex (Fig. 2). Determinations of relative efficiencies of Dicer

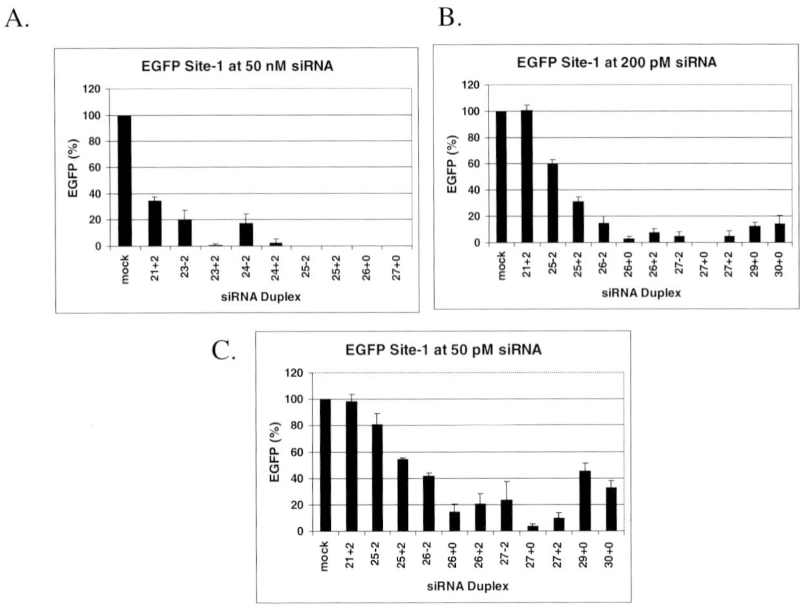

FIG. 1A–C. Longer dsRNAs are more potent effectors of RNAi than a 21+2 siRNA. EGFP expression levels were determined following co-transfection of HEK293 cells with a fixed amount of EGFP expression plasmid and varying concentrations of dsRNAs. Transfections were performed using **A** 50 nM, **B** 200 pM, and **C** 50 pM of the indicated dsRNAs

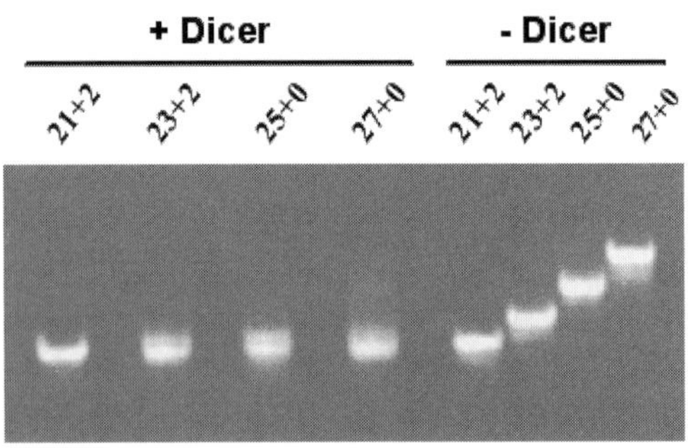

FIG. 2. Cleavage of dsRNAs by recombinant Dicer. Each RNA duplex was incubated in the presence or absence of recombinant human Dicer for 24 h. RNAs were separated using a non-denaturing polyacrylamide gel and visualized by ethidium bromide staining

cleavage were not possible under the in vitro conditions recommended by the supplier due to the slow kinetics of this reaction.

2 Characterization of siRNAs Produced from 27-mers

2.1 *Testing of a Tiled Set of 21-mers*

Cleavage of a 27-mer by Dicer results in a variety of distinct 21-mers, depending on where cleavage occurs; it is possible that one or a mix of these possible 21-mers is sig-

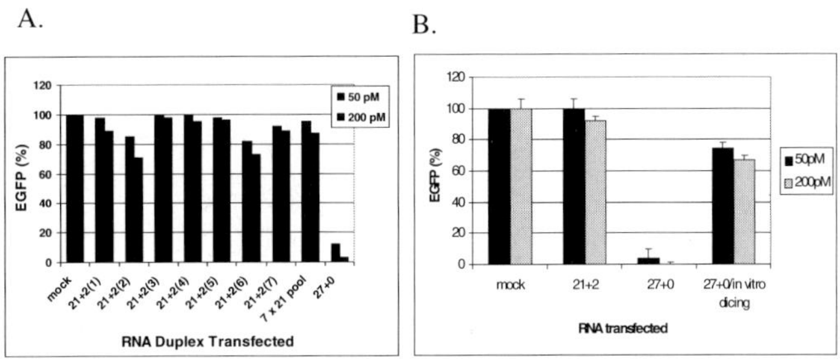

FIG. 3A,B. Testing of a tiled set of individual 21-mer siRNAs versus a pool of siRNAs and the 27-mer dsRNA. **A** Seven possible 21+2 siRNAs predicted from dicing the 27-mer dsRNA were tested individually or as a pool in co-transfection assays with the EGFP reporter construct in HEK293 cells. Each *graph* depicts the average of duplicate experiments. **B** Comparison of in-vitro-diced 27-mer dsRNA versus intact 27-mer dsRNA for RNAi. The respective RNAs were co-transfected as in **A** at the indicated concentrations of dsRNAs. For the diced products, a 1-μM 27-mer dsRNA was incubated in a Dicer reaction buffer without (*column 3*) or with (*column 4*) Dicer at 37°C for 12h. The mixtures were diluted in water and used directly for co-transfection with the EGFP reporter. To control for possible artifacts of residual Dicer in the diluted mixes, the samples in column 4 were phenol extracted and ethanol precipitated prior to transfection (*column 5*)

nificantly more potent than the specific 21-mer employed as our standard for comparison. To test this possibility we synthesized seven different 21-mers that could be derived from the EGFPS1 27+0 duplex, walking in single base steps along the antisense strand and using the traditional 21+2 design (2-base 3′-overhang). These seven duplexes were tested for RNAi activity in the HEK293 cell co-transfection assay individually and as a pool (Fig. 3). At concentrations of 50 or 200 pM, none of the individual 21-mer duplexes or the pooled set of seven 21-mer duplexes showed activity comparable to the 27-mer duplex (Fig. 3). In vitro Dicing of the 27-mers prior to transfection did not significantly enhance efficacy. As an additional control, transfection of a mutated EGFP 27-mer duplex harboring four consecutive, centrally placed mismatched bases was carried out. The mismatches virtually eliminated RNAi efficacy (data not presented).

2.2 Monitoring Dicer Specificity

Holen and others reported that an siRNA with fluorescein conjugated to the 3′-end of the antisense (AS) strand was functional in RNAi silencing (Holen et al. 2002). Harborth and others (Harborth et al. 2003) reported that fluorescein could be placed at either the 5′-end or 3′-end of the sense (S) strand with no adverse effect. However, in contrast to previous findings, this group observed that modification of the 5′-end of the antisense strand was well tolerated but that modification of the 3′-end of the antisense strand affected RNAi activity. The effects of fluorescein end-modification on dicing and ability to trigger RNAi silencing were tested for the 27-mer duplexes. RNA oligonucleotides were synthesized for the EGFPS1 site with 6-carboxyfluorescein (6FAM) attached to the 5′-end of the S strand, 3′-end of the S-strand, 5′-end of the AS strand, and 3′-end of the AS strand. Pairwise combinations were used to make

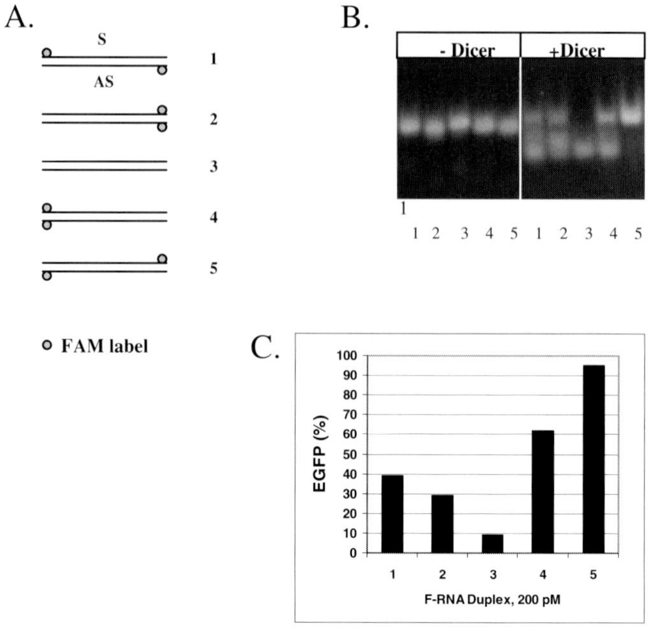

FIG. 4A–C. Dicer processing correlates with RNAi activity. **A** RNA duplexes were employed. Oligonucleotides were conjugated with 6-carboxyfluorescein (6FAM) at the 5′-ends and/or 3′-ends as shown. *Top* and *bottom lines* indicate sense and antisense strands in duplex configuration with the sense in a 5′ to 3′ orientation (*left to right*), and the antisense in a 3′ to 5′ orientation (left to right). **B** 6FAM end-modification affects in vitro dicing. RNA duplexes were incubated with 0.5 U of recombinant human Dicer for 8 h and the products resolved in a 7.5% non-denaturing polyacrylamide gel. The RNAs were visualized by ethidium bromide staining. **C** 6FAM modification affects RNAi activity. RNA duplexes at 200 pM were co-transfected with the EGFP expression plasmid and assayed at 24 h for EGFP fluorescence as described. Reported values for EGFP expression represent the average of two independent experiments. The relative levels of fluorescence were normalized to luciferase

duplex RNAs (Fig. 4A). Duplex #3 was the unmodified wild-type EGFPS1 27+0 duplex (Table 1). RNA duplexes were incubated for 24 h with recombinant human Dicer, separated by non-denaturing PAGE, stained, and visualized using UV excitation (Fig. 4B). Only the unmodified wild-type sequence (#3) was fully cleaved in the in vitro dicing reaction. The duplex bearing a 3′-6FAM on the S strand and a 3′-6FAM on the AS strand (#5) was totally resistant to cleavage under these conditions.

2.3 Intracellular Testing of End-Blocked 27-mers

Functional potencies of these five duplexes were compared in EGFP co-transfection assays (Fig. 4C) using 200-pM RNA concentrations. Parallel to the patterns seen for in vitro dicing, all of the 27-mer duplexes with 6FAM-modified ends were less potent than the unmodified duplex in reducing EGFP expression. Duplexes 1, 2, and 4, which showed partial cleavage with recombinant Dicer, had three- to six-fold reduced RNAi activity. Duplex #5, which showed no cleavage with recombinant Dicer, had minimal RNAi activity, establishing a direct correlation between relative effectiveness of in vitro dicing and RNAi in cell culture. These data are consistent with recent observations that Dicer preferentially recognizes the ends of dsRNA substrates and proceeds to cleave in a processive manner (Zhang et al. 2002). It thus appears that when the 6FAM group is present on the 3′-ends, it interferes with Dicer recognition.

2.4 Making Better Synthetic Substrates for Dicer

Endogenous hairpin RNAs originate in the nucleus and are sequentially processed by two endonucleases, Drosha (nucleus) and Dicer (cytoplasm) (Lee et al. 2003). Thus, the natural pathway for dsRNAs involves two processing steps, and Drosha products are Dicer substrates. We are currently investigating the efficacy of a variety of Dicer substrates with 2-base 3′ overhangs on one end and blunted on the other end. To date, it appears that a 2-base 3′ overhang on one end, with blunt deoxyribonucleotides on the other, generates a predictable 21-mer containing the 2-base 3′ overhang provided (data not presented). This design could facilitate in vivo production of a desired 21-mer with known hyperfunctional activity, perhaps providing even greater potency.

3 Testing 27-mers Against Endogenous Targets

To ensure that the increased potency of the 27-mer dsRNAs was not an artifact of targeting a reporter construct, two endogenous transcripts were targeted. RNA duplexes were synthesized to target randomly chosen sites in the human hnRNP H mRNA

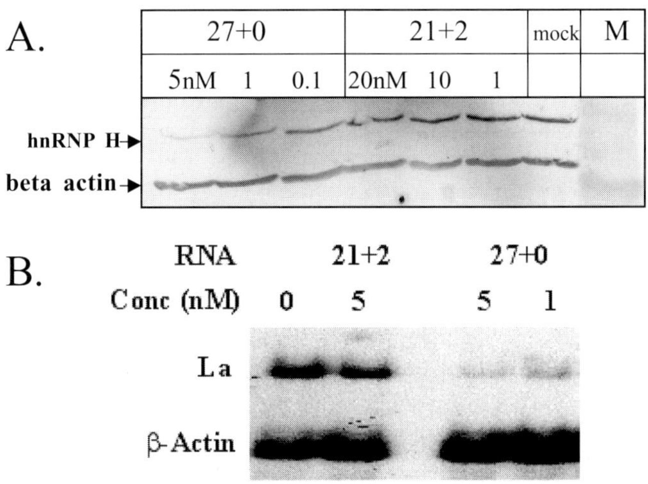

FIG. 5A, B. Comparison of 21-mer siRNA and 27-mer dsRNA in down-regulation of endogenous transcripts. RNAs for a 21+2 siRNA and 27+0 dsRNA were designed to target sites in the hnRNP H mRNA (**A**) or La mRNA (**B**). HnRNP H knockdown was assayed by Western blot and La knockdown by Northern blot analyses. The dsRNAs were used at the indicated concentrations. β-Actin was used as an internal specificity and loading standard in both experiments

(Markovtsov et al. 2000) (analyzed by Western blotting) and La protein encoding RNA (Wolin and Cedervall 2002) (analyzed by Northern blotting (Fig. 5A,B). For both targets, the 27-mer duplex was more potent than the 21-mer siRNAs targeting these messages.

4 Testing for Interferon Pathway Activation and Off-Target Effects of 27-mers

4.1 Interferon and Protein Kinase (PKR) Activation Testing

A potential problem in the use of the longer dsRNAs is activation of protein kinase PKR and induction of interferons (Manche et al. 1992). We thus addressed the question of whether or not a transfected 27-mer dsRNA activates interferon-α or interferon-β (Fig. 6). The indicated RNAs were transfected at a concentration of 20 nM into HEK293 cells and the culture medium was assayed for secreted interferon levels at 24 h post transfection. As a positive control for IFN induction, a triphosphate-containing ssRNA was transfected that potently activated interferon α/β, as reported previously (Kim et al. 2004). Neither cytokine was detected when either the 21+2 siRNA or 27+0 dsRNA was used. We have extended this observation to two other 27-mer sequences targeting different sequences in EGFP (data not presented). PKR activation in cell lysates was also assayed as described previously (Gunnery and Mathews 1998). The lysate was first treated with the indicated RNAs followed by immunoprecipitation. The positive control, long ds RNA elicited PKR activation but none of the shorter RNAs activated PKR (Fig. 6C).

4.2 Testing for Off-Target Effects of 27-mers

Recent studies have shown that the effects of RNAi are not entirely specific and that undesired "off-target" effects can be encountered, the magnitude of which are concentration dependent (Persengiev et al. 2004). It is clear from published data that such effects can be observed in certain cell lines using 21-mer siRNAs (Jackson et al. 2003; Persengiev et al. 2004), but they also can be minimized by using reagents with efficacies in the low to subnanomolar range (Persengiev et al. 2004). To examine this potential problem, DNA microarray analyses were carried out comparing an siRNA 21-mer with a 27-me,r each targeting the EGFP site 1. NIH3T3 cells that stably express EGFP were transfected with concentrations of siRNA that give effective target knockdowns. Total cellular RNAs were prepared from cells 24 and 48 h post-transfection and analyzed by hybridization to an oligonucleotide microarray. Among 16,282 murine genes analyzed, only a small fraction showed up-regulation or down-regulation more than two-fold above or below control values. The 27-mer and 21-mer gave comparable results at their effective RNAi concentrations. The 27-mer showed some increase in the number of transcripts up-regulated when used at the higher 25-nM concentration, but comparisons of the targets affected at 24 versus 48 h and at 5 nM versus 25 nM showed no overlap (Fig. 6D). Rather than specific "off target" effects, these changes are more consistent with statistical scatter within the 16,282 genes examined. The

A.

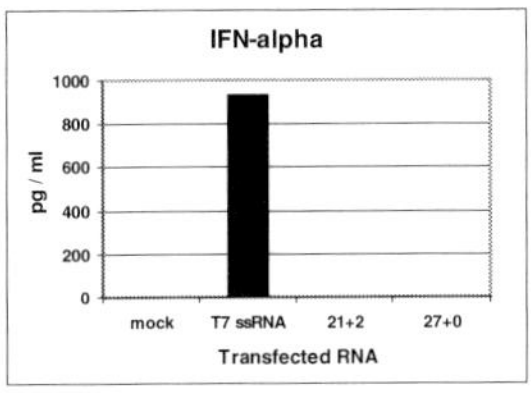

B.

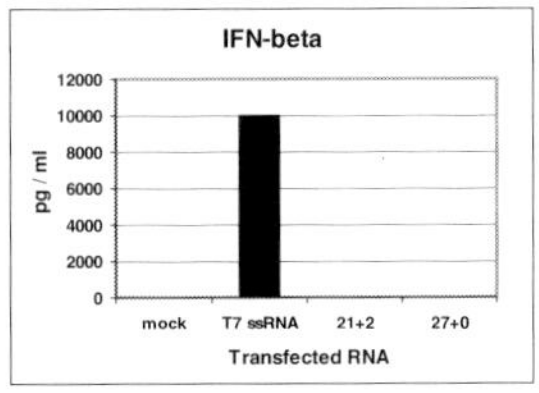

C.

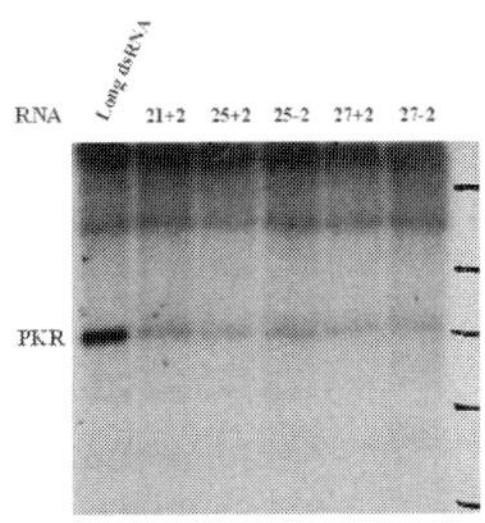

D.

RNA	concentration	hour	X 2 or more/16282	X-2 or less/16282
Site I/21mer	50 nM	24h	11	5
		48h	21	20
Site I/27mer	5 nM	24h	10*	4
		48h	9*	23
	25 nM	24h	71*	3
		48h	22*	0

FIG. 6A–D. Tests for 27-mer non-specific effects. **A,B** Interferon (IFN)-α and β assays. *Column 1*, Positive control for IFN induction (Kim et al. 2004); *column 2*, no RNA; *column 3*, chemically synthesized 21+2 siRNA; *column 4*, chemically synthesized 27+0 dsRNA. **C** PKR activation assay. The long ds RNA used for PKR activation (Kim et al. 2004) and the in vitro PKR activation assay (Manche et al. 1992) have been previously described. Duplex RNAs were transfected as indicated. **D** Summary of results from micro-array analyses for potential off-target effects of 21-mer versus 27-mer in NIH 3T3 cells

same assay was repeated using 27+0 duplex RNA targeting another site in EGFP and comparable results were obtained (data not presented).

Marked differences in potency are best revealed by testing at low nanomolar or picomolar concentrations, which is not routinely done in most labs. Thus far, the 27-mer dsRNA design has shown increased RNAi potency relative to 21+2 siRNAs at every site examined. However within the set of 27-mers studied here, a range of potencies is nevertheless seen between different target sites within the same gene. We have shown that, even in the absence of fully optimized design rules, use of the Dicer-substrate dsRNA approach can increase RNAi potency relative to traditional 21+2 siRNAs. Additionally, the use of 27-mer dsRNAs allows targeting of some sites within a given sequence that are refractory to suppression with traditional 21-mer siRNAs.

Use of Dicer-substrate dsRNAs to trigger RNAi should result in enhanced efficacy and longer duration of RNAi at lower concentrations of RNA than are required for 21+2 applications.

5 Delivery of Synthetic RNAs

Delivery of synthetic siRNAs and Dicer-substrate RNAs in vitro to date has been primarily carried out using cationic lipid carriers(Spagnou et al. 2004). In vivo delivery presents a more difficult challenge. A number of commercial entities are developing siRNA based therapeutics, and as a consequence have focused a great deal of attention on the problem of delivery in vivo. Some backbone modifications on siRNAs appear to stabilize these molecules in serum (Hamada et al. 2002; Chiu and Rana, 2003; Banan and Puri 2004) but these modifications do not necessary facilitate cellular uptake. A breakthrough in this area has recently been achieved by linking cholesterol to backbone-modified siRNAs (Rossi 2004; Soutschek et al. 2004). In a series of very elegant experiments, Soutschek et al. have shown that when cholesterol is covalently linked to the 5′ end of the S strand of an siRNA, the siRNAs can be taken up into a variety of tissues, including liver, jejunum, lung, and kidney. Moreover, the intracellular uptake of the cholesterol-conjugated siRNAs targeting apoB mRNA resulted in sequence-specific knockdown of apoB mRNA and a concomitant reduction in circulating cholesterol as a consequence (Rossi 2004; Soutschek et al. 2004).

Effective delivery of siRNAs in vivo will clearly be a key issue in the coming years. Like its synthetic DNA oligonucleotide predecessor, RNAi technology has the potential for sequence-specific targeting, provided that the molecules can be taken up by tissues in therapeutically effective quantities. The promise of synthetic siRNAs as potent therapeutic agents will only be realized when efficacious, cost-effective delivery is achieved. Given the great scientific and commercial interests in RNAi, this challenge may soon be a memory.

Acknowledgments. D. Kim is a Beckman Fellow. This work was supported by a grant from the Arnold and Mabel Beckman Foundation and the National Institutes of Health (AI29329 and AI42552 and HL074704 to JJR).

References

Banan M, Puri N (2004) The ins and outs of RNAi in mammalian cells. Curr Pharm Biotechnol 5:441–450

Bernstein E, Caudy AA, Hammond SM, Hannon GJ (2001) Role for a bidentate ribonuclease in the initiation step of RNA interference. Nature 409:363–366

Chiu YL, Rana TM (2003) siRNA function in RNAi: a chemical modification analysis. Rna 9:1034–1048

Gunnery S, Mathews MB (1998) RNA binding and modulation of PKR activity. Methods 15:189–198

Hamada M, Ohtsuka T, Kawaida R, Koizumi M, Morita K, Furukawa H, Imanishi T, Miyagishi M, Taira K (2002) Effects on RNA interference in gene expression (RNAi) in cultured mammalian cells of mismatches and the introduction of chemical modifications at the 3′-ends of siRNAs. Antisense Nucleic Acid Drug Dev 12:301–309

Hannon GJ (2002) RNA interference. Nature 418:244–251

Harborth J, Elbashir SM, Vandenburgh K, Manninga H, Scaringe SA, Weber K, Tuschl T (2003) Sequence, chemical, and structural variation of small interfering RNAs and short hairpin RNAs and the effect on mammalian gene silencing. Antisense Nucleic Acid Drug Dev 13:83–105

Holen T, Amarzguioui M, Wiiger MT, Babaie E, Prydz H (2002) Positional effects of short interfering RNAs targeting the human coagulation trigger Tissue Factor. Nucleic Acids Res 30:1757–1766

Hutvagner G, Zamore PD (2002) RNAi: nature abhors a double-strand. Curr Opin Genet Dev 12:225–232

Jackson AL, Bartz SR, Schelter J, Kobayashi SV, Burchard J, Mao M, Li B, Cavet G, Linsley PS (2003) Expression profiling reveals off-target gene regulation by RNAi. Nat Biotechnol 21:635–637

Ketting RF, Fischer SE, Bernstein E, Sijen T, Hannon GJ, Plasterk RH (2001) Dicer functions in RNA interference and in synthesis of small RNA involved in developmental timing in C. elegans. Genes Dev 15:2654–2659

Kim DH, Longo M, Han Y, Lundberg P, Cantin E, Rossi JJ (2004) Interferon induction by siRNAs and ssRNAs synthesized by phage polymerase. Nat Biotechnol 22:321–325

Kim DH, Rossi JJ (2003) Coupling of RNAi-Mediated Target Downregulation with Gene Replacement. Antisense Nucleic Acid Drug Dev 13:151–155

Lee Y, Ahn C, Han J, Choi H, Kim J, Yim J, Lee J, Provost P, Radmark O, Kim S, Kim VN (2003) The nuclear RNase III Drosha initiates microRNA processing. Nature 425:415–419

Lee YS, Nakahara K, Pham JW, Kim K, He Z, Sontheimer EJ, Carthew RW (2004) Distinct roles for Drosophila Dicer-1 and Dicer-2 in the siRNA/miRNA silencing pathways. Cell 117:69–81

Liu Q, Rand TA, Kalidas S, Du F, Kim HE, Smith DP, Wang X (2003) R2D2, a bridge between the initiation and effector steps of the Drosophila RNAi pathway. Science 301:1921–1925

Manche L, Green SR, Schmedt C, Mathews MB (1992) Interactions between double-stranded RNA regulators and the protein kinase DAI. Mol Cell Biol 12:5238–5248

Markovtsov V, Nikolic JM, Goldman JA, Turck CW, Chou MY, Black DL (2000) Cooperative assembly of an hnRNP complex induced by a tissue-specific homolog of polypyrimidine tract binding protein. Mol Cell Biol 20:7463–7479

Persengiev SP, Zhu X, Green MR (2004) Nonspecific, concentration-dependent stimulation and repression of mammalian gene expression by small interfering RNAs (siRNAs). Rna 10:12–18

Pham JW, Pellino JL, Lee YS, Carthew RW, Sontheimer EJ (2004) A Dicer-2-dependent 80s complex cleaves targeted mRNAs during RNAi in Drosophila. Cell 117:83–94

Rossi JJ (2004) Medicine: a cholesterol connection in RNAi. Nature 432:155–156

Sharp PA (1999) RNAi and double-strand RNA. Genes Dev 13:139–141

Soutschek J, Akinc A, Bramlage B, Charisse K, Constien R, Donoghue M, Elbashir S, Geick A, Hadwiger P, Harborth J, John M, Kesavan V, Lavine G, Pandey RK, Racie T, Rajeev KG, Rohl I, Toudjarska I, Wang G, Wuschko S, Bumcrot D, Koteliansky V, Limmer S, Manoharan M, Vornlocher HP (2004) Therapeutic silencing of an endogenous gene by systemic administration of modified siRNAs. Nature 432:173–178

Spagnou S, Miller AD, Keller M (2004) Lipidic carriers of siRNA: differences in the formulation, cellular uptake, and delivery with plasmid DNA. Biochemistry 43:13348–13356

Wolin SL, Cedervall T (2002) The La protein. Annu Rev Biochem 71:375–403

Zamore PD, Tuschl T, Sharp PA, Bartel DP (2000) RNAi: double-stranded RNA directs the ATP-dependent cleavage of mRNA at 21 to 23 nucleotide intervals. Cell 101:25–33

Zhang H, Kolb FA, Brondani V, Billy E, Filipowicz W (2002) Human Dicer preferentially cleaves dsRNAs at their termini without a requirement for ATP. Embo J 21:5875–5885

Subject Index